PT Symmetry
in Quantum and Classical Physics

PT Symmetry
in Quantum and Classical Physics

Carl M. Bender

Washington University in St. Louis, USA

With contributions from

Patrick E. Dorey, Clare Dunning, Andreas Fring, Daniel W. Hook,
Hugh F. Jones, Sergii Kuzhel, Géza Lévai, and Roberto Tateo

We World Scientific

NEW JERSEY · LONDON · SINGAPORE · BEIJING · SHANGHAI · HONG KONG · TAIPEI · CHENNAI · TOKYO

Published by

World Scientific Publishing Europe Ltd.

57 Shelton Street, Covent Garden, London WC2H 9HE

Head office: 5 Toh Tuck Link, Singapore 596224

USA office: 27 Warren Street, Suite 401-402, Hackensack, NJ 07601

Library of Congress Cataloging-in-Publication Data
Names: Bender, Carl M., 1943–   author. | Dorey, P. (Patrick), author.
Title: PT symmetry : in quantum and classical physics / by Carl M. Bender
    (Washington University in St Louis, USA) ; contributions by Patrick E. Dorey
    (Durham University, UK) [and seven others].
Description: Singapore ; Hackensack, NJ : World Scientific Publishing Co. Pte. Ltd., [2018]
Identifiers: LCCN 2018022857| ISBN 9781786345950 (hc ; alk. paper) |
    ISBN 1786345951 (hc ; alk. paper) | ISBN 9781786346681 (pbk. ; alk. paper)
Subjects: LCSH: Symmetry (Physics)--Mathematics. | Space and time--Mathematics. |
    Quantum theory--Mathematics. | Mechanics--Mathematics. | Kreĭn spaces.
Classification: LCC QC174.17.S9 B46 2018 | DDC 539.7/25--dc23
LC record available at https://lccn.loc.gov/2018022857

British Library Cataloguing-in-Publication Data
A catalogue record for this book is available from the British Library.

For any available supplementary material, please visit
https://www.worldscientific.com/worldscibooks/10.1142/Q0178#t=suppl

Desk Editors: Dipasri Sardar/Jennifer Brough/Shi Ying Koe

Typeset by Stallion Press
Email: enquiries@stallionpress.com

Dedicated to Stephen, Piya, and Alexandra

Preface

The condition that a Hamiltonian be $\mathcal{PT}$ symmetric is a generalization of the condition that it be Hermitian. Hamiltonians that are $\mathcal{PT}$ symmetric but not Hermitian can describe actual physical systems. These $\mathcal{PT}$-symmetric systems display a rich array of unexpected features and behaviors and such systems often do not require great knowledge and sophistication to understand. They offer the opportunity to study and explore new and interesting physical theories at both the theoretical and the experimental level. Since 1998, research activity on non-Hermitian $\mathcal{PT}$-symmetric systems has intensified dramatically. As of this writing, over 3500 papers on $\mathcal{PT}$ symmetry have been published and $\mathcal{PT}$ papers have been submitted to two dozen categories of the arXiv. There have been well over 40 international conferences and symposia devoted to $\mathcal{PT}$ symmetry.

The term $\mathcal{PT}$ symmetry refers to combined space-time reflection symmetry. This term first appeared in [Bender and Boettcher (1998b)], but some of the ideas behind $\mathcal{PT}$ symmetry were foreshadowed in earlier work by [Dyson (1956); Cardy and Sugar (1975); Brower *et al.* (1978); Caliceti *et al.* (1980); Harms *et al.* (1980); Scholtz *et al.* (1992); Hollowood (1992); Buslaev and Grecchi (1993); Hatano and Nelson (1996)], and others.

My first exposure to a $\mathcal{PT}$-symmetric Hamiltonian came about in the summer of 1992 in Saclay. In the course of an informal discussion, D. Bessis told me that he and Zinn-Justin had considered the quantum-mechanical analog of the conformal ϕ^3 field theory associated with the Yang-Lee edge singularity. That quantum-mechanical theory is defined by the peculiar non-Hermitian Hamiltonian $\hat{H} = \hat{p}^2 + i\hat{x}^3$. The surprise for me was learning that on the basis of numerical work, they believed that some (and perhaps even all) of the eigenvalues of $\hat{H}$ might be real [Bessis and Zinn-Justin (1992)]. Interestingly, but unknown to us, the numerical reality of the eigenvalues of $\hat{H}$ had already been discovered twelve years earlier in studies of quantum-mechanical analogs of Reggeon field theory [Harms *et al.* (1980)]. Furthermore, in the same year the mathematical properties of perturbation expansions associated with cubic Hamiltonians had been examined [Caliceti *et al.* (1980)].

I would never have worked on PT symmetry had it not be for my previous research on two subjects. First, in collaboration with Wu, Banks, Turbiner, and others, I studied the analytic continuation of eigenvalue problems; that is, the behavior of eigenvalues as functions of a complex coupling constant. This work explained the divergence of perturbation expansions and led to techniques for summing divergent series. Second, in collaboration with Jones, Moshe, and others, I developed the *delta expansion,* a perturbative technique for solving nonlinear problems where the perturbation parameter δ measures the *degree of nonlinearity* of the problem rather than the strength of the coupling. For example, using the delta expansion to solve the Thomas-Fermi equation $y''(x) = [y(x)]^{3/2}/\sqrt{x}$ we consider the problem $y''(x) = y(x)[y(x)/x]^{\delta}$, and to solve a quartic scalar quantum field theory we consider a $\phi^2(\phi^2)^{\delta}$ field theory.[1] We would then seek a perturbation series in powers of δ. This unconventional perturbation procedure is easy to perform and yields great numerical accuracy [Bender *et al.* (1989)].

In 1997 I realized that while the Hamiltonian $H = p^2 + ix^3$ is not Hermitian, it is PT invariant; that is, it remains invariant under $x \to -x$ and $i \to -i$. A possible drawback of a conventional perturbation expansion is that it may violate an invariance (such as gauge invariance) of a Hamiltonian. However, by following the delta expansion and considering the Hamiltonian $\hat{H} = \hat{p}^2 + \hat{x}^2(i\hat{x})^{\varepsilon}$, where ε is real, the PT invariance of the Hamiltonian $\hat{H}$ is preserved for all ε.[2] To my surprise, Boettcher and I found that order by order in powers of ε, the eigenvalues of this Hamiltonian remain real. This was the beginning of the formal study of PT symmetry.

Initial research on PT-symmetric systems was mathematical. This research made use of complex-variable theory, differential equations, and asymptotics. However, since 2009 there has been an avalanche of

---

[1]In this work ϕ^2 and not ϕ is raised to a power δ in order to avoid the appearance of complex numbers when the field ϕ is negative. Amazingly, as explained in this book, it is best to raise $i\phi$ to the power δ.

[2]The cubic Hamiltonian $\hat{H} = \hat{p}^2 + i\hat{x}^3$ is a special case of this Hamiltonian for $\varepsilon = 1$.

beautiful experimental work, mostly but not exclusively in optics. Some of the pioneering optics work is described in [El-Ganainy *et al.* (2018)] and in a special issue of *Nature Photonics* [Feng *et al.* (2017a); Pile (2017); Limonov *et al.* (2017); Horiuchi (2017); Feng *et al.* (2017b)]. Intensive studies have been done on $\mathcal{PT}$-symmetric photonic lattices and graphene [Szameit *et al.* (2011); Regensburger *et al.* (2012); Zhen *et al.* (2015); Weimann *et al.* (2016); Kim *et al.* (2016); Cerjan *et al.* (2016); Zhang *et al.* (2017)], $\mathcal{PT}$-symmetric lasers, coherent perfect absorbers, and unidirectional invisibility [Chong *et al.* (2011); Baranov *et al.* (2017); Hsu *et al.* (2016); Wong *et al.* (2016); Jahromi *et al.* (2017)], $\mathcal{PT}$-symmetric metamaterials [Rechtsman *et al.* (2012); Alaeian and Dionne (2014); Alaeian *et al.* (2016); Weimann *et al.* (2016)], $\mathcal{PT}$-symmetric acoustics [Cummer *et al.* (2016)], $\mathcal{PT}$-symmetric topological insulators [Fleury *et al.* (2016)], $\mathcal{PT}$-symmetric quantum critical phenomena [Ashida *et al.* (2017)], and $\mathcal{PT}$-symmetric excitons and polaritons [Gao *et al.* (2015)].

This experimental work has made the theoretical concepts connected with $\mathcal{PT}$ symmetry clearer and easier to explain. Moreover, it has led to the development of new kinds of metamaterials and devices having remarkable practical applications, such as enhanced sensing [Chen *et al.* (2017)] and wireless power transfer [Assawaworrarit *et al.* (2017)].

To grasp the distinction between $\mathcal{PT}$ symmetry and Hermiticity, we observe that all but one of the axioms of conventional Hermitian quantum theory can be expressed in the language of physics (causality, locality, Lorentz invariance, vacuum-state stability, and so on). However, one axiom is notably different from the others in that it is phrased in the language of mathematics. This is the requirement that the Hamiltonian be Hermitian. In elementary terms, this axiom states that a Hamiltonian matrix $\hat{H}$ remains invariant under *Hermitian conjugation* †, $\hat{H} = \hat{H}^\dagger$, where † denotes combined matrix transposition and complex conjugation.

While Hermitian conjugation sounds more like a mathematical requirement than a physical requirement, the axiom of Hermiticity has significant physical consequences. It guarantees that the eigenvalues of the Hamiltonian are all real, and this is crucial because a measurement of the energy of a physical system will return one of the eigenvalues of the Hamiltonian that describes that system, and such a measurement must have a real outcome.

To show that the eigenvalues of a Hermitian Hamiltonian are real, we take the Hermitian conjugate of the eigenvalue equation $\hat{H}|E\rangle = E|E\rangle$, $\langle E|\hat{H}^\dagger = \langle E|E^*$, and multiply this equation on the right by $|E\rangle$. This gives

$$E\langle E|E\rangle = E^*\langle E|E\rangle,$$

which implies that the eigenvalue $E = E^*$ is real. Furthermore, a Hermitian Hamiltonian generates unitary time evolution, which means that the quantum probability is constant in time. In a closed system it is physically unacceptable for the norm of the state (the quantum probability) to change with time. To see that the norm of a state $|t\rangle$ at time t in a system described by a Hermitian Hamiltonian is conserved in time, we note that $|t\rangle = e^{-i\hat{H}t}|0\rangle$. Thus,

$$\langle t|t\rangle = \langle 0|e^{i\hat{H}t}e^{-i\hat{H}t}|0\rangle = \langle 0|0\rangle.$$

The condition that H be Hermitian is *sufficient* to guarantee the reality of the energy spectrum and the conservation of probability, but it is not necessary. In this book we replace the requirement of Hermiticity by a weaker and more physical condition, namely PT *symmetry* (that is, combined space-time reflection symmetry). Thus, PT symmetry is the condition that if space is reflected *and* time is reversed, the physical system remains invariant. Unlike Hermitian conjugation, space reflection P and time reversal T are physical concepts. Indeed, P and T are elements of the Lorentz group. We will see that a PT-symmetric Hamiltonian has a particularly simple physical interpretation; it represents a nonisolated physical system in contact with its environment in such a way that loss to the environment and gain from the environment is balanced.

The requirement of space-time reflection symmetry is weaker than the condition of Hermiticity, but we show in this book that PT-symmetric Hamiltonians can still have real, positive eigenvalues and can generate unitary time evolution even though they are not Hermitian. Thus, there are new kinds of Hamiltonians that one may consider as possible descriptions of physical systems. Some of these new Hamiltonians are complex, and one can think of these PT-symmetric systems as complex deformations or generalizations of Hermitian Hamiltonians.

The purpose of this book is to acquaint new researchers with the basic ideas and concepts of PT symmetry. The book is not a review or a monograph, but rather an introductory text. The book is organized in two parts, the first presenting a broad and elementary exposition of the fundamentals of PT symmetry, and the second giving a deeper presentation in five selected areas by specialists in the field.

Carl M. Bender
Saint Louis, 2018

About the Authors

Carl M. Bender is the Konneker Distinguished Professor of Physics at Washington University in Saint Louis and the 2017 recipient of the Dannie Heineman Prize in Mathematical Physics awarded jointly by the American Physical Society and the American Institute of Physics. He is a Sloan, Guggenheim, Fulbright, Lady Davis, Rockefeller, Leverhulme, and Ulam Fellow and a Fellow in the American Physical Society and of King's College London. He is coauthor, with S. Orszag, of *Advanced Mathematical Methods for Scientists and Engineers* (Springer). He received his PhD from Harvard University.

Patrick E. Dorey studied in Cambridge and Durham in the UK. After postdocs in Paris and CERN, he returned to Durham as a Lecturer, where (barring the occasional sabbatical) he has stayed ever since. His main research activity has been on exactly-solvable (integrable) quantum field theories in 1+1 dimensions, from which his interest in $\mathcal{PT}$ symmetry developed via the ODE/IM correspondence, discussed in Chapter 6 of this volume.

Clare Dunning is a Reader in Applied Mathematics at the University of Kent. An undergraduate year at CERN, as part of her BSc in Mathematics at the University of Bath, inspired an interest in research. She completed a PhD in Mathematical Physics at Durham University, and then held positions at the University of York and the University of Queensland before coming to the University of Kent.

Andreas Fring studied physics at the Technische Universität München and the University of London and received a PhD in theoretical physics in 1992 from Imperial College London. He joined the Department of Mathematics at City University of London in 2004, where he was promoted to Reader in 2005 and to Professor of Mathematical Physics in 2008. Before joining City University of London, he held postdoc positions at the Universidade de São Paulo (São Carlos, Brazil) and the University of Wales (Swansea, UK). From 1994 until 2004, he was Wissenschaftlicher Mitarbeiter in the Department of Physics at the Freie Universität Berlin.

Daniel W. Hook is CEO of Digital Science, a company that invests in the creation of software to support research. Daniel obtained his PhD in quantum statistical physics in 2007 having studied with Dorje C. Brody at Imperial College London. Since 2005, he has collaborated with Carl M. Bender on PT symmetry and on complex extensions of classical-mechanical systems.

Hugh F. Jones is now retired from the physics department at Imperial College London, where he holds the honorary position of Distinguished Research Fellow. He has published over one hundred papers in particle physics and quantum field theory and is the author of the well-known textbook *Groups, Representations and Physics* (Taylor and Francis). In recent years he has been heavily involved in the development of the theory of $\mathcal{PT}$ symmetry, in particular in its application to classical optics.

Sergii Kuzhel (Sergiusz Kużel), Dr. hab., is a Professor in the Faculty of Applied Mathematics at AGH University of Science and Technology in Kraków, Poland. He earned his postdoctoral degree (habilitation) in mathematics from the Institute of Mathematics of NAS of Ukraine (Kyiv), where he had been working for many years. His research interests include the theory of Kreĭn spaces, mathematical foundations of $\mathcal{PT}$-symmetric quantum mechanics, and scattering theory (Lax–Phillips approach).

With a PhD from the University of Debrecen (Hungary), **Géza Lévai** took a position at the Institute for Nuclear Research (Atomki) in the same city. He visited Oxford University with a Soros Fellowhip and was a Fulbright Fellow at Yale University. He specializes in symmetry-based models in quantum mechanics and nuclear structure theory, paying special attention to exactly-solvable problems. As a Scientific Adviser, he is currently the deputy director of Atomki.

Roberto Tateo is Professor of Physics at the University of Torino in Italy. He is mainly working on integrable models and their applications to the AdS/CFT duality. He received his PhD degree in Physics in 1994 under the supervision of Prof. F. Gliozzi. He worked as a Postdoc at the University of Durham, the University of Amsterdam, and at the Service de Physique Théorique, CEA, Saclay.

Acknowledgment

"The object of knowledge is what exists
and its function to know about reality."
—$\mathcal{P}la\mathcal{T}o$

We are grateful to our many colleagues for their contributions that have helped to advance the field of $\mathcal{PT}$ symmetry and we thank our friends and collaborators for their suggestions and advice concerning the preparation of this book.

We are especially thankful to our colleagues who worked hard and generously to organize the many conferences, meetings, and symposia, and special issues on $\mathcal{PT}$ symmetry. These meetings have helped to bring attention to and inspire people in the scientific community at large to do research in this new area of physics. In particular, we thank M. Znojil, who was the first to organize conferences when the study of $\mathcal{PT}$ symmetry was in its infancy.

We are especially indebted to Jessie for her tireless, persistant, and careful editing work.

In special memory of Boris Samsonov, one of the early contributers to the field.

Contents

Part II. Advanced Topics in $\mathcal{PT}$ Symmetry 173

Part I

Introduction to $\mathcal{PT}$ Symmetry

By Carl M. Bender with Daniel W. Hook

"Numbers have a way of taking a man by the hand and leading him down the path of reason."
—*PyThagoras*

Chapter 1

Basics of $\mathcal{PT}$ Symmetry

> "We consider it a good principle to explain the
> phenomena by the simplest hypothesis possible."
> —$\mathcal{PT}$*olemy*

This chapter introduces the basic ideas of $\mathcal{PT}$-symmetric systems. It begins
with a brief discussion of closed (isolated) and open (non-isolated) systems
and explains that $\mathcal{PT}$-symmetric systems are physical configurations that
may be viewed as intermediate between open and closed systems. The chapter then presents elementary examples of quantum-mechanical and classical $\mathcal{PT}$-symmetric systems. It demonstrates that the Hamiltonians that
describe $\mathcal{PT}$-symmetric systems are complex extensions (deformations) of
conventional real Hamiltonians. Finally, this chapter shows that real systems that are unstable may become stable in the more general complex
setting. Thus, by deforming real systems into the complex domain one
may be able to tame or even eliminate instabilities.

1.1 Open, Closed, and $\mathcal{PT}$-Symmetric Systems

The equations that govern the time evolution of a physical system, whether
it is classical or quantum mechanical, can be derived from the Hamiltonian
for the physical system. However, to obtain a complete physical description of a system, one must also impose appropriate boundary conditions.
Depending on the choice of boundary conditions, physical systems are normally classified as being closed or open; that is, isolated or non-isolated.

A *closed*, or *isolated*, system is one that is not in contact with its environment. In conventional quantum mechanics such a system evolves according
to a Hermitian Hamiltonian. We use the term *Hermitian Hamiltonian* to
mean that if the Hamiltonian H is in matrix form, then H remains invariant

under the combined operations of matrix transposition and complex conjugation. We use the symbol † to represent these combined operations and to indicate that a Hamiltonian is Hermitian we write $H = H^\dagger$. The eigenvalues of a Hermitian Hamiltonian are always real. Moreover, a Hermitian Hamiltonian conserves probability (the norm of a state). When the probability is constant in time, the time evolution is said to be *unitary*.

A closed system may be thought of as *idealized* because its time evolution is not influenced by the external environment. One cannot observe a closed system in a laboratory because making a measurement requires that the system be in contact with the external world. Physically realistic systems, such as scattering experiments, are *open* systems. An open system is subject to external physical influences because energy and/or probability from the outside world flows into and/or out of such a system.

To examine the differences between open and closed systems, we consider a generic nonrelativistic quantum-mechanical Hamiltonian

$$H = \tfrac{1}{2m}\mathbf{p}^2 + V(\mathbf{x}),$$

which describes a particle of mass m subject to a potential $V(\mathbf{x})$ in some region R of space. The function $V(\mathbf{x})$ is assumed to be real. The time-dependent Schrödinger equation associated with this Hamiltonian is

$$i\psi_t(\mathbf{x}, t) = -\tfrac{1}{2}\nabla^2\psi(\mathbf{x}, t) + V(\mathbf{x})\psi(\mathbf{x}, t), \tag{1.1}$$

where we work in units for which $\hbar = 1$ and $m = 1$. If we multiply (1.1) by ψ^*, multiply the complex conjugate of (1.1) by ψ, and subtract the two equations, we obtain the usual quantum-mechanical statement of local conservation of probability:

$$\rho_t(\mathbf{x}, t) + \nabla \cdot \mathbf{J}(\mathbf{x}, t) = 0. \tag{1.2}$$

Here, $\rho = \psi^*\psi$ is the probability density and $\mathbf{J} = \tfrac{i}{2}(\psi\nabla\psi^* - \psi^*\nabla\psi)$ is the probability current. Integrating (1.2) over the region R and applying the divergence theorem,[1] we obtain the equation

$$P'(t) = F(t), \tag{1.3}$$

where $P = \int_R d\mathbf{x}\,\rho$ is the total probability inside the region R and the surface integral $F = \int_S ds\,\mathbf{n}\cdot\mathbf{J}$ represents the net flux of probability passing through the surface S of the region R. (The symbol $\mathbf{n}$ represents a unit vector normal to S.) From (1.3) we can see that if the system is isolated

---

[1]The divergence theorem states that in D-dimensional space if R is a volume, S is the surface of R, ds is an element of surface area, $\mathbf{n}$ is an outward pointing unit vector normal to S, and $\mathbf{F}$ is a vector field, then $\int_R d^D x\,\nabla\cdot\mathbf{F} = \int_S ds\,\mathbf{n}\cdot\mathbf{F}$.

(there is no flow of probability current across any point on the surface of R), then $F = 0$, so the total probability P is conserved (constant in time). However, if the system is open [there is a flow of probability through the surface of R so that $F \neq 0$ (see Fig. 1.1)], then the total probability inside R is not constant. Such a system cannot be in equilibrium.

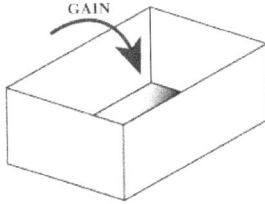

Fig. 1.1 A system with a net flow of probability into it. Such a system is not in equilibrium.

It is possible to construct a physical configuration in which an experimenter makes laboratory measurements and yet no probability flows into or out of the system; we merely include *both* the experiment and the experimenter as components of the physical system. However, then the Hamiltonian for the total system becomes difficult to treat mathematically because there are many degrees of freedom.

In this book we propose a novel way to construct a non-isolated system that has no net flux of probability ($F = 0$): We simply append an identical copy of the original non-isolated system, but with the opposite net flux of probability. (The appended copy is a time-reversed version of the original system.) Thus, the new total physical system consists of two subsystems: (i) the original non-isolated physical system, which has a nonzero net probability flux across the boundary, and (ii) the time-reversed system, which has the opposite flux of probability. Taken together, the two subsystems have no net probability flux because if the original system has gain (or loss), then the time-reversed system has an equal but opposite loss (or gain). Thus the composite system has no net gain or loss (see Fig. 1.2).

The composite loss-gain system exhibits a symmetry called $\mathcal{PT}$ *symmetry*. The symbol $\mathcal{T}$ represents the operation of time reversal, and it has the effect of turning a system with gain into a system with loss (and *vice versa*). The symbol $\mathcal{P}$ is a generic parity (space-reflection) operator; in Fig. 1.2 $\mathcal{P}$ interchanges the gain and loss components of the total system.

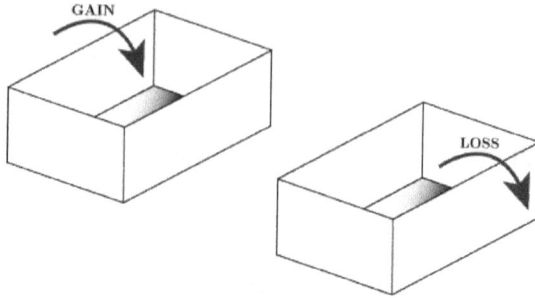

Fig. 1.2　Two subsystems, both in contact with the external environment, one subsystem with gain and the other with loss. One subsystem is the time-reversed version of the other. Taken together, the composite system has no net flux of probability. Even though the composite system is in contact with the environment, it resembles an isolated system because the net flux of probability into or out of the system vanishes. This composite system is $\mathcal{PT}$ symmetric because under time reversal $\mathcal{T}$ LOSS becomes GAIN and GAIN becomes LOSS, and under parity $\mathcal{P}$ the two subsystems are interchanged.

Even though the composite system in Fig. 1.2 has no net probability flux, it differs from a closed system in that the composite system is not in equilibrium. This is because the total probability in one subsystem grows with time, and the total probability in the other subsystem decreases with time. However, if we couple the two components of the $\mathcal{PT}$-symmetric system so that the probability in the subsystem with gain can flow readily into the subsystem with loss (see Fig. 1.3), it may be possible for the composite system to be in dynamic equilibrium. If this system is in dynamic equilibrium, it is said to be in an *unbroken $\mathcal{PT}$-symmetric phase*; if it is not in dynamic equilibrium, it is said to be in a *broken $\mathcal{PT}$-symmetric phase*.

An unbroken $\mathcal{PT}$-symmetric system resembles a closed system because it is in equilibrium; however, it is *not* closed because it is in contact with the external environment, as Fig. 1.3 illustrates. Analogously, a broken $\mathcal{PT}$-symmetric system resembles an open system because it is not in equilibrium, but unlike most open systems the net probability flux vanishes and the system has $\mathcal{PT}$ symmetry. Thus, we can think of $\mathcal{PT}$-symmetric systems as physical systems that are intermediate between open and closed systems.

Typically, the transition to dynamic equilibrium occurs abruptly at a critical value of the coupling strength. The transition from nonequilibrium (broken $\mathcal{PT}$ symmetry) to equilibrium (unbroken $\mathcal{PT}$ symmetry) is called the $\mathcal{PT}$ *phase transition*. An elementary example of this $\mathcal{PT}$ transition is illustrated in Sec. 1.2 by using a 2×2 matrix Hamiltonian.

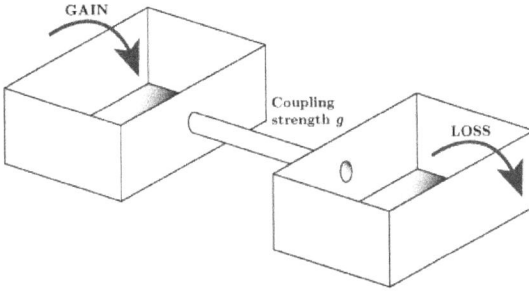

Fig. 1.3   Two coupled subsystems, each in contact with the environment, one subsystem with gain and the other with loss. If the two boxes are coupled so that the probability density in the subsystem with gain can flow rapidly enough into the subsystem with loss, then the composite system can be in dynamic equilibrium.

1.2   Simple $\mathcal{PT}$-Symmetric Matrix Hamiltonians

This section illustrates the qualitative discussion in Sec. 1.1 by using some elementary matrix Hamiltonians. We begin by considering a quantum-mechanical system described by the 1×1 matrix Hamiltonian

$$H = [a + ib],\qquad(1.4)$$

where a and b are real numbers. The physical system described by this Hamiltonian has no spatial dependence and does not move; we can think of the system as being confined to one region of space or even to just one point in space.

The solution to the Schrödinger equation $i\psi_t = (a + ib)\psi$ associated with the Hamiltonian (1.4) is $\psi(t) = Ce^{(-ai+b)t}$, where C is a constant. Observe that if b is nonzero, the Hamiltonian is complex and non-Hermitian. Consequently, for $b \neq 0$ the system is not in equilibrium. Indeed, the probability P, which is conventionally given by

$$P = \psi^*\psi = |C|^2 e^{2bt},$$

grows exponentially in time when b is positive and decays exponentially in time when b is negative. Hence, the system described by H in (1.4) is an instance of the system pictured in Fig. 1.1 for the case when $b > 0$. This system is not $\mathcal{PT}$ symmetric.

As explained in Sec. 1.1, $\mathcal{PT}$-symmetric Hamiltonians can be constructed by combining H with the time-reversed version of H. To do this, we recall the properties of the time-reversal operator $\mathcal{T}$. In the

1930s Wigner showed that the quantum-mechanical time-reversal operator $\mathcal{T}$ involves complex conjugation and thus it replaces i by $-i$. Why does $\mathcal{T}$ change the sign of i? In the time-dependent Schrödinger equation $i\hbar\frac{\partial}{\partial t}\psi = H\psi$, time t is associated with i. Thus, to reverse the sign of t, the time-reversal operator must change the sign of i.

For an alternative explanation of this property of $\mathcal{T}$, recall the Heisenberg algebra $[\hat{x}, \hat{p}] = i\hbar$, which expresses the fundamental operator content of quantum mechanics. The position operator $\hat{x}$ does not change sign under time reversal, $\mathcal{T}\hat{x}\mathcal{T}^{-1} = \hat{x}$, but the momentum operator does change sign, $\mathcal{T}\hat{p}\mathcal{T}^{-1} = -\hat{p}$. Thus, to preserve the Heisenberg algebra under time reversal, $\mathcal{T}$ must change the sign of i: $\mathcal{T}i\mathcal{T}^{-1} = -i$. (Wigner emphasized that because $\mathcal{T}$ changes the sign of i, $\mathcal{T}$ is not a linear operator; rather, it is an *antilinear* operator.) Thus, the time-reversed version of H in (1.4) is

$$H_{\text{time-reversed}} = \mathcal{T}H\mathcal{T}^{-1} = [a - ib].$$

The solution to the Schrödinger equation for the time-reversed version of H behaves as expected; it has the form $\psi(t) = De^{(-ai-b)t}$, where D is a constant. Thus, the quantum-mechanical probability P is given by

$$P = \psi^*\psi = |D|^2 e^{-2bt},$$

which decays in time when b is positive and grows in time when b is negative.

The quantum-mechanical system obtained by combining the subsystem described by H and the subsystem described by $H_{\text{time-reversed}}$ is described by the 2×2 matrix Hamiltonian

$$H_{\text{combined}} = \begin{bmatrix} a + ib & 0 \\ 0 & a - ib \end{bmatrix}.$$

This matrix is diagonal because the two subsystems are side by side but do not interact; one has gain and the other has loss, just as in Fig. 1.2.

The parity operator $\mathcal{P}$, which interchanges the two subsystems, is given by the linear matrix operator

$$\mathcal{P} = \begin{bmatrix} 0 & 1 \\ 1 & 0 \end{bmatrix}.$$

We say that $\mathcal{P}$ is a *reflection operator* because its square is unity, $\mathcal{P}^2 = 1$, and thus applying the parity operator twice leaves the system invariant. Also, $\mathcal{P} = \mathcal{P}^{-1}$. Unlike the parity operator, the time-reversal operator is *antilinear* (that is, it performs complex conjugation), so it does not have a matrix representation. Nevertheless, $\mathcal{T}$ resembles a reflection operator because $\mathcal{T}^2 = 1$. The operations of parity and time reversal are independent of one another so the operators $\mathcal{P}$ and $\mathcal{T}$ commute: $[\mathcal{P}, \mathcal{T}] = 0$.

Note that while the Hamiltonian H_{combined} is not Hermitian, that is, $H_{\text{combined}}^{\dagger} \neq H_{\text{combined}}$, it is $\mathcal{PT}$ symmetric:

$$\mathcal{PT} H_{\text{combined}} (\mathcal{PT})^{-1} = H_{\text{combined}}.$$

The physical system described by the Hamiltonian H_{combined} is not in equilibrium because ψ in one subsystem grows in time and ψ in the other decays in time. However, we can achieve equilibrium by coupling the two subsystems (see Fig. 1.3). To accomplish this coupling, we insert a *coupling parameter* g in the off-diagonal elements of the Hamiltonian:

$$H_{\text{coupled}} = \begin{bmatrix} a + ib & g \\ g & a - ib \end{bmatrix}. \tag{1.5}$$

Now, the subsystems with loss and with gain are coupled. The symmetric matrix Hamiltonian H_{coupled} is not Hermitian, but it is $\mathcal{PT}$ symmetric.

Even though H_{coupled} is not Hermitian, if g becomes sufficiently large, the eigenvalues of H_{coupled} become real, which indicates that the system is in equilibrium. The eigenvalues E are the roots of the secular polynomial

$$\det (H_{\text{coupled}} - IE) = E^2 - 2aE + a^2 + b^2 - g^2, \tag{1.6}$$

where I is the identity matrix. This secular polynomial is *real*.

Solving for the roots of the secular polynomial in (1.6) gives

$$E_{\pm} = a \pm \sqrt{g^2 - b^2}. \tag{1.7}$$

Hence, if the subsystems (represented generically in Fig. 1.3 as boxes) are weakly coupled ($g^2 < b^2$), then the eigenvalues are complex and the system is not in equilibrium; one eigenstate grows in time and the other decays in time. This is the region of *broken $\mathcal{PT}$ symmetry*. However, if the subsystems are strongly coupled ($g^2 > b^2$), then the eigenvalues in (1.7) become real and the entire system is in equilibrium; the eigenstates oscillate and do not grow or decay. This is the region of *unbroken $\mathcal{PT}$ symmetry*.

The eigenvalues $E_{\pm}$ are plotted in Fig. 1.4. At the $\mathcal{PT}$-phase-transition points $g = \pm b$ the real eigenvalues merge and become complex. This merging of eigenvalues can never happen in a system described by a Hermitian Hamiltonian. This is because a Hermitian perturbation of a Hamiltonian system causes the eigenvalues to *repel*; that is, the distance between adjacent eigenvalues *increases*. However, as we can see in (1.7), there is a negative sign in the square root. Thus, the distance between the eigenvalues E_+ and E_- decreases as $|g|$ approaches b. This merging of eigenvalues can only happen in a system described by a non-Hermitian Hamiltonian.

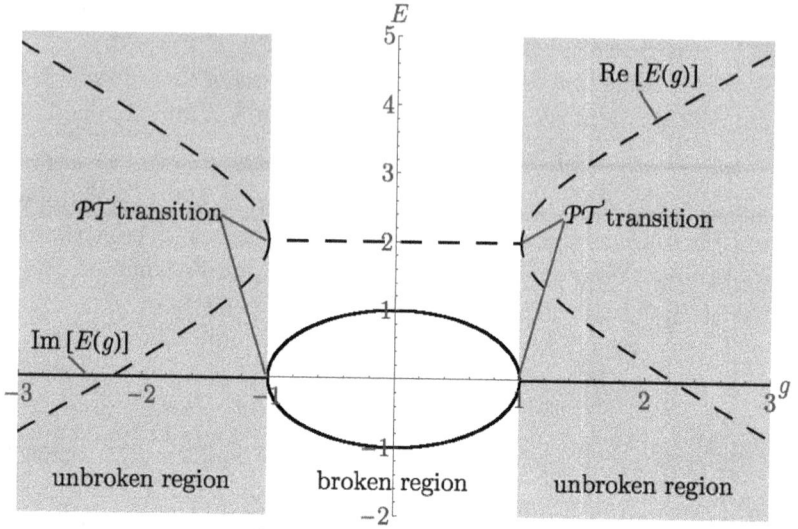

Fig. 1.4 Eigenvalues (1.7) of the $\mathcal{PT}$-symmetric Hamiltonian H_{coupled} (1.5). The eigenvalues are plotted as functions of the coupling g for the parameter choice $a = 2$ and $b = 1$. The real parts of the energies are shown as dashed lines and the imaginary parts of the energies are shown as solid lines. The energy becomes real when $|g|$ exceeds 1; $|g| < 1$ is the region of broken $\mathcal{PT}$ symmetry and $|g| > 1$ is the region of unbroken $\mathcal{PT}$ symmetry. The $\mathcal{PT}$ phase transition occurs at $g = \pm 1$. In the vicinity of this transition both the imaginary and the real parts of the energy split into upper and lower prongs, so the $\mathcal{PT}$ transition is sometimes called a *pitchfork* transition.

The merging of eigenvalues at a $\mathcal{PT}$-phase-transition point has remarkable experimental consequences. Let us suppose that we are trying to determine the value of $g^2 - b^2$ and that this quantity is extremely small. One might think that such a determination would be difficult because it would require a highly sensitive and delicate experiment. However, the quantity to be determined is inside a square root, and *the square root of a very small number is not so small*. Thus, determining the value of $g^2 - b^2$ by making a measurement of the energy difference $E_+ - E_-$ does not in fact require an experiment of great sensitivity; the existence of the $\mathcal{PT}$ phase transition greatly enhances the experimental sensitivity. This elementary observation has enabled experimentalists to make extremely accurate measurements of distances [Liu *et al.* (2016)], angular rotations [Ren *et al.* (2017a)], and temperatures [Hodaei *et al.* (2017)].

1.3   Real Secular Equation for $\mathcal{PT}$-Symmetric Hamiltonians

The eigenvalues E of a Hamiltonian H are the roots of the secular polynomial $f(E) = \det(H - IE)$. We have shown that the Hamiltonian (1.5) has a real secular polynomial [namely, (1.6)]; that is, $f^*(E^*) = f(E)$. If a matrix Hamiltonian is Hermitian, the secular polynomial and all of its roots are real; if a matrix Hamiltonian is not Hermitian, its secular polynomial and its eigenvalues may be complex. However, if the non-Hermitian Hamiltonian is $\mathcal{PT}$ symmetric, the secular polynomial is always real. Hence, *some* of the eigenvalues of a $\mathcal{PT}$-symmetric Hamiltonian (the roots of the secular polynomial) may be real while the remaining eigenvalues must appear as complex-conjugate pairs. Therefore, *the condition of $\mathcal{PT}$ symmetry is weaker and less restrictive than the condition of Hermiticity* [Bender and Mannheim (2010)].

To prove that the secular polynomial of a $\mathcal{PT}$-symmetric Hamiltonian is real, we recall that the determinant of the product of two matrices A and B is independent of the order of multiplication: $\det(AB) = \det(BA)$. Therefore,

$$\det(H - EI) = \det(\mathcal{PT}H\mathcal{T}^{-1}\mathcal{P}^{-1} - \mathcal{P}EI)\mathcal{P}^{-1} = \det(\mathcal{T}H\mathcal{T}^{-1} - EI).$$

The time-reversal operator is the product of some matrix L times the complex-conjugation operator K, so $\mathcal{T} = LK$ and $\mathcal{T}^{-1} = KL^{-1}$, and we get $\det(H - EI) = \det(H^* - EI)$. We conclude that H and H^* have the same set of eigenvalues and that H has a real secular equation. (This conclusion holds for any matrices $\mathcal{P}$ and L.) Thus, we can interpret the condition of $\mathcal{PT}$ symmetry in the generalized sense, where $\mathcal{P}$ is any matrix and $\mathcal{T}$ is antilinear. The generalized $\mathcal{P}$ and $\mathcal{T}$ operators need not even commute with one another.

There is an even shorter proof that the secular polynomial is real. We note that the secular polynomial $f(E)$ has the form $\sum_n a_n E^n = 0$ and that the Hamiltonian matrix itself solves the equation $f(H) = 0$: $\sum_n a_n H^n = 0$. Since the Hamiltonian commutes with the antilinear operator $\mathcal{PT}$, H also obeys $\sum_n a_n^* H^n = 0$. Thus, the secular equation is real. Proving that the secular polynomial is real does not require that $(\mathcal{PT})^2 = 1$.

1.4   Classical $\mathcal{PT}$-Symmetric Coupled Oscillators

The $\mathcal{PT}$ phase transition can occur in classical systems as well as in quantum-mechanical systems. This section discusses a classical

$\mathcal{PT}$-symmetric model, namely, a coupled pair of oscillators. This simple physical system exhibits the features of $\mathcal{PT}$-symmetric systems, as described in Secs. 1.1 and 1.2. This system is especially interesting because there are *two* $\mathcal{PT}$ transitions, a transition from broken to unbroken $\mathcal{PT}$ symmetry when g is small, and a second transition from unbroken to broken $\mathcal{PT}$ symmetry when g is large.

Let us first consider a classical harmonic oscillator of natural frequency ω with a velocity-dependent force of strength γ:

$$\ddot{x} + \omega^2 x + \gamma \dot{x} = 0. \tag{1.8}$$

When $\gamma > 0$ the amplitude of oscillation decays exponentially, and when $\gamma < 0$ the amplitude of oscillation grows exponentially with time. Thus, positive γ simulates the effect of damping, and negative γ corresponds to antidamping.[2] When the parameter γ is nonzero, the quantity $\frac{1}{2}\dot{x}^2 + \frac{1}{2}\omega^2 x^2$ (which we will call the "energy" simply because it looks like the sum of a kinetic energy and a potential energy) is not conserved in time:

$$\frac{d}{dt}\left(\tfrac{1}{2}\dot{x}^2 + \tfrac{1}{2}\omega^2 x^2\right) < 0 \ \ (\gamma > 0), \quad \frac{d}{dt}\left(\tfrac{1}{2}\dot{x}^2 + \tfrac{1}{2}\omega^2 x^2\right) > 0 \ \ (\gamma < 0).$$

The oscillator in (1.8) is not an isolated system because (depending on the sign of γ) the "energy" flows out of or into the system. The case $\gamma < 0$ is an example of the generic system illustrated in Fig. 1.1 because "energy" flows into the oscillator.[3]

Next, we introduce a second oscillator similar to that in (1.8), but with the opposite sign of γ:

$$\ddot{y} + \omega^2 y - \gamma \dot{y} = 0. \tag{1.9}$$

The two oscillators (1.8) and (1.9) are independent of one another, and assuming that $\gamma > 0$, the x oscillator has loss and the y oscillator has corresponding gain. The x and y oscillators taken together comprise a classical system that is $\mathcal{PT}$ symmetric because time reversal $\mathcal{T}$ makes the replacement $t \to -t$ and in this case we define the parity operation $\mathcal{P}$ to

---

[2] *Damping* is a friction force whose direction is opposite to the velocity of the particle and whose magnitude is proportional to the speed of the particle. An *antidamping* force points in the direction of the velocity of the particle and thus it tends to speed up the particle. This kind of force is hard to visualize but one can imagine a grass-powered lawn mower or a spaceship with a vacuum-cleaner drive powered by space dust.

[3] One might think that the classical equation of motion (1.8) cannot be derived from a Hamiltonian because the "energy" is not conserved. In fact, the system (1.8) *does* have a conserved energy (which is a rather complicated expression) and the classical equation of motion can indeed be derived from a time-independent Hamiltonian [Chandrasekar *et al.* (2007); Bender *et al.* (2016b)].

interchange the dynamical variables x and y. The x and y oscillators taken together are an instance of the generic $\mathcal{PT}$-symmetric system illustrated in Fig. 1.2. This system of oscillators is not in equilibrium because the amplitude of one oscillator grows exponentially with t and the amplitude of the other oscillator decays exponentially with t.

Let us now couple the two oscillators (1.8) and (1.9) by introducing the coupling parameter g into the equations of motion:

$$\ddot{x} + \omega^2 x + \gamma \dot{x} = gy, \qquad \ddot{y} + \omega^2 y - \gamma \dot{y} = gx. \tag{1.10}$$

This coupled system is now an instance of the generic balanced-loss-gain system illustrated in Fig. 1.3. This system is not just $\mathcal{PT}$ symmetric; it is also a *Hamiltonian system*, and the equations of motion (1.10) can be derived from the time-independent quadratic Hamiltonian [Bateman (1931); Bender and Gianfreda (2015)]

$$H = pq + \tfrac{1}{2}\gamma(yq - xp) + \left(\omega^2 - \tfrac{1}{4}\gamma^2\right)xy - \tfrac{1}{2}g\left(x^2 + y^2\right). \tag{1.11}$$

To derive (1.10) from H, we apply the standard rules for obtaining Hamilton's equations:

$$\dot{x} = \frac{\partial H}{\partial p} = q - \tfrac{1}{2}\gamma x, \tag{1.12}$$

$$\dot{y} = \frac{\partial H}{\partial q} = p + \tfrac{1}{2}\gamma y, \tag{1.13}$$

$$\dot{p} = -\frac{\partial H}{\partial x} = \tfrac{1}{2}\gamma p - \left(\omega^2 - \tfrac{1}{4}\gamma^2\right)y + gx, \tag{1.14}$$

$$\dot{q} = -\frac{\partial H}{\partial y} = -\tfrac{1}{2}\gamma q - \left(\omega^2 - \tfrac{1}{4}\gamma^2\right)x + gy. \tag{1.15}$$

To obtain the first equation in (1.10) we differentiate (1.12) with respect to t, eliminate $\dot{q}$ by using (1.15), and eliminate q by using (1.12). Similarly, to obtain the second equation in (1.10) we differentiate (1.13) with respect to t, eliminate $\dot{p}$ by using (1.14), and eliminate p by using (1.13). The Hamiltonian (1.11) is $\mathcal{PT}$ symmetric because under parity reflection $\mathcal{P}$ the loss and gain oscillators are interchanged,

$$x \to y, \quad y \to x, \quad p \to q, \quad q \to p,$$

and under time reversal $\mathcal{T}$ the signs of the momenta are reversed,

$$x \to x, \quad y \to y, \quad p \to -p, \quad q \to -q.$$

Note that while the Hamiltonian H in (1.11) is $\mathcal{PT}$ symmetric, it is not invariant under $\mathcal{P}$ or $\mathcal{T}$ separately. Because this balanced-loss-gain system is described by a time-independent Hamiltonian, the energy (that is, the value of H) is conserved. However, the total energy has the complicated form in (1.11) and is not a conventional sum of kinetic and potential energies (which typically would have a form such as $\frac{1}{2}p^2 + \frac{1}{2}q^2 + \frac{1}{2}x^2 + \frac{1}{2}y^2$). If the gain and loss terms in (1.10) were not exactly balanced (that is, if they did not have exactly equal but opposite signs), the system could not be described by a quadratic Hamiltonian [Bender and Gianfreda (2015)].

Let us solve the oscillator equations (1.10). We seek a solution of the form $x(t) = Ae^{i\lambda t}$ and $y(t) = Be^{i\lambda t}$, where A and B are constants. Substituting these expressions into (1.10), we obtain the equations

$$-\lambda^2 + \omega^2 + i\gamma\lambda = Bg/A, \quad -\lambda^2 + \omega^2 - i\gamma\lambda = Ag/B. \quad (1.16)$$

Multiplying the equations in (1.16) gives an equation for λ:

$$\left(\lambda^2 - \omega^2\right)^2 + \gamma^2\lambda^2 = g^2.$$

Hence, the frequencies λ satisfy the fourth-degree polynomial equation

$$P(\lambda) = \lambda^4 + \lambda^2\left(\gamma^2 - 2\omega^2\right) + \omega^4 - g^2 = 0. \quad (1.17)$$

This polynomial equation, like the quadratic secular equation (1.6), is *real* because the oscillator system (1.10) is PT symmetric.

Depending on how many times $P(\lambda)$ crosses the λ axis, the polynomial may have four, two, or no real roots. A complex root implies that the solution $e^{i\lambda t}$ grows or decays exponentially in t, and a real root implies that the solution just oscillates in time and does not grow or decay. To determine whether the roots of (1.17) are real or complex, we differentiate $P(\lambda)$ with respect to λ and use the condition that the polynomial is stationary to get

$$P'(\lambda) = 4\lambda^3 + 2\lambda\left(\gamma^2 - 2\omega^2\right) = 0. \quad (1.18)$$

Thus, stationary points are located at $\lambda = 0$ and at $\lambda^2 = \omega^2 - \frac{1}{2}\gamma^2$.

Let us make some notational simplifications: First, we rescale the time variable t so that without loss of generality we may assume that $\omega = 1$. Second, we take the damping parameter $\gamma > 0$ to be small so that $\omega^2 - \frac{1}{2}\gamma^2$ is positive; a simple choice is $\gamma = \frac{1}{10}$. With this choice of parameters, the stationary points are located at 0 and $\pm\sqrt{199/200}$. At these points the polynomial $P(\lambda)$ has the maximum value $1 - g^2$ and the minimum values $\frac{399}{40000} - g^2$. Thus, there are three regions of g depending on whether the polynomial crosses the horizontal axis no times, four times, or two times.

In the *weak-coupling* region $0 \leq g^2 < \frac{399}{40000}$ there are no real roots; this is a region of broken PT symmetry. In this region the solutions $x(t)$ and

$y(t)$ grow exponentially, so the oscillator system is not in equilibrium. In the *intermediate-coupling* region $\frac{399}{40000} < g^2 < 1$ there are four real roots; this is the region of unbroken $\mathcal{PT}$ symmetry. In this region all solutions $x(t)$ and $y(t)$ oscillate and do not grow or decay exponentially, so the physical system is in equilibrium. Finally, in the *strong-coupling* region $g > 1$, there are two real roots and two complex roots. Once again, this is a region of broken $\mathcal{PT}$ symmetry. The three regions are separated by the two $\mathcal{PT}$ phase transitions at $g = \frac{399}{40000}$ and at $g = 1$.

For typical values of ω and γ the polynomial $P(\lambda)$ in (1.17) is plotted in Fig. 1.5 as a function of λ. The three curves illustrate the behavior of $P(\lambda)$ in the weak-, intermediate-, and strong-coupling regions.

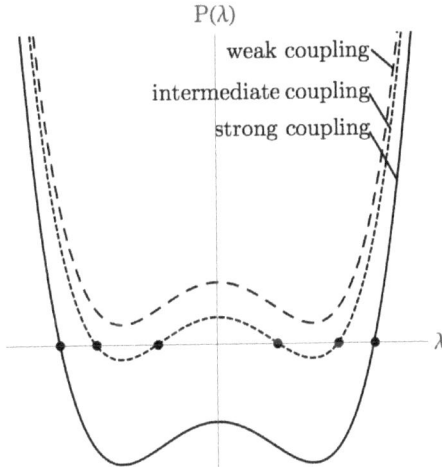

Fig. 1.5 Typical plots of the polynomial $P(\lambda)$ in (1.18) for fixed ω and γ and for three different values of the coupling strength g. For g in the weak-coupling regime (long dashes) there are no real roots. This is a region of broken $\mathcal{PT}$ symmetry. For g in the intermediate-coupling region (short dashes), there are four real roots, which are shown as dots. This is the region of unbroken $\mathcal{PT}$ symmetry. For g in the strong-coupling regime (solid line) there are two real roots. This is another region of broken $\mathcal{PT}$ symmetry.

In Fig. 1.6 we plot the solutions $x(t)$ and $y(t)$ to the coupled-oscillator equations (1.12)-(1.15) as functions of t. Note that these solutions have characteristically different behaviors in the weak-, intermediate-, and strong-coupling regimes and that there are two $\mathcal{PT}$ transitions. In the weak-coupling regime the solutions oscillate and grow exponentially in time. In the intermediate-coupling (unbroken-$\mathcal{PT}$) regime the solutions oscillate

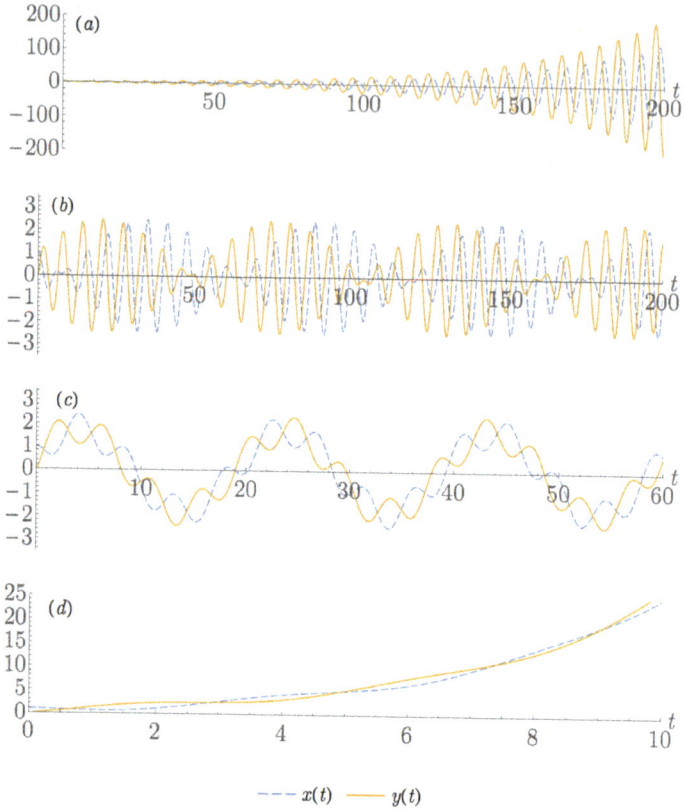

Fig. 1.6   Solutions $x(t)$ (dashed blue line) and $y(t)$ (solid orange line) to the coupled-oscillator equations (1.12)-(1.15) plotted as functions of t. The initial values are $x(0) = 1.0$, $p(0) = 1.0$, $y(0) = 0.0$, $q(0) = 0.0$. For all plots the parameter values $\omega = 1.0$ and $\gamma = 0.1$ have been used. Plots are shown for four different values of the coupling strength g: (a) $g = 0.09$ (in the weak-coupling region of broken PT symmetry), (b) $g = 0.15$ (in the lower range of the intermediate-coupling region of unbroken PT symmetry), (c) $g = 0.9$ (in the upper range of the intermediate-coupling region), (d) $g = 1.1$ (in the strong-coupling region of broken PT symmetry). The behavior in the two regions of broken PT symmetry is qualitatively different. In the weak-coupling broken-PT-symmetric region both $x(t)$ and $y(t)$ oscillate and grow exponentially. However, in the strong-coupling regime the solutions grow so rapidly that they do not oscillate about the horizontal axis. In the unbroken regime the envelopes of the solutions exhibit long-time oscillations (Rabi oscillations). The Rabi oscillations become harder to see near the upper end of the unbroken region. In all plots the value of H in (1.11) remains constant and independent of time.

and remain bounded. In this regime the envelopes of the $x(t)$ and the $y(t)$ solutions grow and decay periodically and there is a lag time between the maxima. These long-time vibrations are called *Rabi oscillations.* The behavior in the strong-coupling (broken-$\mathcal{PT}$) regime is different from the behavior in the weak-coupling (broken-$\mathcal{PT}$) regime. In both cases, the solutions grow exponentially, but in the strong-coupling regime, the solutions grow so rapidly that they do not oscillate about the horizontal axes.

The remarkable behaviors illustrated in Fig. 1.6 have been seen in many laboratory experiments. It has been observed in coupled electronic oscillators [Schindler *et al.* (2011)], mechanical oscillators [Bender *et al.* (2013a)], and more recently in optical whispering-gallery cavity resonators [Peng *et al.* (2014)], acoustical systems [Aurégan and Pagneux (2017)], and optomechanical systems [Lü *et al.* (2015); Li *et al.* (2016)].

1.5 Complex Deformation of Real Physical Theories

The study of $\mathcal{PT}$ symmetry has deep connections to the theory of complex variables and this section explores these connections. A common feature of $\mathcal{PT}$-symmetric quantum-mechanical Hamiltonians, such as H_{coupled} in (1.5), is that they have complex matrix elements that violate the condition of Hermiticity. Consequently, we say that $\mathcal{PT}$-symmetric Hamiltonians are *deformations, extensions,* or *generalizations* of conventional Hermitian Hamiltonians into the complex domain. For example, the Hamiltonian H_{coupled} is Hermitian when $b = 0$ but if we deform this Hamiltonian into the complex domain by allowing the parameter b to take on nonzero real values, the Hamiltonian ceases to be Hermitian but it remains $\mathcal{PT}$ symmetric.

The discovery of complex numbers as a generalization of real numbers was of immense importance in mathematics. It allowed one to solve some real equations, such as $x^4 + 1 = 0$, which were thought at first to have no solutions at all. (This equation has four complex solutions $x = e^{\pm i\pi/4}, e^{\pm 3i\pi/4}$.) Complex-variable theory helps to explain the properties of real functions. For example, it explains why the function $f(x) = 1/(1 + x^4)$, which is smooth and infinitely differentiable for *all* real x (see Fig. 1.7), has a Taylor expansion that converges only in a *finite* real interval. The Taylor series for $f(x)$,

$$f(x) = \frac{1}{1 + x^4} = \sum_{n=0}^{\infty} (-1)^n x^{4n}, \qquad (1.19)$$

converges only for $-1 < x < 1$ because there are *complex* zeros of the denominator $x^4 + 1$ that give rise to singularities in the complex-x plane.

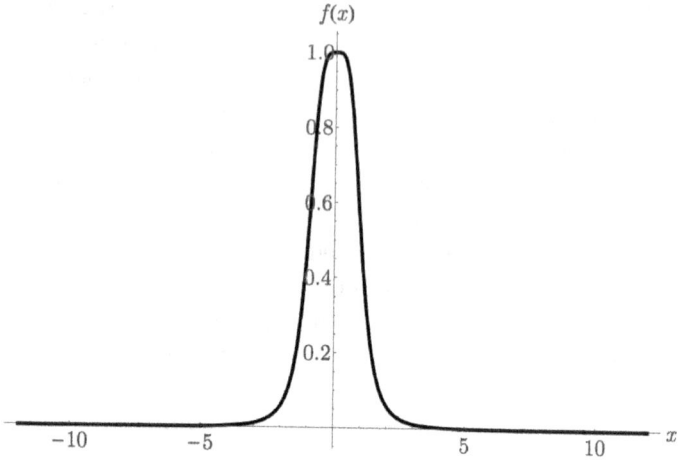

Fig. 1.7   Plot of the function $f(x) = 1/\left(1 + x^4\right)$. This function is smooth and infinitely differentiable for all x and yet the Taylor series (1.19) for $f(x)$ converges only if $|x| < 1$. Complex variable theory explains why this is so.

Complex analysis also enables one to understand multivalued functions. The simplest example of such a function is the square-root function $\sqrt{x}$: When one takes the square root of a real positive number, one obtains two possible answers, one having a positive sign and the other having a negative sign. This seems to contradict the term square-root *function* because a function must have a unique point in its range corresponding to each point in its domain. To resolve this problem we must extend the definition of the square-root function from the real-positive axis to a complex two-sheeted Riemann surface. For each point on this complex domain the complex square-root function has a unique value.

A striking illustration of the insights provided by complex analysis is provided by the integral

$$I(x) = \int_{-\infty}^{\infty} dt \, \cos(\pi x t) e^{-x(\cosh t + t^2/2)}. \tag{1.20}$$

This integral is a real function of x and the path of integration is the real-t axis. Nevertheless, to determine the asymptotic behavior of $I(x)$ as $x \to \infty$, it is essential to make use of complex techniques. It is almost impossible to deduce the large-x behavior of $I(x)$ without using complex analysis, and without using complex variables one may even be led to an incorrect result. Indeed, it is tempting to argue that because the function

$\exp[-x(\cosh t + t^2/2)]$ is sharply localized at $t = 0$ when x is large, we may approximate $\cosh t + t^2/2$ by the first two terms in its Taylor series, $1 + t^2$, and evaluate the resulting Gaussian integral for $I(x)$. However, if we do so, we obtain the wrong asymptotic behavior. This is because the contribution to the integral is *not* localized on the real-t axis at $t = 0$ when x is large and positive. Thus, such a *local* approximation of the integrand is not valid.

To find the correct asymptotic behavior of (1.20) we *deform* the original integration contour, which lies on the real axis, into the complex plane. The appropriate complex contour is determined by the *method of steepest descents* [Bender and Orszag (1999)] because only on a steepest-descent contour can one perform a local approximation to the integrand for large positive x. We rewrite $I(x)$ as

$$I(x) = \text{Re} \int_{-\infty}^{\infty} dt \, e^{x\phi(t)},$$

where $\phi(t) = it + \cosh t + t^2/2$. The condition $\phi'(t) = 0$ implies that there is a saddle point at $t = i\pi$. The steepest-descent path through this saddle point obeys the equation $\text{Im}\,\phi(t) = 0$ and is shown in Fig. 1.8. The advantage of using this steepest-descent contour is that for large positive x the contribution to the integral is *localized* at $t = i\pi$. It is then straightforward [Bender and Orszag (1999)] to show that the asymptotic behavior of $I(x)$ has the simple form

$$I(x) \sim \tfrac{1}{2}(6/x)^{1/4}\Gamma\left(\tfrac{1}{4}\right) e^{x(1-\pi^2/2)} \quad (x \to \infty).$$

Complex-variable theory is also used to make sense of an *improper* real integral whose region of integration contains a simple pole. To interpret such an integral one introduces the concept of a *Cauchy principal part*, which relies on symmetrically truncating the range of integration. For example, if the integrand has a simple pole at $x = 0$, we replace the function $1/x$ using the well known identity

$$\lim_{\epsilon \to 0} \frac{1}{x \pm i\epsilon} \equiv \text{P}\frac{1}{x} \pm i\pi\delta(x).$$

(For a discussion of this formula see, for example, [Saff and Snider (1993)].) Note that the above identity is not invariant under parity reflection $x \to -x$. However, it *is* invariant under PT reflection $x \to -x$, $i \to -i$.

An even more striking example of the power of complex-variable theory is in its application to summation theory, which is a technique for assigning a unique and meaningful sum to a divergent series. Two elementary examples of divergent series are

$$1 - 1 + 1 - 1 + 1 - 1 + \dots \quad \text{and} \quad 1 + 2 + 4 + 8 + 16 + 32 + \dots .$$

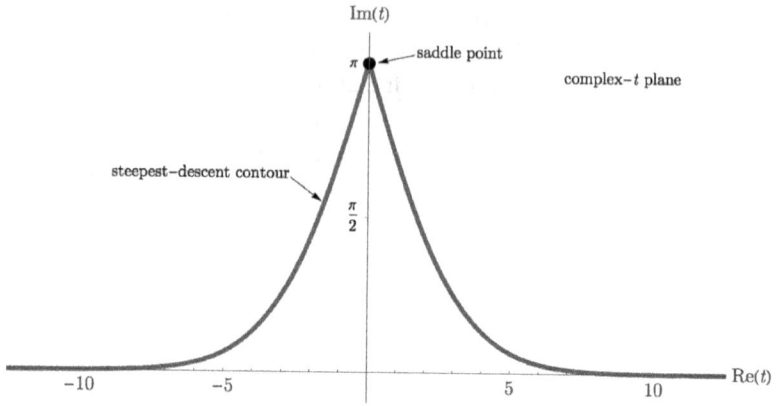

Fig. 1.8   Plot of the steepest-descent contour for the integral $I(x)$ in (1.20). This path passes through the saddle point at $t = i\pi$. Although $I(x)$ is a real integral, one must deform the path of integration into the complex plane to find the asymptotic behavior of $I(x)$ for large real x.

It is relatively easy to show that the unique sum of the first series is $\frac{1}{2}$ and the unique sum of the second series is -1 [Bender and Orszag (1999)] and obtaining this result only requires the use of real-variable theory. However, some more sophisticated divergent series are

$$1 + 1 + 1 + 1 + 1 + \dots \quad \text{and} \quad 1 + 2 + 3 + 4 + 5 + \dots .$$

Real-variable techniques are not powerful enough to assign values to the sums of these series, but complex-variable techniques can be used to show that the first sum has the value $-\frac{1}{2}$ and the second sum has the value $-\frac{1}{12}$. To obtain these results one must analytically continue the zeta function $\zeta(x)$, which is defined as the series $\sum_{n=1}^{\infty} n^{-x}$ for $\operatorname{Re} x > 1$, to negative values of x. Divergent series like these arise in physics when one is performing a sum over eigenmodes, such as when one is calculating the quantum Casimir force between a pair of uncharged conducting parallel plates [Milton (2001)]. The validity of using complex-variable techniques to sum divergent series can be established by measuring the strength of the Casimir force experimentally.

The theory of complex variables also enhances our understanding of the conventional tools that are used to study real physical theories. For example, let us see what happens if we perform a perturbative calculation of the eigenvalues of the quantum anharmonic-oscillator Hamiltonian $\hat{H} = \hat{p}^2 + \hat{x}^2 + \hat{x}^4$. We insert a perturbation parameter ε and expand the eigenvalues and eigenfunctions of the Schrödinger eigenvalue problem

$$- \psi''(x) + x^2 \psi(x) + \varepsilon x^4 \psi(x) = E\psi(x), \qquad \psi(\pm\infty) = 0, \qquad (1.21)$$

as formal series in powers of ε:

$$E = \sum_{n=0}^{\infty} a_n \varepsilon^n, \qquad \psi(x) = \sum_{n=0}^{\infty} \psi_n(x) \varepsilon^n. \qquad (1.22)$$

To our surprise and dismay we discover that the perturbation series for $E(\varepsilon)$ diverges for all $\varepsilon \neq 0$ because for $n \gg 1$ the nth coefficient in the perturbation series grows roughly like $n!$ [Bender and Wu (1973)].

A quick and superficial way to explain the divergence of this perturbation expansion comes from the observation that the nth perturbation coefficient a_n is the sum of Feynman graphs having n vertices. The number of such graphs grows like $n!$ and the graphs all add in phase [Bender and Wu (1969)] and this implies that the perturbation coefficients grow like $n!$. This observation, while technically correct, does not provide a fundamental explanation for the divergence of the perturbation expansion.

To understand the rapid growth of the perturbation coefficients we turn to the methods of complex analysis. We extend the Schrödinger eigenvalue problem (1.21) into the complex domain and examine its behavior for *complex* ε. We discover that the function $E(\varepsilon)$ has singularities in the complex-ε plane that lie arbitrarily close to the origin $\varepsilon = 0$ [Bender and Wu (1969, 1968)]. Consequently, the perturbation series for $E(\varepsilon)$ has a vanishing radius of convergence and it diverges for any nonzero ε.

Typical perturbation expansions often have zero radii of convergence, and this may be discouraging because it suggests that perturbation theory is ineffective for calculating the eigenvalues of a Hamiltonian. However, complex variable theory saves the day! A powerful result of complex analysis is that if one converts the formal perturbation series (1.22) representing the eigenvalue $E(\varepsilon)$ to a sequence of *diagonal Padé approximants* [Bender and Orszag (1999)], the Padé sequence converges to the exact eigenvalue for all positive real values of ε even though the perturbation series diverges [Simon and Dicke (1970)].

In order to study the properties of the complex-valued function $E(\varepsilon)$, we must deform the real differential-equation eigenvalue problem (1.21) into the complex-ε domain. To perform this deformation we must apply the ideas of asymptotic approximations to differential equations, and this requires an elementary understanding of WKB theory and the notion of Stokes sectors. We defer a detailed discussion of these techniques to Chap. 2; our objective in this chapter is just to illustrate the nature of solutions to complex eigenvalue problems.

Consider the simple two-dimensional matrix eigenvalue problem $H\psi = E\psi$, where H is the 2×2 matrix

$$H = \begin{bmatrix} a & g \\ g & b \end{bmatrix}. \tag{1.23}$$

This Hamiltonian, whose structure is similar to that in (1.5), describes a coupled two-state Hermitian system, whose energy eigenvalues are a and b when the coupling constant g vanishes. The eigenvalues E of the coupled system are the zeros of the quadratic secular equation

$$0 = \det(H - IE) = \det \begin{bmatrix} a - E & g \\ g & b - E \end{bmatrix} = E^2 - (a+b)E + ab - g^2.$$

The solution to this equation is

$$E_{\pm}(g) = \tfrac{1}{2}a + \tfrac{1}{2}b \pm \tfrac{1}{2}\sqrt{(a-b)^2 + g^2}. \tag{1.24}$$

If we set $g = 0$, the eigenvalues $E_{\pm}$ reduce to b and a. Note that had we performed a perturbative calculation of $E_{\pm}(g)$, the perturbation series would converge only for $|g| < |a - b|$. This is because there are two square-root singularities in the complex-g plane.

These square-root singularities have a remarkable physical interpretation. To simplify the discussion we choose $a = 1$ and $b = 2$ so that

$$E_{\pm} = \tfrac{3}{2} \pm \tfrac{1}{2}\sqrt{1 + g^2}. \tag{1.25}$$

Now, the eigenvalues $E_{\pm}(g)$ in (1.25) have square-root branch-point singularities at $g = i$ and at $g = -i$. These singularities lie at a distance of 1 from the origin in the complex-g plane; if we calculate the eigenvalues as perturbation series in powers of g, the perturbation series will have a radius of convergence of 1. These square-root singularities are often called *exceptional points*. In general, eigenvalues *cross* (that is, they become degenerate) at exceptional points. When two eigenvalues cross at an exceptional point, the exceptional point is said be of order two.[4] An exceptional point also marks the boundary between broken and unbroken PT-symmetric phases. For example, Fig. 1.4 shows that there are exceptional points at the PT-phase-transition points $g = \pm 1$.

The radius of convergence of the perturbation series for an energy level of a quantum-mechanics problem is determined by the locations of the exceptional points. In the case of the quantum anharmonic oscillator (1.21)

---

[4]It is uncommon but possible for $n > 2$ eigenvalues to become degenerate at an exceptional point. When this happens, the exceptional point is said to be of order n [Bender *et al.* (1992)].

there are *infinitely* many exceptional points in the complex-ε plane, all of order 2. To find the locations of these exceptional points we must first calculate the secular equation $S(E, g) = \det(H - IE) = 0$ for the eigenvalues. (This can be done approximately for the quantum anharmonic oscillator.) We then determine the exceptional values of ε for which there are two-fold degeneracies by solving the pair of simultaneous equations

$$S(E, \varepsilon) = 0 \quad \text{and} \quad \frac{\partial}{\partial E} S(E, \varepsilon) = 0. \tag{1.26}$$

When we do this, we find that there are exceptional points arbitrarily close to the origin, and that this is the underlying reason that the perturbation series in (1.22) diverges.

By deforming a quantum-mechanical eigenvalue problem into the complex domain and locating the exceptional points, we can determine the radius of convergence of the perturbation series. However, by doing so, we achieve much more than that: The complex deformation provides a deep understanding of the nature of quantization. The conventional picture of quantum mechanics is that the energy levels are discrete and quantized functions of the real coupling constant. However, from the more general complex-variable perspective in which the coupling constant is complex, we can see that the energy levels of H in (1.23) are in fact *continuous* functions of the complex coupling constant g. Indeed, the function $E(g) = \frac{3}{2} + \frac{1}{2}\sqrt{1 + g^2}$ is a continuous function of complex g on the two-sheeted Riemann surface in Fig. 1.9.

This behavior of $E(g)$ reveals that quantization has a topological interpretation. Note that there are two eigenvalues $E_\pm(g)$ simply because there are two sheets in the Riemann surface. Figure 1.9 shows that if we follow a closed continuous path in the complex-g plane that does not wind around an exceptional point, the energy levels themselves follow closed paths in the complex-energy plane (see Fig. 1.10). However, on a continuous path that encircles an exceptional point, the energy levels exchange their identities because they are analytic continuations of one another (see Fig. 1.11). This exchange phenomenon, called *level crossing*, is not just a mathematical concept; the analytic continuation and crossing of energy levels has been observed experimentally in PT-symmetric microwave cavities [Bittner *et al.* (2012)] and more recently in beautiful experiments involving optical waveguides [Xu *et al.* (2016); Doppler *et al.* (2016)].

In general, when we extend the parameters (the coupling constants) of a Hamiltonian into the complex domain, we find that the energy eigenvalues are all branches of a multivalued function; they are analytic continuations of

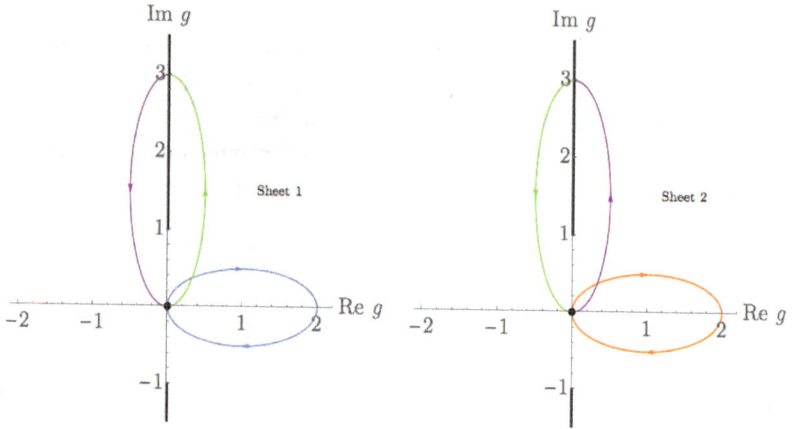

Fig. 1.9   Two-sheeted Riemann surface in the complex variable g for the energy levels (1.25) of the two-state system (1.23) with $a = 1$ and $b = 2$. The energy levels correspond to the two branches of the function $E(g) = \frac{3}{2} + \frac{1}{2}\sqrt{1 + g^2}$. The two sheets of the Riemann surface are connected by square-root branch cuts that we have chosen to run from i to $+i\infty$ and from $-i$ to $-i\infty$ along the $\mathrm{Im}(g)$ axis. These cuts are indicated by heavy solid lines on the vertical axes. On the first sheet $E(0) = 1$ and on the second sheet $E(0) = 2$. If we follow a closed elliptical path (orange) that begins and ends at $g = 0$ on sheet 2, the energy $E_+(g)$ follows a closed curve (also orange) that begins and terminates at $E = 2$ in the complex-E plane as shown on Fig. 1.10. Similarly, if we follow the equivalent path (blue) on the first sheet, then $E_-(g)$ follows a closed path in Fig. 1.10 (also blue) that begins and ends at $E = 1$. On the other hand, if we follow a continuous elliptical path (green) that begins at $g = 0$ on the first sheet, winds around a branch-point singularity (exceptional point), and terminates at $g = 0$ in the second sheet, the energy levels continuously deform and *exchange* their identities, as shown by the green curve in Fig. 1.11. Similarly, if we follow a path (purple) from $g = 0$ on the second sheet to $g = 0$ on the first sheet, the $E = 2$ energy level deforms to the $E = 1$ energy level as shown by the purple curve in Fig. 1.11.

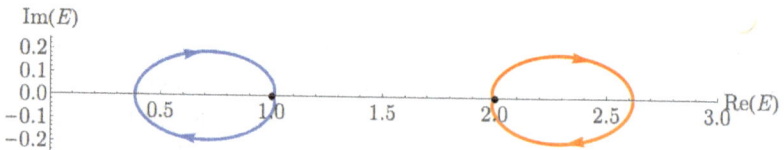

Fig. 1.10   Closed loops in the complex-energy $E(g)$ plane. These paths are images of closed elliptical paths that begin and end at $g = 0$ in the upper sheet (orange) and the lower sheet (blue) in the Riemann surface pictured in Fig. 1.9.

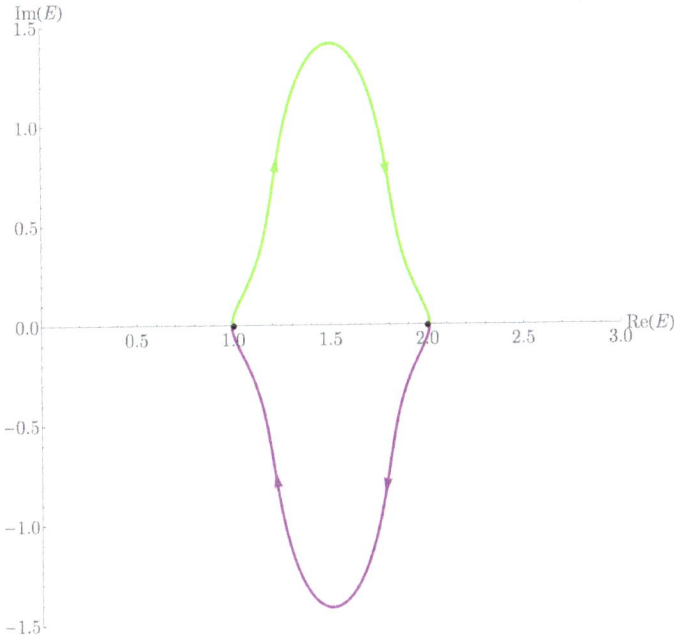

Fig. 1.11 Level crossing around exceptional points on the Riemann surface in Fig. 1.9. The closed paths (green and purple) are images in the complex-E plane of the green and purple paths in Fig. 1.9.

one another and can be continuously deformed into one another by varying the parameters. Thus, complex analysis provides a remarkable topological picture of quantization; the energy levels are in one-to-one correspondence with the sheets of a Riemann surface. From the narrow perspective of real variables this interpretation of quantization is invisible; if we had insisted that the coupling constant g had to be kept real in order that the Hamiltonian (1.23) be Hermitian, we would never have discovered this beautiful picture of the nature of quantization.

1.6 Classical Mechanics in the Complex Domain

Having shown in Sec. 1.5 that extending quantum-mechanical theories into the complex domain yields new insights regarding the mathematical structure of quantization, our objective in this section is to show that extending *classical mechanics* into the complex domain yields valuable insights as well.

To illustrate, we consider the elementary problem of the classical harmonic oscillator, which is described by the Hamiltonian

$$H = \tfrac{1}{2m}p^2 + \tfrac{k}{2}x^2. \tag{1.27}$$

Here, m is the mass of a particle subject to a spring force and k is the spring constant. Note that this Hamiltonian is PT symmetric if we interpret parity reflection as changing the signs of x and p and time reversal as changing the sign of p.

The Hamiltonian (1.27) gives rise to two coupled equations of motion for the coordinate variable $x(t)$ and the momentum variable $p(t)$:

$$\dot{x} = \frac{\partial H}{\partial p} = p/m, \tag{1.28}$$

$$\dot{p} = -\frac{\partial H}{\partial x} = -kx. \tag{1.29}$$

Combining these two equations, we get the harmonic-oscillator equation of motion $m\ddot{x} + kx = 0$. The general solution to this differential equation is

$$x(t) = A\sin(\omega t) + B\cos(\omega t), \tag{1.30}$$

where A and B are arbitrary constants and the frequency is $\omega = \sqrt{k/m}$.

The conventional way to explain the harmonic oscillator is as follows: Multiply (1.28) by kx, multiply (1.29) by p/m, and add the resulting equations. This verifies that the Hamiltonian (1.27) is time independent; that is, $\frac{d}{dt}H = 0$. Next, use (1.28) to eliminate p from H to establish that

$$E \equiv \tfrac{1}{2}m\dot{x}^2 + \tfrac{1}{2}kx^2 \tag{1.31}$$

is a constant of the motion. Observe that E is a sum of two terms, which are recognizable as (i) the kinetic energy $\tfrac{1}{2}m\dot{x}^2$ of a particle of mass m, and (ii) the potential energy $\tfrac{1}{2}kx^2$ of the particle in a quadratic potential well. The negative derivative of the potential is the force $-kx$ on the particle, and this force is proportional to the displacement of the particle from its equilibrium (rest) position at $x = 0$. This force, called *Hooke's law*, acts to restore the particle to its equilibrium position. As the particle moves away from $x = 0$, its kinetic energy decreases and its potential energy increases. When $x = \pm\sqrt{2E/k}$, the potential energy reaches its maximum value E and the kinetic energy of the particle vanishes. At these points, called *turning points*, the particle stops, turns around, and heads back towards its equilibrium position. This back-and-forth motion of the particle in the quadratic potential well is called *simple harmonic motion*. This motion is described explicitly in (1.30).

One might think that this is all there is to say about the harmonic oscillator. However, if we extend the harmonic oscillator into the complex domain [that is, to complex values of $x(t)$], we discover a more elaborate and intricate behavior. Let us begin by asking, What happens if we choose the initial position $x(0)$ of a particle of energy E to be outside of the real interval bounded by the turning points at $\pm\sqrt{2E/k}$; that is, what happens if $x(0) > \sqrt{2E/k}$? The conventional answer to this question is that this choice of initial condition is classically impossible because it would imply that the potential energy exceeds the total energy $\frac{1}{2}k[x(0)]^2 > E$, and thus the kinetic energy would be negative. One would think that because the kinetic energy $\frac{1}{2}m\dot{x}^2$ is a squared quantity, it cannot be negative. Thus, the real values of x satisfying $|x| > \sqrt{2E/k}$ are conventionally designated as the *classically forbidden* region and it is generally understood that this region is only accessible at the quantum level; particles can only enter the classically forbidden region via the process of quantum-mechanical tunneling.

Of course, the square of an imaginary number is negative. Thus, while the total energy E in (1.31) remains real and positive, the kinetic energy can be negative if the momentum $\dot{x}$ is *imaginary*. We generalize from real classical mechanics to complex classical mechanics by allowing for complex solutions to Hamilton's equations (1.28) and (1.29). Indeed, if the classical particle is initially in the classically forbidden region at $t = 0$ and the total energy of the particle is real and positive, we can show that the classical trajectory of the particle is an *ellipse* in the complex-x plane.

The calculation goes as follows: Let the energy E of the particle be real and let the initial position $x(0)$ of the particle lie to the right of the turning point at $\sqrt{2E/k}$. Then, from (1.31) we obtain a first-order ordinary differential equation for the position of the particle:

$$\dot{x} = \pm\sqrt{(2E - kx^2)/m}. \tag{1.32}$$

This equation is separable, so we substitute $x(t) = z(t)\sqrt{2E/k}$ and write the differential equation in separated form as $dz/\sqrt{1 - z^2} = \pm dt\sqrt{k/m}$. The general solution to this equation is $z = \cos\left(t\sqrt{k/m} + C\right)$, where C is an arbitrary constant. Thus,

$$x(t) = \sqrt{2E/k}\,\cos\left(t\sqrt{k/m} + C\right).$$

To determine C we incorporate the initial condition

$$x(0) = \sqrt{2E/k}\,\cos C = \sqrt{2E/k}\,\cos(i\beta) = \sqrt{2E/k}\,\cosh(\beta)$$

and obtain the following equation for the complex classical path of the particle as a function of time t:

$$x(t) = \sqrt{2E/k}\,[\cos(tk/m)\cosh\beta - i\sin(tk/m)\sinh\beta]. \tag{1.33}$$

To see that this path is an ellipse in the complex-x plane, we note that $R = \mathrm{Re}\, x(t)$ and $I = \mathrm{Im}\, x(t)$ satisfy the equation

$$\frac{R^2}{\cosh^2 \beta} + \frac{I^2}{\sinh^2 \beta} = \frac{2E}{k}. \qquad (1.34)$$

Thus, the semi-major axis of the ellipse is $\sqrt{2E/K}\cosh\beta$ and the semi-minor axis of the ellipse is $\sqrt{2E/K}\sinh\beta$. (Some elliptical classical paths are shown in Fig. 1.12.) Interestingly, the foci of the ellipses lie on the real axis at $\pm\sqrt{2E/K}$, which are precisely the positions of the classical turning points. Note that conventional back-and-forth harmonic motion is merely a limiting degenerate form of the ellipse in which the semi-minor axis vanishes (because $\beta = 0$).

The elliptical classical paths shown in Fig. 1.12 are left-right symmetric. This is a consequence of the $\mathcal{PT}$ symmetry of the harmonic-oscillator Hamiltonian in (1.27). Under parity reflection $x \to -x$ even if x is complex, and under time reversal $x \to x^*$. When these transformations are combined, the net result is just a reflection through the vertical (imaginary) axis.

The most obvious feature of the elliptical classical paths in Fig. 1.12 is that while the classical particle passes through the classically forbidden regions to the left of the left turning point and to the right of the right turning point, the particle is moving vertically when it does so. In general, a classical particle can never travel parallel to the real axis in a classically forbidden region because the kinetic energy is negative.

There is a simple optical analogy for this behavior involving the phenomenon of total internal reflection. Figure 1.13 shows a light ray in a slab of glass. This ray reflects off the glass-air interface, so no light is transmitted across the boundary of the two media. However, there *is* an electromagnetic field to the right of the interface, and the magnitude of this field decays exponentially with the distance to the right of the boundary [Jackson (1999)]. In a steady-state configuration no energy flows in the horizontal direction in the air region. If one calculates the Poynting vector, one finds that the energy flow to the right of the glass-air interface is entirely vertical, in analogy with the classical trajectories in Fig. 1.12.

While the classical orbits shown in Fig. 1.12 are ellipses, the orbits are not Keplerian. In Keplerian dynamics the more distant planets take longer to go around the sun than the inner planets. (Pluto's year is much longer than an Earth year.) In contrast, as a consequence of the Cauchy integral theorem, *all* complex-harmonic-oscillator orbits of energy E have exactly the same period. To see why this is so, we calculate the period T of a

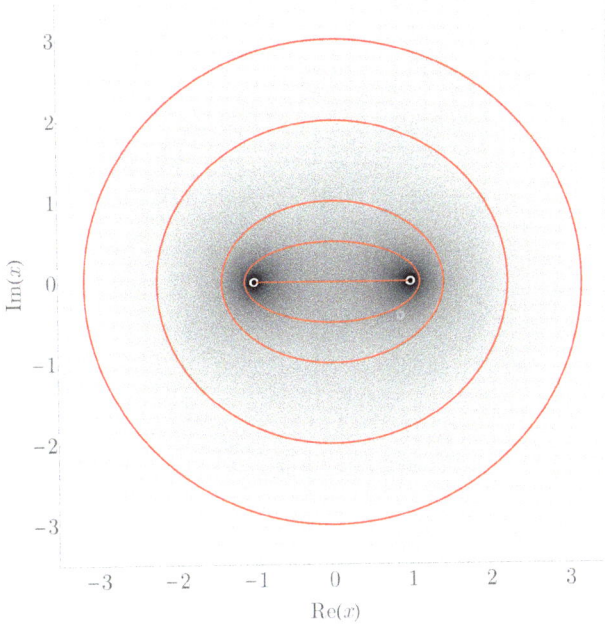

Fig. 1.12   Five complex trajectories of a classical particle obeying the complex-harmonic-oscillator equation (1.33). For this plot $m = 1/2$, $k = 2$, and the energy of the particle is $E = 1$. For each of the four initial conditions $x(0) = i/2$, i, $2i$, $3i$, the orbit of the particle is an ellipse whose foci are located at the turning points ± 1. When the particle starts at the origin $x = 0$, its orbit is a degenerate ellipse and it executes simple real harmonic motion between the turning points at ± 1. Note the background shading: Light shading indicates that the particle moves rapidly and dark shading indicates that the particle moves slowly. The speed of the particle is slow when it is near a turning point and fast when it is far from a turning point.

classical orbit as a path integral and use the chain rule:

$$T = \oint dt = \oint dx \frac{dt}{dx}. \tag{1.35}$$

Next, we substitute for $\frac{dx}{dt}$ by using (1.32) and get

$$T = \oint \frac{dt}{\sqrt{(2E - kx^2)/m}} = \sqrt{\frac{m}{k}} \oint \frac{ds}{\sqrt{1 - s^2}} = 2\pi \sqrt{\frac{m}{k}}. \tag{1.36}$$

The integral on the right encloses the branch cut running from the left turning point to the right turning point. Since there are no other singularities, it follows by path independence that the orbital period T is the same for any orbit.

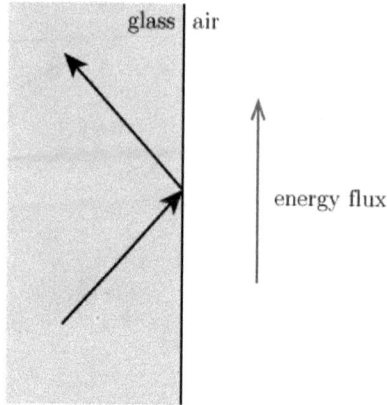

Fig. 1.13   Total internal reflection, an optical analogy with the motion of the complex classical particle motion shown in Fig. 1.12. A light ray in the slab of glass is totally reflected off the glass-air interface. In the steady state no electromagnetic energy flows in the air region in the horizontal direction. The energy flow in this region is entirely vertical, just as the motion of the classical particle is vertical as it crosses the real axis in the classically forbidden region.

In complex classical mechanics the orbital period T is determined by the energy only and it is independent of the initial condition. This explains why the speed of the particle in the larger ellipses in Fig. 1.12 is so large. The particle in the outer ellipses must travel faster in order for the orbital period to remain constant.

1.7   Complex Deformed Classical Harmonic Oscillator

A nontrivial model of a deformation is the complex-deformed harmonic oscillator. The Hamiltonian for this oscillator model is

$$H = p^2 + x^2(ix)^\varepsilon \qquad (\varepsilon \text{ real}). \qquad (1.37)$$

This model is historically important because the quantized version of this Hamiltonian began the field known as $\mathcal{PT}$-symmetric quantum mechanics [Bender and Boettcher (1998b)]. At the quantum level, when $\varepsilon \neq 0$, this Hamiltonian is complex and not Hermitian, but it remains $\mathcal{PT}$ invariant for all real ε because the combination $i\hat{x}$ is $\mathcal{PT}$ symmetric. We show how to calculate the eigenvalues of the quantized version of this Hamiltonian in Chap. 2 and we investigate its spectrum at a mathematically rigorous level in Chap. 6. We will see that $\varepsilon \geq 0$ is the region of unbroken $\mathcal{PT}$ symmetry (where the eigenvalues are all real, positive, and discrete) and that $\varepsilon < 0$ is the region of broken $\mathcal{PT}$ symmetry (where a finite number of eigenvalues are real and the remaining eigenvalues form complex-conjugate pairs).

In this section we restrict our attention to the classical version of H. (A further study of this classical Hamiltonian is presented in Chap. 4). When $\varepsilon = 0$, H in (1.37) reduces to the harmonic-oscillator Hamiltonian, but as ε assumes nonreal values, H deforms smoothly into the complex domain. We will see that the classical trajectories of a particle described by H undergo a dramatic topological transition at $\varepsilon = 0$. When $\varepsilon \geq 0$ the classical trajectories are closed orbits, but when $\varepsilon < 0$ the classical trajectories become open. This transition in the character of the classical orbits is the underlying cause of the quantum $\mathcal{PT}$ phase transition.

Let us begin by finding the locations of the classical turning points. A classical particle has no kinetic energy when it is at a turning point so $p^2 = 0$. Thus, if the total energy of the particle is E, the turning points obey the equation

$$x^2(ix)^\varepsilon = E, \tag{1.38}$$

which has many complex solutions. However, we assume that the energy E is real and ask what happens to the two turning points that lie on the real axis at $\pm\sqrt{E}$ when $\varepsilon = 0$ (the harmonic-oscillator case) as ε moves away from 0. We seek a solution to (1.38) in polar form and find that the left and right turning points are located in the complex plane at

$$x_{\text{left}} = E^{1/(2+\varepsilon)}e^{i\theta_{\text{left}}} \quad \text{and} \quad x_{\text{right}} = E^{1/(2+\varepsilon)}e^{i\theta_{\text{right}}}, \tag{1.39}$$

where

$$\theta_{\text{left}} = -\frac{4+\varepsilon}{4+2\varepsilon}\pi \quad \text{and} \quad \theta_{\text{right}} = -\frac{\varepsilon}{4+2\varepsilon}\pi. \tag{1.40}$$

Thus, the two turning points for the classical harmonic oscillator move downward into the complex plane as ε increases from 0 and they reach the negative-imaginary axis as $\varepsilon \to +\infty$. They move upward as ε decreases below 0. Note that for all ε the positions of these turning points are symmetric with respect to reflections through the imaginary axis. This is a consequence of the $\mathcal{PT}$ symmetry of the Hamiltonian.

Let us consider the *cubic* $\mathcal{PT}$-symmetric oscillator obtained by setting $\varepsilon = 1$. The Hamiltonian for this oscillator is

$$H = p^2 + ix^3. \tag{1.41}$$

There are altogether three turning points located in the complex plane, the two turning points given in (1.39) and (1.40) and an additional turning point on the positive-imaginary axis. Eight classical trajectories for the complex cubic oscillator are shown in Fig. 1.14. We can see in this figure that the complex ellipses in Fig. 1.12 are now deformed but that the classical orbits are still closed and left-right symmetric.

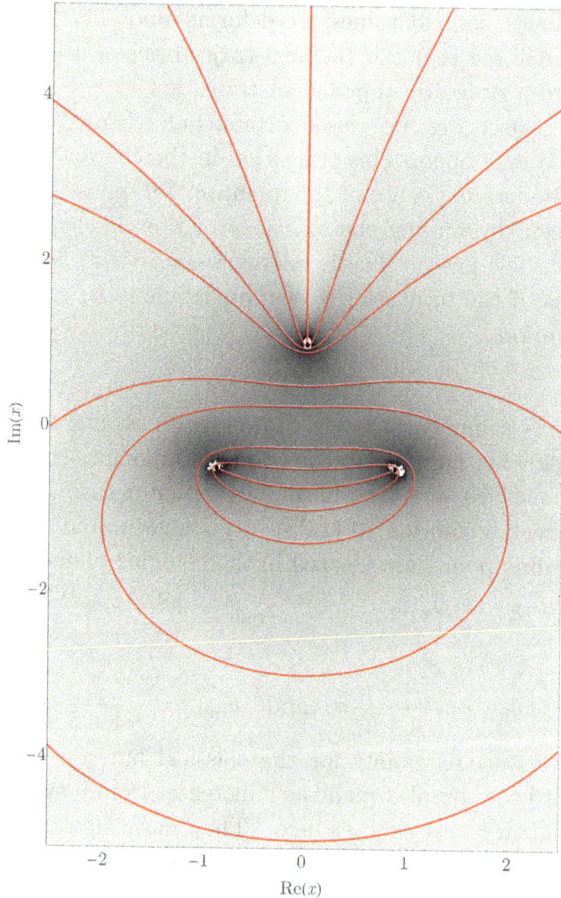

Fig. 1.14   Nine classical orbits for the cubic oscillator in (1.41) for a classical particle having energy $E = 1$. These orbits are all PT symmetric (left-right symmetric). There is one special trajectory that oscillates between the left and right turning points in the lower-half complex plane; this trajectory begins at $x = -i$. Seven classical orbits begin at $x = -i/4$, $i/4$, $i/2$, $9i/10$, $95i/100$, $99i/100$, $-73i/100$. Finally, there is a special singular trajectory that oscillates between the upper turning point and $i\infty$. Like the trajectories shown in Fig. 1.12, the classical particle moves fast in the lighter regions and slower in the darker regions.

Next, let us consider the quartic $\mathcal{PT}$-symmetric oscillator obtained by setting $\varepsilon = 2$ in (1.37). The resulting potential $V(x) = -x^4$ is *upside down* and the classical system described by this potential is apparently unstable on the real axis. The conventional view is that one should reject theories based on unstable potentials such as $-x^4$ because a classical particle on the real axis subject to this potential will zoom off to $\pm\infty$ (unless it is precariously balanced at rest at $x = 0$). However, this conclusion is simplistic because, as we show below, a particle of energy $E > 0$ reaches infinity in finite time and so we must answer the question, Where does the particle go next? Real analysis does not provide an answer to this question. However, complex-variable theory is able to answer the question, and the answer is most remarkable [Kevrekidis *et al.* (2017)].

Let us calculate the time for a classical particle of energy $E > 0$ that is initially at the origin to travel to infinity. The time of flight T from $x = 0$ to $x = \infty$ for a particle described by the Hamiltonian $H = p^2 - x^4$ is *finite*:

$$T = \int_{t=0}^{T} dt = \int_{x=0}^{\infty} dx/\dot{x} = \tfrac{1}{2}\int_{x=0}^{\infty} dx/\sqrt{E + x^4},$$

where we have used Hamilton's equation $\dot{x} = \frac{\partial H}{\partial p} = 2p$. This integral can be evaluated in terms of the beta function B [Abramowitz and Stegun (1965)]:

$$T = \tfrac{1}{8}E^{-1/4}\int_{u=0}^{\infty} du\, u^{-3/4}(1+u)^{-1/2}$$

$$= \tfrac{1}{8}E^{-1/4}B\left(\tfrac{1}{4},\tfrac{1}{4}\right) = \tfrac{1}{8\sqrt{\pi}}E^{-1/4}\left[\Gamma\left(\tfrac{1}{4}\right)\right]^2. \tag{1.42}$$

Thus, the time T is finite for any positive value of the classical energy E.

To find out where the particle goes next, we approach the real axis from the complex plane. Figure 1.15 shows eight classical trajectories in the complex-x plane for $E = 1$. There are now four turning points. One trajectory represents a particle that oscillates between the upper pair of turning points and another trajectory represents a particle that oscillates between the lower pair of turning points. The other six classical trajectories are all closed periodic orbits. The time for a particle of energy $E = 1$ to execute one closed orbit is

$$2T = \frac{1}{4\sqrt{\pi}}\left[\Gamma\left(\tfrac{1}{4}\right)\right]^2,$$

regardless of which path it follows. As the orbits approach the real axis, the particle moves faster, and in the limiting case we can see that a particle that begins at the origin reaches $+\infty$ in finite time T, zooms around to $-\infty$ in no time at all because it travels infinitely fast, and then it *returns to the origin* in time T. The particle continues to follow this limiting closed path and

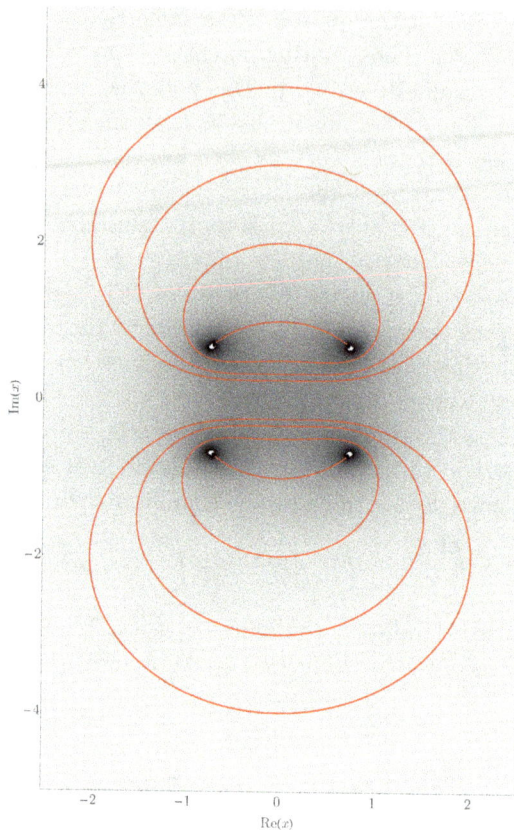

Fig. 1.15   Eight closed periodic trajectories for $H = p^2 - x^4$ in the complex-x plane. The energy is $E = 1$. Turning points located at $x = e^{\pm i\pi/4}$ and at $x = e^{\pm 3i\pi/4}$ are indicated by dots. Initial conditions are $x_0 = \pm i, \pm i/2, \pm i/4, \pm i/8$. A key feature of $\mathcal{PT}$-symmetric systems is that while the classical orbits may appear to be open from the narrow perspective of the real axis, they are seen to be closed when examined in the complex plane.

repeatedly returns to the origin. Thus, a classical particle does not simply disappear at $x = \pm\infty$. Rather, when the particle reaches $\pm\infty$, it instantly reappears at $\mp\infty$. Thus, the classical particle *periodically* completes the transit from $\pm\infty$ to $\mp\infty$. (This periodic motion is equivalent to a $\mathcal{PT}$-symmetric flux of particles on the real axis with a source at $\pm\infty$ balanced by a sink at $\mp\infty$.)

We may now ask, As it executes this periodic motion on the real axis, where is the classical particle most likely to be found? To answer this

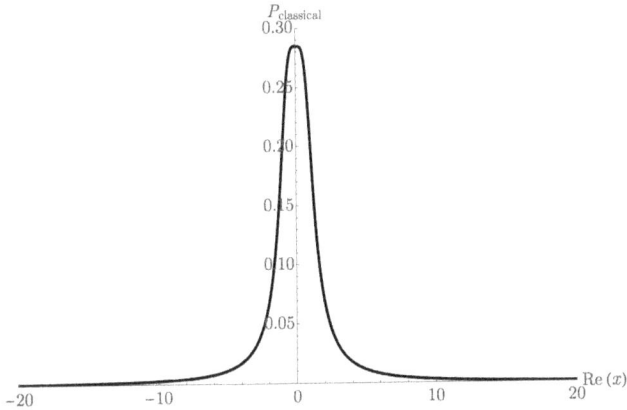

Fig. 1.16   Normalized classical probability $P_{\text{classical}}(x)$ for the quartic oscillator described by the Hamiltonian $H = p^2 - x^4$. The classical particle has energy $E = 1$. This plot shows that while the classical particle repeatedly zooms off to infinity, it spends almost all its time (and is most likely to be found) near the origin $x = 0$. Thus, while $x = 0$ is a point of static instability, the particle motion is dynamically stable!

question, we note that the classical probability density $P_{\text{classical}}(x)$ is proportional to the inverse of the speed of the particle at the point x; the faster the particle moves, the less likely it is to be there. In Fig. 1.16 we plot the normalized classical probability density of finding a particle of energy $E = 1$ on the real axis.[5] The key feature of this plot is that it is highly peaked at the origin. This shows that the particle is most likely to be found very near the origin because it spends most of its time there. Thus, we can interpret the $-x^4$ potential as being *dynamically stable* and not unstable because this potential confines classical particles to the vicinity of the origin. More precisely, while this potential pushes classical particles off to infinity, the chance of ever finding particles there is zero!

In Fig. 1.17 we give a three-dimensional plot of the absolute value of the complex classical probability density. Note that as the particle executes a closed orbit in the complex plane, it is most likely to be found near the origin. Thus, in a time-averaged sense, we can think of the classical particle as being *bound* to the origin; the classical particle is always sliding down the potential hill and whizzing off to infinity and yet at any given instant it is most likely to be very near the origin.

---

[5]The classical probability density is normalized by dividing $P_{\text{classical}}(x)$ by $\int_{-\infty}^{\infty} dx\, P_{\text{classical}}(x)$. Of course, it is essential that this integral converge; otherwise, the normalized classical probability density will vanish.

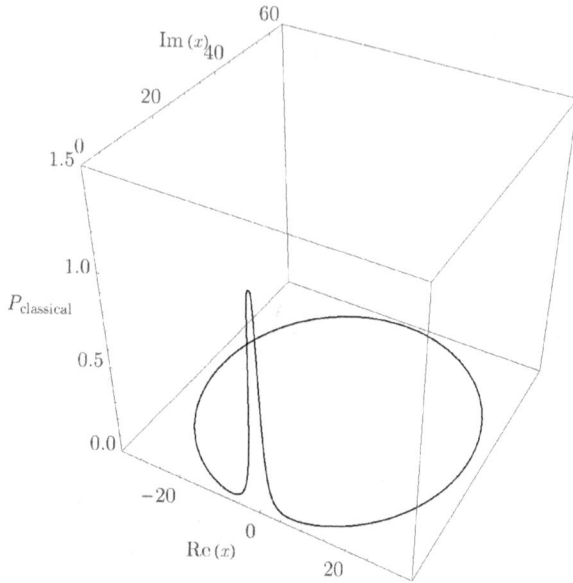

Fig. 1.17   Three-dimensional plot of the absolute value of the complex classical probability $1/|\dot{x}|$ for a closed orbit beginning at $x_0 = i/64$ in the complex plane. The classical energy is $E = 1$. The classical particle executes a rapid loop in the complex plane but goes very slowly near the origin.

    We conclude that potentials that one might think of as being unstable from a purely real perspective may no longer be unstable in a complex regime. The fundamental reason for this is that when we extend the real number system to the complex number system, we lose the property of *ordering*; that is, one can say that $a > b$ if a and b are real, but not if they are complex. In physics it is conventional to say that a potential is unstable if it is unbounded below. However, the assertion that a potential $V(x)$ is unbounded below becomes meaningless if it is complex. Thus, the complex domain provides a setting for the taming of physical instabilities by making static instabilities become dynamically stable.

    In Chap. 2 we extend this notion of a particle being bound to the origin from classical mechanics to quantum mechanics, and we give a simple and rigorous proof that a quantum particle subject to the repulsive $-x^4$ potential is also in a bound state and confined to the vicinity of the origin. Our classical study suggests a heuristic explanation for why the energy levels of the *quantized* Hamiltonian $H = \hat{p}^2 - \hat{x}^4$ are all real and positive. One can obtain quantum mechanics by summing over all possible classical paths [Feynman and Hibbs (1965)]. If one extends this path integral to include a sum over all *complex* classical paths and not just real paths, then all classical paths in the sum are stable closed orbits and the seemingly unstable motion on the real-x axis becomes infinitesimally rare (with probabilistic measure zero).

Not all complex potentials are stable. The PT-symmetric potential $x^2(ix)^\varepsilon$ has stable closed orbits when $\varepsilon \geq 0$, but it becomes unstable when $\varepsilon < 0$. For negative values of ε the classical trajectories are PT symmetric (left-right symmetric), but the orbits are no longer closed and a classical particle under the influence of the potential drifts off to infinity in infinite (not finite) time. As ε decreases below 0, the turning points in (1.37) and (1.38), which lie on the real axis in Fig. 1.12, move into the upper-half complex plane. Classical paths wind around these turning points and eventually veer off to infinity. Because the trajectories in this case are open and nonperiodic, the system is no longer in equilibrium, and we say that the PT symmetry is broken. Figure 1.18 shows that as ε approaches 0 from below, the trajectories wind more and more before veering off to infinity. Trajectories for three different negative values of ε are shown in Fig. 1.18.

We show in Chap. 2 that the difference between closed and open classical trajectories is crucial for the quantized version of the theory. If the classical trajectory is closed, we can use the Bohr–Sommerfeld quantization formula

$$\oint_C dx \sqrt{E - V(x)} \sim \left(n + \tfrac{1}{2}\right)\pi, \qquad (1.43)$$

where C is the closed classical trajectory, to obtain an approximate formula for the quantized energy levels E_n. One may think of the closed classical orbits as representing complex classical atoms. The semiclassical interpretation of the Bohr–Sommerfeld formula is that in quantum mechanics we think of a particle as a wave. If a particle is in a closed orbit and the orbit consists of an integer number n of wavelengths, then the wave interferes with itself constructively. This is precisely the meaning of (1.43), which is an approximate condition for an energy level. Thus, an allowed energy is a resonant effect, like an echo chamber or a *whispering gallery* [Strutt (1878)].

As a consequence of PT symmetry, this approximate quantization formula gives *real* values for the energies E_n [Bender *et al.* (2007c)]. However, if the classical trajectories are open, the quantization integral (1.43) cannot

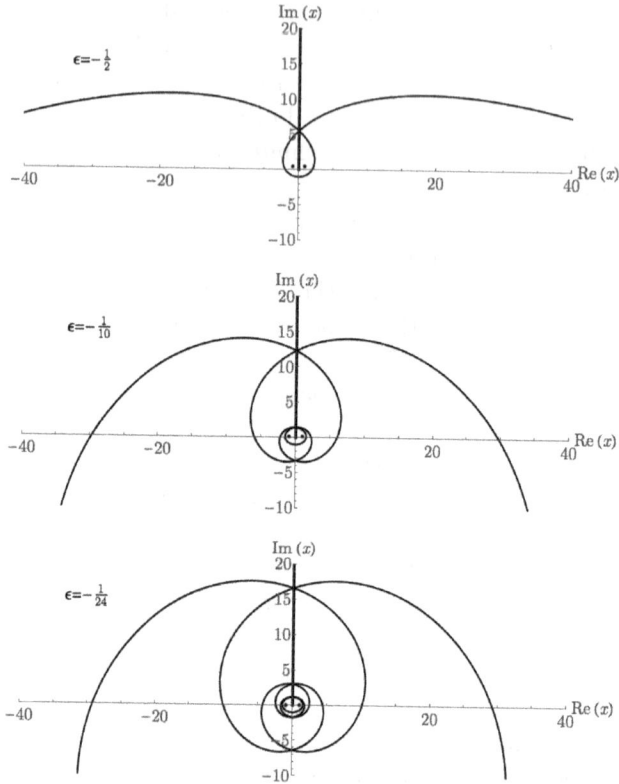

Fig. 1.18   Three plots of the classical trajectories for negative ε. For these plots $E = 1$. The classical trajectories begin at $x = -i$, and while they appear to cross, they cannot do so. Rather, the trajectories lie on an infinite-sheeted Riemann surface because the potential function $x^2(ix)^\varepsilon$ has a logarithmic branch-point singularity at $x = 0$. The sheets are joined by branch cuts (indicated by a solid line), which we take to lie on the positive-imaginary axis. Turning points on the principal sheet are indicated by dots. The trajectories shown represent the path of a classical particle that goes both forward and backward in time. These trajectories are PT symmetric (left-right symmetric). The curves wind around the origin and eventually approach infinity.

be used because, as the symbol $\oint$ indicates, the integration contour must be *closed*. Indeed, we will see that when the classical trajectories are open, the PT symmetry is broken and the eigenvalues of the quantum Hamiltonian become complex. For the potential $x^2(ix)^\varepsilon$ the orbits are open when $\varepsilon < 0$ because the orbital period is infinite. An echo chamber cannot resonate if the wave does not have enough time to interfere constructively with itself.

Chapter 2

$\mathcal{PT}$-Symmetric Eigenvalue Problems

"The difficulty, as in all this work, is to find a
notation which is both concise and intelligible to
at least two people of whom one may be the author."

—$\mathcal{P}\mathcal{T}$ *Matthews*

Perhaps the most remarkable feature of a $\mathcal{PT}$-symmetric quantum-
mechanical Hamiltonian is that its eigenvalues can be all real even if the
Hamiltonian is not Hermitian. The purpose of this chapter is to explain the
mathematics (complex asymptotic methods, WKB approximation, Stokes
sectors) needed to define and solve the eigenvalue problems associated with
$\mathcal{PT}$-symmetric Hamiltonians. We apply these techniques to a wide variety
of examples that illustrate the parametric deformation (analytic continuation) of differential-equation eigenvalue problems.

In reading this chapter one should keep in mind that it is quite nontrivial
to prove that a non-Hermitian $\mathcal{PT}$-symmetric Hamiltonian has an entirely
real spectrum even though at first glance it might seem that spectral reality
would be an elementary consequence of $\mathcal{PT}$ invariance. A Hamiltonian has
an invariance if the operator representing that invariance commutes with
the Hamiltonian. For example, if a Hamiltonian is parity invariant, then $\mathcal{P}$
commutes with the Hamiltonian, and as a consequence, the eigenstates of
the Hamiltonian are also eigenstates of $\mathcal{P}$. If a Hamiltonian is $\mathcal{PT}$ symmetric ($\mathcal{PT}$ invariant), then the $\mathcal{PT}$ operator commutes with the Hamiltonian.
However, $\mathcal{PT}$ symmetry is more subtle than parity symmetry because the
$\mathcal{PT}$ operator is *not* linear. Because of this nonlinearity, the eigenstates of
the Hamiltonian *may or may not* be eigenstates of $\mathcal{PT}$.

Let us see how we are led astray if we (wrongly) assume that a given
eigenstate ψ of a $\mathcal{PT}$-invariant Hamiltonian $\hat{H}$ is necessarily an eigenstate
of the $\mathcal{PT}$ operator. Denoting the eigenvalue of the $\mathcal{PT}$ operator by λ, we

see that the eigenvalue equation reads

$$\mathcal{P}\mathcal{T}\psi = \lambda\psi.$$

We multiply this equation on the left by $\mathcal{P}\mathcal{T}$ and substitute $(\mathcal{P}\mathcal{T})^2 = 1$:

$$\psi = (\mathcal{P}\mathcal{T})\lambda(\mathcal{P}\mathcal{T})^2\psi.$$

Since $\mathcal{T}$ is antilinear, we get

$$\psi = \lambda^*\lambda\psi = |\lambda|^2\psi.$$

Thus, $|\lambda|^2 = 1$, so the eigenvalue λ of the $\mathcal{P}\mathcal{T}$ operator is a pure phase:

$$\lambda = e^{i\alpha}.$$

This result is certainly valid; an eigenvalue λ of the $\mathcal{P}\mathcal{T}$ operator always has the form $e^{i\alpha}$.

Next, we multiply the eigenvalue equation

$$\hat{H}\psi = E\psi$$

on the left by $\mathcal{P}\mathcal{T}$ and again use the property that $(\mathcal{P}\mathcal{T})^2 = 1$:

$$(\mathcal{P}\mathcal{T})\hat{H}\psi = (\mathcal{P}\mathcal{T})E(\mathcal{P}\mathcal{T})^2\psi.$$

Using the eigenvalue equation $\mathcal{P}\mathcal{T}\psi = \lambda\psi$ and recalling that $\mathcal{P}\mathcal{T}$ commutes with $\hat{H}$, we get

$$\hat{H}\lambda\psi = (\mathcal{P}\mathcal{T})E(\mathcal{P}\mathcal{T})\lambda\psi.$$

Finally, we again use the property that $\mathcal{T}$ is antilinear to obtain

$$E\lambda\psi = E^*\lambda\psi.$$

Since λ is a pure phase, it is nonzero, so we divide this equation by λ and conclude that the eigenvalue E is real: $E = E^*$. But this conclusion is wrong because, as we saw in Chap. 1, an eigenvalue of a $\mathcal{P}\mathcal{T}$-invariant Hamiltonian may well be complex.

The chain of reasoning above is invalid because it was not correct to assume at the start that the $\mathcal{P}\mathcal{T}$ invariance of the Hamiltonian implies that an eigenstate ψ of $\hat{H}$ is also an eigenstate of $\mathcal{P}\mathcal{T}$. However, we have discovered something useful: *If an eigenfunction of a $\mathcal{P}\mathcal{T}$-symmetric Hamiltonian is also an eigenfunction of $\mathcal{P}\mathcal{T}$, then the associated energy eigenvalue is real.* Therefore, if *every* eigenfunction of a $\mathcal{P}\mathcal{T}$-symmetric Hamiltonian is also an eigenfunction of the $\mathcal{P}\mathcal{T}$ operator, then the spectrum of $\hat{H}$ is entirely real. To use the terminology introduced in Chap. 1, if every eigenfunction of the Hamiltonian is also an eigenstate of the $\mathcal{P}\mathcal{T}$ operator, the $\mathcal{P}\mathcal{T}$ symmetry of $\hat{H}$ is *unbroken*; if some of the eigenfunctions of a PT-symmetric Hamiltonian are not simultaneously eigenfunctions of the PT operator, the PT symmetry of $\hat{H}$ is *broken*.

Proving that the spectrum of a PT-symmetric Hamiltonian is real is not easy. It took several years after the discovery of the family of PT-symmetric Hamiltonians $\hat{H} = \hat{p}^2 + \hat{x}^2(i\hat{x})^\varepsilon$ ($\varepsilon \geq 0$) before a rigorous proof that the eigenvalues of $\hat{H}$ are real was finally published [Dorey *et al.* (2001a, 2007b)].[1] Many others contributed to the early rigorous mathematical development of the theory of PT symmetry. These include [Shin (1998, 2001, 2002, 2004, 2005a,b); Delabaere and Pham (1998); Delabaere and Trinh (2000); Weigert (2003a,b, 2005, 2006); Davies (2010); Ralston (2007); Scholtz and Geyer (2006a,b); Mostafazadeh (2002b,c,d); Bender *et al.* (2016a)].

2.1 Examples of PT-Deformed Eigenvalue Problems

Ordinarily, we construct a PT-symmetric Hamiltonian from a Hermitian Hamiltonian by introducing a continuous parameter, usually denoted by ε, that *deforms* (analytically continues) the Hermitian Hamiltonian into the complex (non-Hermitian) domain. For example, for the Hamiltonian H_{coupled} in (1.5) this parameter is b, and for the quantized version of the Hamiltonian H in (1.37) this parameter is ε.[2] These PT-symmetric Hamiltonians are Hermitian only for the special cases $b = 0$ and $\varepsilon = 0$ and they become non-Hermitian when $b \neq 0$ and $\varepsilon \neq 0$. To someone with a background in conventional quantum mechanics it is surprising that there actually exist parametric regions of b and ε for which the eigenvalues are all real even though these Hamiltonians are not Hermitian.

An elementary class of complex PT-symmetric Hamiltonians is

$$\hat{H} = \hat{p}^2 + \hat{x}^2 + i\varepsilon\hat{x}. \tag{2.1}$$

This Hamiltonian is a complex deformation of the Hermitian harmonic-oscillator Hamiltonian $\hat{p}^2 + \hat{x}^2$. If we take the deformation parameter ε to be real, the Hamiltonian $\hat{H}$ is PT symmetric for all ε because, as we have seen in Chap. 1, the combination $i\hat{x}$ is invariant under PT reflection.

---

[1]The proof by Dorey *et al.* draws from many areas of theoretical and mathematical physics and uses spectral determinants, the Bethe *ansatz*, the Baxter TQ relation, the monodromy group, and an array of techniques used in conformal quantum field theory. This proof is a *tour de force* and is described in Chap. 6. The proof establishes a correspondence, known as the *ODE/IM correspondence*, between ordinary differential equations and integrable models [Dorey *et al.* (2007b, 2008)].

[2]In Chap. 9 we study the complex deformation of integrable classical systems.

To formulate the coordinate-space Schrödinger equation associated with $\hat{H}$ in (2.1), we make the usual transcriptions

$$\hat{x} \to x \quad \text{and} \quad \hat{p} \to -i\frac{d}{dx}. \tag{2.2}$$

The formal Schrödinger eigenvalue equation $\hat{H}\psi = E\psi$ then takes the form of the differential equation

$$-\psi''(x) + x^2\psi(x) + i\varepsilon x\psi(x) = E\psi(x) \tag{2.3}$$

with the accompanying boundary conditions that $\psi(x)$ vanishes as x approaches $\pm\infty$ on the real-x axis:

$$\psi(\pm\infty) = 0. \tag{2.4}$$

The novelty of PT-symmetric quantum mechanics is that we treat the coordinate variable x in a Schrödinger eigenvalue problem as *complex*. This does not affect the transcription (2.2) because the Heisenberg algebra $[\hat{x}, \hat{p}] = i$ continues to hold even if x is complex.

To solve the eigenvalue problem (2.3) analytically we make the change of independent variable $x \to x - i\varepsilon/2$, which reduces the eigenvalue equation (2.3) to the eigenvalue equation for the quantum harmonic oscillator:

$$-\psi''(x) + x^2\psi(x) = \left(E - \varepsilon^2/4\right)\psi(x). \tag{2.5}$$

This example is elementary because the change of variable does not affect the boundary conditions on the eigenfunctions; the boundary conditions (2.4) remain the same for any value of ε. Since the eigenvalues of $\hat{p}^2 + \hat{x}^2$ are $E_n = 2n + 1$ ($n = 0, 1, 2, 3, \ldots$), the eigenvalues of $\hat{H}$ in (2.1) are

$$E_n = 2n + 1 + \varepsilon^2/4 \quad (n = 0, 1, 2, 3, \ldots). \tag{2.6}$$

In summary, while $\hat{H}$ is Hermitian only for $\varepsilon = 0$, its eigenvalues remain real for all real values of ε; there is no PT transition between unbroken- and broken-PT-symmetric regions of ε.

A fancier deformation of the harmonic-oscillator Hamiltonian $\hat{p}^2 + \hat{x}^2$ is

$$\hat{H} = \hat{p}^2 + \hat{x}^2(i\hat{x})^\varepsilon, \tag{2.7}$$

where the deformation parameter ε is again real. The eigenvalue differential equation $\hat{H}\psi = E\psi$ associated with this Hamiltonian is

$$-\psi''(x) + x^2(ix)^\varepsilon\psi(x) = E\psi(x). \tag{2.8}$$

Unlike the boundary conditions in (2.4), the boundary conditions for this differential equation depend on the deformation parameter ε. Thus, it is crucial to understand how to continue analytically an eigenvalue problem into the complex domain. This is a principal objective of this chapter.

We emphasize that while a classical dynamical system is *local*, a differential-equation eigenvalue problem is inherently *nonlocal*. The solution to the differential equation(s) for a dynamical system is determined once the initial conditions are specified, and these initial conditions are local. (They are given at a point, say at $t = 0$.) However, the eigenvalues of a differential equation are determined by widely separated boundary conditions. It is because of this nonlocality that quantum-mechanical systems, and especially complex ones, must be treated with great care.

We cannot solve the differential equation (2.8) exactly and thus we cannot calculate the eigenvalues of the Hamiltonian (2.7) analytically unless $\varepsilon = 0$.[3] However, numerical calculations (see Sec. 2.6) reveal that there is a PT transition at $\varepsilon = 0$, as Fig. 2.1 shows. This transition resembles the PT transition in Fig. 1.4 for the Hamiltonian in (1.5), but the transition illustrated in Fig. 2.1 is more elaborate; when $\varepsilon < 0$, there is not just one pair of merging eigenvalues but rather an infinite number of merging pairs. (Recall from Chap. 1 that the point at which a pair of eigenvalues merges is called an *exceptional point*.) The sequence of special real values of ε at which pairs of eigenvalues merge converges monotonically upwards to $\varepsilon = 0$.

The PT transition at $\varepsilon = 0$ has an underlying classical correspondence. In Sec. 1.7 we saw that when $\varepsilon > 0$, the classical trajectories associated with the classical Hamiltonian are closed and that the periods of the classical orbits are finite. Because these orbits are closed, the Bohr–Sommerfeld quantization formula (1.43) applies, and as shown in Sec. 2.6, this semiclassical approximation predicts that the eigenvalues are real and positive. However, when $\varepsilon < 0$, the classical trajectories are open and the classical particles drift off to infinity, as shown in Fig. 1.18. [The classical trajectories $x(t)$ reach infinity in infinite time, so they cannot close.] Thus, the Bohr–Sommerfeld formula is not applicable when $\varepsilon < 0$, so it is not surprising that the eigenvalues become complex.

It *is* surprising that the eigenvalues of $\hat{H}$ in (2.7) are all real when $\varepsilon \geq 0$, and it is particularly counterintuitive that, as we can see in Fig. 2.1, the eigenvalues are all real even when $\varepsilon = 2$. This is because for $\varepsilon = 2$ the resulting Hamiltonian $\hat{H} = \hat{p}^2 - \hat{x}^4$ has an upside-down potential. As explained in Sec. 1.7, classical potentials that are unstable on the real axis may actually become stable in the complex domain; instabilities on the real

---

[3] The limit $\varepsilon \to \infty$ also gives an exactly solvable potential and the potential obtained in this limit is the complex analog of a square-well potential. See [Bender *et al.* (1999a)].

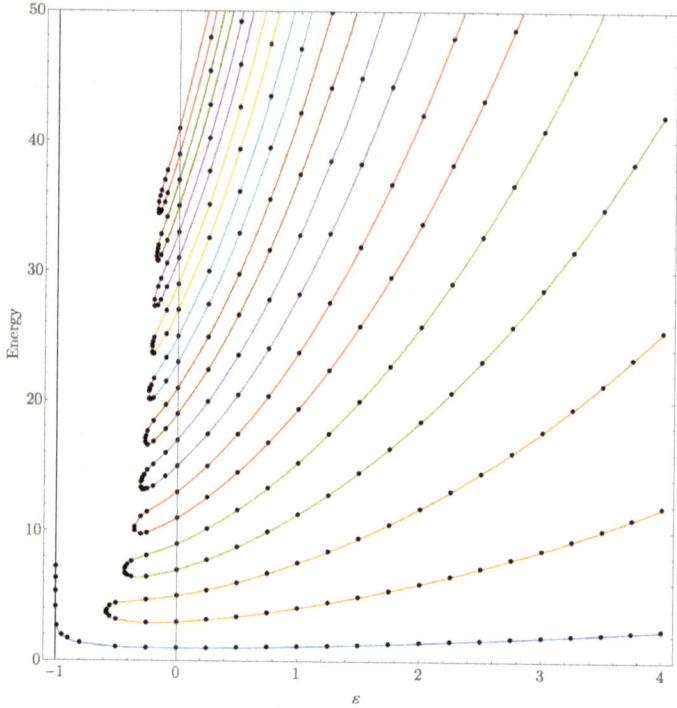

Fig. 2.1   Eigenvalues of $\hat{H} = \hat{p}^2 + \hat{x}^2(i\hat{x})^\varepsilon$ plotted as functions of real ε. This Hamiltonian is a complex $\mathcal{PT}$-symmetric deformation of the harmonic-oscillator Hamiltonian. The eigenvalues are all real when $\varepsilon \geq 0$. This is the region of unbroken $\mathcal{PT}$ symmetry. At the transition $\varepsilon = 0$ the eigenvalues $(1, 3, 5, 7, \ldots)$ are those of the harmonic oscillator. When ε becomes negative, pairs of eigenvalues merge and become degenerate, and like the eigenvalues in Fig. 1.4, they veer into the complex plane as complex-conjugate pairs. (The complex eigenvalues are not shown on this graph.) The merging pairs of eigenvalues are color coded; with increasing energy the merging pairs are colored yellow, followed by green, red, purple, brown, and so on. As ε becomes more negative, eventually only one real eigenvalue (blue) remains, and this eigenvalue becomes infinite as ε approaches -1. At $\varepsilon = -1$ the spectrum is *null* (there are no real or complex eigenvalues at all). This disappearance of eigenvalues is explained in Sec. 2.7. Numerical values of the first five eigenvalues for $\varepsilon = 0, 1/2, 1, 3/2, 2, 5/2, 3, 7/2, 4$ are listed in Table 2.1.

axis can become tamed in the complex domain. In Sec. 2.4 we prove that this stabilization effect persists at the quantum-mechanical level as well as at the classical level. Of course, it is startling that an upside-down potential can have *localized* bound states and *discrete positive* eigenvalues because this seems to be in conflict with the conventional interpretation that a $-x^4$ potential has a continuum of energy levels that is unbounded below. The

Table 2.1   Numerical values of the first five eigenvalues, E_0, E_1, E_2, E_3, and E_4, of the Hamiltonian $\hat{H} = \hat{p}^2 + \hat{x}^2(i\hat{x})^\varepsilon$ for various values of ε.

	E_0	E_1	E_2	E_3	E_4
$\varepsilon=0$	1	3	5	7	9
$\varepsilon=1/2$	1.048956	3.434539	6.051737	8.791012	11.620695
$\varepsilon=1$	1.156267	4.109229	7.562274	11.314422	15.291554
$\varepsilon=3/2$	1.301514	4.969791	9.480030	14.530476	19.997745
$\varepsilon=2$	1.477150	6.003386	11.802434	18.458819	25.791792
$\varepsilon=5/2$	1.679907	7.208428	14.540831	23.134243	32.741996
$\varepsilon=3$	1.908265	8.587221	17.710809	28.595103	40.918891
$\varepsilon=7/2$	2.161511	10.143518	21.328941	34.879469	50.390825
$\varepsilon=4$	2.439346	11.881565	25.411553	42.023722	61.222419

explanation of this discrepancy is that *depending on the choice of boundary conditions, a Hamiltonian may have several completely different spectra.* In order to determine the appropriate boundary conditions for the $-x^4$ potential, we must first understand how to deform an eigenvalue problem as a function of a deformation parameter.

We can obtain an eigenvalue problem for the Hamiltonian $\hat{H} = \hat{p}^2 - \hat{x}^4$ in two completely different ways by using two different deformation procedures. First, we can begin with the conventional Hermitian eigenvalue problem for the quartic quantum anharmonic oscillator

$$- \psi''(x) + \varepsilon x^4 \psi(x) = E\psi(x), \qquad \psi(\pm\infty) = 0, \qquad (2.9)$$

where $\varepsilon > 0$. We then deform this problem by allowing ε to vary from 1 to -1 along a path in the complex plane that follows the unit semicircle; that is, we take $\varepsilon = e^{i\theta}$ and let $\theta : 0 \to \pi$. Second, we can begin with the conventional Hermitian eigenvalue problem for the quantum-harmonic-oscillator equation and then deform this problem by allowing ε in

$$- \psi''(x) + x^2(ix)^\varepsilon \psi(x) = E\psi(x), \qquad \psi(\pm\infty) = 0 \qquad (2.10)$$

to increase continuously from 0 to 2 through real values. In both cases one arrives at the same eigenvalue differential equation that is associated with the upside-down potential $-x^4$:

$$- \psi''(x) - x^4 \psi(x) = E\psi(x). \qquad (2.11)$$

However, in the first case the eigenvalue spectrum ends up being complex while in the second case the spectrum ends up being discrete, real, and positive, as shown in Fig. 2.1. The differences between the two spectra are due to the different boundary conditions that one obtains from the two different deformation procedures. These boundary conditions depend on the choice of deformation procedure.

To illustrate the subtle issues involved in the deformation of eigenvalue problems, we consider two paradoxical examples that were first discussed in [Bender and Turbiner (1993)].

Example 1: *Deformed harmonic oscillator.*

Consider the eigenvalue problem for the quantum harmonic oscillator in which the square of a deformation parameter ε appears in the coefficient of the x^2 potential:

$$-\psi''(x) + \varepsilon^2 x^2 \psi(x) = E\psi(x), \qquad \psi(\pm\infty) = 0. \qquad (2.12)$$

Let us initially take ε to be a positive real parameter. We know that the eigenvalues are linear functions of ε:

$$E_n = (2n+1)\varepsilon \qquad (n = 0, 1, 2, 3, \ldots). \qquad (2.13)$$

What happens if we deform this eigenvalue problem by analytically continuing from $\varepsilon = 1$ to $\varepsilon = -1$ along the unit semicircle in the complex-ε plane? To be precise, what happens if we let $\varepsilon = e^{i\theta}$ and let θ range smoothly from 0 to π. Of course, at the end of the deformation process the differential equation (2.12) reappears unchanged. However, the eigenvalues (2.13) become negative under this analytic continuation. Can the eigenvalues of the quantum-harmonic-oscillator Hamiltonian really be negative?

Example 2: *Deformed sextic anharmonic oscillator.*

Consider the eigenvalue problem for the sextic quantum anharmonic oscillator

$$-\psi''(x) + \varepsilon^2 x^6 \psi(x) - 3\varepsilon x^2 \psi(x) = E\psi(x), \qquad \psi(\pm\infty) = 0. \qquad (2.14)$$

Let us assume that ε is initially a positive parameter $\varepsilon > 0$, so on the real-x axis the potential $\varepsilon^2 x^6 - 3\varepsilon x^2$ has a double well. We cannot calculate all the eigenvalues of this potential analytically, but there is a surprisingly simple exact formula for the *ground-state* energy: $E_0(\varepsilon) = 0$. What happens if we deform (analytically continue) this problem from positive ε to negative ε along the unit semicircle in the complex-ε plane? That is, suppose we let $\varepsilon = e^{i\theta}$ and allow θ to vary smoothly from 0 to π. When the deformation is completed and $\varepsilon < 0$, the potential has a *single well* as a function of real x and the minimum of the potential on the real axis is 0. However, under this analytic continuation the ground-state energy remains unchanged; that is, $E_0(\varepsilon)$ is still 0. Does this contradict the standard result in quantum mechanics that because of zero-point-energy fluctuations the lowest energy level in a potential well must lie above the minimum of the potential?

To resolve the nontrivial and paradoxical results in these two examples requires that we understand what happens to the boundary conditions when an eigenvalue problem is deformed (or analytically continued). This in turn requires that we introduce the concept of Stokes sectors. We explain these ideas in the next section.

2.2   Deformed Eigenvalue Problems and Stokes Sectors

The purpose of this section is to explain how to formulate the Schrödinger differential-equation eigenvalue problem for a deformed Hamiltonian such as that in (2.7). (Detailed discussions may be found in [Bender and Wu (1968, 1969); Bender and Turbiner (1993)].) The conceptual objective here is to ascertain what happens to the boundary conditions on coordinate-space eigenfunctions as the eigenvalue problem is analytically deformed. We begin our discussion by explaining the paradoxes raised in Examples 1 and 2 in the previous section.

2.2.1   *Resolution of puzzles in examples of Sec. 2.1*

Because we can solve the eigenvalue problems (2.12) and (2.14) exactly and can find closed-form expressions for the eigenfunctions, it is not hard to explain the paradoxical results that we discovered in the examples in the previous section when we deformed these eigenvalue problems. We examine first the quantum-harmonic-oscillator differential equation (2.12). When ε is real and positive, the nth eigenfunction $\psi_n(x)$ has the form of the gaussian $e^{-\varepsilon x^2/2}$ multiplied by the nth Hermite polynomial. These eigenfunctions vanish exponentially as $x \to \infty$ and as $x \to -\infty$ along the real-x axis. The usual way to find these eigenfunctions, either numerically or analytically, is to integrate the differential equation along a straight-line contour that follows the real axis from $-\infty$ to $+\infty$ (see Fig. 2.2a), and to require that the solution vanish at both ends of this contour.

Next, we observe that these eigenfunctions do not vanish on the real axis only; they vanish exponentially as $|x| \to \infty$ in the complex-x plane in two angular sectors that contain the real axis:

$$-\tfrac{1}{4}\pi < \arg x < \tfrac{1}{4}\pi \quad \text{and} \quad \tfrac{3}{4}\pi < \arg x < \tfrac{5}{4}\pi.$$

These angular sectors are called *Stokes sectors* or *Stokes wedges* [Bender and Orszag (1999)] and are shown in Fig. 2.2a. To find the eigenfunctions we need not restrict the integration to a contour along the real axis. Indeed, since the eigenvalue differential equation has no singular points [Bender and Orszag (1999)] in the finite complex plane, its solutions are entire (that is, analytic in the finite complex plane). Thus, we can deform the original

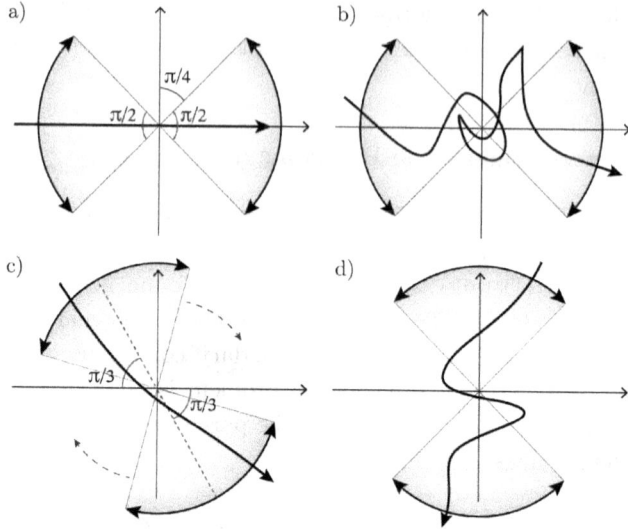

Fig. 2.2   Stokes sectors for the deformed quantum-harmonic-oscillator eigenvalue problem (2.12), where $\varepsilon = e^{i\theta}$. The purpose of the shading in the Stokes sectors is to indicate that the crucial feature is the angular opening of the sector at complex infinity, where boundary conditions are imposed; the Stokes sectors play no role in the finite complex plane. The configuration of Stokes sectors for the undeformed problem $\theta = 0$ is shown in panel (a). These Stokes sectors have angular opening $\pi/2$ and are centered about the positive-real and the negative-real axes. The integration contour in panel (a) follows the real axis from $-\infty$ to $+\infty$. However, the integration contour can be deformed into the complex-x plane as long as the endpoints of the contour remain in the Stokes sectors. One such deformed contour is shown in panel (b). The deformed contour must be continuous but it need not be smooth and may even cross itself, as shown. As θ increases, the Stokes sectors rotate clockwise. Panel (c) shows the sectors for $\theta = 2\pi/3$ and panel (d) shows the sectors for $\theta = \pi$. The integration contours for these panels can no longer terminate on the real axis.

integration contour, which lies on the real-x axis, into the complex-x plane and follow *any* contour as long as the contour terminates inside the left and right Stokes sectors. One such path is shown in Fig. 2.2b. Because the eigenfunctions (the solutions to the differential equation) are analytic functions of x, the eigenvalues are independent of the choice of contour.

What happens to the orientation of the two Stokes sectors in which the eigenfunctions vanish exponentially as we rotate the parameter $\varepsilon = e^{i\theta}$ into the complex plane? As θ increases, the angular openings of the Stokes sectors remain constant and equal to $\frac{1}{2}\pi$, but the sectors rotate *clockwise*. The centers of the Stokes sectors for $\theta > 0$ are located at $-\frac{1}{2}\theta$ and at $\pi - \frac{1}{2}\theta$.

(Figure 2.2c shows the orientation of the sectors for $\theta = \frac{2}{3}\pi$.) Thus, when θ reaches the value π, the Stokes sectors are centered about the *imaginary-x* axes rather than the real-x axes. When $\theta = \pi$, the original differential equation at $\theta = 0$ (2.26) reappears, but this deformation has generated a new eigenvalue problem in which the eigenfunctions vanish exponentially along the imaginary axis rather than along the real axis (see Fig. 2.2d). The eigenvalues for this new unconventional eigenvalue problem are *negative*,

$$E_n = -(2n + 1) \qquad (n = 0, 1, 2, 3, \ldots), \qquad (2.15)$$

and the same as we obtain from a direct analytic continuation of the eigenvalues of the harmonic oscillator (2.12) from $\varepsilon = 1$ to $\varepsilon = -1$. It is remarkable that the *eigenvalues have changed sign* even though the replacement $\varepsilon \to -\varepsilon$ leaves the Hamiltonian $\hat{H} = \hat{p}^2 + \varepsilon^2 \hat{x}^2$ invariant. This demonstrates that the eigenspectrum of a Hamiltonian depends crucially on the boundary conditions that are imposed on the eigenfunctions.

The conclusion is simply this: For each pair of Stokes sectors we can pose a coordinate-space eigenvalue problem for which the boundary conditions are that the eigenfunctions vanish exponentially in the two Stokes sectors. However, it is crucial that the eigenvalue problem be posed in *two separated and nonadjacent* Stokes sectors. If the sectors were adjacent, we could deform the entire integration contour out to infinity, and there would be no eigenvalue problem at all! In order to prevent the integration contour from being deformed out to infinity the endpoints of the contour must be pinned at infinity in nonadjacent Stokes sectors.

For the harmonic-oscillator Hamiltonian $\hat{H} = \hat{p}^2 + \hat{x}^2$ there are four Stokes sectors, each having an angular opening angle of $\frac{1}{2}\pi$ but there are only two distinct nonadjacent pairs of Stokes sectors. Therefore, for this Hamiltonian there are *two* different eigenvalue problems, one having the conventional positive spectrum and the other having a negative spectrum. This explains the paradox associated with the deformation of (2.12).

Next, we turn to the sextic-oscillator eigenvalue problem in (2.14). For $\varepsilon > 0$ the potential has a double well. The exact ground-state eigenfunction is $\psi_0(x) = e^{-\varepsilon x^4/4}$ and the exact ground-state energy is $E_0 = 0$. This eigenfunction vanishes exponentially as $|x| \to \pm\infty$ in two Stokes sectors of angular opening $\frac{1}{4}\pi$ centered about the positive-real and negative-real axes (see Fig. 2.3a). If we let $\varepsilon = e^{i\theta}$ and allow θ to increase from 0, these sectors rotate clockwise and are centered about $-\frac{1}{4}\theta$ and at $\pi - \frac{1}{4}\theta$. Thus, when $\theta = \pi$ these two sectors lie $22\frac{1}{2}^\circ$ below the positive-real axis and $22\frac{1}{2}^\circ$ above the negative-real axis (see Fig. 2.3b).

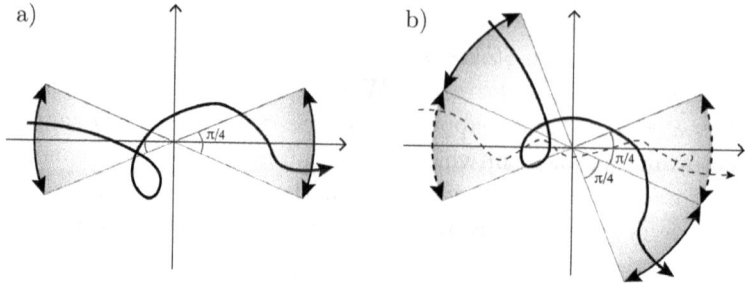

Fig. 2.3   Stokes sectors for the deformed quantum-harmonic-oscillator eigenvalue problem (2.14) with $\varepsilon = e^{i\theta}$. The Stokes sectors for the undeformed problem $\theta = 0$ are shown in panel (a). These Stokes sectors have angular opening $\pi/4$ and are centered about the positive-real and the negative-real axis. The integration contour can follow the real axis from $-\infty$ to $+\infty$, and this contour can then be deformed into the complex-x plane as long as the endpoints of the contour remain in the Stokes sectors. One such contour (heavy line) is shown. As θ increases, the Stokes sectors rotate clockwise. Panel (b) shows the rotated sectors for $\theta = \pi$; the original Stokes sectors have rotated clockwise by 45° so the integration contour (solid heavy line) for these sectors can no longer terminate on the real axis. However, as shown in panel (b), one can formulate a totally new and different eigenvalue problem with a new set of boundary conditions. The integration contours for the new problem (dashed lines) terminate in Stokes sectors of angular opening $\pi/4$ that are centered about the positive-real and negative-real axes.

Like the case of the deformed harmonic oscillator, the key point here is that with this deformation the differential equation is once again real when $\theta = \pi$, even though we have obtained a potential that has a single well on the real axis. However, the Stokes sectors inside of which the boundary conditions are imposed have rotated into the complex plane; the boundary conditions can no longer be imposed on the real-x axis. [If we were to impose exponentially vanishing boundary conditions on the real-x axis by solving the differential equation along the dashed contour in Fig. 2.3b, we would of course find that the ground-state energy (the zero-point energy) is greater than 0.] However, in the rotated problem the eigenfunction, which now has the form $\psi_0(x) = e^{x^4/4}$, still has eigenvalue 0. Note that this eigenfunction vanishes in the rotated Stokes sector of the deformed problem but blows up exponentially as $|x| \to \infty$ on the real axis. This resolves the puzzle regarding the eigenvalue problem (2.14).

2.2.2   *Analytic deformation of eigenvalue problems*

Fortunately, to determine the locations of the Stokes sectors for general eigenvalue problems it is not necessary to solve the differential equation exactly. [This is good news because, for example, it is hopeless to try to solve the eigenvalue differential equation (2.8) exactly!] To understand what happens to the boundary conditions of a deformed eigenvalue problem, it is only necessary to determine the possible asymptotic behaviors of the solutions for large $|x|$, where x is complex. To do this we can use the WKB approximation [Bender and Orszag (1999)].

To apply WKB analysis to a general Schrödinger differential equation

$$y''(x) = Q(x)y(x), \tag{2.16}$$

we need only assume that $|Q(x)| \to \infty$ as $|x| \to \infty$ in the complex-x plane. For all the deformed Hamiltonians considered in Sec. 2.1 this assumption is valid. For large $|x|$ the leading-order WKB approximation to $y(x)$ is

$$y(x) \sim Q(x)^{-1/4} \exp\left[\pm \int^x ds\sqrt{Q(s)}\right]. \tag{2.17}$$

This approximation is called the *physical-optics approximation*. In fact, to study the deformation of eigenvalue problems, we only need the *exponential component* of the WKB approximation to $y(x)$ for large $|x|$,

$$\exp\left[\pm \int^x ds\sqrt{Q(s)}\right], \tag{2.18}$$

which is called the *geometrical-optics approximation* to $y(x)$.

The eigenvalue differential equation (2.8) has the general form (2.16). To identify the appropriate boundary conditions to impose on $\psi(x)$, we consider first the *undeformed* problem, that is, the Hermitian harmonic-oscillator problem for which $\varepsilon = 0$. From (2.18) we see immediately that the possible geometrical-optics approximations to the solutions $\psi(x)$ are $\exp\left(\pm x^2/2\right)$. The conventional requirement that the eigenfunction be square integrable on the real-x axis implies that we must choose the *negative* sign in the exponent. Therefore, the eigenfunctions $\psi(x)$ are Gaussian-like for large $\pm x$. This Gaussian behavior analytically continues from the real-x axis into the complex-x plane: As shown in Fig. 2.2, because the eigenfunctions vanish exponentially on the real-x axis for large $\pm x$, they also vanish exponentially in two Stokes sectors of opening angle $\frac{1}{2}\pi$ in the complex-x plane, a right sector centered about the positive-real axis and a left sector centered about the negative-real axis. The eigenfunctions in these Stokes sectors vanish most rapidly at the centers of the sectors; at the edges of the sectors the eigenfunctions just oscillate and no longer vanish exponentially.

What happens to these sectors as ε increases from 0? As soon as $\varepsilon > 0$, a logarithmic branch-point singularity appears at the origin $x = 0$. We must therefore introduce a branch cut, and we are free to choose the branch cut to run up the positive-imaginary axis from $x = 0$ to $x = i\infty$. In this cut plane the solutions to the differential equation (2.8) are analytic and single-valued. From the WKB geometric-optics formula (2.18) we see that for large $|x|$ the possible exponential behaviors of solutions to (2.8) are

$$\exp\left[\pm\frac{2}{4+\varepsilon}i^{\varepsilon/2}x^{2+\varepsilon/2}\right].$$

If we let $x = re^{i\theta}$, we can express the angular orientations of the Stokes sectors in the complex-x plane in terms of θ. For example, for the decaying solution the location of the center of the right sector is determined by the phase condition $\theta(2 + \varepsilon/2) + \pi\varepsilon/4 = 0$. Thus,

$$\theta_{\text{center, right}} = -\frac{\pi\varepsilon}{8 + 2\varepsilon}. \tag{2.19}$$

The center of the left sector is a reflection about the imaginary axis and thus is located at

$$\theta_{\text{center, left}} = -\pi + \frac{\pi\varepsilon}{8 + 2\varepsilon}. \tag{2.20}$$

This left-right reflection symmetry is the coordinate-space realization of $\mathcal{PT}$ symmetry. Note that if we choose any point x in the complex-x plane and perform a parity reflection, then $x \to -x$. Moreover, since time reversal replaces i by $-i$, we see that $\mathcal{T}$ replaces $-x$ by its complex conjugate $-x^*$. Thus, in the coordinate representation $\mathcal{PT}$ symmetry is just left-right symmetry. (We have already seen this left-right symmetry in Sec. 1.6, where we studied classical $\mathcal{PT}$-symmetric trajectories.)

The upper edge of the right Stokes sector is determined by the phase condition $\theta(2 + \varepsilon/2) + \pi\varepsilon/4 = \pi/2$. Thus,

$$\theta_{\text{upper, right}} = \frac{\pi(2 - \varepsilon)}{8 + 2\varepsilon}, \tag{2.21}$$

and by left-right reflection the upper edge of the left sector is at

$$\theta_{\text{upper, left}} = -\pi - \frac{\pi(2 - \varepsilon)}{8 + 2\varepsilon}. \tag{2.22}$$

The lower edge of the right Stokes sector is determined by the phase condition $\theta(2 + \varepsilon/2) + \pi\varepsilon/4 = -\pi/2$. Thus,

$$\theta_{\text{lower, right}} = -\frac{\pi(2 + \varepsilon)}{8 + 2\varepsilon} \tag{2.23}$$

and

$$\theta_{\text{lower, left}} = -\pi + \frac{\pi(2 + \varepsilon)}{8 + 2\varepsilon}. \tag{2.24}$$

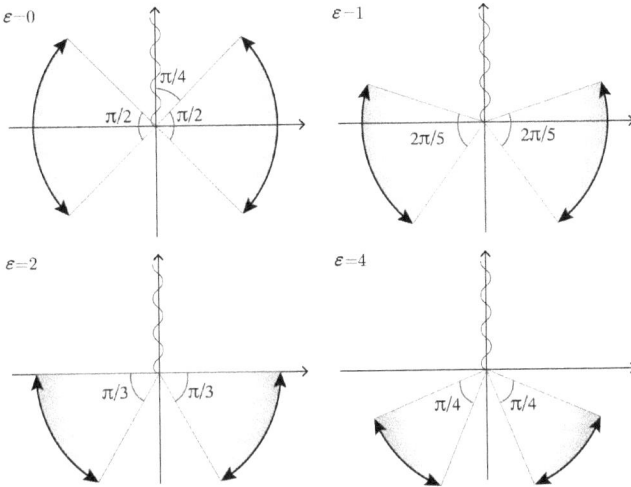

Fig. 2.4   Stokes sectors for the deformed eigenvalue problem (2.8), where $\varepsilon = 0$, 1, 2, 4. [The Stokes sectors for the undeformed Hermitian quantum-harmonic-oscillator problem ($\varepsilon = 0$) have already been examined in Fig. 2.2.] These Stokes sectors have angular opening $\pi/2$ and are centered about the positive-real and the negative-real axis. When $\varepsilon = 1$ the opening angles of the Stokes sectors shrink to $2\pi/5$ and the sectors rotate downward. When $\varepsilon = 2$ the opening angles of the sectors shrink even further to $\pi/3$ and the real axis is no longer in the interior of the sectors; it lies along the upper edge of each sector. At $\varepsilon = 4$ the opening angle of the sectors is $\pi/4$. For all values of ε the Stokes sectors are mirror images of one another through the imaginary axis. The logarithmic branch cut on the imaginary axis is indicated by a wiggly line.

The opening angle Δ of the left and right Stokes sectors is given by

$$\Delta = \theta_{\text{upper, right}} - \theta_{\text{lower, right}} = \frac{2\pi}{4 + \varepsilon}. \qquad (2.25)$$

To summarize, the two Stokes sectors inside of which $\psi(x) \to 0$ as $|x| \to \infty$ rotate downward into the complex-x plane and the opening angles of the sectors decrease (the sectors become thinner) as ε increases. As ε approaches ∞, the centers of the sectors approach $-\frac{1}{2}\pi$ and the opening angles approach 0. (The configuration of the Stokes sectors for some positive values of ε is shown in Fig. 2.4.) On the other hand, if ε decreases below 0, the sectors rotate upwards and become wider. This region of ε is particularly interesting because as soon as ε becomes negative, the $\mathcal{PT}$ symmetry becomes broken and the eigenspectrum is no longer entirely real. This behavior is discussed in detail in Sec. 2.7, where we describe the behavior of eigenvalues in the broken-$\mathcal{PT}$-symmetric region.

In general, if the parameters of an eigenvalue differential equation are varied smoothly, the Stokes sectors of this deformed problem rotate continuously in the complex-x plane. Indeed, we can see from (2.19)–(2.25) that the opening angles and orientations of the Stokes sectors in Fig. 2.4 are smooth functions of ε. However, the Stokes sectors shown in this figure are not the only Stokes sectors associated with the differential equation (2.8). Indeed, *there are many possible pairs of sectors in which one can require that* $\psi(x) \to 0$ *as* $|x| \to \infty$. Thus, there are many different eigenvalue problems associated with the differential equation (2.8). However, in (2.19)–(2.25) we have specifically chosen to seek those eigenvalues associated with the PT-symmetric pair of sectors that rotate *smoothly* away from the harmonic-oscillator sectors at $\varepsilon = 0$.

To emphasize this point we note that the eigenvalues shown in Fig. 2.1 for $\varepsilon = 4$ are associated with the Hamiltonian $\hat{H} = \hat{p}^2 + \hat{x}^6$ and that the boundary conditions for this value of ε are associated with the Stokes sectors displayed in Fig. 2.4. However, while these eigenvalues are discrete, real, and positive, they differ numerically from the eigenvalues of the conventional Hermitian sextic-anharmonic-oscillator eigenvalue problem

$$-\psi''(x) + x^6\psi(x) = E\psi(x), \qquad \psi(\pm\infty) = 0. \tag{2.26}$$

The first five eigenvalues of (2.26) are $E_0 = 1.144802$, $E_1 = 4.338599$, $E_2 = 9.073085$, $E_3 = 14.935170$, and $E_4 = 21.714165$. These eigenvalues are shown in Fig. 2.8 for $\varepsilon = 0$ and these numbers are different from the eigenvalues of $\hat{H}$ in (2.7) at $\varepsilon = 4$ (see Table 2.1). This is because the boundary conditions satisfied by $\psi(x)$ are imposed in a different pair of Stokes sectors.

A crucial distinction between the eigenvalue problem for the sextic oscillator associated with the Stokes sectors in Fig. 2.4 and the conventional eigenvalue problem with Stokes sectors of angular opening $\frac{1}{4}\pi$ centered about the positive-real and negative-real axes is that in the conventional case the eigenfunctions are states of definite parity while in the former case the eigenfunctions are PT symmetric but are *not* states of definite parity. To see this, observe that in Fig 2.4 the eigenfunctions vanish exponentially fast in the southeast direction. If the eigenfunctions were states of parity, they would also vanish in the *northwest* direction, but these eigenfunctions blow up exponentially in this direction! Instead, the eigenfunctions vanish in the *southwest* direction.

2.3   Proof of Spectral Reality for $-x^4$ Potential

We can now use what we have learned about Stokes sectors to prove that
the wrong-sign quartic Hamiltonian

$$\hat{H} = \hat{p}^2 - g\hat{x}^4 \tag{2.27}$$

has a real positive discrete spectrum. It is not easy to prove that the
eigenvalues plotted in Fig. 2.1 are real for *all* $\varepsilon \geq 0$. (For detailed discussion
of the proof see Chap. 6.) However, there is an easy low-brow proof that
the eigenvalues of the quartic Hamiltonian $\hat{H}$ in (2.27) [which is obtained
by setting $\varepsilon = 2$ in $\hat{H} = \hat{p}^2 + g\hat{x}^2(i\hat{x})^\varepsilon$] are all positive and discrete.

The proof does not require that we calculate the eigenvalues; an exact
solution to the eigenvalue problem for the quartic Hamiltonian has never
been found. However, it is quite easy to show that the eigenvalues of this
quartic $\mathcal{PT}$-symmetric Hamiltonian are identical to those of the conventional Hermitian right-sign quartic Hamiltonian

$$\hat{h} = \hat{p}^2 + 4g\hat{x}^4 - 2\hbar\sqrt{g}\,\hat{x}. \tag{2.28}$$

We follow the procedure described in [Bender *et al.* (2006a)] and use a
simple transformation of variables to establish that the two Hamiltonians
$\hat{H}$ and $\hat{h}$ are *isospectral*; that is, they have the same eigenvalues.

Let us consider the one-dimensional Schrödinger eigenvalue problem

$$-\hbar^2\psi''(x) - gx^4\psi(x) = E\psi(x) \tag{2.29}$$

for the non-Hermitian Hamiltonian (2.27). We have included a factor of $\hbar$
in the transcription $\hat{p} \to -i\hbar\frac{d}{dx}$ here because a term containing $\hbar$ appears
in the equivalent Hamiltonian $\hat{h}$ in (2.28). Such a term is called a *quantum anomaly* because there is no corresponding linear term in the classical
version of the theory (see Subsec. 4.2.4).

As we learned in Sec. 2.2, the boundary conditions on $\psi(x)$ in (2.29)
are that $\psi(x) \to 0$ as $|x| \to \infty$ in the two Stokes sectors

$$-\tfrac{1}{3}\pi < \arg x < 0 \quad \text{and} \quad -\pi < \arg x < -\tfrac{2}{3}\pi.$$

These Stokes sectors are adjacent to but do not include the real-x axis.
The differential equation (2.29) must be solved along a contour whose ends
lie in these Stokes sectors in the complex-x plane (see Fig. 2.5). We use
here the special complex contour that Jones and Mateo employed in their
operator analysis of the Hamiltonian (2.27) [Jones and Mateo (2006)]:

$$x = -2i\sqrt{1 + it}. \tag{2.30}$$

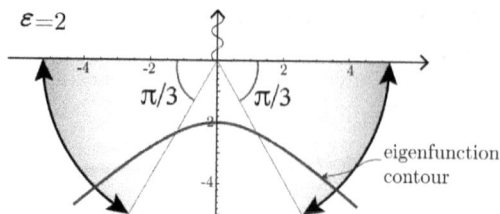

Fig. 2.5 Stokes sectors in the lower-half complex-x plane for the Schrödinger equation (2.29) arising from the quartic Hamiltonian $\hat{H}$ in (2.27). The eigenfunctions of H decay exponentially as $|x| \to \infty$ inside these sectors. The integration contour in (2.30) is shown as a solid line.

While x is complex, t is a real parameter that runs from $-\infty$ to ∞. Such a contour is acceptable because as $t \to \pm\infty$, $\arg x$ approaches $-45°$ and $-135°$, so the contour lies inside the Stokes sectors. When we change the independent variable in (2.29) from x to t according to (2.30), the Schrödinger equation (2.29) becomes

$$-\hbar^2(1+it)\phi''(t) - \tfrac{1}{2}i\hbar^2\phi'(t) - 16g(1+it)^2\phi(t) = E\phi(t). \quad (2.31)$$

Next, we perform a Fourier transform of (2.31). We define

$$\tilde{f}(p) \equiv \int_{-\infty}^{\infty} dt\, e^{-itp/\hbar} f(t), \quad (2.32)$$

so the Fourier transform of $f'(t)$ is $ip\tilde{f}(p)/\hbar$ and the Fourier transform of $tf(t)$ is $i\hbar\tilde{f}'(p)$. Then, the Fourier transform of the Schrödinger equation (2.31) is

$$\left(1 - \hbar\frac{d}{dp}\right)p^2\tilde{\phi}(p) + \frac{\hbar}{2}p\tilde{\phi}(p) - 16g\left(1 - \hbar\frac{d}{dp}\right)^2\tilde{\phi}(p) = E\tilde{\phi}(p). \quad (2.33)$$

Expanding and simplifying this equation, we obtain

$$-16g\hbar^2\tilde{\phi}''(p) - \hbar\left(p^2 - 32g\right)\tilde{\phi}'(p)$$
$$+ \left(p^2 - \tfrac{3}{2}p\hbar - 16g\right)\tilde{\phi}(p) = E\tilde{\phi}(p). \quad (2.34)$$

[Note that the variable p used here is not the same as the variable p used in (2.27). Here, considered as an operator, p represents $-i\hbar\frac{d}{dt}$, whereas in (2.27) p represents $-i\hbar\frac{d}{dx}$.]

Equation (2.34) is not a Schrödinger equation because it has a one-derivative term. However, we can eliminate this term by performing the simple transformation

$$\tilde{\phi}(p) = e^{Q(p)/2}\Phi(p). \quad (2.35)$$

The condition on $Q(p)$ for which the equation satisfied by $\Phi(p)$ has no one-derivative term is a first-order differential equation whose solution is

$$Q(p) = \frac{2}{\hbar}p - \frac{1}{48g\hbar}p^3. \tag{2.36}$$

It is interesting that $e^{Q(p)}$ is precisely the operator found in [Jones and Mateo (2006)].[4] Substituting this expression for Q gives the Schrödinger equation satisfied by $\Phi(p)$:

$$-16g\hbar^2\Phi''(p) + \left(-\frac{\hbar p}{2} + \frac{p^4}{64g}\right)\Phi(p) = E\Phi(p). \tag{2.37}$$

Finally, we make the scaling substitution

$$p = z\sqrt{16g} \tag{2.38}$$

to replace the p variable, which has units of momentum, by z, which is a coordinate variable having units of length. The resulting eigenvalue equation, posed on the real-z axis, is

$$-\hbar^2\Phi''(z) + \left(-2\hbar\sqrt{g}\,z + 4gz^4\right)\Phi(z) = E\Phi(z). \tag{2.39}$$

We emphasize that z is not a conventional coordinate variable because it is *odd* under the discrete transformation of time reversal.

Observe that the eigenvalue problem (2.39) is similar in structure to that in (2.29). [Equation (2.39) is not *dual* to (2.29) because it is still weakly coupled.] However, the potential has acquired a linear term, and since this linear term is proportional to $\hbar$, we regard this term as a quantum anomaly. The linear term has no classical analog because the classical equations of motion are parity symmetric. The breaking of parity symmetry occurs at large values of x where the boundary conditions on the wave function $\psi(x)$ are imposed. Because we have taken a Fourier transform to obtain the Schrödinger equation (2.39), this parity anomaly now manifests itself at small values of z.

The Hamiltonian $\tilde{H}$ for which (2.39) is the eigenvalue problem is

$$\hat{h} = \hat{p}^2 - 2\hbar\sqrt{g}\,\hat{z} + 4g\hat{z}^4. \tag{2.40}$$

This Hamiltonian is Hermitian in the Dirac sense and is *bounded below on the real-z axis*. Furthermore, it is also PT symmetric. This is because at

---

[4]Note that $e^{Q(p)}$ is *not* the CP operator, which will be discussed in Chap. 3 and which could at least in principle be used in a similarity transformation to produce the Hermitian Hamiltonian from the non-Hermitian PT-symmetric Hamiltonian (2.27). One can only use $e^{Q(p)}$ to transform the non-Hermitian Hamiltonian $\hat{H}$ to Hermitian form if $\hat{H}$ has first been written in terms of the real variable y.

every stage in the sequence of transformations above, PT symmetry is exactly preserved. However, while z and $\tilde{p}$ are canonically conjugate operators satisfying $[z, \tilde{p}] = i$, the new variable z behaves like a momentum rather than a coordinate variable because z changes sign under time reversal.

Since the Hamiltonian $\hat{h}$ in (2.40) has a positive discrete spectrum and $\hat{h}$ is isospectral to the Hamiltonian $\hat{H}$ in (2.27), the eigenvalues E_n of the PT-symmetric upside-down quartic potential are positive and discrete. A simple physical interpretation for the positive eigenvalues of $\hat{H}$ is given in [Ahmed *et al.* (2005)]: If we rotate the eigenfunction contour in the left and right Stokes sectors in Fig. 2.5 upward towards the real-x axis, the asymptotic behavior of the eigenfunction $\psi(x)$ in (2.29) becomes purely oscillatory in both sectors instead of exponentially decaying. On both the negative-real-x axis and positive-real-x axis, $\psi(x)$ represents *right-going waves*; $\psi(x)$ represents an *incoming wave* from $x = -\infty$ and an *outgoing wave* to $x = +\infty$. Thus, at the eigenvalues E_n of $\hat{H}$ the upside-down $-x^4$ potential becomes *reflectionless*; that is, $\psi(x)$ represents a *steady-state scattering process in which the incident wave from $-\infty$ is transmitted to $+\infty$ and there is no reflected wave.* At all other energies the incident wave is partially reflected.

If we generalize the Hamiltonian (2.27) to include a quadratic term in the potential,

$$\hat{H} = \hat{p}^2 + \mu^2 \hat{x}^2 - g\hat{x}^4, \tag{2.41}$$

then the same differential-equation analysis used above straightforwardly yields the following equivalent isospectral Hermitian Hamiltonian:

$$\hat{h} = \hat{p}^2 - 2\hbar\sqrt{g}\,\hat{z} + 4g\left(\hat{z}^2 - \frac{\mu^2}{4g}\right)^2. \tag{2.42}$$

This result was found in [Andrianov (1982); Buslaev and Grecchi (1993); Jones and Mateo (2006)]. For this more general Hamiltonian the form of the linear anomaly term remains unchanged from that in (2.39).

One may wonder whether we can use the transformation procedure explained in this subsection to establish isospectral equivalences between other pairs of PT-symmetric Hamiltonians, one having a form like that in (2.27) and the other having a form containing anomalies like that in (2.28). The answer is that there is an *infinite* tower of such pairs of Hamiltonians, but they are rather complicated [Bender and Hook (2008)]. The nth PT-symmetric Hamiltonian has the general form

$$\hat{H}_n = \hat{p}^n - g(i\hat{x})^{(n^2)} \quad (n = 2, 3, 4, \ldots),$$

where (2.27) corresponds to the case $n = 2$. The equivalent Hermitian Hamiltonians $\hat{h}_n$ become more elaborate as n increases and they contain $n-1$ powers of $\hbar$. For example, for $n = 3$, $\hat{H}_3 = \hat{p}^3 - ig\hat{x}^9$ and the equivalent Hamiltonian,

$$\hat{h}_3 = i\hat{x}^3 + i\left(-27g^{1/3}\hbar\hat{p}^2 + 243g^{2/3}\hat{p}^6\right)\hat{x}$$
$$+ \left(972g^{2/3}\hbar\hat{p}^5 - 6g^{1/3}\hbar^2\hat{p} + 1458g\hat{p}^9\right),$$

contains a quadratic anomaly. For $n = 4$, $\hat{H}_4 = \hat{p}^4 - g\hat{x}^{16}$ and the equivalent Hamiltonian,

$$\hat{h}_4 = -\hat{x}^4 + 4^{12}3g\hat{p}^{16} + 4^9g^{3/4}\left(8i\hat{p}^{12}\hat{x} + 54\hbar\hat{p}^{11}\right)$$
$$+ 4^5\sqrt{g}\left(-24\hat{p}^8\hat{x}^2 + 240\hbar i\hat{p}^7\hat{x} + 483\hbar^2\hat{p}^6\right)$$
$$- 8g^{1/4}\left(48\hbar\hat{p}^3\hat{x}^2 - 6i\hbar^2\hat{p}^2\hat{x} + 87\hbar^3\hat{p}\right),$$

has an anomaly proportional to $\hbar^3$.

2.4 Additional PT-Deformed Eigenvalue Problems

In this section we present additional examples of Hermitian eigenvalue problems for which the eigenvalues remain real after a non-Hermitian PT-symmetric deformation. To begin we ask, What happens if instead of performing a PT-symmetric deformation of the quantum *harmonic* oscillator, we perform a PT-symmetric deformation of the quartic quantum *anharmonic* oscillator $\hat{H} = \hat{p}^2 + \hat{x}^4$? The deformed Hamiltonian is

$$\hat{H} = \hat{p}^2 + \hat{x}^4(i\hat{x})^\varepsilon. \tag{2.43}$$

The eigenvalues of this Hamiltonian are plotted in Fig. 2.6 as functions of real ε. These eigenvalues behave similarly to those in Fig. 2.1 for the Hamiltonian in (2.7). The non-Hermitian deformed Hamiltonian (2.43) has a region of unbroken PT symmetry ($\varepsilon \geq 0$) in which the eigenvalues are all real and a region of broken PT symmetry ($\varepsilon < 0$) in which the eigenvalues are almost all complex. However, there is a remarkable new feature — at one isolated point $\varepsilon = -1$ in the broken-PT-symmetric region the eigenvalues are all real (see Fig. 2.7). In the neighborhood of this point, all of the complex eigenvalues pass through exceptional points and become real pairs. At $\varepsilon = -2$ the spectrum is null; there are no eigenvalues at all.

Next, let us perform a PT-symmetric deformation of the sextic quantum anharmonic oscillator $\hat{p}^2 + \hat{x}^6$:

$$\hat{H} = \hat{p}^2 + \hat{x}^6(i\hat{x})^\varepsilon \qquad (\varepsilon \text{ real}). \tag{2.44}$$

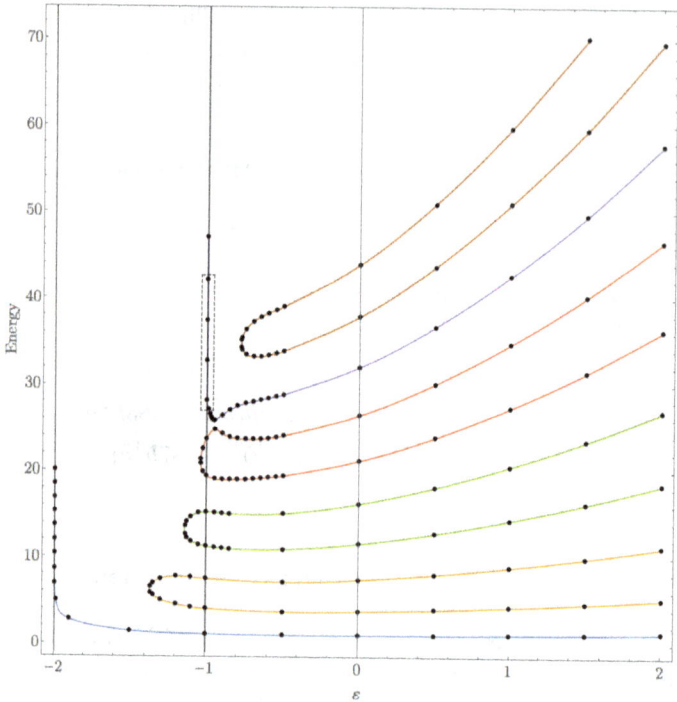

Fig. 2.6   Eigenvalues of the $\mathcal{PT}$-symmetric non-Hermitian Hamiltonian $\hat{H} = \hat{p}^2 + \hat{x}^4(i\hat{x})^\varepsilon$ plotted as functions of ε. The eigenvalues behave similarly to and are color coded like those in those in Fig. 2.1. They are all real, positive, and discrete for $\varepsilon \geq 0$. Below the transition at $\varepsilon = 0$ the real eigenvalues disappear as they merge to become complex-conjugate pairs. Interestingly, when ε is exactly -1, the eigenvalues are all real again. The area within the dashed line is blown up in Fig. 2.7, which details the process by which these complex eigenvalues re-emerge from the complex plane and become real. At $\varepsilon = -2$ there are no eigenvalues whatsoever, real or complex.

The eigenvalues of this Hamiltonian are plotted in Fig. 2.8.   Like the non-Hermitian $\mathcal{PT}$-symmetric Hamiltonians in (2.7) and (2.43), this non-Hermitian deformed Hamiltonian has a region of unbroken $\mathcal{PT}$ symmetry for $\varepsilon \geq 0$ and a region of broken $\mathcal{PT}$ symmetry for $\varepsilon < 0$. The eigenvalues are all real at the isolated values $\varepsilon = -1$ and $\varepsilon = -2$ (see [Bender et $al.$ (1999b)], Fig. 20). When $\varepsilon = -3$ there are no eigenvalues at all.

From Figs. 2.1, 2.6, and 2.8 we can see a pattern emerging. Based on these figures, we expect that the eigenvalues of the Nth class of $\mathcal{PT}$-symmetric Hamiltonians

$$\hat{H} = \hat{p}^2 + \hat{x}^{2N}(i\hat{x})^\varepsilon \qquad (N = 1, 2, 3, \ldots) \qquad (2.45)$$

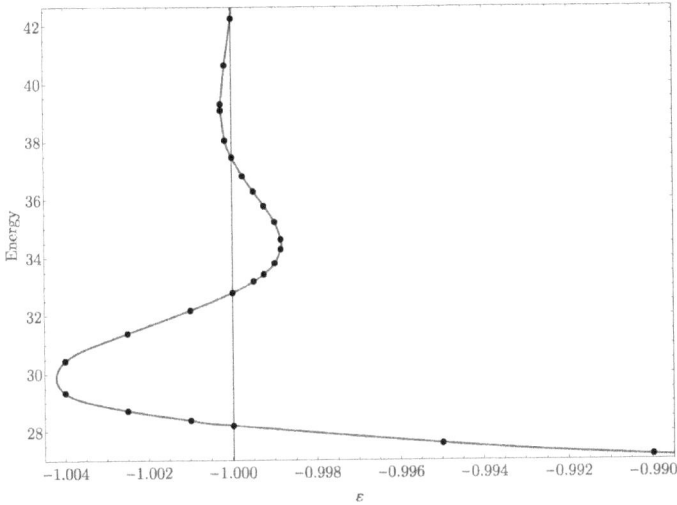

Fig. 2.7   Blow up of Fig. 2.6 showing the detailed behavior of the eigenvalues of the $\mathcal{PT}$-symmetric non-Hermitian Hamiltonian $\hat{H} = \hat{p}^2 + \hat{x}^4(i\hat{x})^\varepsilon$ near $\varepsilon = -1$. In the neighborhood of this value of ε complex-conjugate pairs of eigenvalues emerge from the complex plane and combine to become real.

are all real, positive, and discrete when $\varepsilon \geq 0$ and that there is a $\mathcal{PT}$ transition at $\varepsilon = 0$. In the region of broken $\mathcal{PT}$ symmetry $\varepsilon < 0$, the eigenvalues are almost all complex. However, in the region of broken $\mathcal{PT}$ symmetry there are isolated points $\varepsilon = -1, -2, \ldots, -N+1$ at which the eigenvalues are all real. There are no eigenvalues at all when $\varepsilon = -N$.

For the Nth Hamiltonian in (2.45) the eigenvalue differential equation reads

$$-\psi''(x) + x^{2N}(ix)^\varepsilon \psi(x) = E\psi(x). \qquad (2.46)$$

Using WKB theory, we find that the geometrical-optics approximation to $\psi(x)$ is $\exp\left(\pm i^{\varepsilon/2} \int^x dt\, t^{N+\varepsilon/2}\right)$. Thus, if we make the *ansatz* $x = re^{i\theta}$, we find that $\theta(N+1+\varepsilon/2) + \pi\varepsilon/4 = 0$ at the center of the right wedge, where the phase of the WKB approximation vanishes. At the upper edge of the right wedge we have $\theta(N+1+\varepsilon/2) + \pi\varepsilon/4 = \pi/2$, and at the lower edge of the right wedge we have $\theta(N+1+\varepsilon/2) + \pi\varepsilon/4 = -\pi/2$. Therefore, the centers of the right and left sectors are at

$$\theta_{\text{center, right}} = -\frac{\pi\varepsilon}{4N+4+2\varepsilon}, \qquad \theta_{\text{center, left}} = -\pi + \frac{\pi\varepsilon}{4N+4+2\varepsilon}, \qquad (2.47)$$

the upper edges of the sectors are at

$$\theta_{\text{upper, right}} = \frac{\pi(2-\varepsilon)}{4N+4+2\varepsilon}, \qquad \theta_{\text{upper, left}} = -\pi - \frac{\pi(2-\varepsilon)}{4N+4+2\varepsilon}, \qquad (2.48)$$

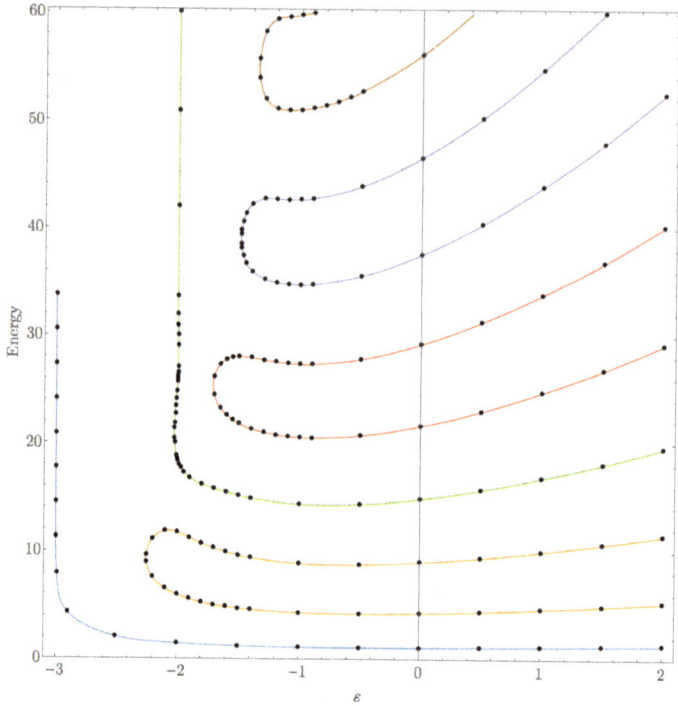

Fig. 2.8   Eigenvalues of the non-Hermitian $\mathcal{PT}$-symmetric Hamiltonian $\hat{H} = \hat{p}^2 + \hat{x}^6 (i\hat{x})^\varepsilon$ in (2.44). The eigenvalues are color coded like those in Figs. 2.1 and 2.6. There is a $\mathcal{PT}$ transition at $\varepsilon = 0$. All eigenvalues are real at $\varepsilon = -1$ and at $\varepsilon = -2$.

and the lower edges of the sectors are at

$$\theta_{\text{lower, right}} = -\frac{\pi(2+\varepsilon)}{4N+4+2\varepsilon}, \qquad \theta_{\text{lower, left}} = -\pi + \frac{\pi(2+\varepsilon)}{4N+4+2\varepsilon}. \quad (2.49)$$

Also, the opening angle Δ of each sector is

$$\Delta = \theta_{\text{upper}} - \theta_{\text{lower}} = \frac{4\pi}{4N+4+2\varepsilon}. \quad (2.50)$$

As for the case $N = 1$, we note that as $\varepsilon \to \infty$, the center of each sector approaches $-\frac{1}{2}\pi$ and the opening angle of each sector tends to 0. Also, the upper edge of each wedge approaches $\frac{1}{2}\pi$ when $\varepsilon = -N$. As explained in Sec. 2.7, these Stokes sectors are no longer separated when $\varepsilon = -N$, so the eigenvalue problem is no longer well posed and the eigenspectrum is null. This disappearance of eigenvalues is evident in Figs. 2.1, 2.6, and 2.8.

One can also deform Hermitian Hamiltonians containing higher powers of $\hat{p}$. For example, in Figs. 2.9 and 2.10 we plot the eigenvalues of the deformed Hamiltonians ([Bender and Hook (2012)])

$$\hat{H} = \hat{p}^4 + \hat{x}^2(i\hat{x})^\varepsilon \qquad (\varepsilon \text{ real}) \qquad (2.51)$$

and

$$\hat{H} = \hat{p}^6 + \hat{x}^2(i\hat{x})^\varepsilon \qquad (\varepsilon \text{ real}). \qquad (2.52)$$

Again, we see an eigenvalue behavior that strongly resembles that shown in Figs. 2.1, 2.6, and 2.8.

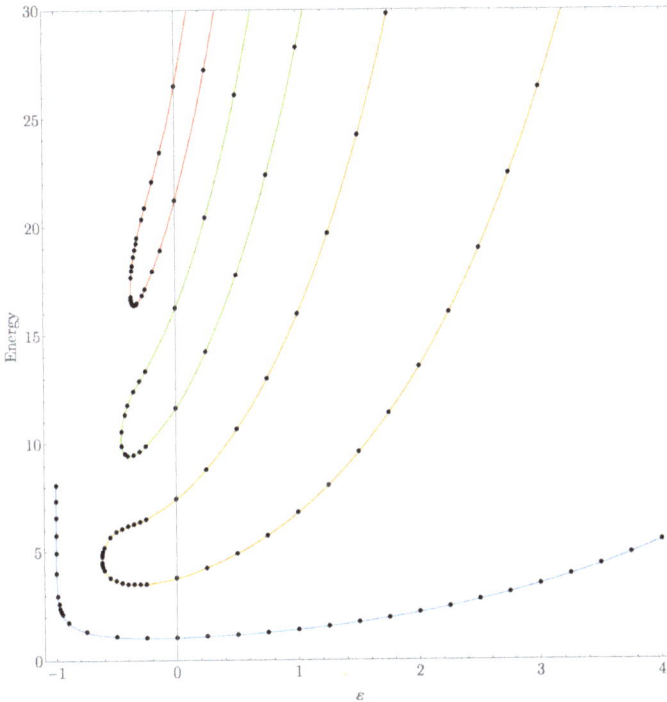

Fig. 2.9 Real eigenvalues of $\hat{H} = \hat{p}^4 + \hat{x}^2(i\hat{x})^\varepsilon$ in (2.51). The eigenvalues are color coded like those in Fig. 2.1.

Next, let us consider some more elaborate classes of PT-symmetric deformed Hamiltonians. For example, the eigenvalues of the class of logarithmic PT-symmetric Hamiltonians [Bender *et al.* (2016c)]

$$\hat{H} = \hat{p}^2 + \hat{x}^2(i\hat{x})^\varepsilon \log(i\hat{x}) \qquad (\varepsilon \text{ real}) \qquad (2.53)$$

behave quite similarly to those of the Hamiltonians in (2.7) (see Fig. 2.11). It is interesting that one can deform the *non*-Hermitian PT-symmetric

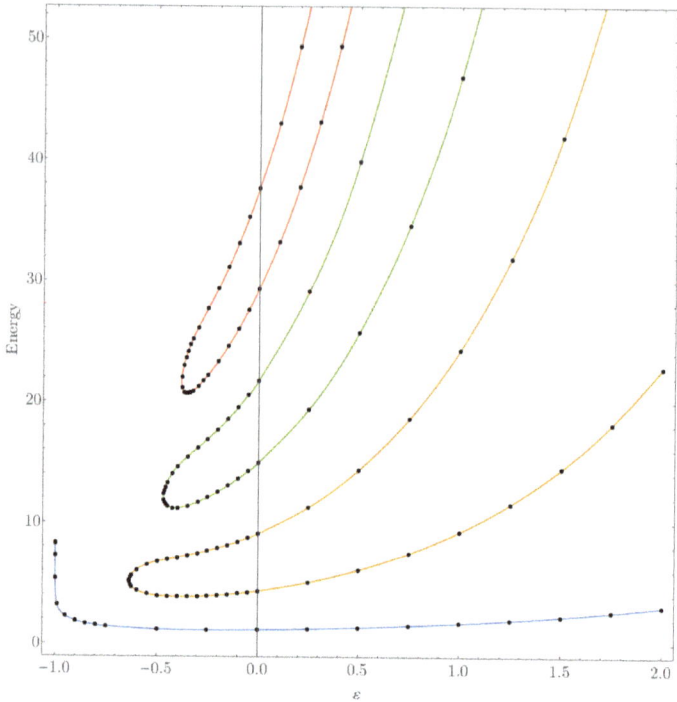

Fig. 2.10 Real eigenvalues of $\hat{H} = \hat{p}^6 + \hat{x}^2(i\hat{x})^\varepsilon$ in (2.52). The eigenvalues are color coded like those in Fig. 2.1.

Hamiltonian $\hat{H} = \hat{p}^2 + \hat{x}^2 \log(i\hat{x})$ and still obtain a class of Hamiltonians that have the usual broken- and unbroken-PT-symmetric regions of ε. Although there is no proof of reality, it appears that in the unbroken region of PT symmetry $\varepsilon \geq 0$ the eigenvalues of the quantum Hamiltonians (2.53) appear to be real. Correspondingly, in this region the underlying classical theory has closed trajectories (see Fig. 2.12).

Not all PT-symmetric deformations of Hermitian Hamiltonians have unbroken-PT-symmetric regions. For example, while the Hermitian Hamiltonian $\hat{H} = \hat{p}^2 + \hat{x}^2 \log(x^2)$ has real eigenvalues, we can see from Fig. 2.13 that the class of deformed PT-symmetric Hamiltonians

$$\hat{H} = \hat{p}^2 + \hat{x}^2(i\hat{x})^\varepsilon \log(\hat{x}^2) \qquad (\varepsilon \text{ real}) \qquad (2.54)$$

has complex eigenvalues if $\varepsilon \neq 0$. The nonreality of the eigenvalues correlates with the property that the classical trajectories are not closed (see Fig. 2.14).

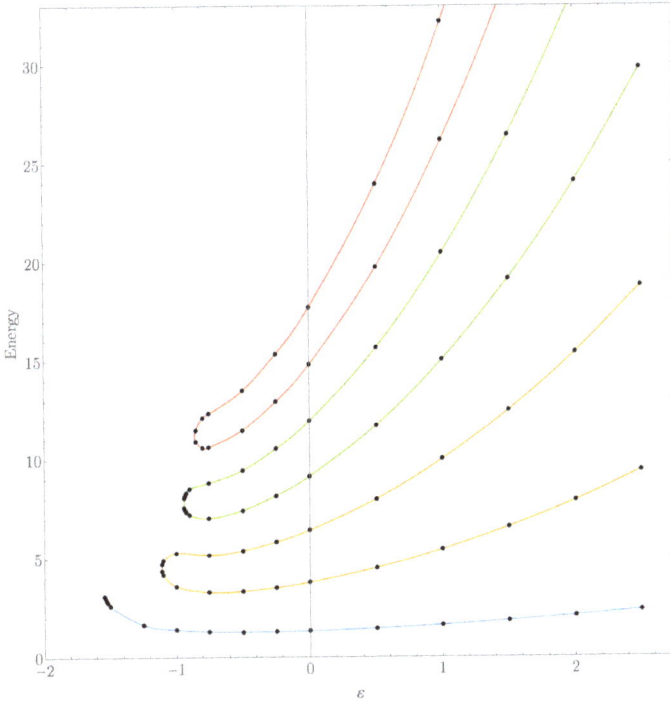

Fig. 2.11   Eigenvalues of the class of Hamiltonians $\hat{H} = \hat{p}^2 + \hat{x}^2(i\hat{x})^\varepsilon \log(i\hat{x})$ plotted as functions of real ε. Note that the eigenvalues of these rather complicated $\mathcal{PT}$-symmetric Hamiltonians appear to be entirely real in the unbroken region $\varepsilon \geq 0$.

The Hamiltonian (2.54) suggests that it may be easy to find classes of $\mathcal{PT}$-symmetric Hamiltonians that do not have parametric regions in which the eigenvalues are all real and this is indeed so. For example, let us consider the class of $\mathcal{PT}$-symmetric Hamiltonians

$$\hat{H} = \hat{p}^2 + |\hat{x}|^b(i\hat{x})^\varepsilon \quad (b, \ \varepsilon \text{ real and positive}). \qquad (2.55)$$

An examination of the Hamiltonians in this class shows that there is never a region of unbroken $\mathcal{PT}$ symmetry except when b is a positive even integer [Bender *et al.* (2008b)]. Let us examine the special exactly solvable case that is obtained by setting $b = 1$ and $\varepsilon = 1$:

$$\hat{H} = \hat{p}^2 + i|\hat{x}|\hat{x}. \qquad (2.56)$$

The Schrödinger eigenvalue differential equation associated with this Hamiltonian is

$$-\psi''(x) + i|x|x\psi(x) = E\psi(x), \qquad (2.57)$$

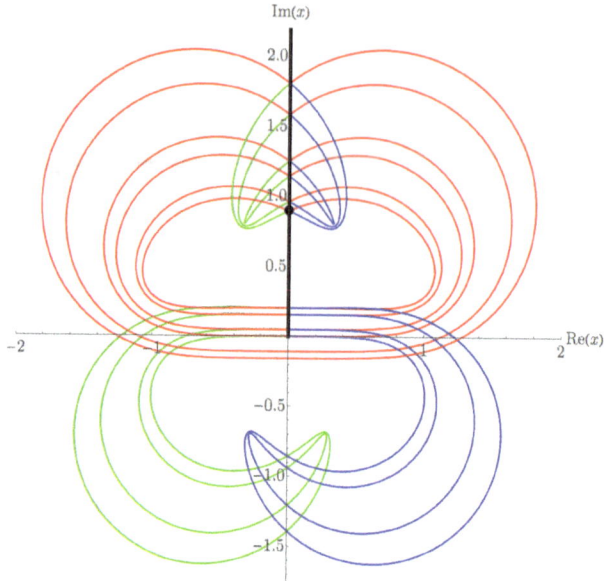

Fig. 2.12   Closed classical trajectory for a Hamiltonian belonging to the class of logarithmic Hamiltonians (2.53) with $\varepsilon = 2$. The classical energy is $E = 2.07734$ for this
plot. The trajectory enters three sheets of the complex-x Riemann surface. The heavy
solid black line indicates the location of the branch cuts that separate the sheets. On
the principal sheet (sheet 0) ($-\frac{3}{2}\pi < \arg x \le \frac{1}{2}\pi$) the classical trajectory is shown in
red. On sheet 1 ($\frac{1}{2}\pi < \arg x \le \frac{5}{2}\pi$) the trajectory is indicated in green and on sheet
-1 ($-\frac{7}{2}\pi < \arg x \le -\frac{3}{2}\pi$) it is indicated in blue. The trajectory begins (and ends) at
the black dot, which is located on sheet 0 at $0.9i$. It then continues in an east-northeast
direction, curves downward, crosses the branch cut, enters sheet -1, and changes its
color to green. (This color change is hard to see because another part of the trajectory
lies very close to this line.) Note that the trajectory never crosses itself; at the apparent
crossing points the curves lie on different Riemann sheets, as indicated by the color.

where x is real and where the eigenfunction $\psi(x)$ is required to obey the
boundary conditions that $\psi \to 0$ as $x \to \pm\infty$.

   To obtain the eigenvalues we solve this differential equation in each of
two regions, $x > 0$ and $x < 0$, and then patch the two solutions together.
In the region $x > 0$ the exact solution is

$$\psi(x) = c_1 D_\nu\big(xe^{i\pi/8}\sqrt{2}\big) + c_2 D_\nu\big(-xe^{i\pi/8}\sqrt{2}\big),$$

where D_ν is the parabolic cylinder function with index $\nu = \frac{1}{2}Ee^{-i\pi/4} -
\frac{1}{2}$ and c_1 and c_2 are arbitrary constants. The boundary condition
$\lim_{x \to +\infty} \psi(x) = 0$ implies that $c_2 = 0$. Thus, for $x > 0$ we find that

$$\psi(x) = c_1 D_\nu\big(xe^{i\pi/8}\sqrt{2}\big). \tag{2.58}$$

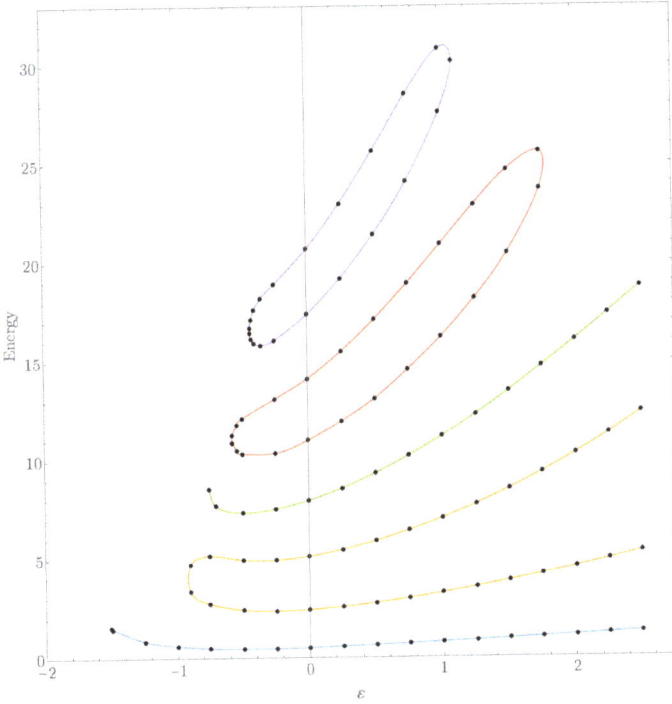

Fig. 2.13 Eigenvalues of $\hat{H} = \hat{p}^2 + \hat{x}^2(i\hat{x})^\varepsilon \log(\hat{x}^2)$. Note that when $\varepsilon \neq 0$ only a few eigenvalues are real and the rest are complex. Apart from the point $\varepsilon = 0$, this class of $\mathcal{PT}$-symmetric Hamiltonians does not have a region of unbroken $\mathcal{PT}$ symmetry.

However, if $x < 0$ the exact solution to the differential equation (2.57) is

$$\psi(x) = d_1 D_\mu\big(xe^{-i\pi/8}\sqrt{2}\big) + d_2 D_\mu\big(-xe^{-i\pi/8}\sqrt{2}\big),$$

where $\mu = \frac{1}{2}e^{i\pi/4}E - \frac{1}{2}$ and d_1 and d_2 are arbitrary constants. The boundary condition $\lim_{x \to -\infty} \psi(x) = 0$ gives $d_1 = 0$. Thus, for $x < 0$ we get

$$\psi(x) = d_2 D_\mu\big(-xe^{-i\pi/8}\sqrt{2}\big). \qquad (2.59)$$

We patch the two solutions (2.58) and (2.59) together at the origin $x = 0$. Continuity of $\psi(x)$ at $x = 0$ implies that $c_1 D_\nu(0) = d_2 D_\mu(0)$, and continuity of $\psi'(x)$ at $x = 0$ implies that $c_1 e^{i\pi/8} D'_\nu(0) = -d_2 e^{-i\pi/8} D'_\mu(0)$. We then combine these two equations and eliminate the constants c_1 and d_2 to obtain an exact equation for the eigenvalues:

$$\frac{e^{i\pi/8} D'_\nu(0)}{D_\nu(0)} = -\frac{e^{-i\pi/8} D'_\mu(0)}{D_\mu(0)}. \qquad (2.60)$$

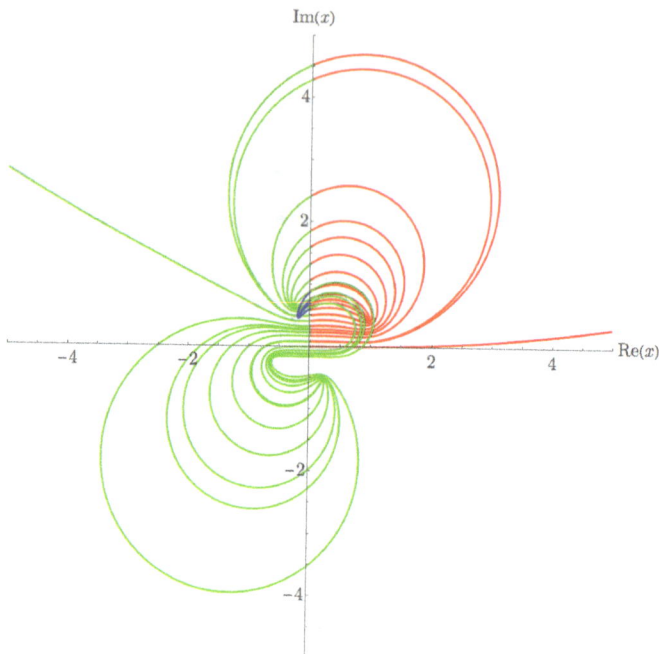

Fig. 2.14   Open classical trajectory for the Hamiltonian (2.54) for $\varepsilon = 2$. The trajectory begins at the point $0.9i$ on the zeroth Riemann sheet (red), and as t runs from $t = 0$ to $t = 30$, the trajectory repeatedly enters the first (green) and second (blue) sheets. The classical energy is $E = 1.10543$. The trajectory never crosses itself and it never closes. The absence of closed classical orbits correlates with the absence of a region of unbroken PT symmetry in which all the eigenvalues of the Hamiltonian are real.

The condition (2.60) can be rewritten simply in terms of gamma functions:

$$e^{i\pi/8}\frac{\Gamma\left(\frac{3}{4} - \frac{1}{4}Ee^{-i\pi/4}\right)}{\Gamma\left(\frac{1}{4} - \frac{1}{4}Ee^{-i\pi/4}\right)} + e^{-i\pi/8}\frac{\Gamma\left(\frac{3}{4} - \frac{1}{4}Ee^{i\pi/4}\right)}{\Gamma\left(\frac{1}{4} - \frac{1}{4}Ee^{i\pi/4}\right)} = 0. \qquad (2.61)$$

As required by PT symmetry, the secular equation (2.61) is a *real* function of E. (We verify this by observing that it is the sum of two terms that are complex conjugates of one another.)

We can solve (2.61) numerically for E by substituting $E = \operatorname{Re} E + i \operatorname{Im} E$ and taking the real and imaginary parts of the resulting equation. We then plot in Fig. 2.15 the curves in the complex-E plane along which the real part of (2.61) vanishes (solid line) and the imaginary part of (2.61) vanishes (dotted line). Note that the condition of PT symmetry requires that the dotted line lie along the real-E axis. However, note that the real-E axis is not the only curve along which the imaginary part of (2.61) vanishes.

The intersections of the solid and dotted lines in Fig. 2.15 are the eigenvalues of the Hamiltonian (2.56). Evidently, there is only one real eigenvalue; all other intersections occur in complex-conjugate pairs. The numerical value of this real eigenvalue is 1.258092 .... For this real energy the corresponding eigenfunction is PT symmetric. This can be seen by examining the eigenfunction in (2.58) and (2.59): If we reverse the sign of x and simultaneously take the complex conjugate, we see that the eigenfunction remains invariant.

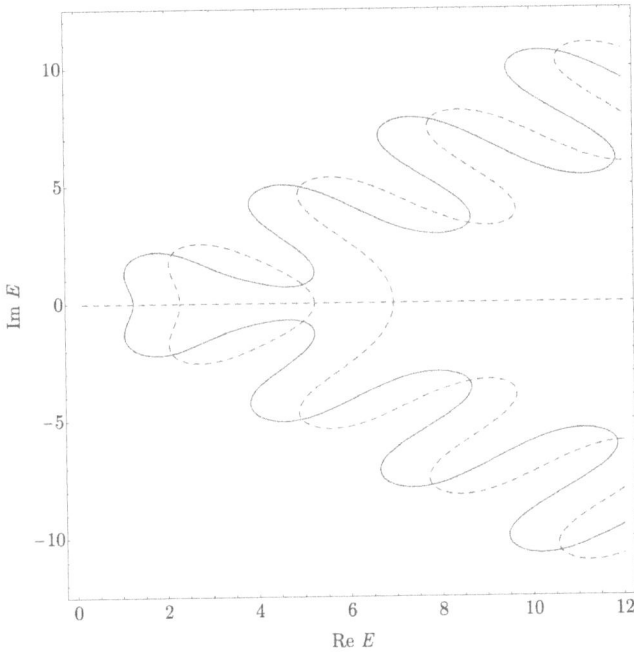

Fig. 2.15   Numerical solution to the secular equation (2.61) for the eigenvalues of the Hamiltonian $\hat{H} = \hat{p}^2 + i|\hat{x}|\hat{x}$. The real part of the secular equation vanishes on the solid line in the complex-E plane, and the imaginary part of the secular equation vanishes on the dotted line. The eigenvalues lie at the intersections of the solid and dotted lines. Observe that there is just one real eigenvalue, which is located at $E = 1.258092$. All other eigenvalues are complex and occur in complex-conjugate pairs.

A numerical study of the eigenvalues of $\hat{H}$ in (2.55) for a range of values of b and ε was done in [Bender *et al.* (2008b)] and these results are listed in Tables 2.2 and 2.3. Of course, when b is an even integer, $\hat{H}$ in (2.55) reduces to the classes of Hamiltonians whose eigenvalues are plotted in Figs. 2.1,

2.6, and 2.8 and the spectrum is entirely real. However, for other values of b and for $\varepsilon \neq 0$ there are only a finite number of real eigenvalues. The number of real eigenvalues appears to decrease as ε increases and appears to increase as b increases, but it is not clear from these tables that this behavior continues to hold when ε is small. Therefore, we examine the case $b = \frac{1}{2}$ for some additional values of ε and find that when $\varepsilon = \frac{1}{4}$ there is one real eigenvalue, when $\varepsilon = \frac{1}{8}$ there are *two* real eigenvalues at 1.064 07 and 2.058 27, and when $\varepsilon = \frac{1}{16}$, there are *three* eigenvalues at 1.0582, 2.3488, and 3.4132. As ε decreases further, the number of real eigenvalues grows until at $\varepsilon = 0$ there are an infinite number of real eigenvalues.

Table 2.2   Real eigenvalues for the Hamiltonian $\hat{H} = \hat{p}^2 + |\hat{x}|^b (i\hat{x})^\varepsilon$, where $\varepsilon = \frac{1}{2}$. The underlined eigenvalues are the largest of all the real eigenvalues. The symbol ... indicates that the spectrum is entirely real and infinite.

	$b = 1/2$	$b = 1$	$b = 3/2$	$b = 2$	$b = 5/2$	$b = 3$
$\varepsilon = 1/2$	<u>1.180777</u>	1.08693	1.05583	1.04896	1.05404	1.06568
		3.19578	3.27843	3.43454	3.59460	3.74791
		<u>4.4220</u>	5.36421	6.05174	6.64515	7.17496
			7.67568	8.79101	9.91884	10.9735
			<u>9.53919</u>	11.6207	13.4256	15.1112
				14.5219	17.0514	19.4889
				17.4829	20.8691	24.1139
				20.4952	24.7239	28.9111
				23.5529	28.8137	33.9218
				26.6504	32.7868	39.0482
				29.7848	37.2141	44.3936
				32.9526	41.0803	49.7770
				36.1511	<u>46.2256</u>	55.4476
				39.3784		60.9963
				42.6321		67.0561
				45.9112		72.5848
				...		79.2958
						84.3126
						<u>92.7345</u>

In [Bender *et al.* (2008b)] it was conjectured that the absence of a region of unbroken $\mathcal{PT}$ symmetry for the Hamiltonian (2.55) was due to the nonanalyticity of the function $|x|$ that is being deformed into the complex domain. Indeed, the function $|x|$ is analytic nowhere in the complex plane. We can replace this function in (2.55) by the function $\log(2 \cosh x)$, which behaves like $|x|$ for large real $|x|$ and is analytic in the strip $|\mathrm{Im}\, x| < \frac{1}{2}\pi$, which includes the real axis. Doing so yields a greater number of real eigenvalues. While the Hamiltonian (2.56) has only one real eigenvalue,

Table 2.3   Same as in Table 2.2 except that $\varepsilon = 1$ and $\varepsilon = 3/2$.

	$b = 1/2$	$b = 1$	$b = 3/2$	$b = 2$	$b = 5/2$	$b = 3$
$\varepsilon = 1$	1.446448	1.25809	1.18627	1.15627	1.14615	1.14685
			4.21683	4.10923	4.13051	4.19436
			6.93323	7.56227	7.95153	8.30206
				11.3144	12.0844	12.9101
				15.2916	16.8072	18.1062
				19.4515	21.3065	23.5322
				23.76667	27.4779	29.6147
				28.2175	30.3268	35.3873
				32.7891		42.9034
				37.4698		47.4048
				42.2504		
				. . .		
$\varepsilon = 3/2$	1.791941	1.48873	1.36338	1.30151	1.26993	1.2550
			5.52801	4.96979	4.80096	4.7494
			8.50818	9.48003	9.60759	9.7042
				14.5305	14.6672	15.2406
				19.9977	21.7069	21.891
				25.8103	24.9567	28.1147
				31.9205		
				38.2938		
				44.904		
				. . .		

the deformed Hamiltonian

$$\hat{H} = \hat{p}^2 + i \log(2 \cosh x) x$$

has *five* real eigenvalues, which are located at 1.46908, 4.23578, 6.97342, 10.1872, and 11.8161. However, the remaining eigenvalues all appear as complex-conjugate pairs. It may be that the spectra plotted in Figs. 2.1, 2.6, and 2.8 are entirely real because the functions x^2, x^4, x^6 that are being deformed into the complex plane are entire (have no singularities).

2.5   Numerical Calculation of Eigenvalues

In this section we explain some of the numerical techniques that may be used to calculate the eigenvalues of a $\mathcal{PT}$-deformed Hamiltonian and to produce eigenvalue plots like that in Fig. 2.1. We mention some of the original techniques that were used in [Bender and Boettcher (1998b)] to calculate eigenvalues. In Sec. 2.7 we describe a more powerful numerical technique, known as the *Arnoldi algorithm*, that can be used to calculate the complex eigenvalues in the broken-$\mathcal{PT}$-symmetric region $\varepsilon < 0$ as well as the real eigenvalues in the unbroken-$\mathcal{PT}$-symmetric region $\varepsilon \geq 0$.

2.5.1 *Shooting method*

The simplest and most direct numerical method for solving a time-independent Schrödinger eigenvalue equation like that (2.8) is to integrate the differential equation directly by using a Runge–Kutta integration package. To do so we make a change of variable that allows us to integrate radially down the center of a Stokes sector. For example, for the right Stokes sector we make the substitution $x = re^{i\theta_{\text{right}}}$, where $\theta_{\text{right}} = -\pi\varepsilon/(8 + 2\varepsilon)$ is given in (2.19), and we obtain the differential equation

$$-\psi''_{\text{right}}(r) + r^{2+\varepsilon}\psi_{\text{right}}(r) = e^{-i\pi\varepsilon/(4+\varepsilon)}E\psi_{\text{right}}(r). \qquad (2.62)$$

Similarly, for the left Stokes sector we refer to (2.20) and substitute $x = re^{i\theta}$, where $\theta = -\pi + \pi\varepsilon/(8 + 2\varepsilon)$:

$$-\psi''_{\text{left}}(r) + r^{2+\varepsilon}\psi_{\text{left}}(r) = e^{i\pi\varepsilon/(4+\varepsilon)}E\psi_{\text{left}}(r). \qquad (2.63)$$

We seek solutions to (2.62) and (2.63) that vanish as $r \to \infty$, and to find these solutions we take a large value of r, say $r_0 = 10$ or $r_0 = 20$, and use WKB theory to determine the initial conditions there. For simplicity we choose $\psi_{\text{right}}(r_0) = 1$. Then, differentiating the WKB approximation in (2.17), we find that for the decaying solution, we must choose

$$\psi'_{\text{right}}(r_0) = -\sqrt{Q(r_0)}, \text{ where } Q(r) = r^{2+\varepsilon} + Ee^{2i\theta_{\text{right}}}.$$

Then we use Runge–Kutta to integrate from r down to 0. The advantage of integrating *down* the center of the Stokes sector is that the solution that we seek grows most rapidly as we approach the origin, while the unwanted solution (the one that grows exponentially as $r \to \infty$) is most suppressed. Having integrated both the left and right solutions down to $r = 0$, we now know the values of ψ_{right} and ψ_{left} and their derivatives at $r = 0$.

Finally, we *patch* these two solutions together at $r = 0$ by requiring both their continuity and the continuity of their first derivatives. The patching condition is that the Wronskian-like quantity

$$W \equiv \psi_{\text{right}}(0)\psi'_{\text{left}}(0)e^{i\theta_{\text{right}}} - \psi_{\text{left}}(0)\psi'_{\text{right}}(0)e^{i\theta_{rmleft}}$$

must vanish [Bender and Hook (2012)]. We then *shoot* by iterating and adjusting E to make $|W|$ as small as possible. This procedure gives accurate results for E if E is real.

The shortcoming of this procedure is that it is not effective in the region of broken $\mathcal{PT}$ symmetry, where E is complex. This is because a two-dimensional search for the optimal complex value of E that minimizes $|W|$ is much more difficult to carry out than a one-dimensional search for a real value of E. A more widely applicable numerical method for computing eigenvalues, whether they are real or complex, is the Arnoldi method. This fancy discretization technique is described in Sec. 2.7.

2.5.2 Variational techniques

An completely different way to compute the eigenvalues is to use a variational method. To do so we must introduce a complete set of basis functions (for example, the harmonic-oscillator eigenfunctions) and express the Hamiltonian $\hat{H}$ in (2.7) as an infinite-dimensional numerical matrix. We then truncate this matrix to a finite-dimensional $N \times N$ matrix and numerically compute the eigenvalues of this matrix. As $N \to \infty$ we hope that the eigenvalues of the $N \times N$ matrix approach those of $\hat{H}$.

The drawback of this numerical scheme is while it works well for small ε (it is quite effective when $\varepsilon = 1$ [Bender et $al.$ (1999b)]), it rapidly becomes inaccurate as ε increases. This is because the asymptotic behaviors of the harmonic-oscillator basis functions cease to match the asymptotic behaviors of the eigenfunctions of $\hat{H}$. As we see in Fig. 2.4, when ε increases, the Stokes sectors associated with the eigenfunctions of $\hat{H}$ rotate downward and cease to overlap with the sectors for $\varepsilon = 0$.

A further drawback of this scheme is that because $\hat{H}$ is not Hermitian, the eigenvalues of the $N \times N$ matrix do not approach the eigenvalues of $\hat{H}$ monotonically. Rather, the eigenvalues of the $N \times N$ matrix converge in an irregular fashion, and some of the eigenvalues of the $N \times N$ matrix are even complex [Bender et $al.$ (1999b)]).

An alternative variational approach is to determine the stationary points of the functional

$$\langle H \rangle(a,b,c) \equiv \frac{\int_C dx \, \psi(x) H \psi(x)}{\int_C dx \, \psi^2(x)}, \tag{2.64}$$

where the trial wave function

$$\psi(x) = (ix)^c \exp\left[a(ix)^b\right] \tag{2.65}$$

is a three-parameter class of $\mathcal{PT}$-invariant trial wave functions [Bender et $al.$ (1999c)]. The integration contour C used to define $\langle H \rangle(a,b,c)$ must lie inside the Stokes sectors in the complex-x plane in which the wave function $\psi(x)$ decays exponentially at infinity. Again, because $\hat{H}$ is not Hermitian, the expectation value (2.64) does not have a local minimum; rather this functional has a $saddle$ $point$ in (a,b,c)-space. At this saddle point the numerical prediction for the ground-state energy is accurate for a wide range of ε. This method also determines approximate eigenfunctions and eigenvalues of the excited states of $\hat{H}$.

Yet another variational approach is to solve the coupled-moment problem [Handy (2001)]. This technique produces accurate results for the eigenvalues and is the quantum-mechanical analog of solving the Dyson–Schwinger equations in quantum field theory (see Chap. 5) by successive truncation.

2.6   Approximate Analytical Calculation of Eigenvalues

It is impossible to solve the differential-equation eigenvalue problem (2.8) analytically and in closed form except in two special cases, namely, for $\varepsilon = 0$ (the harmonic oscillator) and for $\varepsilon \to \infty$. The latter is the $\mathcal{PT}$-symmetric version of the square-well potential and the solution is given in [Bender *et al.* (1999a)]). However, WKB theory provides an accurate analytical way to calculate the real eigenvalues of the Hamiltonian $\hat{H}$ in (2.7) in the unbroken-$\mathcal{PT}$-symmetric region $\varepsilon \geq 0$.

The WKB calculation is interesting because it must be performed in the complex plane rather than on the real-x axis. The turning points $x_\pm$ are those roots of $E = x^2(ix)^\varepsilon$ that *analytically continue* off the real axis as ε increases from 0. These turning points,

$$x_- = E^{\frac{1}{\varepsilon+2}} e^{i\pi\left(\frac{3}{2}-\frac{1}{\varepsilon+2}\right)}, \quad \text{and} \quad x_+ = E^{\frac{1}{\varepsilon+2}} e^{-i\pi\left(\frac{1}{2}-\frac{1}{\varepsilon+2}\right)}, \qquad (2.66)$$

lie in the lower-half x plane in Fig. 2.4 when $\varepsilon > 0$ (and they lie in the upper-half plane when $\varepsilon < 0$).

The leading-order WKB phase-integral quantization condition is

$$(n + 1/2)\pi = \int_{x_-}^{x_+} dx\, \sqrt{E - x^2(ix)^\varepsilon}. \qquad (2.67)$$

When $\varepsilon > 0$ this path lies entirely in the lower-half x plane, and when $\varepsilon = 0$ (the case of the harmonic oscillator) the path lies on the real axis. However, when $\varepsilon < 0$ the path lies in the upper-half x plane and crosses the cut on the positive-imaginary-x axis. In this case there is *no continuous path joining the turning points*. Hence, WKB fails when $\varepsilon < 0$.

When $\varepsilon \geq 0$, we deform the phase-integral contour so that it follows the rays from x_- to 0 and from 0 to x_+:

$$\left(n + \tfrac{1}{2}\right)\pi = 2\sin\left(\frac{\pi}{\varepsilon+2}\right) E^{(\varepsilon+4)/(2\varepsilon+4)} \int_0^1 ds\, \sqrt{1 - s^{\varepsilon+2}}. \qquad (2.68)$$

We then solve for E_n:

$$E_n \sim \left[\frac{\Gamma\left(\frac{3}{2} + \frac{1}{\varepsilon+2}\right)\sqrt{\pi}\left(n + \frac{1}{2}\right)}{\sin\left(\frac{\pi}{\varepsilon+2}\right)\Gamma\left(1 + \frac{1}{\varepsilon+2}\right)}\right]^{\frac{2\varepsilon+4}{\varepsilon+4}} \qquad (n \to \infty). \qquad (2.69)$$

This formula gives a very accurate approximation to the eigenvalues plotted in Fig. 2.1, and it shows, at least in the WKB approximation, that the eigenvalues of $\hat{H}$ in (2.7) are real and positive (see Table 2.4). We can, in addition, perform a higher-order WKB calculation by replacing the phase integral by a *closed contour* that encircles the path joining the turning points [Bender *et al.* (1999b); Bender and Orszag (1999)].

Table 2.4 Comparison of the exact eigenvalues (obtained with Runge–Kutta) and the WKB result in (2.69).

ε	n	E_{exact}	E_{WKB}	ε	n	E_{exact}	E_{WKB}
1	0	1.156 267 072	1.0943	2	0	1.477 149 753	1.3765
	1	4.109 228 752	4.0895		1	6.003 386 082	5.9558
	2	7.562 273 854	7.5489		2	11.802 433 593	11.7690
	3	11.314 421 818	11.3043		3	18.458 818 694	18.4321
	4	15.291 553 748	15.2832		4	25.791 792 423	25.7692
	5	19.451 529 125	19.4444		5	33.694 279 298	33.6746
	6	23.766 740 439	23.7606		6	42.093 814 569	42.0761
	7	28.217 524 934	28.2120		7	50.937 278 826	50.9214
	8	32.789 082 922	32.7841		8	60.185 767 651	60.1696
	9	37.469 824 697	37.4653		9	69.795 703 031	69.7884

2.7 Eigenvalues in the Broken-$\mathcal{PT}$-Symmetric Region

In the region of broken $\mathcal{PT}$ symmetry, $\varepsilon < 0$, the eigenvalues shown in Fig. 2.1 begin to form complex-conjugate pairs. In this section we show how to compute the eigenvalues in this region and we give a detailed description of their behavior as ε ranges from 0 down to -2.

When ε goes below 0, (2.19)–(2.25) are still valid and they continue to describe the rotation and angular opening of the Stokes sectors. The Stokes sectors for four values of ε are shown in Fig. 2.16. Note that at $\varepsilon = -1$ the upper edges of the left and right Stokes sectors lie at the angle $\frac{1}{2}\pi$. The sectors touch and fuse at $\varepsilon = -1$, so the integration contour for the eigenvalue problem can be deformed out to infinity. At this value of ε we no longer have an eigenvalue problem; this explains why the eigenvalues in Fig. 2.1 disappear when $\varepsilon = -1$. For a proof that the eigenvalue equation $-y''(x) + ixy(x) = Ey(x)$ has a null spectrum see [Herbst (1979)].

How do we calculate the eigenvalues in the region of broken $\mathcal{PT}$ symmetry? The eigenvalue differential equation (2.8) may be integrated along any path in the complex-x plane *so long as the ends of the path approach complex infinity inside the left sector and inside the right sector.* A particularly effective numerical technique for computing eigenvalues in the broken-$\mathcal{PT}$-symmetric region $\varepsilon < 0$ begins by parametrizing the integration path so

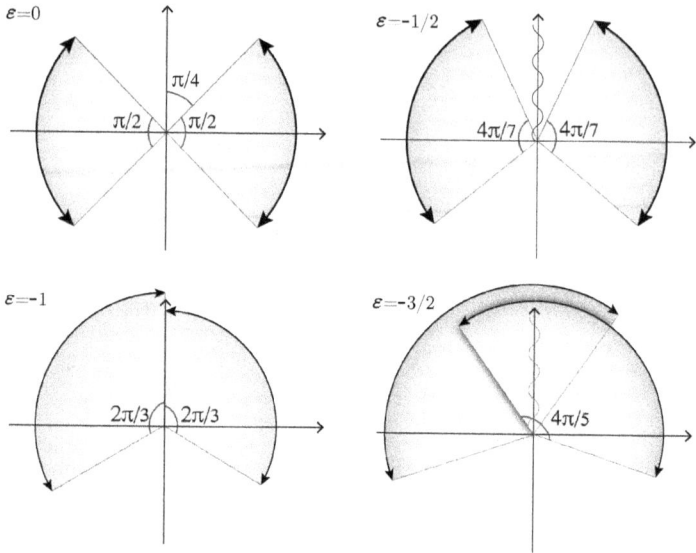

Fig. 2.16   Stokes sectors for the deformed quantum-harmonic-oscillator eigenvalue problem (2.8) for four values of ε. The plot begins with $\varepsilon = 0$, where the Stokes sectors have angular opening $\frac{1}{2}\pi$. As ε decreases, the Stokes sectors rotate upward and become wider. At $\varepsilon = -1$ the branch cut disappears and the sectors touch and fuse. Below $\varepsilon = -1$ the sectors pass through the branch cut on the positive-imaginary axis and enter different sheets of the Riemann surface.

that the eigenvalue problem is reformulated on a *finite real interval*. The appropriate change of variable is

$$x(t) = -i\frac{1}{1 - t^2} \exp\frac{2\pi it}{4 + \varepsilon}. \tag{2.70}$$

With this change of variable, the parameter t ranges from -1 to $+1$ on the real-t axis; the complex eigenvalue problem along an infinitely long curving path in the complex-x plane is thus reduced to an eigenvalue problem on a finite interval on the real-t axis.

To be specific, as t ranges from -1 to 0, the path $x(t)$ in the complex-x plane comes in from infinity in the center of the left Stokes sector [see (2.20)], avoids (loops around) the logarithmic branch-point singularity at the origin, and arrives at the point $x = -i$ as t reaches 0. Then, as t ranges from 0 to $+1$, $x(t)$ follows a PT reflection of the path from $t = -1$ to $t = 0$; that is, $x(t)$ begins at $-i$ and then goes back out to infinity in the center of the right Stokes sector. Figure 2.17 shows the curves $x(t)$ in the complex-x plane for three values of ε. These curves all originate and terminate in

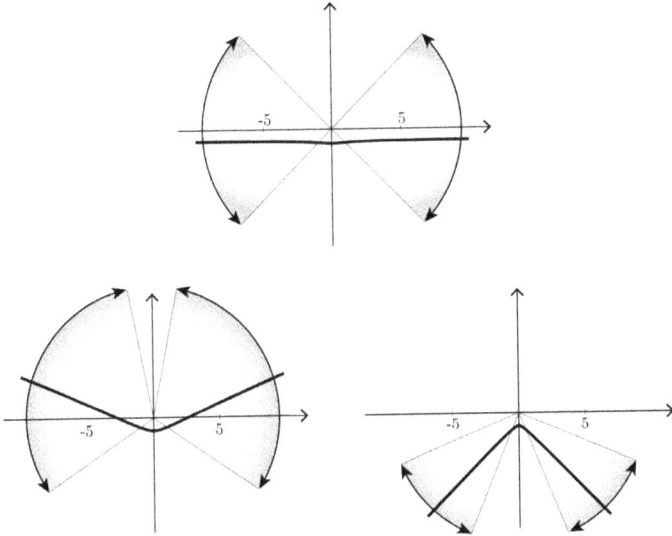

Fig. 2.17   Curves $x(t)$ in (2.70) in the complex x plane for three values of ε: $\varepsilon = -0.8$ (left panel), $\varepsilon = 0$ (center panel), and $\varepsilon = 4.0$ (right panel). By construction, the curves terminate in the centers of the left and right Stokes sectors when $t = -1$ and $t = 1$. When $t = 0$ the curves pass through $x = -i$.

the centers of the Stokes sectors corresponding to the choice of ε and pass through the point $-i$. All curves are left-right symmetric ($\mathcal{PT}$ symmetric).

In terms of the t variable the eigenvalue equation (2.8) has the form

$$-\frac{1}{[x'(t)]^2}\psi''(t) + \frac{x''(t)}{[x'(t)]^3}\psi'(t) + [x(t)]^2[ix(t)]^\varepsilon\psi(t) = E\psi(t), \qquad (2.71)$$

where $\psi(t)$ satisfies the boundary condition $\psi(-1) = \psi(1) = 0$. We then apply the Arnoldi algorithm, which uses the method of finite elements to discretize the t interval and thereby reduce the problem of finding the eigenvalues to that of diagonalizing a finite matrix. An efficient matrix diagonalization determines the low-lying eigenvalues. This method is accurate whether or not the eigenvalues are real. The Arnoldi algorithm is available on Mathematica [Wolfram Research (2015); Trefethen and Bau (1997)].

Since the differential equation (2.71) is singular at $t = \pm 1$, we apply the Arnoldi algorithm to (2.71) subject to the homogeneous Dirichlet boundary conditions $\psi(-1 + \eta) = \psi(1 - \eta) = 0$ and then let η approach 0^+. As $\eta \to 0$, the eigenvalues rapidly approach limiting values. This method is particularly effective for finding the complex energies in the broken-$\mathcal{PT}$-symmetric region $\varepsilon < 0$. A plot of the real and complex eigenvalues of (2.8)

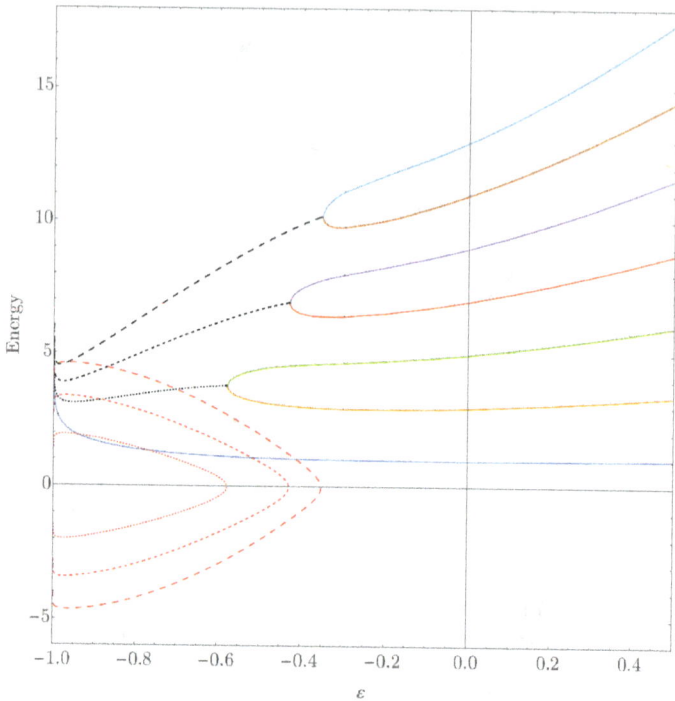

Fig. 2.18   Eigenvalues of the Hamiltonian $H = p^2 + x^2(ix)^\varepsilon$ plotted as functions of the parameter ε for $-1.0 < \varepsilon < 0.5$. This graph is a continuation of that in Fig. 2.1. As ε decreases below 0 and enters the region of broken $\mathcal{PT}$ symmetry, the real eigenvalues, which are plotted according to the color scheme in Fig. 2.1, become pairwise degenerate (except for the lowest eigenvalue, which remains a solid blue curve) and then split into complex-conjugate pairs. The real parts of these pairs of eigenvalues (black lines) initially decrease slightly as ε decreases but blow up abruptly as ε gets close to -1. The imaginary parts of the eigenvalue pairs (red lines) remain finite and abruptly fall to 0 when ε gets close to -1. A blow up of the region around $\varepsilon = -1$ is given in Fig. 2.20, which shows that the real parts of the eigenvalues come down from ∞ and the imaginary parts of the eigenvalues rise back up from 0 as ε passes through -1.

for $-1.1 < \varepsilon < 0$ is depicted in Fig. 2.18. The Arnoldi algorithm gives these eigenvalues to an accuracy of better than one part in 10^{10}.

The Arnoldi algorithm also gives highly accurate plots of the eigenfunctions. Plots of the first two eigenfunctions are given in Fig. 2.19. Note that there are no nodes shown in the plots. This is because the eigenfunctions have no zeros anywhere along the curve $x(t)$ in the complex-x plane. For $\varepsilon = 0$ (the quantum harmonic oscillator) the zeros of the eigenfunctions

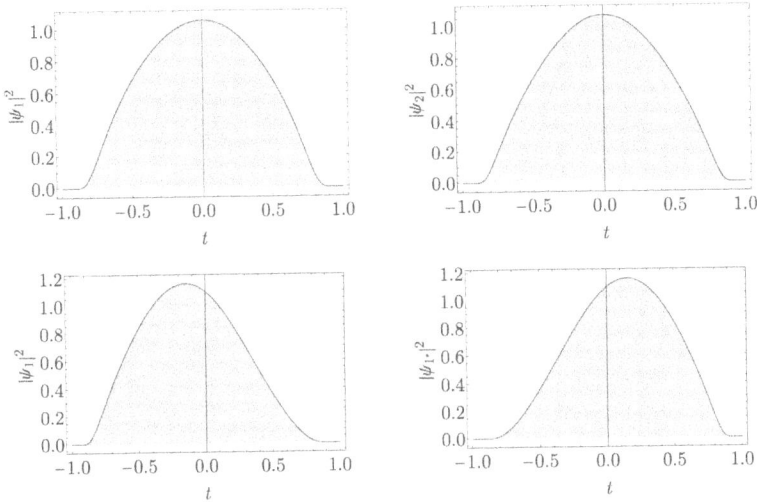

Fig. 2.19 Plot of the absolute squares of two eigenfunctions associated with the eigenvalues in Fig. 2.18 for two values of ε plotted as functions of real t ($-1 \leq t << 1$). In the top row are the eigenfunctions for the first and second excited states at $\varepsilon = -0.4$; that is, $\psi_1(t)$, whose energy is $E = 3.01916$, and $\psi_2(t)$, whose energy is $E = 4.58511$. Note that these plots are left-right symmetric (PT symmetric). This is because the energies are real. In the second row are the analytic continuations of the eigenfunctions in the top row from $\varepsilon = -0.4$ to $\varepsilon = -0.8$. At this value of ε the energies are complex conjugates of one another: $E = 3.4218 \pm 1.4709i$. These plots are no longer left-right symmetric, but the left plot is the mirror image (PT reflection) of the right plot.

are on the real-z axis. For $\varepsilon \neq 0$ the situation is more complicated. For example, when $\varepsilon = 1$, each eigenfunction has an infinite number of zeros on the positive-imaginary-x axis and a finite number of zeros in the lower-half x plane. The zeros in the lower-half plane exhibit an elaborate interlacing pattern [Bender *et al.* (2000a)].

In [Bender *et al.* (2017b)] the asymptotic behavior of the eigenvalues as $\varepsilon \to -1$ was determined analytically. It was found that if $\varepsilon = -1 + \delta$, where δ is small, then as $\delta \to 0$, the real parts of the eigenvalues are logarithmically divergent but the imaginary parts of the eigenvalues are small:

$$\operatorname{Re} E \sim \left(-\frac{3}{4} \ln|\delta|\right)^{2/3}, \quad \operatorname{Im} E \sim \frac{n\pi}{2} \left(-\frac{3}{4} \ln|\delta|\right)^{-1/3}. \tag{2.72}$$

Here, n is an even integer for $\delta > 0$ and n is an odd integer for $\delta < 0$. The imaginary parts of the eigenvalues vary rapidly as ε passes through -1 because there is a *logarithmic* singularity at $\varepsilon = -1$. A detailed description of the region $-1.01 < \varepsilon < -0.99$ is shown in Fig. 2.20.

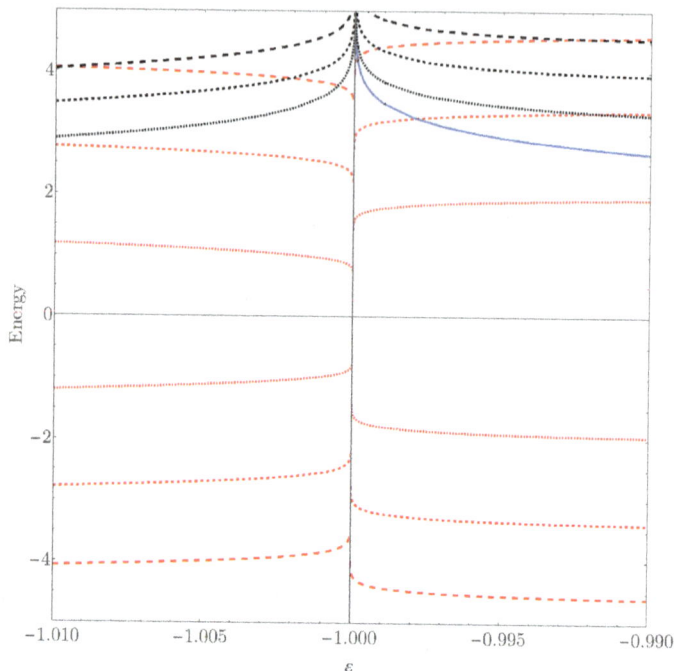

Fig. 2.20   Detailed view of Fig. 2.18 showing the behavior of the eigenvalues of the Hamiltonian $H = p^2 + x^2(ix)^\varepsilon$ plotted as functions of the parameter ε for $-1.01 \leq \varepsilon \leq -0.99$. There is one real eigenvalue for $\varepsilon > -1$ (solid blue line). The real parts of the complex eigenvalues (black lines) and also the real eigenvalue diverge at $\varepsilon = -1$. The complex eigenvalues occur in complex-conjugate pairs and the imaginary parts of the eigenvalues (red lines) abruptly fall to 0 at ε approaches -1 from above and then the imaginary parts bounce back up when ε goes below -1. The eigenvalue behavior near $\varepsilon = -1$ is expressed quantitatively in (2.72). Note that when $\varepsilon > 0$, the zeroth eigenvalue is real and the eigenvalues 1-2, 3-4, 5-6, and so on, form complex-conjugate pairs. However, when $\varepsilon < 0$, the pairing is switched and now the pairing is 0-1, 2-3, 4-5, and so on.

To visualize the behavior of the eigenvalues near $\varepsilon = -1$ more clearly, we have plotted the imaginary and real parts of the eigenvalues in the *complex-ε* plane in the left and right panels of Fig. 2.21. Observe that the imaginary parts of the eigenvalues lie on a helix and that the real parts of the eigenvalues lie on a *double helix* as ε winds around the logarithmic singularity at $\varepsilon = -1$. This logarithmic singularity is an *infinite-order exceptional point*, which one finds only rarely in studies of the analytic structure of eigenvalue problems.

In Fig. 2.22 we plot the first three complex-conjugate pairs of eigenvalues in the range $-2.0 \leq \varepsilon \leq -1.1$. Note that the eigenvalues E_k coalesce to the value -1 as ε approaches -2. As ε decreases towards -2, $\mathrm{Re}\,E_k$ becomes more negative as k increases, and the spectrum becomes *inverted*; that

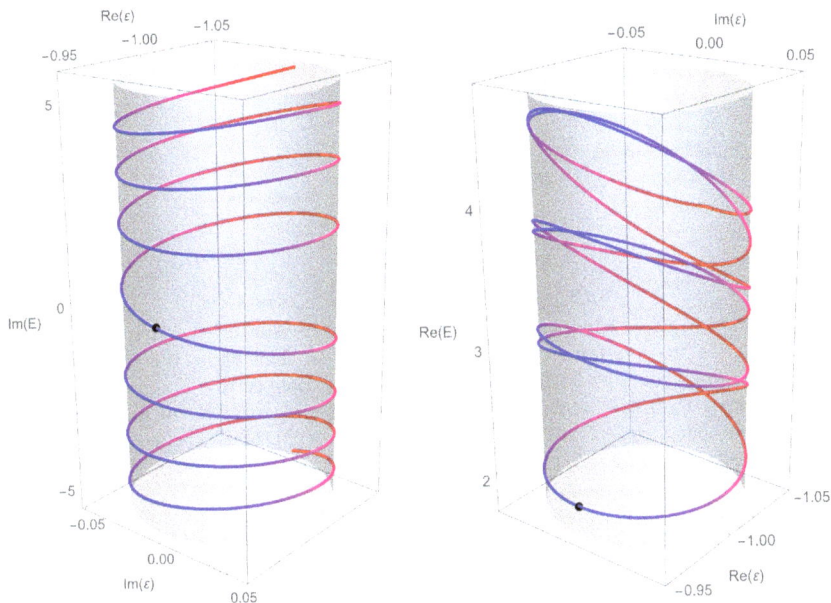

Fig. 2.21   Behavior of the eigenvalues of the Hamiltonian $H = p^2 + x^2(ix)^\varepsilon$ as the parameter ε winds around the exceptional point at $\varepsilon = -1$ in a circle of radius 0.05 in the complex-ε plane. This singular point is an infinite-order exceptional point, and all of the complex eigenvalues analytically continue into one another as we encircle this exceptional point. The lines are shaded blue when $\mathrm{Re}\,\varepsilon > 0$ and red when $\mathrm{Im}\,\varepsilon < 0$. The behavior of the imaginary parts of the eigenvalues (left panel) are easier to visualize because they exhibit a simple logarithmic spiral. The dot shows that the imaginary part of an eigenvalue (the eigenvalue shown in black in Figs. 2.18 and 2.20) vanishes (the eigenvalue is real) when $\mathrm{Re}\,\varepsilon > 0$. As we wind in one direction the imaginary parts of the eigenvalues increase in a helical fashion, and as we wind in the opposite direction the imaginary parts of the eigenvalues decrease in a helical fashion. As we pass the real-ε axis, we pass through the values plotted on the red dashed lines shown in Figs. 2.18 and 2.20. A shaded cylinder has been drawn to assist the eye in following this helix. The behavior of the real parts of the eigenvalues (right panel) is complicated because the curves form a *double* helix. The two helices intersect *four* times each time the singular point at $\varepsilon = -1$ is encircled, and they intersect at $90°$ intervals. If we begin at the dot, we see that the real parts of the eigenvalues increase as we rotate about $\varepsilon = -1$ in either direction. Each time ε crosses the real axis in the complex-ε plane, the curves pass through the values shown at the left and right edges of Fig. 2.20.

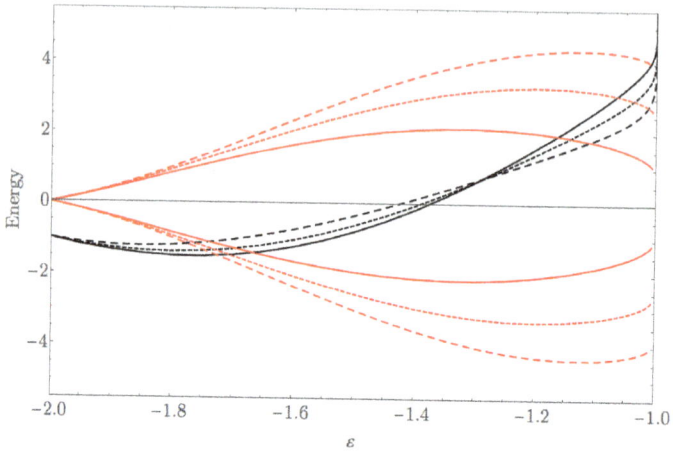

Fig. 2.22   First three complex-conjugate pairs of eigenvalues of the Hamiltonian $H = p^2 + x^2(ix)^\varepsilon$ plotted as functions of ε for $-2 \leq \varepsilon \leq -1$. This figure is a continuation of Fig. 2.20. Note that the real parts of the eigenvalues coalesce to -1 and the imaginary parts coalesce to 0 as ε approaches -2. The results of a WKB calculation of these eigenvalues near $\varepsilon = -2$ is given in [Bender *et al.* (2017b)]. The real parts of the eigenvalues cross near $\varepsilon = -1.3$, but they do not all cross at the same point.

is, the higher-lying real parts of the eigenvalues when ε is near -1.7 (for example) decrease as ε decreases and they cross when $\varepsilon \approx -1.3$.

Chapter 3

$\mathcal{PT}$-Symmetric Quantum Mechanics

> "It's only performance art, you know. Rhetoric,
> they used to teach it in ancient times, like $\mathcal{PT}$."
> —*Tom STOPPard, Arcadia*

We have seen in Chap. 2 that a $\mathcal{PT}$-symmetric Hamiltonian can have an unbroken region of $\mathcal{PT}$ symmetry in which its eigenvalues are all real and positive even though the Hamiltonian is not Hermitian. This raises an obvious question: Can a non-Hermitian Hamiltonian such as $\hat{H}$ in (2.7) define an actual physical theory of quantum mechanics or is the reality and positivity of its spectrum just a mathematical curiosity having no important physical consequences? A Hamiltonian that defines a physical quantum theory must have a real energy spectrum because an energy measurement must return a real number. (The energy spectrum must also be bounded below because a physical theory must have a stable ground state.) Second, the theory must possess a Hilbert space of state vectors that is endowed with an inner product having a positive norm; this is because the norm of a state is interpreted as a probability, and a probability is a positive number. Third, the Hamiltonian of the theory must generate unitary time evolution because probability must be preserved in time.

A simple condition on the Hamiltonian $\hat{H}$ that guarantees that the quantum theory satisfies these three requirements is that $\hat{H}$ be real and symmetric. However, this condition is too restrictive; $\hat{H}$ may be complex as long as it satisfies a more elaborate symmetry property, namely that it be *Dirac Hermitian*: $\hat{H}^{\dagger} = \hat{H}$. In this chapter we explain why we can replace this condition of Hermiticity by the less restrictive condition that $\hat{H}$ have an unbroken $\mathcal{PT}$ symmetry and yet still satisfy the above requirements for a physical quantum theory.[1]

---

[1]Most of the $\mathcal{PT}$-symmetric Hamiltonians considered in this chapter are symmetric

This chapter begins with a review of how to construct a conventional quantum theory when one is given its (Hermitian) Hamiltonian. It then explains the corresponding procedure for the case of a non-Hermitian PT-symmetric Hamiltonian. The principal novelty associated with a PT-symmetric Hamiltonian is that one must calculate a new linear operator called the C operator. This operator is needed to have a probabilistic interpretation of a PT-symmetric quantum theory. The construction of C is nontrivial and several techniques for calculating C are presented.

3.1   Hermitian Quantum Mechanics

For purposes of comparison, we review the conventional textbook procedure that one follows in developing a theory that is defined by a Hermitian quantum-mechanical Hamiltonian $\hat{H}$. For simplicity, we will assume that $\hat{H}$ describes a particle in a one-dimensional confining potential (a potential that rises as $|x| \to \infty$) so that the energy levels are discrete. Using the discussion in this section as a guide, we then explain in the next section the corresponding procedures for a non-Hermitian PT-symmetric Hamiltonian.

(a) *Eigenfunctions and eigenvalues of H.* Given the Hamiltonian $\hat{H}$, one can at least in principle construct and solve the time-independent Schrödinger differential equation $\hat{H}\psi = E\psi$ and obtain the eigenvalues E_n and the eigenfunctions $\psi_n(x)$. Such a calculation is usually difficult to perform analytically, but it may be done with great accuracy by using numerical techniques. We will be interested primarily in performing these calculations in coordinate space, and in coordinate space the zeros of the eigenfunctions of Hermitian Hamiltonians exhibit the remarkable phenomenon of *interlacing*; this means that $\psi_n(x)$ has exactly n zeros and between every pair of successive zeros of $\psi_n(x)$ is exactly one zero of $\psi_{n-1}(x)$.

(b) *Orthogonality of eigenfunctions.* Because $\hat{H}$ is Hermitian, its eigenfunctions must be orthogonal with respect to the standard Hermitian coordinate-space inner product:

$$(\psi, \phi) \equiv \int dx \, [\psi(x)]^* \phi(x). \tag{3.1}$$

---

under matrix transposition. This matrix symmetry condition is not necessary, but it has the simplifying virtue that the we do not need to introduce a biorthogonal set of basis states. We can consider PT-symmetric Hamiltonians that are not symmetric under matrix transposition, but only if we introduce a biorthogonal basis [Weigert (2003a); Curtright and Mezincescu (2007); Brody (2014)].

Orthogonality means that the inner product of eigenfunctions $\psi_m(x)$ and $\psi_n(x)$ associated with eigenvalues $E_m \neq E_n$ vanishes: $(\psi_m, \phi_n) = 0.$[2]

(c) *Orthonormality of eigenfunctions.* Because the Hamiltonian $\hat{H}$ is Hermitian, the norm of any vector is positive. Therefore, we can *normalize* the eigenfunctions of $\hat{H}$ so that the norm of every eigenfunction is unity:

$$(\psi_n, \psi_n) = 1. \tag{3.2}$$

In coordinate space the orthonormality condition reads

$$\int dx \, \psi_m^*(x)\psi_n(x) = \delta_{m,n}, \tag{3.3}$$

where $\delta_{m,n}$ is the Kroneker delta function.

(d) *Completeness of eigenfunctions.* A profound result of the theory of linear operators on Hilbert spaces is that the eigenfunctions of a Hermitian Hamiltonian are *complete*. This means that a given (finite-norm) vector χ in the Hilbert space can be approximated by linear combinations of the eigenfunctions of $\hat{H}$: $\chi = \lim_{m \to \infty} \sum_{n=0}^{m} a_n^m \psi_n$. At a formal level, completeness is expressed by representing the *unit operator* (the delta function) as a sum over the eigenfunctions: $\sum_{n=0}^{\infty} [\psi_n(x)]^* \psi_n(y) = \delta(x - y)$. This result can be restated using bra and ket notation as $\sum_n |n\rangle\langle n| = \mathbb{1}$.

(e) *Coordinate-space reconstruction of the Hamiltonian.* The Hamiltonian $\hat{H}$ may be represented as a matrix $H(x, y)$ in coordinate space by a formal sum over eigenfunctions:

$$\sum_{n=0}^{\infty} [\psi_n(x)]^* \psi_n(y) E_n = H(x, y). \tag{3.4}$$

In terms of this coordinate-space matrix representation, matrix multiplication takes the form of integration. Thus, we express the eigenvalue equation $\hat{H}\psi_n = E_n\psi_n$ as the integral $\int dy \, H(x, y)\psi_n(y) = E_n\psi_n(y)$, where we have used the orthogonality condition (3.3). (We have assumed here that it is valid to interchange the operations of summation and integration.)

(f) *Coordinate-space representation of the Green's function.* The Green's function $G(x, y)$ is the matrix inverse of the Hamiltonian: $\int dy \, H(x, y)G(y, z) = \delta(x - z)$. In matrix form, the Green's function is a formal sum over the eigenfunctions:

$$\sum_{n=0}^{\infty} [\psi_n(x)]^* \psi_n(y) E_n^{-1} = G(x, y). \tag{3.5}$$

Equations (3.4) and (3.5) may be rewritten using bra and ket notation as

$$\sum_n |n\rangle E_n \langle n| = H \quad \text{and} \quad \sum_n |n\rangle E_n^{-1} \langle n| = G.$$

---

[2] We do not discuss here the technical problems that arise if the spectrum contains degenerate eigenvalues.

The formula for the Green's function in (3.5) allows us to calculate the sum of the reciprocals of the energy eigenvalues: We set $x = y$ in (3.5), integrate with respect to x (assuming that this integral exists), and then use the normalization condition (3.2) to obtain

$$\int dx\, G(x, x) = \sum_{n=0}^{\infty} E_n^{-1}. \tag{3.6}$$

The summation on the right side of (3.6) is called the *spectral zeta function*. This sum converges if the energy levels E_n grow like n^α, where $\alpha > 1$. The spectral zeta function for the harmonic oscillator diverges, but the WKB approximation (2.64) shows that the sum exists for the $\mathcal{PT}$-symmetric $x^2(ix)^\varepsilon$ potential if $\varepsilon > 0$.[3]

(g) *Time evolution and unitarity.* For a Hermitian Hamiltonian the time-evolution operator $e^{-i\hat{H}t}$ is *unitary*: it preserves the inner product

$$\langle \chi(t)|\chi(t)\rangle = \langle \chi(0)e^{i\hat{H}t}|e^{-i\hat{H}t}\chi(t)\rangle = \langle \chi(0)|\chi(0)\rangle.$$

(h) *Observables.* An *observable* may be represented as a linear Hermitian operator O. The outcome of a measurement of O is one of the *real* eigenvalues of this operator. (Observables in $\mathcal{PT}$-symmetric quantum mechanics are discussed in Sec. 3.4.)

(i) *Additional topics.* A course on quantum mechanics addresses a number of additional topics, such as the classical and semiclassical limits of the quantum theory, probability density and currents, perturbative and nonperturbative calculations, and so on. For $\mathcal{PT}$-symmetric Hamiltonians some of these topics are discussed in Chaps. 2 and 4.

3.2 $\mathcal{PT}$-Symmetric Quantum Mechanics

Let us apply the recipe in Sec. 3.1 to a non-Hermitian $\mathcal{PT}$-symmetric Hamiltonian having an unbroken $\mathcal{PT}$ symmetry. The novelty here is that, unlike the case of conventional quantum mechanics, we do not know *a priori* the definition of the inner product. We will discover the appropriate inner product in the course of our analysis and we will see that this inner product is determined by the Hamiltonian. Thus, unlike conventional quantum mechanics, $\mathcal{PT}$-symmetric quantum mechanics is a kind of *bootstrap* theory in the sense that the Hamiltonian operator $\hat{H}$ chooses the Hilbert space (and associated inner product) in which it prefers to live.

---

[3]The exact closed-form expression for the spectral zeta function for the $\mathcal{PT}$-symmetric Hamiltonian $\hat{H}$ in (2.7) is given in (3.13).

(a) *Eigenfunctions and eigenvalues of $\hat{H}$.* In Chap. 2 we discussed the numerical and analytic techniques that are available for finding the eigenvalues and corresponding coordinate-space eigenfunctions of a non-Hermitian Hamiltonian. Therefore, we assume that we have already found the eigenvalues E_n by using these methods and that these eigenvalues are all real; that is, we assume that we are in a region of *unbroken PT symmetry*. This assumption is crucial, and as we showed at the beginning of Chap. 2, this is equivalent to assuming that all of the eigenfunctions $\psi_n(x)$ of $\hat{H}$ are also eigenfunctions of PT. The zeros of non-Hermitian PT-symmetric Hamiltonians do not interlace; however, the eigenfunctions of PT-symmetric Hamiltonians exhibit the phenomenon of *winding*, which is the non-Hermitian analog and generalization of Hermitian interlacing [Bender *et al.* (2000a); Eremenko *et al.* (2008); Hezaro (2008); Schindler and Bender (2018)].

(b) *Orthogonality of eigenfunctions.* To test the orthogonality of the eigenfunctions, we must specify an inner product. (A pair of vectors can be orthogonal with respect to one inner product and not orthogonal with respect to another.) We do not yet know an appropriate inner product, so let us simply guess the inner product. Arguing by analogy, one might think that since the inner product in (3.1) is appropriate for Hermitian Hamiltonians ($\hat{H} = \hat{H}^\dagger$), a possible choice for the inner product associated with a PT-symmetric Hamiltonian ($\hat{H} = \hat{H}^{PT}$) might be

$$(\psi, \phi) \equiv \int_C dx \, [\psi(x)]^{PT} \phi(x) = \int_C dx \, [\psi(-x)]^* \phi(x),$$

where C is a contour that terminates in the Stokes sectors in which we impose the boundary conditions on the eigenvalue equation associated with the Hamiltonian. With this inner-product definition we can use integration by parts to show that pairs of eigenfunctions of $\hat{H}$ associated with different eigenvalues are orthogonal. The fact that we can establish orthogonality is good news, but unfortunately, this guess for an inner product is not acceptable for formulating a physical quantum theory because the norm of a state is not necessarily positive.[4]

(c) *The CPT inner product.* It is possible to construct an inner product with a positive norm for a complex non-Hermitian Hamiltonian having an *unbroken PT* symmetry. To do so we invent a new linear operator C that commutes with both $\hat{H}$ and PT. The existence of such an operator C that commutes with the Hamiltonian reveals that the Hamiltonian has a completely unanticipated symmetry. This symmetry embodies our assumption that the PT symmetry of the Hamiltonian is unbroken; that is,

---

[4] Nonpositive inner products and their use in Kreĭn spaces are discussed in Chap. 8.

that all of the eigenstates of the Hamiltonian are also eigenstates of $\mathcal{PT}$ [Bender *et al.* (2002)]. We use the symbol $\mathcal{C}$ to represent this new symmetry because the properties of $\mathcal{C}$ are mathematically similar to those of the charge-conjugation operator in particle physics. The inner product with respect to $\mathcal{CPT}$ conjugation is thus defined as

$$\langle\psi|\chi\rangle^{\mathcal{CPT}} \equiv \int dx \, \psi^{\mathcal{CPT}}(x)\chi(x), \qquad (3.7)$$

where $\psi^{\mathcal{CPT}}(x) = \int dy \, \mathcal{C}(x,y)\psi^*(-y)$. We will see shortly that this inner product satisfies the requirements for the quantum theory defined by $\hat{H}$ to have a Hilbert space with a positive norm and to be a unitary theory of quantum mechanics and we will show how to represent the $\mathcal{C}$ operator as a sum over the eigenfunctions of H [just as we represented the Hamiltonian in (3.4) and the Green's function in (3.5)], but before doing so we must first show how to normalize the eigenfunctions of a $\mathcal{PT}$-symmetric Hamiltonian.

(d) $\mathcal{PT}$-*symmetric normalization and completeness of the eigenfunctions.* Recall that at the beginning of Chap. 2 we showed that the eigenfunctions $\psi_n(x)$ of $\hat{H}$ are also eigenfunctions of the $\mathcal{PT}$ operator with the eigenvalue $\lambda = e^{i\alpha}$, where λ and α depend on n. Thus, we choose to define the $\mathcal{PT}$-normalized eigenfunctions $\phi_n(x)$ as follows:

$$\phi_n(x) \equiv e^{-i\alpha/2}\psi_n(x). \qquad (3.8)$$

Note that $\phi_n(x)$ is an eigenfunction of $\hat{H}$, but it is also an eigenfunction of $\mathcal{PT}$ with eigenvalue 1. One can show both numerically and analytically that for $\hat{H}$ in (2.7) the algebraic sign of the $\mathcal{PT}$ norm in (3.7) of $\phi_n(x)$ is $(-1)^n$ for all n and for all $\varepsilon > 0$ [Bender *et al.* (2002, 2004b)]. Thus, we *define* the eigenfunctions so that their $\mathcal{PT}$ norms are exactly $(-1)^n$:

$$\int_C dx \, [\phi_n(x)]^{\mathcal{PT}}\phi_n(x) = \int_C dx \, [\phi_n(-x)]^*\phi_n(x) = (-1)^n, \qquad (3.9)$$

where the contour C lies in the Stokes sectors shown in Fig. 2.4. In terms of these $\mathcal{PT}$-normalized eigenfunctions there is a simple but unusual [because of the factor of $(-1)^n$ in the sum] statement of completeness:

$$\sum_{n=0}^{\infty}(-1)^n\phi_n(x)\phi_n(y) = \delta(x-y). \qquad (3.10)$$

This unusual statement of completeness has been verified for all $\varepsilon > 0$ [Mezincescu (2000); Bender and Wang (2001); Weigert (2003a)]. Using (3.9) one can verify that the left side of (3.10) satisfies the integration rule for delta functions: $\int dy \, \delta(x-y)\delta(y-z) = \delta(x-z)$.

Example: *PT-symmetric normalization of harmonic-oscillator eigen-functions.* For the Hamiltonian $\hat{H} = \hat{p}^2 + \hat{x}^2$ the eigenfunctions are Gaussians multiplied by Hermite polynomials:

$$\psi_0(x) = \exp\left(-\tfrac{1}{2}x^2\right),$$
$$\psi_1(x) = x \exp\left(-\tfrac{1}{2}x^2\right),$$
$$\psi_2(x) = (2x^2 - 1) \exp\left(-\tfrac{1}{2}x^2\right),$$
$$\psi_3(x) = (2x^3 - 3x) \exp\left(-\tfrac{1}{2}x^2\right),$$

and so on. By an appropriate choice of phase, we normalize these eigenfunctions so that they are eigenfunctions of PT with eigenvalue 1:

$$\phi_0(x) = a_0 \exp\left(-\tfrac{1}{2}x^2\right),$$
$$\phi_1(x) = a_1 ix \exp\left(-\tfrac{1}{2}x^2\right),$$
$$\psi_2(x) = a_2(2x^2 - 1) \exp\left(-\tfrac{1}{2}x^2\right),$$
$$\psi_3(x) = a_3 i(2x^3 - 3x) \exp\left(-\tfrac{1}{2}x^2\right),$$

where the real numbers a_n are such that the integral in (3.9) evaluates to $(-1)^n$ for all n. If the eigenfunctions $\phi_n(x)$ are substituted into (3.10) and the summation is performed, then the Dirac delta function on the right side of (3.10) is obtained.

(e) *Coordinate-space representation of $\hat{H}$ and $\hat{G}$ and the spectral zeta function.* From the statement of completeness in (3.10) we can construct coordinate-space representations of the linear operators. For example, in coordinate space the parity operator is $P(x, y) = \delta(x + y)$, so

$$P(x, y) = \sum_{n=0}^{\infty} (-1)^n \phi_n(x)\phi_n(-y). \tag{3.11}$$

Also, the coordinate-space representations of the Hamiltonian and the Green's function are

$$H(x, y) = \sum_{n=0}^{\infty} (-1)^n E_n \phi_n(x)\phi_n(y),$$
$$G(x, y) = \sum_{n=0}^{\infty} (-1)^n E_n^{-1} \phi_n(x)\phi_n(y).$$

By using (3.9), we can show that $\hat{G}$ is the functional inverse of H:

$$\int dy \, H(x, y)G(y, z) = \delta(x - z).$$

For the PT-symmetric Hamiltonian (2.7) this equation is a differential equation satisfied by $G(x, y)$ in coordinate space:

$$- G_{xx}(x, y) + x^2 (ix)^\varepsilon G(x, y) = \delta(x - y). \tag{3.12}$$

We solve (3.12) in terms of associated Bessel functions in each of two regions, $x > y$ and $x < y$. We then patch the solutions together at $x = y$ to obtain a closed-form expression for $G(x, y)$ [Mezincescu (2000); Bender and Wang (2001)]. Finally, we use (3.6) to obtain an exact formula for the spectral zeta function for all values of $\varepsilon > 0$:

$$\sum_n \frac{1}{E_n} = \left[1 + \frac{\cos\left(\frac{3\varepsilon\pi}{2\varepsilon+8}\right) \sin\left(\frac{\pi}{4+\varepsilon}\right)}{\cos\left(\frac{\varepsilon\pi}{4+2\varepsilon}\right) \sin\left(\frac{3\pi}{4+\varepsilon}\right)} \right]$$

$$\times \frac{\Gamma\left(\frac{1}{4+\varepsilon}\right) \Gamma\left(\frac{2}{4+\varepsilon}\right) \Gamma\left(\frac{\varepsilon}{4+\varepsilon}\right)}{(4+\varepsilon)^{\frac{4+2\varepsilon}{4+\varepsilon}} \Gamma\left(\frac{1+\varepsilon}{4+\varepsilon}\right) \Gamma\left(\frac{2+\varepsilon}{4+\varepsilon}\right)}. \tag{3.13}$$

(f) *Construction of the $\mathcal{C}$ operator.* Note that half of the energy eigenstates of $\hat{H}$ have positive $\mathcal{PT}$ norm and half have negative $\mathcal{PT}$ norm [see (3.9)]. This situation is analogous to the problem that Dirac encountered in formulating the spinor wave equation in relativistic quantum theory [Dirac (1942)]. To solve this problem Dirac sought a physical interpretation of the negative-norm states in terms of antiparticles.

For our problem we observe that any Hamiltonian $\hat{H}$ having an unbroken $\mathcal{PT}$ symmetry has an additional symmetry of $\hat{H}$ represented by the linear operator $\mathcal{C}$, and this operator arises because there are equal numbers of positive-norm and negative-norm states all having real energy eigenvalues. In coordinate space the operator $\mathcal{C}$ that embodies this symmetry is a sum over the $\mathcal{PT}$-normalized eigenfunctions of the $\mathcal{PT}$-symmetric Hamiltonian:

$$\mathcal{C}(x, y) = \sum_{n=0}^{\infty} \phi_n(x)\phi_n(y). \tag{3.14}$$

This equation is identical to the statement of completeness in (3.10) except that *the factor of* $(-1)^n$ *is absent*. We use (3.9) and (3.10) to see that the square of $\mathcal{C}$ is unity ($\mathcal{C}^2 = \mathbb{1}$): $\int dy\, \mathcal{C}(x, y)\mathcal{C}(y, z) = \delta(x - z)$. Observe that the eigenvalues of $\mathcal{C}$ are ± 1 and thus we interpret $\mathcal{C}$ as a *reflection operator*. Also, we can verify that $\mathcal{C}$ commutes with $\hat{H}$. Consequently, since $\mathcal{C}$ is linear, the eigenstates of $\hat{H}$ have definite values of $\mathcal{C}$; that is, each state of $\hat{H}$ has a definite $\mathcal{C}$ parity:

$$\mathcal{C}\phi_n(x) = \int dy\, \mathcal{C}(x, y)\phi_n(y) = \sum_{m=0}^{\infty} \phi_m(x) \int dy\, \phi_m(y)\phi_n(y) = (-1)^n \phi_n(x).$$

The $\mathcal{C}$ operator resembles the charge-conjugation operator in quantum field theory. However, here $\mathcal{C}$ represents the measurement of the sign of the $\mathcal{PT}$ norm (3.9). The operators $\mathcal{P}$ and $\mathcal{C}$ are distinct square roots of the unity operator $\delta(x - y)$. That is, $\mathcal{P}^2 = \mathcal{C}^2 = 1$, but $\mathcal{P} \neq \mathcal{C}$; indeed $\mathcal{P}$ is real, while $\mathcal{C}$ is complex. The parity operator in coordinate space is real $\mathcal{P}(x, y) = \delta(x + y)$, but the $\mathcal{C}(x, y)$ operator in coordinate space is complex because it is a sum of products of complex functions. The two operators $\mathcal{P}$ and $\mathcal{C}$ do not commute, but $\mathcal{C}$ *does* commute with $\mathcal{PT}$.

(g) *Positive norm and unitarity in* $\mathcal{PT}$*-symmetric quantum mechanics.* Having constructed the $\mathcal{C}$ operator, we now use the $\mathcal{CPT}$ inner product defined in (3.8). Like the $\mathcal{PT}$ inner product, this new inner product is phase independent. Also, because the time-evolution operator (as in ordinary quantum mechanics) is e^{-iHt} and because H commutes with $\mathcal{PT}$ and with $\mathcal{CPT}$, both the $\mathcal{PT}$ inner product and the $\mathcal{CPT}$ inner product remain time independent as the states evolve. However, unlike the $\mathcal{PT}$ inner product, the $\mathcal{CPT}$ inner product is *positive definite* because $\mathcal{C}$ contributes a factor of -1 when it acts on states with negative $\mathcal{PT}$ norm. In terms of the $\mathcal{CPT}$ conjugate, the completeness condition reads

$$\sum_{n=0}^{\infty} \phi_n(x)[\mathcal{CPT}\phi_n(y)] = \delta(x - y).$$

3.3 Comparison of Hermitian and $\mathcal{PT}$-Symmetric Theories

In Secs. 3.1 and 3.2 we showed how to construct quantum theories based on Hermitian and on non-Hermitian Hamiltonians. For a conventional quantum theory defined by a Hermitian Hamiltonian, the Hilbert space of physical states is specified even before the Hamiltonian is known. The inner product (3.1) in this vector space is defined with respect to Dirac Hermitian conjugation (combined complex conjugation and matrix transposition). The Hamiltonian is then chosen and the eigenvectors and eigenvalues of the Hamiltonian are determined. In contrast, the inner product for a quantum theory defined by a non-Hermitian $\mathcal{PT}$-symmetric Hamiltonian depends on the Hamiltonian itself and thus it is determined *dynamically*. We must solve for the eigenvalues and eigenstates of $\hat{H}$ and verify that the eigenvalues are real (and that we have an unbroken region of $\mathcal{PT}$ symmetry) before we can know the Hilbert space and the associated inner product of the theory. This is because the $\mathcal{C}$ operator is defined and constructed in terms of the eigenstates of the Hamiltonian. The Hilbert space, which consists of all complex linear combinations of the eigenstates of $\hat{H}$, and the $\mathcal{CPT}$ inner product are determined by these eigenstates.

The $\mathcal{C}$ operator does not exist as a distinct entity in conventional Hermitian quantum mechanics. For example, if we let the parameter ε in (2.7) tend to 0, the $\mathcal{C}$ operator in this limit becomes identical to $\mathcal{P}$ and the $\mathcal{CPT}$

operator becomes $\mathcal{T}$, which performs complex conjugation. Hence, the inner product defined with respect to $\mathcal{CPT}$ conjugation reduces to the inner product of conventional quantum mechanics and (3.10) reduces to the usual statement of completeness $\sum_n \phi_n(x)\phi_n^*(y) = \delta(x - y)$ when $\varepsilon \to 0$.

The $\mathcal{CPT}$ inner product is independent of the choice of integration contour C as long as C terminates inside the Stokes sectors associated with the boundary conditions for the eigenvalue problem. In contrast, in ordinary quantum mechanics, where the positive-definite inner product has the form $\int dx\, f^*(x)g(x)$, the integral is taken along the real axis and the path of integration cannot be deformed into the complex plane because the integrand is not analytic. The $\mathcal{PT}$ inner product shares with the $\mathcal{CPT}$ inner product the advantage of analyticity and path independence, but the $\mathcal{PT}$ inner product suffers from nonpositivity. It is surprising that we can construct a positive-definite metric by using $\mathcal{CPT}$ conjugation without disturbing the path independence of the inner-product integral.

Time evolution is expressed by the operator $e^{-i\hat{H}t}$ regardless of whether the theory is determined by a $\mathcal{PT}$-symmetric Hamiltonian or a conventional Hermitian Hamiltonian. To verify that time evolution is unitary we show that as a state vector evolves, its norm does not change in time. If $\psi_0(x)$ is any given initial vector belonging to the Hilbert space spanned by the energy eigenstates, then it evolves into the state $\psi_t(x)$ at time t according to $\psi_t(x) = e^{-i\hat{H}t}\psi_0(x)$. With respect to the $\mathcal{CPT}$ inner product the norm of $\psi_t(x)$ does not change in time simply because $i\hat{H}$ commutes with $\mathcal{CPT}$.

3.4 Observables

In conventional quantum mechanics the condition for a linear operator A to be an observable is that it be Hermitian $A = A^\dagger$. This condition guarantees that the expectation value of A in a state is real. Operators in the Heisenberg picture evolve in time according to $A(t) = e^{i\hat{H}t}A(0)e^{-i\hat{H}t}$, so this Hermiticity condition is maintained in time. In $\mathcal{PT}$-symmetric quantum mechanics the equivalent condition is that at time $t = 0$ the operator A must obey the condition $A^{\mathrm{T}} = \mathcal{CPT}A\mathcal{CPT}$, where A^{T} is the *transpose* of A [Bender *et al.* (2002, 2004b)]. If this condition holds at $t = 0$, then it continues to hold at later times because we have assumed that $\hat{H}$ is symmetric $(\hat{H} = \hat{H}^{\mathrm{T}})$. This condition also guarantees that the expectation value of A in any state is real.[5]

---

[5]The requirement for A to be an observable normally involves matrix transposition. This condition is more restrictive than is necessary because we are assuming here that the

The operator $\mathcal{C}$ itself is an observable because it satisfies this requirement. The Hamiltonian is also an observable. However, the $\hat{x}$ and $\hat{p}$ operators are not observables. Indeed, the expectation value of $\hat{x}$ in the ground state for the theory defined by $\hat{H}$ in (2.7) is a negative imaginary number [Bender *et al.* (2001c)]. Thus, there is no position operator in PT-symmetric quantum mechanics.[6]

In this sense PT-symmetric quantum mechanics is similar to fermionic quantum field theories. In such theories the fermion field corresponds to the $\hat{x}$ operator. The fermion field is complex and does not have a classical limit. One cannot measure the position of an electron; one can only measure the position of the *charge* or of the *energy* of the electron.

In relativistic quantum field theory there is no position operator at all. This is because if we could measure the position of a particle very accurately, say by confining it to a small box, then the uncertainty in the momentum would be large. Hence, the energy in the interior of the box would be gigantic. As a consequence, particle production would occur and the resulting particles would no longer be confined to the box! Thus, the notion of a position operator is of limited applicability and only applies to the nonrelativistic limit of a relativistic quantum theory.

To see why the expectation value of the $\hat{x}$ operator in PT-symmetric quantum mechanics is a negative imaginary number, we examine a classical trajectory like those shown in Fig. 1.14. Note that the classical trajectories have left-right (PT) symmetry, but not up-down symmetry. The classical paths favor (spend more time in) the lower-half complex-x plane. Thus, it is not surprising that the average classical position is a negative imaginary number. Just as the classical particle moves about in the complex plane, the quantum probability current also flows in the complex plane. A plausible interpretation of PT-symmetric quantum mechanics is that it describes extended rather than pointlike objects, where these objects extend in the imaginary as well as in the real direction.

3.5 Pseudo-Hermiticity and Quasi-Hermiticity

The underlying idea of PT symmetry (that is, replacing the mathematical condition of Hermiticity by the more physical condition of PT symmetry)

Hamiltonian is symmetric. Nonsymmetric Hamiltonians are discussed in [Mostafazadeh and Batal (2004); Mostafazadeh (2005); Jones (2005)].

[6]Nevertheless, it may be possible to formulate a meaningful description of the probability density and probability current of a quantum-mechanical particle in the complex plane. See Sec. 4.3 for a discussion of this idea.

can be placed in a more general mathematical setting known as *pseudo-Hermiticity*. A linear operator A is *pseudo-Hermitian* if there is a Hermitian operator η such that

$$A^\dagger = \eta A \eta. \tag{3.15}$$

The operator η is called an *intertwining* operator. The condition (3.15) reduces to ordinary Hermiticity when η is the identity $\mathbf{1}$ and it reduces to $\mathcal{PT}$ symmetry when $\eta = \mathcal{P}$. The concept of pseudo-Hermiticity was introduced in the 1940s by Dirac and Pauli, and later discussed by Lee, Wick, and Sudarshan, who were interested in resolving the problems that arise in quantizing electrodynamics and other quantum field theories in which negative-norm states appear as a consequence of renormalization [Pauli (1943); Gupta (1950a,b); Bleuler (1950); Sudarshan (1961); Lee and Wick (1969)]. These problems are illustrated clearly by the Lee model, which is discussed in Sec. 5.6.

The related notion of quasi-Hermiticity was discussed in [Scholtz *et al.* (1992)]. This deep paper is relevant to $\mathcal{PT}$ symmetry because it was the first to show how to construct a similarity transformation that maps Hermitian operators onto the corresponding quasi-Hermitian operators and also the first to consider the corresponding transformations of infinite-dimensional Hilbert-space inner products.

Mostafazadeh first pointed out that because the parity operator $\mathcal{P}$ is Hermitian, it may be used as an intertwining operator. The class of Hamiltonians $\hat{H}$ in (2.7) is pseudo-Hermitian because the parity operator $\mathcal{P}$ changes the sign of $\hat{x}$ while Dirac Hermitian conjugation changes the sign of i [Mostafazadeh (2002a,b,c,d,e, 2003a,c, 2005)]: $\hat{H}^\dagger = \mathcal{P}\hat{H}\mathcal{P}$. Further references on the generalization of $\mathcal{PT}$ symmetry to pseudo-Hermiticity are[7] [Bagchi and Quesne (2002); Ahmed (2002); Japaridze (2002); Ahmed (2003a,c); Ahmed and Jain (2003b,a); Ahmed (2003b,d); Blasi *et al.* (2004); Bagchi *et al.* (2005)].

3.6   Model 2×2 $\mathcal{PT}$-Symmetric Matrix Hamiltonian

In order to illustrate the formal procedures discussed in Sec. 3.2 we examine the exactly solvable $\mathcal{PT}$-symmetric matrix model Hamiltonian in (1.5), which we rewrite in polar form as [Bender *et al.* (2003a)]

$$\hat{H} = \begin{pmatrix} re^{i\theta} & s \\ s & re^{-i\theta} \end{pmatrix}. \tag{3.16}$$

---

[7]The term *crypto-Hermiticity* has also been introduced [Smilga (2008a,b)].

The three parameters r, s, and θ are real. Of course, the Hamiltonian (3.16) is not Hermitian, but we can see that it is PT symmetric if we define the parity operator as

$$P = \begin{pmatrix} 0 & 1 \\ 1 & 0 \end{pmatrix} \tag{3.17}$$

and we define the operator T to perform complex conjugation.

The first step in developing the quantum-mechanical theory defined by the Hamiltonian (3.16) is to calculate its two eigenvalues:

$$E_\pm = r\cos\theta \pm (s^2 - r^2 \sin^2\theta)^{1/2}. \tag{3.18}$$

As we explained in Sec. 1.2, there are two parametric regions, one for which the square root in (3.18) is real and the other for which it is imaginary. When $s^2 < r^2 \sin^2\theta$, the energy eigenvalues form a complex-conjugate pair. This is the region of broken PT symmetry. However, when $s^2 \geq r^2 \sin^2\theta$, the eigenvalues $E_\pm = r\cos\theta \pm (s^2 - r^2 \sin^2\theta)^{1/2}$ are real. This is the region of unbroken PT symmetry. In the unbroken region the simultaneous eigenstates of the operators $\hat{H}$ and PT are

$$|E_+\rangle = \frac{1}{\sqrt{2\cos\alpha}} \begin{pmatrix} e^{i\alpha/2} \\ e^{-i\alpha/2} \end{pmatrix} \quad \text{and} \quad |E_-\rangle = \frac{i}{\sqrt{2\cos\alpha}} \begin{pmatrix} e^{-i\alpha/2} \\ -e^{i\alpha/2} \end{pmatrix}, \tag{3.19}$$

where $\sin\alpha = \frac{r}{s}\sin\theta$. The PT inner product gives

$$(E_\pm, E_\pm) = \pm 1 \quad \text{and} \quad (E_\pm, E_\mp) = 0, \tag{3.20}$$

where $(u, v) = (PT u) \cdot v$. As we predicted in Sec. 3.2, with respect to the PT inner product the vector space spanned by the energy eigenstates has a metric of signature $(+, -)$. If the condition $s^2 > r^2 \sin^2\theta$ for an unbroken PT symmetry is violated, the states (3.19) are no longer eigenstates of PT because α becomes imaginary. When PT symmetry is broken, the PT norm of the energy eigenstate vanishes.

Following the recipe in (3.14), we construct the C operator:

$$C = \frac{1}{\cos\alpha} \begin{pmatrix} i\sin\alpha & 1 \\ 1 & -i\sin\alpha \end{pmatrix}. \tag{3.21}$$

A short calculation shows that, as we predicted, the C operator commutes with H and satisfies $C^2 = 1$. Also, as expected, the eigenvectors of $\hat{H}$ are also eigenvectors of C with eigenvalues $+1$ and -1: $C|E_\pm\rangle = \pm|E_\pm\rangle$. Thus, the eigenvalues of C are exactly the signs of the PT norms of the corresponding eigenstates.

Next, we construct the new CPT inner product $\langle u|v\rangle = (CPT u) \cdot v$ by using the operator C. This inner product is positive definite because

$\langle E_\pm | E_\pm \rangle = 1$. The two-dimensional Hilbert space spanned by $|E_\pm\rangle$ and with inner product $\langle \cdot | \cdot \rangle$ has the signature $(+, +)$.

Finally, we demonstrate that the $\mathcal{CPT}$ norm of *any* nonzero vector is positive. For the vector $\psi = \binom{a}{b}$, where a and b are arbitrary complex numbers, we calculate that

$$\mathcal{T}\psi = \begin{pmatrix} a^* \\ b^* \end{pmatrix}, \quad \mathcal{PT}\psi = \begin{pmatrix} b^* \\ a^* \end{pmatrix}, \quad \mathcal{CPT}\psi = \frac{1}{\cos\alpha} \begin{pmatrix} a^* + ib^* \sin\alpha \\ b^* - ia^* \sin\alpha \end{pmatrix}.$$

Hence, $\langle \psi | \psi \rangle = (\mathcal{CPT}\psi) \cdot \psi = \frac{1}{\cos\alpha}[a^*a + b^*b + i(b^*b - a^*a)\sin\alpha]$. Thus, if we take $a = x + iy$ and $b = u + iv$ with x, y, u, and v real, then

$$\langle \psi | \psi \rangle = \frac{1}{\cos\alpha} \left(x^2 + v^2 + 2xv\sin\alpha + y^2 + u^2 - 2yu\sin\alpha \right).$$

This inner product is positive and vanishes only if $x = y = u = v = 0$.

We denote the $\mathcal{CPT}$-conjugate of $|u\rangle$ as $\langle u|$. The completeness condition is then

$$|E_+\rangle\langle E_+| + |E_-\rangle\langle E_-| = \begin{pmatrix} 1 & 0 \\ 0 & 1 \end{pmatrix}.$$

Furthermore, using the $\mathcal{CPT}$ conjugate $\langle E_\pm|$, we can represent the $\mathcal{C}$ operator as $\mathcal{C} = |E_+\rangle\langle E_+| - |E_-\rangle\langle E_-|$. In the limit as θ approaches 0, the Hamiltonian (3.16) becomes Hermitian and the $\mathcal{C}$ operator reduces to the parity operator $\mathcal{P}$. Thus, the $\mathcal{CPT}$ invariance reduces to the conventional condition of Hermiticity for a symmetric matrix: $\hat{H} = \hat{H}^*$.

There is another exactly solvable $\mathcal{PT}$-symmetric Hamiltonian that has been used to study and illustrate the properties of $\mathcal{PT}$-symmetric quantum theory. This beautiful Hamiltonian, known as the *Swanson Hamiltonian*, has been widely examined in many papers on $\mathcal{PT}$ symmetry [Swanson (2004); Graefe *et al.* (2015a)]. The Swanson Hamiltonian is exactly solvable because it is quadratic in $\hat{x}$ and $\hat{p}$.

3.7 Calculating the $\mathcal{C}$ Operator

The distinguishing feature of $\mathcal{PT}$-symmetric quantum mechanics is the $\mathcal{C}$ operator; there is no such operator in Hermitian quantum mechanics. Only a non-Hermitian $\mathcal{PT}$-symmetric Hamiltonian possesses a $\mathcal{C}$ operator that is distinct from the parity operator $\mathcal{P}$. If we were to sum the series in (3.14) for a *Hermitian* $\mathcal{PT}$-symmetric Hamiltonian, we would obtain the parity operator $\mathcal{P}$, which is simply $\delta(x + y)$ in coordinate space [see (3.11)].

The $\mathcal{C}$ operator is represented as a formal infinite series in (3.14), but it is not easy to calculate the sum of this series directly. Attempting to obtain $\mathcal{C}$ by a direct brute-force evaluation of the sum in (3.14) would be difficult because it requires that we find analytic expressions for all the eigenfunctions $\phi_n(x)$ of $\hat{H}$. Worse yet, such a procedure would be impossible in quantum field theory because there is no simple analog of the Schrödinger eigenvalue equation.

An early attempt to calculate $\mathcal{C}$ proposed a perturbative approach [Bender *et al.* (2003b)]. This paper considers the PT-symmetric Hamiltonian

$$\hat{H} = \tfrac{1}{2}\hat{p}^2 + \tfrac{1}{2}\hat{x}^2 + i\varepsilon\hat{x}^3 \qquad (3.22)$$

and treats ε as a small real parameter. When $\varepsilon = 0$, the Hamiltonian $\hat{H}$ reduces to that of the quantum harmonic oscillator, all of whose eigenfunctions can be calculated exactly. Thus, one can express each of the eigenfunctions of $\hat{H}$ in (3.22) as a perturbation series in powers of ε. In this paper the first few terms of these perturbation series are calculated, the perturbation series are substituted into (3.14), and the summation over the nth eigenfunction is done for the first few powers of ε. The result is a formal perturbation expansion of the $\mathcal{C}$ operator in powers of ε.

This calculation is complicated and difficult and the final result is quite messy. However, the calculation is of interest because a careful examination of the series for $\mathcal{C}$ reveals that in coordinate space the result simplifies dramatically if the $\mathcal{C}$ operator is written as the exponential of a derivative operator Q multiplying the parity operator $\mathcal{P}$:

$$\mathcal{C}(x,y) = \exp\left[Q(x, -i\,d/dx)\right]\delta(x+y). \qquad (3.23)$$

Thus, while the perturbative expression for $\mathcal{C}$ is a series in *all* positive integer powers of ε, when $\mathcal{C}$ is written in the form $\mathcal{C} = e^Q \mathcal{P}$, Q becomes a series in *odd* powers of ε only. Also, since Q is a series in odd powers of ε, we see that in the limit $\varepsilon \to 0$ the function Q vanishes. Thus, in this limit e^Q tends to 1 and the $\mathcal{C}$ operator collapses to the parity operator $\mathcal{P}$.

In fact, the expression in (3.23) need not be limited to coordinate space. A completely general way to represent the $\mathcal{C}$ operator is to express it in terms of the basic quantum-mechanical operators $\hat{x}$ and $\hat{p}$ as

$$\mathcal{C} = e^{Q(\hat{x},\hat{p})}\mathcal{P}. \qquad (3.24)$$

Thus, the problem of finding the operator $\mathcal{C}$ is reduced to the problem of finding the operator Q as a *real* function of the two fundamental conjugate dynamical variables $\hat{x}$ and $\hat{p}$.

Given the representation of $\mathcal{C}$ in (3.24), it is possible to devise powerful analytical tools for calculating it. However, before doing so, we illustrate this representation by using two elementary Hamiltonians. First, we consider the shifted harmonic oscillator $\hat{H} = \frac{1}{2}\hat{p}^2 + \frac{1}{2}\hat{x}^2 + i\varepsilon\hat{x}$. This Hamiltonian has an unbroken PT symmetry for all real ε and its eigenvalues $E_n = n + \frac{1}{2} + \frac{1}{2}\varepsilon^2$ are all real. The exact formula for C in this theory is $C = e^Q P$, where $Q = -\varepsilon\hat{p}$. Thus, in the limit $\varepsilon \to 0$, where the Hamiltonian $\hat{H}$ becomes Hermitian, C reduces to P.

Second, we consider the non-Hermitian 2×2 matrix Hamiltonian (3.16). The C operator in (3.21) that is associated with this Hamiltonian can also be written in the form $C = e^Q P$, where

$$Q = \frac{1}{2}\sigma_2 \log\left(\frac{1 - \sin\alpha}{1 + \sin\alpha}\right), \tag{3.25}$$

and σ_2 is the Pauli matrix

$$\sigma_2 = \begin{pmatrix} 0 & -i \\ i & 0 \end{pmatrix}. \tag{3.26}$$

Again, we see that in the limit $\theta \to 0$, in which the Hamiltonian becomes Hermitian, the C operator reduces to the parity operator P.

3.8 Algebraic Equations Satisfied by C

There is a straightforward algebraic procedure for calculating the C operator, and this procedure avoids the difficult problem of evaluating the sum in (3.14). The technique also generalizes easily from quantum mechanics to quantum field theory (see Sec. 5.3). In this section we illustrate this technique by calculating C for the PT-symmetric cubic Hamiltonian (3.22). We calculate C to high order as a perturbation series in powers of ε. Note that calculating C for other kinds of interactions may still be difficult and may require sophisticated techniques such as semiclassical approximations [Bender and Jones (2004)].

To calculate C we invoke its three fundamental algebraic properties. First, while C does not, in general, commute with P or T separately, it does commute with the space-time reflection operator PT:

$$[C, PT] = 0. \tag{3.27}$$

Second, C is a reflection operator; that is, the square of C is the identity

$$C^2 = \mathbb{1}. \tag{3.28}$$

Third, C commutes with the Hamiltonian $\hat{H}$,

$$[C, \hat{H}] = 0, \tag{3.29}$$

and thus $\mathcal{C}$ is time independent. In short, $\mathcal{C}$ is a time-independent PT-symmetric reflection operator.[8]

We begin by substituting the operator representation in (3.24) into the three algebraic equations (3.27)–(3.29). We then solve the resulting equations for Q. Substituting (3.24) into (3.27), we get

$$e^{Q(\hat{x},\hat{p})} = PTe^{Q(\hat{x},\hat{p})}PT = e^{Q(-\hat{x},\hat{p})},$$

from which we conclude that $Q(\hat{x},\hat{p})$ is an *even* function of $\hat{x}$.

Next, we substitute (3.24) into (3.28) and find that

$$e^{Q(\hat{x},\hat{p})}Pe^{Q(\hat{x},\hat{p})}P = e^{Q(\hat{x},\hat{p})}e^{Q(-\hat{x},-\hat{p})} = 1.$$

This equation implies that $Q(\hat{x},\hat{p}) = -Q(-\hat{x},-\hat{p})$ and since $Q(\hat{x},\hat{p})$ is an even function of $\hat{x}$, we conclude that it is an *odd* function of $\hat{p}$.

The third condition (3.29) is that the operator $\mathcal{C}$ commutes with $\hat{H}$. While the first two conditions above are *kinematic* conditions on Q that are true for any Hamiltonian, the condition (3.29) is equivalent to imposing the dynamics specified by the particular Hamiltonian that defines the quantum theory. Thus, substituting $\mathcal{C} = e^{Q(\hat{x},\hat{p})}P$ into (3.29), we get

$$e^{Q(\hat{x},\hat{p})}[P,\hat{H}] + [e^{Q(\hat{x},\hat{p})},H]P = 0. \tag{3.30}$$

This equation is difficult to solve in general, so it is probably best to use perturbative methods, as we explain in the next subsection.

3.8.1   *Perturbative calculation of* $\mathcal{C}$

To solve (3.30) for the Hamiltonian (3.22), we express the Hamiltonian in the form $\hat{H} = \hat{H}_0 + \varepsilon \hat{H}_1$, where $\hat{H}_0 = \frac{1}{2}\hat{p}^2 + \frac{1}{2}\hat{x}^2$ is the harmonic-oscillator Hamiltonian, which commutes with the parity operator P. The operator $\hat{H}_1 = i\hat{x}^3$ *anticommutes* with P. Thus, the condition (3.30) becomes

$$2\varepsilon e^{Q(\hat{x},\hat{p})}\hat{H}_1 = [e^{Q(\hat{x},\hat{p})},\hat{H}].$$

Next, we expand $Q(\hat{x},\hat{p})$ as a perturbation series in *odd* powers of ε,

$$Q(\hat{x},\hat{p}) = \varepsilon Q_1(\hat{x},\hat{p}) + \varepsilon^3 Q_3(\hat{x},\hat{p}) + \varepsilon^5 Q_5(\hat{x},\hat{p}) + \cdots , \tag{3.31}$$

and substitute this expansion into the exponential $e^{Q(\hat{x},\hat{p})}$. After some algebra, we obtain a sequence of equations that can be solved systematically for the operator-valued functions $Q_n(\hat{x},\hat{p})$ ($n = 1,3,5,\ldots$) subject to the

---

[8]Although the $\mathcal{C}$ operator satisfies these three equations, it is unfortunately not the unique solution to these equations. Another solution is simply 1. The solution $\mathcal{C}$ that we seek has the property that $\mathcal{C}P = e^Q$ is a positive operator.

symmetry constraints that ensure the conditions (3.27) and (3.28). The first three of these equations are

$$[\hat{H}_0, Q_1] = -2\hat{H}_1,$$
$$[\hat{H}_0, Q_3] = -\tfrac{1}{6}[Q_1, [Q_1, \hat{H}_1]],$$
$$[\hat{H}_0, Q_5] = \tfrac{1}{360}[Q_1, [Q_1, [Q_1, [Q_1, \hat{H}_1]]]] - \tfrac{1}{6}[Q_1, [Q_3, \hat{H}_1]]$$
$$+\tfrac{1}{6}[Q_3, [Q_1, \hat{H}_1]]. \tag{3.32}$$

Let us solve these equations for the Hamiltonian $\hat{H}$ in (3.22) for which $\hat{H}_0 = \tfrac{1}{2}\hat{p}^2 + \tfrac{1}{2}\hat{x}^2$ and $\hat{H}_1 = i\hat{x}^3$. The procedure is generic: We substitute a general polynomial form for Q_n using arbitrary coefficients and then solve for these coefficients. For example, to solve $[\hat{H}_0, Q_1] = -2i\hat{x}^3$, the first of the equations in (3.32), we take as an *ansatz* for Q_1 the most general Hermitian cubic polynomial that is both *even* in $\hat{x}$ and *odd* in p:

$$Q_1(\hat{x}, \hat{p}) = M\hat{p}^3 + N\hat{x}\hat{p}\hat{x}. \tag{3.33}$$

Here, M and N are numerical coefficients to be determined. The operator equation for Q_1 is clearly satisfied if $M = -\tfrac{4}{3}$ and $N = -2$.

It is tedious but straightforward to continue this process. To represent the solutions for $Q_n(\hat{x}, \hat{p})$ ($n > 1$), we introduce the following convenient notation: Let $S_{m,n}$ represent the *totally symmetrized* sum over all terms containing m factors of $\hat{p}$ and n factors of $\hat{x}$. For example,

$$S_{0,0} = 1, \quad S_{0,3} = \hat{x}^3, \quad S_{1,1} = \tfrac{1}{2}(\hat{x}\hat{p} + \hat{p}\hat{x}), \quad S_{1,2} = \tfrac{1}{3}(\hat{x}^2\hat{p} + \hat{x}\hat{p}\hat{x} + \hat{p}\hat{x}^2),$$

and so on. The properties of the operators $S_{m,n}$ are summarized in [Bender and Dunne (1989a,b)].

Expressed in terms of the symmetrized operators $S_{m,n}$, the first three functions Q_{2n+1} are given by

$$Q_1 = -\tfrac{4}{3}\hat{p}^3 - 2S_{1,2},$$
$$Q_3 = \tfrac{128}{15}\hat{p}^5 + \tfrac{40}{3}S_{3,2} + 8S_{1,4} - 12\hat{p},$$
$$Q_5 = \tfrac{24\,736}{45}\hat{p}^3 - \tfrac{320}{3}\hat{p}^7 - \tfrac{544}{3}S_{5,2} - \tfrac{512}{3}S_{3,4} - 64S_{1,6} + \tfrac{6\,368}{15}S_{1,2}. \tag{3.34}$$

Equations (3.24), (3.31), and (3.34) comprise a perturbative expansion of $\mathcal{C}$ in terms of $\hat{x}$ and $\hat{p}$. This expansion is accurate to order ε^6.

In summary, we can use the *ansatz* (3.24) to calculate $\mathcal{C}$ to high order in perturbation theory. This perturbative procedure avoids the necessity of calculating the $\mathcal{PT}$-normalized eigenfunctions $\phi_n(x)$. We use these same techniques for cubic quantum-field-theoretic Hamiltonians in Sec. 5.3.

3.8.2   Calculation of C for other Hamiltonians

The C operator has been calculated perturbatively for a variety of quantum-mechanical models. For example, consider the case of the Hamiltonian

$$\hat{H} = \tfrac{1}{2}\left(\hat{p}^2 + \hat{q}^2\right) + \tfrac{1}{2}\left(\hat{x}^2 + \hat{y}^2\right) + i\varepsilon\hat{x}^2\hat{y}, \qquad (3.35)$$

which has two degrees of freedom. The energies of this complex *Hénon-Heiles* theory were studied in Ref. [Bender *et al.* (2001b)], and a perturbative calculation of the C operator for this Hamiltonian was given in [Bender *et al.* (2004a,c,d)]. The coefficients in the series $Q = Q_1\varepsilon + Q_3\varepsilon^3 + \dots$ are

$$Q_1(\hat{x}, \hat{y}, \hat{p}, \hat{q}) = -\tfrac{4}{3}\hat{p}^2\hat{q} - \tfrac{1}{3}S_{1,1}y - \tfrac{2}{3}\hat{x}^2\hat{q},$$

$$Q_3(\hat{x}, \hat{y}, \hat{p}, \hat{q}) = \tfrac{512}{405}\hat{p}^2\hat{q}^3 + \tfrac{512}{405}\hat{p}^4\hat{q} + \tfrac{1088}{405}S_{1,1}T_{2,1} - \tfrac{256}{405}\hat{p}^2T_{1,2}$$
$$+ \tfrac{512}{405}S_{3,1}\hat{y} + \tfrac{288}{405}S_{2,2}\hat{q} - \tfrac{32}{405}\hat{x}^2\hat{q}^3 + \tfrac{736}{405}\hat{x}^2T_{1,2}$$
$$- \tfrac{256}{405}S_{1,1}\hat{y}^3 + \tfrac{608}{405}S_{1,3}\hat{y} - \tfrac{128}{405}\hat{x}^4\hat{q} - \tfrac{8}{9}q. \qquad (3.36)$$

$T_{m,n}$ is a totally symmetric product of m factors of $\hat{q}$ and n factors of $\hat{y}$.

Next, consider the Hamiltonian

$$\hat{H} = \tfrac{1}{2}\left(\hat{p}^2 + \hat{q}^2 + \hat{r}^2\right) + \tfrac{1}{2}\left(\hat{x}^2 + \hat{y}^2 + \hat{z}^2\right) + i\varepsilon\hat{x}\hat{y}\hat{z}, \qquad (3.37)$$

which has three degrees of freedom. For this Hamiltonian we get [Bender *et al.* (2004a,d,c)]

$$Q_1(\hat{x}, \hat{y}, \hat{z}, \hat{p}, \hat{q}, \hat{r}) = -\tfrac{2}{3}(\hat{y}\hat{z}\hat{p} + \hat{x}\hat{z}\hat{q} + \hat{x}\hat{y}\hat{r}) - \tfrac{4}{3}\hat{p}\hat{q}\hat{r},$$

$$Q_3(\hat{x}, \hat{y}, \hat{z}, \hat{p}, \hat{q}, \hat{r}) = \tfrac{128}{405}\left(\hat{p}^3\hat{q}\hat{r} + \hat{q}^3\hat{p}\hat{r} + \hat{r}^3\hat{q}\hat{p}\right)$$
$$+ \tfrac{136}{405}[\hat{p}\hat{x}\hat{p}(\hat{y}\hat{r} + \hat{z}\hat{q}) + \hat{q}\hat{y}\hat{q}(\hat{x}\hat{r} + \hat{z}\hat{p}) + \hat{r}\hat{z}\hat{r}(\hat{x}\hat{q} + \hat{y}\hat{p})]$$
$$- \tfrac{64}{405}(\hat{x}\hat{p}\hat{x}\hat{q}\hat{r} + \hat{y}\hat{q}\hat{y}\hat{p}\hat{r} + \hat{z}\hat{r}\hat{z}\hat{p}\hat{q}) + \tfrac{184}{405}(\hat{x}\hat{p}\hat{x}\hat{y}\hat{z} + \hat{y}\hat{q}\hat{y}\hat{x}\hat{z} + \hat{z}\hat{r}\hat{z}\hat{x}\hat{y})$$
$$- \tfrac{32}{405}[\hat{x}^3(\hat{y}\hat{r} + \hat{z}\hat{q}) + \hat{y}^3(\hat{x}\hat{r} + \hat{z}\hat{p}) + \hat{z}^3(\hat{x}\hat{q} + \hat{y}\hat{p})]$$
$$- \tfrac{8}{405}\left(\hat{p}^3\hat{y}\hat{z} + \hat{q}^3\hat{x}\hat{z} + \hat{r}^3\hat{x}\hat{y}\right). \qquad (3.38)$$

In the examples discussed so far the coordinate-space representation of the C operator is a combination of integer powers of x and integer numbers of derivatives multiplying the parity operator $\mathcal{P}$. Thus, the Q operator in (3.24) is a polynomial in the operators $\hat{x}$ and $\hat{p} = -i\frac{d}{dx}$. However, for the deceptively simple-looking $\mathcal{PT}$-symmetric square-well Hamiltonian discussed below, C contains *integrals* of $\mathcal{P}$, as we will show. Thus, the Q operator is *not* just a polynomial in $\hat{x}$ and $\hat{p}$, and therefore C cannot be found by the algebraic perturbative methods that we have used above.

In coordinate space the $\mathcal{PT}$-symmetric square-well Hamiltonian is defined on the domain $0 < x < \pi$ and is given by

$$\hat{H} = \hat{p}^2 + V(\hat{x}), \qquad (3.39)$$

where $V(x) = \infty$ for $x < 0$ and $x > \pi$ and

$$V(x) = \begin{cases} i\varepsilon & \text{for } \frac{\pi}{2} < x < \pi, \\ -i\varepsilon & \text{for } 0 < x < \frac{\pi}{2}. \end{cases}$$

The $\mathcal{PT}$-symmetric square-well Hamiltonian was introduced in [Znojil (2001)] and it has been heavily studied by many researchers including [Znojil and Lévai (2001); Bagchi et al. (2002); Mostafazadeh and Batal (2004); Znojil (2005b)]. This Hamiltonian is a complex deformation of the conventional Hermitian square-well Hamiltonian, and it reduces to the conventional square-well Hamiltonian as the deformation parameter ε tends to 0. For $\hat{H}$ in (3.39) the parity operator $\mathcal{P}$ performs a reflection about $x = \frac{\pi}{2}$ rather than about $x = 0$; that is, $\mathcal{P} : \hat{x} \to \pi - \hat{x}$.

In [Bender and Tan (2006)] the $\mathcal{C}$ operator for this square-well Hamiltonian was obtained by using the brute-force approach of calculating the $\mathcal{PT}$-normalized eigenfunctions $\phi_n(x)$ of $\hat{H}$ in (3.39) and summing over these eigenfunctions according to (3.14). In that paper the eigenfunctions $\phi_n(x)$ were obtained as perturbation series to second order in powers of ε. The eigenfunctions were then normalized according to (3.9) and substituted into the summation formula (3.14). The sum was then evaluated directly to obtain the $\mathcal{C}$ operator accurate to order ε^2. The advantage of using the domain $0 < x < \pi$ instead of the domain $-\frac{1}{2}\pi < x < \frac{1}{2}\pi$ is that this sum reduces to a set of Fourier sine and cosine series that can be evaluated in closed form. After evaluating the sum, one can translate the expression for $\mathcal{C}$ to the symmetric region $-\frac{1}{2}\pi < x < \frac{1}{2}\pi$. Recall that on the symmetric domain the parity operator in coordinate space is $\mathcal{P}(x, y) = \delta(x + y)$.

The final step is to show that the $\mathcal{C}$ operator to order ε^2 has the form $e^Q \mathcal{P}$ and then to evaluate the function Q to order ε^2. On the symmetric domain the $Q(x, y)$ operator has the relatively simple structure

$$Q(x, y) = \tfrac{1}{4} i\varepsilon [x - y + \Theta(x - y)\, (|x + y| - \pi)] + \mathcal{O}(\varepsilon^3), \qquad (3.40)$$

where $\Theta(x)$ is the step function defined as

$$\Theta(x) \equiv \begin{cases} 1 & (x > 0), \\ 0 & (x = 0), \\ -1 & (x < 0). \end{cases} \qquad (3.41)$$

It may seem surprising to say that the formula for $Q(x, y)$ in (3.40) has a relatively simple structure, but this structure is indeed simple in comparison with the expression for the $\mathcal{C}$ operator as a series in powers of

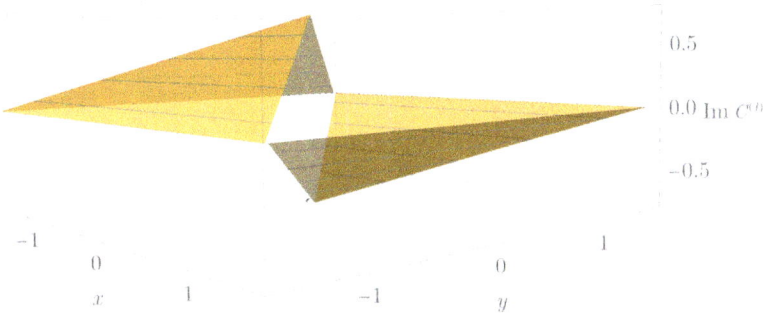

Fig. 3.1 Three-dimensional plot of the imaginary part of $\mathcal{C}^{(1)}(x,y)$, the first-order perturbative contribution in (3.42) to the $\mathcal{C}$ operator in coordinate space. The plot is on the symmetric square domain $-\frac{1}{2}\pi < (x,y) < \frac{1}{2}\pi$. Note that $\mathcal{C}^{(1)}(x,y)$ vanishes on the boundary of this square domain because the eigenfunctions $\phi_n(x)$ are required to vanish at $x = 0$ and at $x = \pi$.

ε: $C(x,y) = C^{(0)} + \varepsilon C^{(1)} + \varepsilon^2 C^{(2)} + \mathcal{O}(\varepsilon^3)$. The formulas for the first three coefficients in this series for $\mathcal{C}$ are

$$\mathcal{C}^{(0)}(x,y) = \delta(x+y),$$
$$\mathcal{C}^{(1)}(x,y) = \tfrac{1}{4}i[x+y+\Theta(x+y)\,(|x-y|-\pi)],$$
$$\mathcal{C}^{(2)}(x,y) = \tfrac{1}{96}\pi^3 - \tfrac{1}{24}(x^3+y^3)\,\Theta(x+y) - \tfrac{1}{24}(y^3-x^3)\,\Theta(y-x)$$
$$+\tfrac{1}{8}xy\pi - \tfrac{1}{16}\pi^2(x+y)\,\Theta(x+y) + \tfrac{1}{8}\pi(x|x|+y|y|)\,\Theta(x+y)$$
$$-\tfrac{1}{4}xy\{|x|[\theta(x-y)\,\theta(-x-y) + \theta(y-x)\,\theta(x+y)]$$
$$+|y|[\theta(y-x)\,\theta(-x-y) + \theta(x-y)\,\theta(x+y)]\}, \qquad (3.42)$$

where $\theta(x) = \frac{1}{2}[1+\Theta(x)]$ is the half-step function.

We plot the imaginary part of $\mathcal{C}^{(1)}(x,y)$ in Fig. 3.1 and plot $\mathcal{C}^{(2)}(x,y)$ in Fig. 3.2. These three-dimensional plots show $\mathcal{C}^{(1)}(x,y)$ and $\mathcal{C}^{(2)}(x,y)$ on the symmetric domain $-\frac{1}{2}\pi < (x,y) < \frac{1}{2}\pi$.

The noteworthy property of the $\mathcal{C}$ operator for the square-well model is that the associated operator Q is a nonpolynomial function, and this kind of structure had not been seen in previous studies of $\mathcal{C}$. It was originally believed that for such a simple $\mathcal{PT}$-symmetric Hamiltonian it would be possible to calculate the $\mathcal{C}$ operator in closed form. It is a surprise that even for this elementary model the $\mathcal{C}$ operator is so complicated.

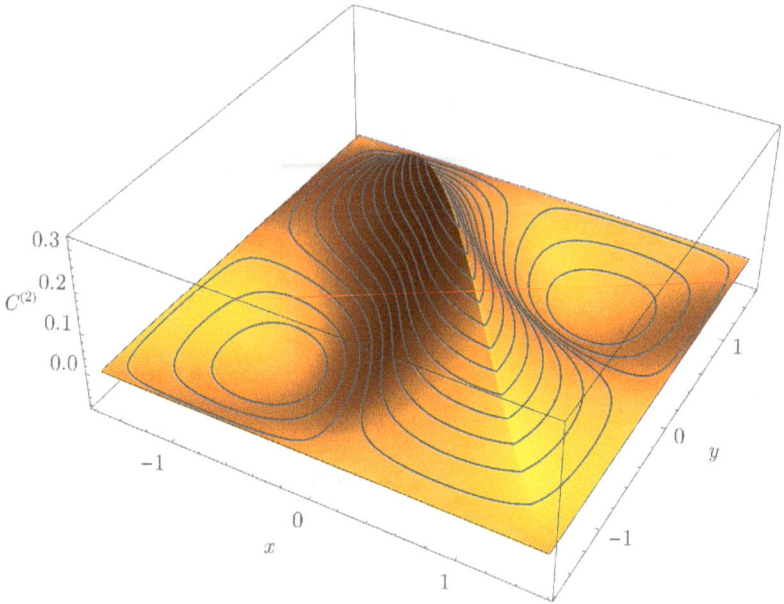

Fig. 3.2   Three-dimensional plot of $\mathcal{C}^{(2)}(x,y)$ in (3.42) on the symmetric square domain $-\frac{1}{2}\pi < (x,y) < \frac{1}{2}\pi$. The function $\mathcal{C}^{(2)}(x,y)$ vanishes on the boundary of this square domain because the eigenfunctions $\phi_n(x)$ from which it was constructed vanish at the boundaries of the square well.

3.9   Mapping $\mathcal{PT}$-Symmetric to Hermitian Hamiltonians

The square root of the positive operator e^Q can be used to construct a similarity transformation from a non-Hermitian $\mathcal{PT}$-symmetric Hamiltonian $\hat{H}$ to an equivalent Hermitian Hamiltonian $\hat{h}$ [Mostafazadeh (2002b, 2003b)]:

$$\hat{h} = e^{-Q/2}\hat{H}e^{Q/2}. \tag{3.43}$$

We say that the Hamiltonian $\hat{h}$ is equivalent to $\hat{H}$ in the sense that it has the same eigenvalues as $\hat{H}$; that is, $\hat{h}$ and $\hat{H}$ are *isospectral*.

To explain the similarity transformation (3.43) we recall from (3.24) that the $\mathcal{C}$ operator has the general form $\mathcal{C} = e^Q\mathcal{P}$, where $Q = Q(\hat{x},\hat{p})$ is a real function of the fundamental dynamical operator variables $\hat{x}$ and $\hat{p}$ of the quantum theory. We then multiply $\mathcal{C}$ on the right by $\mathcal{P}$ to obtain an expression for e^Q:

$$e^Q = \mathcal{C}\mathcal{P}. \tag{3.44}$$

This formula shows that $\mathcal{C}\mathcal{P}$ is a positive and invertible operator.

To verify that $\hat{h}$ in (3.43) is Hermitian, we take the Hermitian conjugate of (3.43): $\hat{h}^\dagger = e^{Q/2}\hat{H}^\dagger e^{-Q/2}$. We rewrite this as $\hat{h}^\dagger = e^{-Q/2}e^Q\hat{H}^\dagger e^{-Q}e^{Q/2}$. We then use (3.44) to replace e^Q by $\mathcal{CP}$ and e^{-Q} by $\mathcal{PC}$,

$$\hat{h}^\dagger = e^{-Q/2}\mathcal{CP}\hat{H}^\dagger\mathcal{PC}e^{Q/2}, \tag{3.45}$$

and recall from (3.16) that $\hat{H}^\dagger$ can be replaced by $\mathcal{P}\hat{H}\mathcal{P}$. This gives

$$\hat{h}^\dagger = e^{-Q/2}\mathcal{CPP}\hat{H}\mathcal{PPC}e^{Q/2} = e^{-Q/2}\mathcal{C}\hat{H}\mathcal{C}e^{Q/2}.$$

Finally, we recall that $\mathcal{C}$ commutes with $\hat{H}$ [see (3.29)] and that the square of $\mathcal{C}$ is unity [see (3.28)] in order to reduce the right side of (3.45) to the right side of (3.43). Thus, $\hat{h}$ is Hermitian in the Dirac sense.

This calculation demonstrates that for every non-Hermitian $\mathcal{PT}$-symmetric Hamiltonian $\hat{H}$ whose $\mathcal{PT}$ symmetry is unbroken, it is possible, at least in principle, to construct via (3.43) a Hermitian Hamiltonian $\hat{h}$ that has exactly the same eigenvalues as $\hat{H}$. The assumption that $\hat{H}$ has an unbroken $\mathcal{PT}$ symmetry is crucial because this allows us to construct $\mathcal{C}$, which then allows us to construct the similarity operator $e^{Q/2}$.

This construction poses a profound and as yet unanswered question: Are $\mathcal{PT}$-symmetric Hamiltonians physically new and distinct from ordinary Hermitian Hamiltonians, or do they describe exactly the same physical processes that ordinary Hermitian Hamiltonians describe? That is, is $\mathcal{PT}$-symmetric quantum mechanics physically new and different from conventional Hermitian quantum mechanics or are $\mathcal{PT}$-symmetric Hamiltonians merely just complicated transformations of conventional Hamiltonians.

There are two answers to this question, the first being technical and practical and the second being an answer in principle. First, while there is a formal similarity transformation that maps a $\mathcal{PT}$-symmetric Hamiltonian $\hat{H}$ to a Hermitian Hamiltonian $\hat{h}$ whose spectrum is identical to that of $\hat{H}$, it cannot be done except at the perturbative level because the transformation is horribly complicated. (See, for example, the discussion of the square well in Subsec. 3.8.2!) Furthermore, while the $\mathcal{PT}$-symmetric Hamiltonian $\hat{H}$ is simple in structure and easy to use in calculations because the interaction term is local, it is shown in [Bender *et al.* (2006c)] that $\hat{h}$ is in general *nonlocal*; that is, its interaction term has arbitrarily high powers of the variables $\hat{x}$ and $\hat{p}$. Thus, it is not merely difficult to calculate using $\hat{h}$; it is virtually *impossible* because the usual regulation schemes to handle the interaction term are hopelessly inadequate.

Apart from finite-dimensional matrix systems [Brody (2016)], there is only one known nontrivial example continuum for which there is actually a closed-form expression for both $\hat{H}$ and $\hat{h}$: This is the case of the quartic Hamiltonian (2.27) discussed in Sec. 2.3. However, even in this case it is not possible to obtain $\mathcal{C}$ in closed form because this operator is nonlocal (it contains a Fourier transform) and it performs a transformation in the complex plane. This is the only simple example for which it is practical to calculate with both $\hat{H}$ and $\hat{h}$. Hence, while the mapping from $\hat{H}$ to $\hat{h}$ is of theoretical interest, in general it does not have any practical value.

The second answer is of much greater significance because it has crucial physical implications, and Chap. 8 provides a basis for a discussion of this question. The point is that the mapping between $\hat{H}$ and $\hat{h}$ in (3.43) is a *similarity* transformation and not a *unitary* transformation. A unitary transformation is *bounded*, but a similarity transformation operator e^Q is typically unbounded. Therefore, while the Hamiltonians $\hat{H}$ and $\hat{h}$ are *formally* isospectral, the mapping between these two Hamiltonians is of questionable significance because it does not map *all* the vectors in the Hilbert space associated with $\hat{h}$ onto *all* the vectors in the Hilbert space associated with $\hat{H}$. The mapping between these Hilbert spaces is not 1-to-1 because under the similarity mapping some vectors may be mapped to infinity. Of course, for a finite-dimensional matrix Hamiltonian like that in (3.16), the operator e^Q is bounded. However, for infinite-dimensional Hamiltonians like those in (2.7) the mapping e^Q is unbounded. Thus, there may be experimentally measurable physical differences between $\hat{H}$ and $\hat{h}$, perhaps as a sort of Bohm–Aharanov effect. This is an issue of great significance that deserves much future investigation because it addresses the question of whether $\mathcal{PT}$-symmetric quantum mechanics is physically different at a fundamental level from ordinary quantum mechanics.

There are many other areas in which quantum-mechanical studies of $\mathcal{PT}$-symmetric systems have been and should be further pursued. These include $\mathcal{PT}$-symmetric wave packets, density matrices, and coherent states [Brody and Graefe (2012, 2013); Graefe and Schubert (2012, 2011)], $\mathcal{PT}$-symmetric Bose–Einstein condensates [Graefe (2012)], $\mathcal{PT}$-symmetric quantum noise [Schomerus (2010)], $\mathcal{PT}$-symmetric random matrices [Ahmed and Jain (2003c); Graefe *et al.* (2015b); Srivastava and Jain (2013)], $\mathcal{PT}$-symmetric nonlinear eigenvalue problems [Bender *et al.* (2014a); Bender and Komijani (2015); Bender *et al.* (2017c)], and $\mathcal{PT}$-symmetric exceptional points [Goldzak *et al.* (2018)].

Chapter 4

$\mathcal{PT}$-Symmetric Classical Mechanics

> "The great ambition should be to excel all
> others engaged in the same occupation."
>
> —$\mathcal{P}\,\mathcal{T}$ *Barnam*

The requirement that a physical system be $\mathcal{PT}$ symmetric (that is, invariant under combined space and time reflection) can be imposed on classical as well as on quantum systems. Elementary examples of classical $\mathcal{PT}$-symmetric systems, such as coupled linear oscillators with balanced gain and loss, are used in Chap. 1 to illustrate the basic concepts of $\mathcal{PT}$ symmetry. Classical $\mathcal{PT}$-symmetric Hamiltonians arise as the classical limit ($\hbar \to 0$) of quantum-mechanical $\mathcal{PT}$-symmetric Hamiltonians. We saw in Chap. 1 that it is useful to study such classical systems because closed classical trajectories (see Figs. 1.14, 1.15, and 2.12) indicate that the corresponding quantum system has an unbroken $\mathcal{PT}$ symmetry (with a real spectrum), and open classical trajectories (see Figs. 1.18 and 2.14) indicate that the corresponding quantum system has a broken $\mathcal{PT}$ symmetry (with a partially or entirely complex spectrum).

This chapter discusses in detail the properties of some typical $\mathcal{PT}$-symmetric classical systems. It is interesting that many nonlinear classical systems, such as the simple pendulum, the Lotka–Volterra equations (which describe competing biological populations), and the Euler equations (which describe a rotating rigid body) are $\mathcal{PT}$ symmetric. Moreover, many classical nonlinear wave equations, such as the Korteweg–de Vries and the generalized Korteweg–de Vries equations, the Camassa–Holm equation, the Sine–Gordon equation, and the Boussinesq equation are also $\mathcal{PT}$ symmetric [Bender *et al.* (2007c)].

This chapter begins with a study of the remarkable classical motion of a particle described by the $\mathcal{PT}$-symmetric Hamiltonian $H = p^2 + x^2(ix)^\varepsilon$.

(This classical Hamiltonian was discussed in Sec. 1.7 but only for integer values of ε.) We then examine some dynamical systems whose equations of motion are $\mathcal{PT}$ symmetric. Next, we construct a $\mathcal{PT}$-symmetric classical random walk and discuss the problem of having a probabilistic interpretation of a complex physical system. We conclude with some brief remarks concerning $\mathcal{PT}$-symmetric nonlinear wave equations. (Chapter 9 discusses $\mathcal{PT}$-symmetric deformations of nonlinear wave equations.)

4.1   Classical Trajectories for Noninteger ε

This section describes the properties of the $\mathcal{PT}$-symmetric classical-mechanical theory that underlies the quantum-mechanical theory defined by the Hamiltonian (2.7); namely, $\hat{H} = \hat{p}^2 + \hat{x}^2(i\hat{x})^\varepsilon$. We discuss the motion of a particle that feels complex forces and responds by moving about in the complex plane. Among the many early papers that have been published on this topic are [Bender *et al.* (1999b); Nanayakkara (2004a,b); Bender *et al.* (2006b, 2007b); Bender and Darg (2007); Fedorov and Gomez-Ullate (2007)] and we summarize here just a few of the remarkable discoveries.

Chapter 1 examines the complex classical trajectories of the Hamiltonian $\hat{H}$ for *integer* values of ε: $\varepsilon = 0, 1, 2$ (see [Bender *et al.* (1999b)]). However, when ε is noninteger, the classical trajectories exhibit new and surprising behaviors: Trajectories can visit many sheets of a Riemann surface. Also, in [Bender *et al.* (2006b); Bender and Darg (2007)] it was shown that the classical trajectories can exhibit both broken and unbroken $\mathcal{PT}$ symmetry and can even give rise to strange fractal-like structures.

Without loss of generality, we may take the energy E of a classical particle whose motion is determined by the Hamiltonian $p^2 + x^2(ix)^\varepsilon$ to be 1. As ε increases from 0, the turning points at $x = 1$ (and at $x = -1$) rotate downward and clockwise (anticlockwise) into the complex-x plane. These turning points are solutions to the equation

$$1 + (ix)^{2+\varepsilon} = 0.$$

When ε is noninteger, this equation has many solutions that all lie on the unit circle and have the form

$$x = \exp\left(i\pi\frac{4N - \varepsilon}{4 + 2\varepsilon}\right) \quad (N \text{ integer}). \tag{4.1}$$

The turning points occur in $\mathcal{PT}$-symmetric pairs (pairs that are symmetric when reflected through the negative-imaginary axis), which correspond to the N values ($N = -1$, $N = 0$), ($N = -2$, $N = 1$), ($N = -3$, $N = 2$),

($N = -4$, $N = 3$), and so on. We label these pairs by the integer K ($K = 0, 1, 2, 3, \ldots$) so that the Kth pair corresponds to $N = -K - 1$, $N = K$. The pair of turning points on the real-x axis for $\varepsilon = 0$ deforms continuously into the $K = 0$ pair of turning points when $\varepsilon > 0$. Clearly, if ε is rational there are a finite number of turning points on the complex-x algebraic Riemann surface, but if ε is not rational there are an infinite number of turning points on the complex-x logarithmic Riemann surface.

On each sheet of the Riemann surface we take the branch cut to run up the positive-imaginary axis from $x = 0$ to $x = i\infty$. The principal sheet is labeled sheet 0. On the principal sheet $\arg x$ ranges from $-\frac{3}{2}\pi$ up to $\frac{1}{2}\pi$. Sheets 1 and -1 are PT-symmetric reflections of one another; on sheet 1, $\frac{1}{2}\pi < \arg x < \frac{5}{2}\pi$, and on sheet -1, $-\frac{7}{2}\pi < \arg xi < -\frac{3}{2}\pi$. Sheets 2 and -2, 3 and -3, and so on, all occur in PT-symmetric pairs.

To illustrate, let us choose $\varepsilon = \sqrt{2}$. The turning points on the sheets 0, ± 1, and ± 2 of the Riemann surface are shown in Fig. 4.1. There are altogether nine pairs of turning points shown, two pairs on sheet 0, three pairs on sheets ± 1, and four pairs on sheets ± 2.

As ε increases from 0, the elliptical complex trajectories shown in Fig. 1.12 for the harmonic oscillator begin to deform. However, these deformed classical trajectories $x(t)$ remain closed and continue to be left-right symmetric. Three of these trajectories are shown in Fig. 4.2. Each trajectory begins at a point on the negative-imaginary axis. The first such trajectory begins at $x(0) = -0.873i$ and oscillates between the $N = -1$ and $N = 0$ turning points. The second and third trajectories begin lower down on the imaginary axis at $x(0) = -2.000i$ and $x(0) = -3.378i$. Note that as the starting points of the trajectories move down the negative-imaginary axis, the trajectories become larger and remain nested.

Figure 4.2 shows that the top of each trajectory comes closer to the branch cut on the positive-imaginary axis as the starting point moves downward along the imaginary axis. When the starting point passes a critical value near $x(0) = -3.378i$, the top of the orbit crosses the branch cut, the trajectory enters sheets ± 1, and the topology of the trajectories changes abruptly. However, the trajectories remain closed and left-right symmetric. One such trajectory, which is shown in Fig. 4.3, starts at $x(0) = -4.000i$. Note that a classical path cannot cross itself; the apparent self-intersections are paths lying on different sheets of the Riemann surface.

To demonstrate that the trajectory shown on Fig. 4.3 does not cross itself, we have plotted in Fig. 4.4 the angular rotation of the classical particle as a function of elapsed time. The color scheme is the same as in Fig. 4.3.

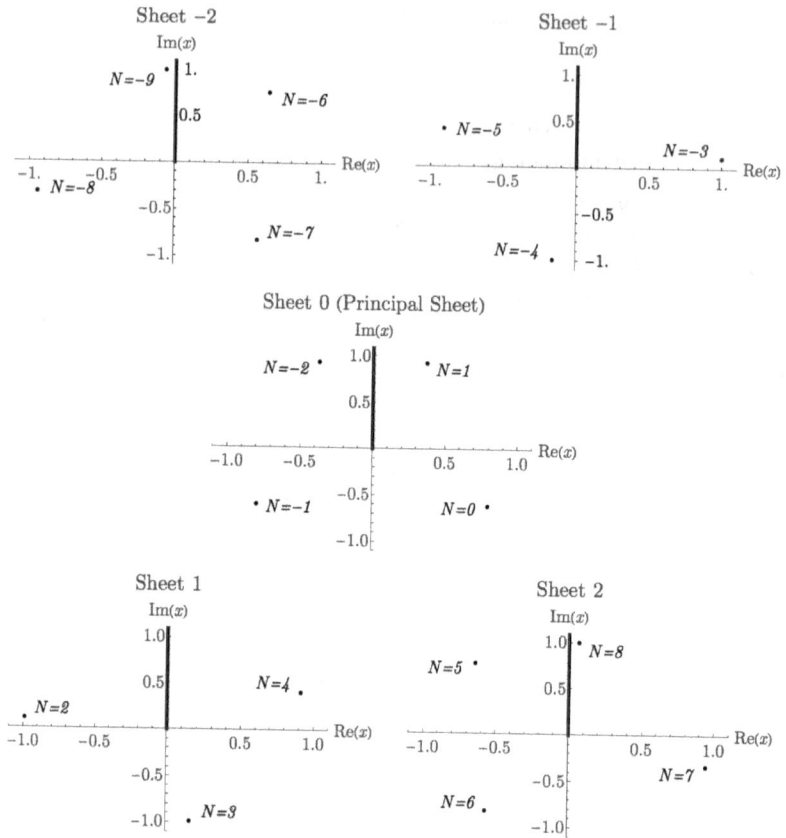

Fig. 4.1   Locations of the turning points (dots) for $\varepsilon = \sqrt{2}$ on sheets 0, ± 1, and ± 2 of the complex-x logarithmic Riemann surface. There are two $\mathcal{PT}$-symmetric pairs of turning points on sheet 0, three $\mathcal{PT}$-symmetric pairs of turning points on sheets ± 1, and four $\mathcal{PT}$-symmetric pairs of turning points on sheets ± 2. On each sheet a branch cut (heavy line) runs up the positive-imaginary axis from the origin to $i\infty$.

Figure 4.5 redisplays Fig. 4.4 in a three-dimensional perspective. Rather than plotting the rotation of $x(t)$ as a function of elapsed time t, it plots the complex location $x(t)$ of the classical particle as a function of the rotation angle. This plot displays the topological structure of the Riemann surface and shows how the classical trajectory avoids crossing itself.

As we can see in Figs. 4.3–4.5, the complex classical trajectory enters sheets ± 1 but does not cross the branch cuts on the positive-imaginary axes on these sheets. However, as the starting point of the trajectory $x(0)$

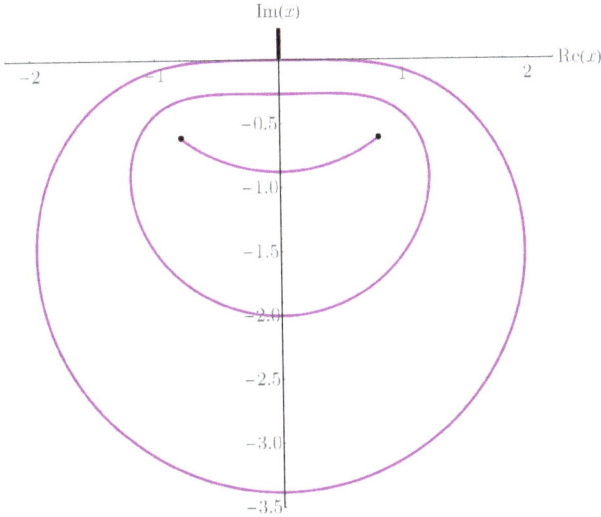

Fig. 4.2 Three classical trajectories $x(t)$ of the Hamiltonian $H = \hat{p}^2 + x^2(ix)^\varepsilon$ with $\varepsilon = \sqrt{2}$. The starting points of the trajectories are at $x(0) = -0.873i$, $-2.000i$, and $-3.378i$. All three trajectories are closed and $\mathcal{PT}$ symmetric and lie entirely on the principal sheet (sheet 0) on the Riemann surface. The first trajectory oscillates between the $N = 0$ and $N = -1$ turning points. As the starting point moves down the imaginary axis, the resulting trajectories become larger and remain nested. The third trajectory begins very close to a critical point; the top part of this trajectory just grazes the origin. For starting points below this critical point the trajectory crosses the branch point and enters sheets ± 1 on the Riemann surface (see Fig. 4.3).

continues to move down the negative-imaginary axis on sheet 0, we en-counter a second critical point. If the trajectory starts below this critical point, the trajectory not only enters sheets ± 1 but also crosses the branch cuts on these sheets and enters sheets ± 2. An example of such a five-sheeted trajectory is shown in Fig. 4.6. This trajectory begins at $x(0) = -12.000i$.

Figures 4.7 and 4.8 are analogs of Figs. 4.5 and 4.6: The first is a plot of the complex rotation angle of $x(t)$ as a function of elapsed time t and the second is a three-dimensional plot of the complex value of $x(t)$ as a function of the rotation angle. Observe that the trajectory briefly dips into sheets ± 2 but does not cross the branch cuts on these sheets.

The orbits in Figs. 4.1–4.8 for $H = p^2 + x^2(ix)^\varepsilon$ with $\varepsilon = \sqrt{2}$ are closed and $\mathcal{PT}$ symmetric. Usually, when $\varepsilon > 0$, the complex trajectories are closed and periodic except for special isolated trajectories that run off to complex infinity. (One such singular trajectory begins at the turning point

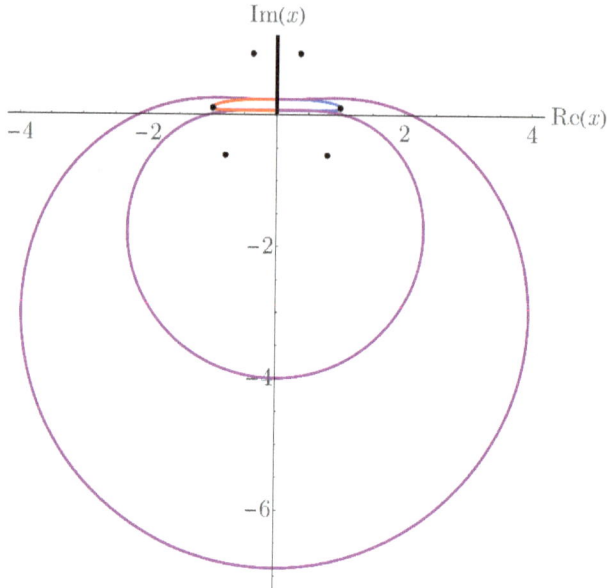

Fig. 4.3　A closed classical trajectory that visits three sheets of the Riemann surface. The trajectory $x(t)$ originates at $x(0) = -4.000i$, and while it lies mostly on the principal sheet of the Riemann surface (purple), it crosses the branch cut and briefly dips into the ± 1 sheets of the Riemann surface. On sheet 1 the trajectory is colored red and on the sheet -1 it is colored blue. Note that the trajectory does not cross itself. Rather, this plot is a projection of the three-sheeted trajectory onto the principal sheet. The dots indicate the locations of turning points: The lower ($N = -1$ and $N = 0$) and upper ($N = -2$ and $N = 1$) pair of turning points lie on sheet 0, while the middle pair ($N = -3$ and $N = 2$) lie on sheets -1 and 1.

in Fig. 1.14 and runs up the positive-imaginary axis to ∞. Such singular trajectories only occur for integer ε.) It was originally thought that all closed periodic orbits are PT (left-right) symmetric. However, this is not quite true. As shown in [Bender *et al.* (2006b); Bender and Darg (2007)], closed periodic *non-PT*-symmetric orbits exist, but only for rational values ε of the form $\varepsilon = \frac{a}{b}$, where b is an odd integer and a is an even integer that is evenly divisible by 4. For such values of ε one can find failed PT-symmetric orbits that attempt to form a PT-symmetric figure, but exactly half way along the orbit the classical particle bangs into a turning point in the complex plane and is reflected back to its starting point. These surprising orbits are not left-right symmetric, but instead are up-down symmetric. This is because a non-PT-symmetric orbit must join or encircle a pair of

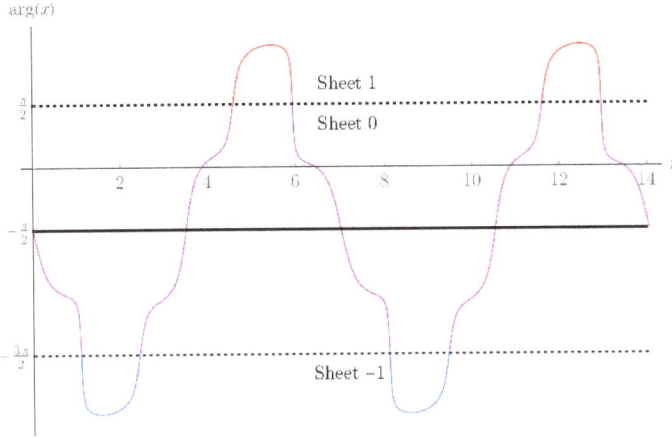

Fig. 4.4 Angular evolution of the classical orbit shown in Fig. 4.3. The angular rotation of $x(t)$ is plotted as a function of elapsed time t. The color scheme is the same as that in Fig. 4.3. The trajectory begins at an angle of $-\frac{1}{2}\pi$ on the principal sheet (sheet 0). It then briefly visits sheet -1 and returns back to sheet 0 and finally makes a short visit to sheet 1 and again returns to sheet 0. Two orbit cycles are shown.

complex-conjugate turning points. If we change the value of ε slightly, $\mathcal{PT}$ symmetry is restored and all orbits are left-right symmetric.

4.2 Some $\mathcal{PT}$-Symmetric Classical Dynamical Systems

This section examines the beautiful complex trajectories of some well known dynamical systems whose equations of motion are $\mathcal{PT}$ symmetric.

4.2.1 *Lotka–Volterra equations for predator-prey models*

The Lotka–Volterra equations

$$\dot{x} = x - xy, \quad \dot{y} = -y + xy \tag{4.2}$$

are a pair of nonlinear equations whose positive real solutions describe an ecological system of two competing species, where $x(t)$ represents the population of the prey species and $y(t)$ represents the population of the predator species. For example, we can take $x(t)$ to be the number of rabbits and $y(t)$ to be the number of foxes. We see immediately from (4.2) that if there are no rabbits initially [that is, $y(0) = 0$], then the population of rabbits grows exponentially with time t, and if there are no rabbits initially

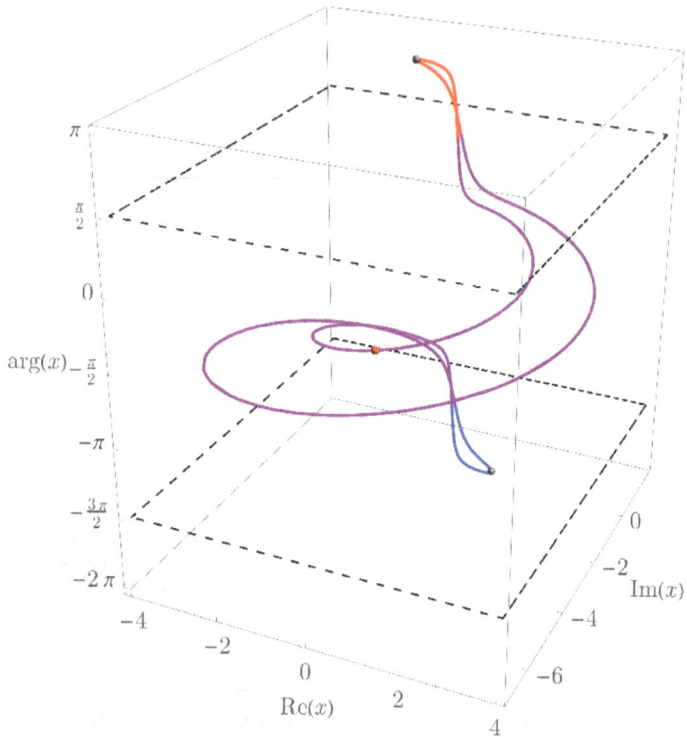

Fig. 4.5   Three-dimensional plot of the complex trajectory $x(t)$ as a function of the angular rotation. The color scheme is the same as that in Fig. 4.3. This plot shows why the complex classical trajectory in Fig. 4.3 does not cross itself.

[that is, $x(0) = 0$], then the population of foxes decays exponentially. However, if $x(0)$ and $y(0)$ are both nonzero, then the populations of rabbits and foxes oscillates in time. Figure 4.9 illustrates this oscillation.

The Lotka–Volterra equations provide a nice two-dimensional example of a nonlinear $\mathcal{PT}$-symmetric dynamical system whose complex solutions are generally nonperiodic but whose $\mathcal{PT}$-symmetric complex solutions are periodic [Bender *et al.* (2007c)]. To see that these equations are $\mathcal{PT}$ symmetric, we must generalize slightly the definition of $\mathcal{P}$ reflection to mean that $\mathcal{P}$ interchanges the roles of x and y:

$$\mathcal{P} : (x, y) \rightarrow (y, x).$$

This is the same definition of $\mathcal{P}$ that was used for the coupled linear oscillator system (1.10).

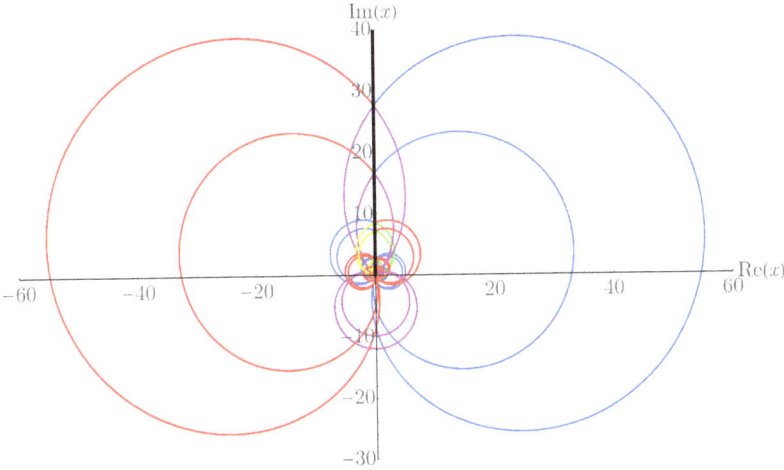

Fig. 4.6   Five-sheeted complex orbit that begins at $x(0) = -12.000i$. The color scheme is the same as that in Fig. 4.3 except that on sheet 2 the trajectory is colored lime green and on sheet -2 the trajectory is colored turquoise. As we can see in Figs. 4.7 and 4.8, the trajectory is PT symmetric and does not cross itself.

There is one constant of the motion for the Lotka–Volterra equations:

$$x + y - \log(xy) = C. \tag{4.3}$$

If we define the time-reversal operator T to reverse the sign of t and also to take the complex conjugate (because time reversal has this effect for quantum-mechanical systems), then for PT-symmetric complex solutions C must be real. Note that we always treat time t as a real parameter; the possibility of complex time is not considered in this book.

Let us compare the complex trajectories that arise for real C and for complex C. We first choose the initial conditions $x(0) = 1 + i$ and $y(0) = 0.0765 + 0.0181i$ for which $C = 3.273$ is real. [This is the same value of C that one gets for the real initial conditions $x(0) = \frac{1}{4}$, $y(0) = \frac{1}{4}$.] The resulting complex trajectories are closed and periodic (see Fig. 4.10). Next, we choose the complex initial conditions $x(0) = 1 + i$ and $y(0) = 0.380 - 0.022i$ for which the conserved quantity $C = 2 + i/4$ is complex. These initial conditions give a nonperiodic classical trajectory (see Fig. 4.11).

There are many possible PT-symmetric generalizations of the Lotka–Volterra equations that describe more complicated predator-prey systems having multiple species. One such system has an interpretation as a biological model of immune response [Bender *et al.* (2017a)].

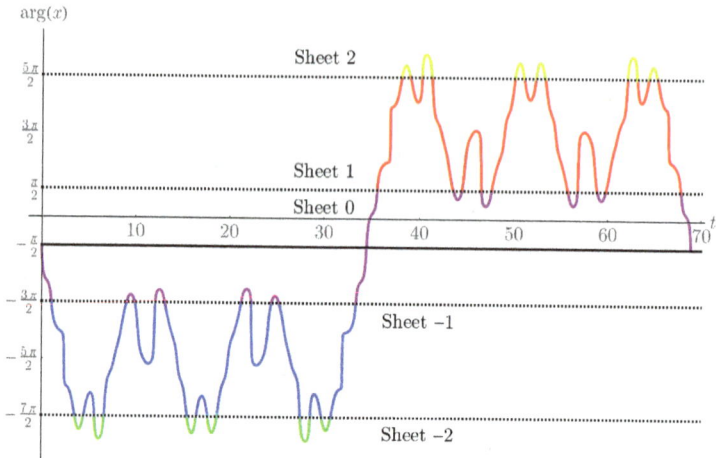

Fig. 4.7   Complex rotation angle of $x(t)$ in Fig. 4.6 plotted as a function of elapsed time t. The starting point of the orbit is $x(0) = -12.000i$. Observe that the orbit dips into each of the Riemann sheets ± 2 six times per cycle. However, it does not cross the branch cut on either of these sheets. This plot is a five-sheeted analog of Fig. 4.4. The color scheme is the same as that in Fig. 4.6.

4.2.2   Euler's equation for a rotating rigid body

Euler's differential equations govern the free three-dimensional rotation of a rigid body about its center of mass. In dimensionless form these equations may be written very simply as

$$\dot{x} = yz, \quad \dot{y} = -2xz, \quad \dot{z} = xy. \tag{4.4}$$

This system of equations is $\mathcal{PT}$ symmetric, where the effect of a parity reflection is to change the signs of x, y, z,

$$\mathcal{P}: x \to -x, \, y \to -y, \, z \to -z,$$

and the effect of time reversal $\mathcal{T}$ is to change the sign of t: $t \to -t$. We also assume that $\mathcal{T}$ performs complex conjugation.

Euler's equations possess two conserved quantities,

$$R^2 = x^2 + y^2 + z^2 \quad \text{and} \quad B = z^2 - x^2.$$

Thus, the real solutions to Euler's equations are trajectories confined to the surface of a sphere of radius R and each trajectory is the intersection of this sphere and a hyperboloid of revolution about the y axis. Figure 4.12 displays seven such trajectories for the case $R = 1$.

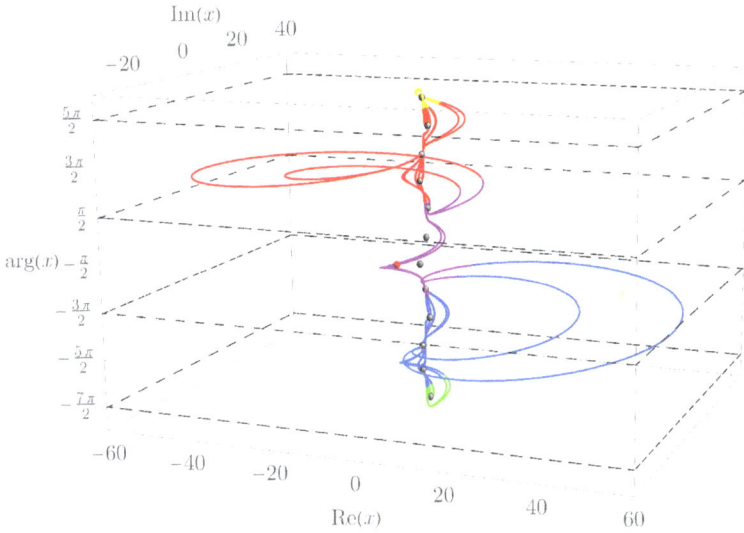

Fig. 4.8   Analog of Fig. 4.5 for the complex orbit shown in Fig. 4.6. This figure makes it clear that the orbit shown in Fig. 4.6 does not intersect itself.

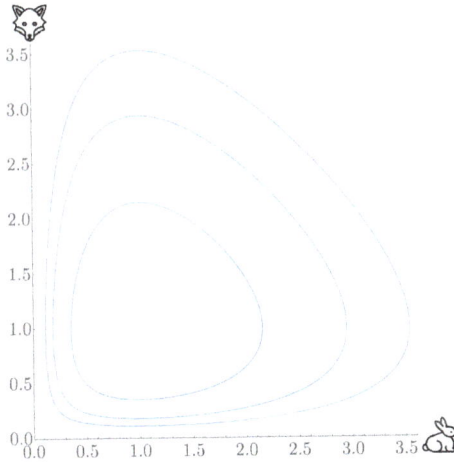

Fig. 4.9   Three periodic orbits for the Lotka–Volterra equations (4.2). The initial values are $\left[x(0) = \frac{1}{4}, y(0) = \frac{1}{4}\right]$, $\left[x(0) = \frac{1}{3}, y(0) = \frac{1}{3}\right]$, and $\left[x(0) = \frac{1}{2}, y(0) = \frac{1}{2}\right]$. The orbits are all periodic and $\mathcal{PT}$ symmetric; that is, they are invariant under $x \leftrightarrow y$, $t \to -t$.

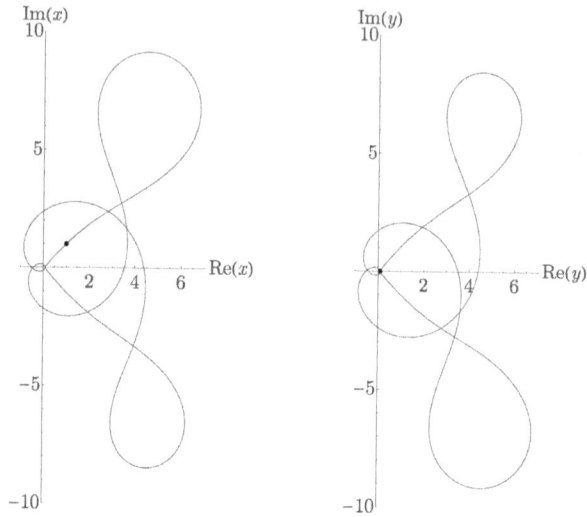

Fig. 4.10 Classical orbits for the Lotka–Volterra equation for the complex initial conditions $x(0) = 1 + i$ and $y(0) = 0.0765 + 0.0181i$, which give a real conserved quantity $C = 3.273$. The classical trajectories are closed and PT symmetric; that is, they are invariant under combined $x \leftrightarrow y$ and complex conjugation.

The condition of PT symmetry constrains the conserved quantities R and B to be real. As we did for the Lotka–Volterra equations, we consider two possibilities, real and nonreal values of the conserved quantities. First, we take as initial conditions $x(0) = 0.75$, $y(0) = -\sqrt{3/8}$, and $z(0) = -0.25$. For these initial conditions $R = 1$ and $B = -0.5$. The resulting trajectory is periodic and PT symmetric (see Fig. 4.13).

Second, we take as initial conditions $x(0) = 1.5$, $y(0) = -0.00721682 + 1.73207i$, and $z(0) = -\sqrt{7/4}$. This complex initial condition has the value $B = -1/2$, but the resulting complex trajectory is not PT symmetric because the value of R is complex: $R = \sqrt{1 - i/40}$. Consequently, the trajectory is not closed (see Fig. 4.14). Complex solutions to Euler's equations are studied in [Bender *et al.* (2007c)].

4.2.3 *Simple pendulum*

The simple pendulum is a particularly elementary dynamical system whose classical Hamiltonian

$$H = \tfrac{1}{2}p^2 - \cos x. \tag{4.5}$$

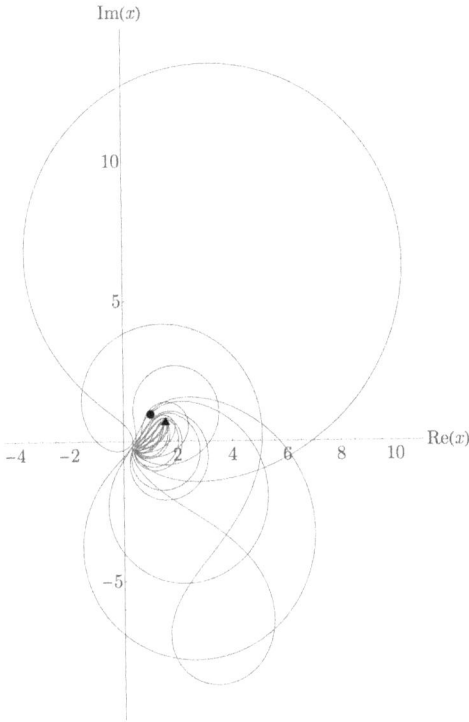

Fig. 4.11   Here the conserved quantity for the Lotka–Volterra equations is $2 + i/4$ and the initial conditions are $x(0) = 1 + i$ and $y(0) = 0.380 - 0.022i$. The trajectory begins at the dot and the final plotted point is the triangle. Observe that the trajectory is not $\mathcal{PT}$ symmetric and does not close.

is $\mathcal{PT}$ symmetric.[1] In fact, this Hamiltonian is independently $\mathcal{P}$ symmetric and $\mathcal{T}$ symmetric. The complex trajectories for this classical system have been studied in [Bender *et al.* (2007b)] and the corresponding quantized system has been analyzed in [Bender *et al.* (1999d)].

Let us examine some classical trajectories for the Hamiltonian (4.5). For simplicity, we take the classical energy to be 0. The turning points are then located at $\pm\frac{1}{2}\pi$, $\pm\frac{3}{2}\pi$, $\pm\frac{5}{2}\pi$, and so on. Real classical trajectories then connect adjacent pairs of turning points, as shown in Fig. 4.15. However, this figure also shows that there are nested families of complex trajectories that enclose these real trajectories. Each such trajectory is confined to a vertical strip of width 2π in the complex-x plane.

---

[1]Note that if we replace $\cos x$ by $\sin x$, the resulting Hamiltonian is still $\mathcal{PT}$ symmetric, but now we must define $\mathcal{P}$ to perform a reflection about the point $x = \frac{1}{2}\pi$.

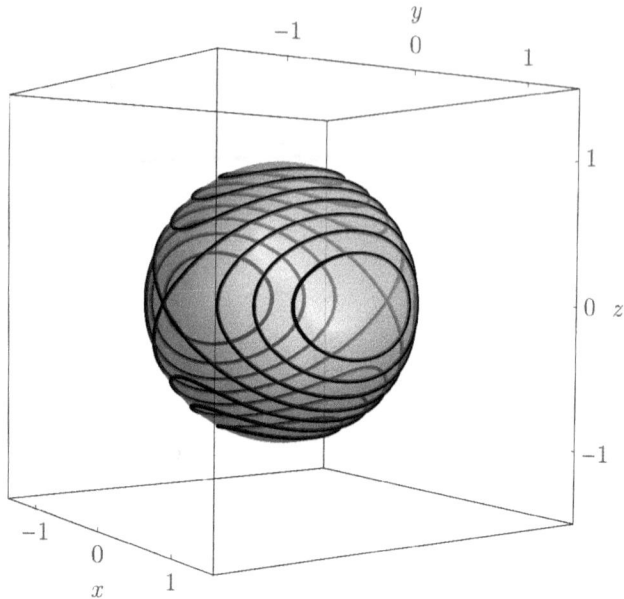

Fig. 4.12   Solution curves to Euler's equations (4.4) for the case $R = 1$. Each solution corresponds to a different value of the conserved quantity B.

Observe that there is no trajectory that goes from one strip to another. This is because the cosine potential consists of periodic potential wells separated by potential barriers. It is not possible for a classical particle of energy less than 1 to leap over the barriers that separate the strips, even if the classical particle travels through the complex plane. However, if the energy is allowed to be complex, the complex path of a classical particle is no longer closed. Instead of executing periodic motion, the particle follows an open nonintersecting trajectory; this trajectory is a deterministic random walk in which the particle repeatedly visits all the different strips (see Fig. 4.16). This classical behavior might be viewed as a kind of tunneling process that is analogous to a quantum particle in a periodic potential hopping from well to well. However, the classical particle does not really tunnel through the barriers separating the potential wells but rather follows a complex path that circumvents the barriers [Bender *et al.* (2008a); Arpornthip and Bender (2009); Bender *et al.* (2010a); Anderson *et al.* (2011a)].

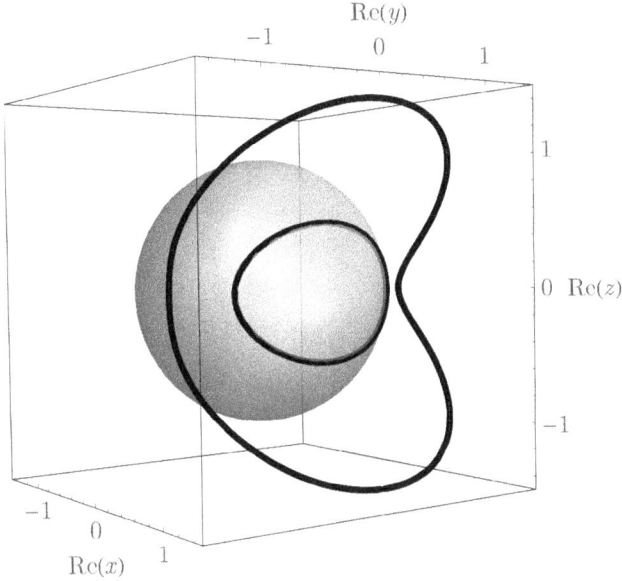

Fig. 4.13   Complex $\mathcal{PT}$-symmetric trajectory for the Euler equations (4.4). The initial condition is $x(0) = 1.5$, $y(0) = -i\sqrt{3}$, and $z(0) = -\sqrt{7/4}$, which gives the real values $R = 1$ and $B = -0.5$. The resulting butterfly-shaped trajectory is closed and $\mathcal{PT}$ symmetric. For reference, the ball in Fig. 4.12 is included in this picture and a trajectory with initial conditions $x(0) = 0.75$, $y(0) = -\sqrt{3/8}$, and $z(0) = -0.25$ is shown. This trajectory has the same values of $R = 1$ and $B = -0.5$.

Let us see what happens if we modify the classical pendulum Hamiltonian (4.5) by making the gravitational field *imaginary*:

$$H = p^2 + i \sin x. \tag{4.6}$$

This Hamiltonian is still $\mathcal{PT}$ symmetric. However, the classical trajectories are no longer confined to vertical strips in the complex plane. Rather, there are classical trajectories that run vertically from the turning points to infinity in the positive-imaginary and negative-imaginary directions and there are wavy horizontal trajectories that run from negative-real infinity to positive-real infinity (see Fig. 4.17).

4.2.4   Classical trajectories for isospectral Hamiltonians

In Sec. 2.3 we proved that the $\mathcal{PT}$-symmetric quantum-mechanical Hamiltonian $\hat{H} = \hat{p}^2 - g\hat{x}^4$ in (2.27) has a positive real spectrum by establishing that its eigenvalues are identical to those of the conventional Hermitian

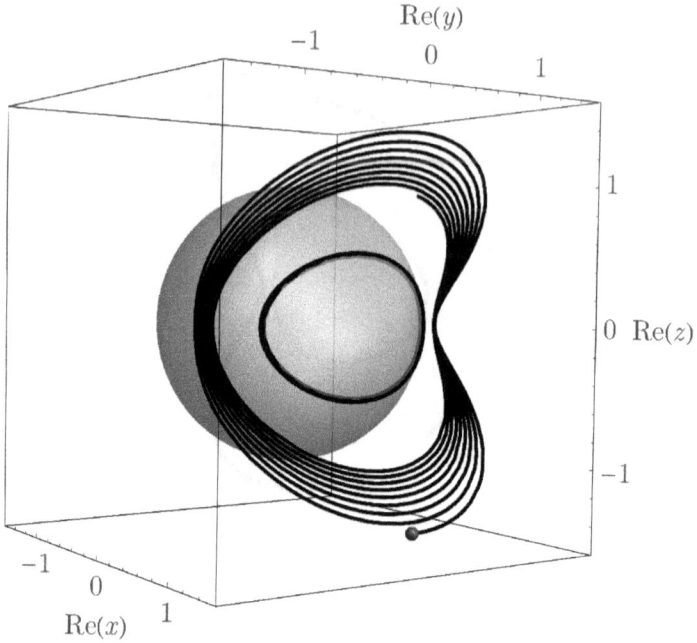

Fig. 4.14   Complex trajectory for the Euler equations (4.4). The initial condition is $x(0) = 1.5$, $y(0) = -0.00721682 + 1.73207i$, and $z(0) = -\sqrt{7/4}$. For this initial condition $R = \sqrt{1 - i/40}$ and $B = -1/2$. Since the conserved quantities are not real the orbit is not closed and not $\mathcal{PT}$ symmetric.

quantum-mechanical quartic Hamiltonian $\hat{h} = \hat{p}^2 + 4g\hat{x}^4 \pm 2\hbar\sqrt{g}\hat{x}$ in (2.28). We argued there that the linear term in $\hat{h}$ is a quantum anomaly that arises because the boundary conditions on $\hat{H}$, which are imposed in Stokes sectors, violate parity invariance.

How is the isospectrality of these two quantum-mechanical Hamiltonians reflected at the classical level? To answer this interesting question we first take $g = 1$ and plot the orbits of the $\mathcal{PT}$-symmetric classical Hamiltonian $H = p^2 - x^4$ for the classical energy $E = 1$. These orbits are shown in Fig. 1.15. The classical period of the orbits is $T = 1.854$. Next, we plot the classical orbits of the corresponding spectrally equivalent Hamiltonian $h = p^2 + 4x^4$ for $E = 1$. (It is crucial here that we drop the anomaly term containing $\hbar$. This is because $\hbar = 0$ in the classical limit.) A peanut-shaped classical orbit of this Hamiltonian is shown in Fig. 4.18 (upper panel). The period of this orbit is also exactly $T = 1.854$. Thus, the two Hamiltonians

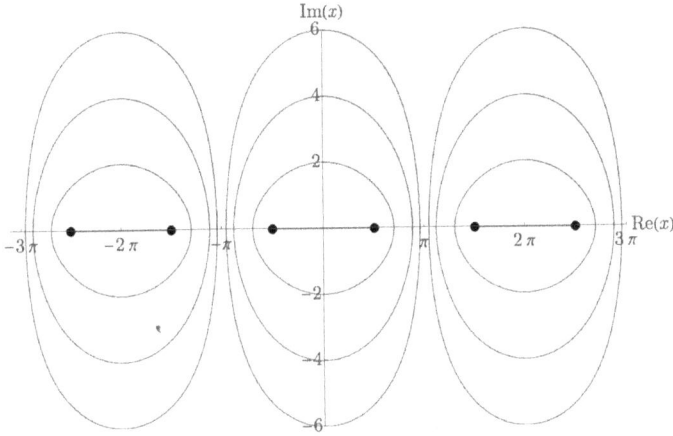

Fig. 4.15 Twelve closed trajectories for the simple pendulum Hamiltonian $H = \frac{1}{2}p^2 - \cos x$. The energy for all trajectories is 0. The complex-x plane is decomposed into vertical strips of width 2π and three such strips are shown. In each strip are two turning points (indicated by dots) on the real axis. Each classical path is confined to a given strip; no path goes from one strip to another. The trajectories in each strip are topologically equivalent to (and are just squeezed versions of) those in Fig. 1.12.

are *classically* equivalent. However, if we do not discard the anomaly term, the period of the classical orbit *changes* to $T = 1.669$. We plot in Fig. 4.18 (lower panel) a classical orbit of the Hamiltonian $h = p^2 + 4x^4 + 2x$ for the case $E = 1$. We find the surprising result that despite the spectral equivalence of the two Hamiltonians at the quantum-mechanical level, the classical periods for the same energy are different.

4.2.5 A more complicated oscillatory system

Consider the rather complicated $\mathcal{PT}$-symmetric Hamiltonian that arises in the context of Liouville quantum field theory [Bender *et al.* (2014b)]:

$$H = \tfrac{1}{2}p^2 - xe^{ix}. \tag{4.7}$$

The classical trajectories for this Hamiltonian are remarkable because they divide into an infinite number of classes depending on how many pairs of turning points they encircle before they close. Three classes of orbits are shown in Fig. 4.19, two orbits enclosing one pair of turning points, one orbit enclosing two pairs of turning points, and one orbit enclosing three pairs of turning points. The different classes of orbits are separated by separatrix paths, which do not close but rather approach negative-imaginary infinity.

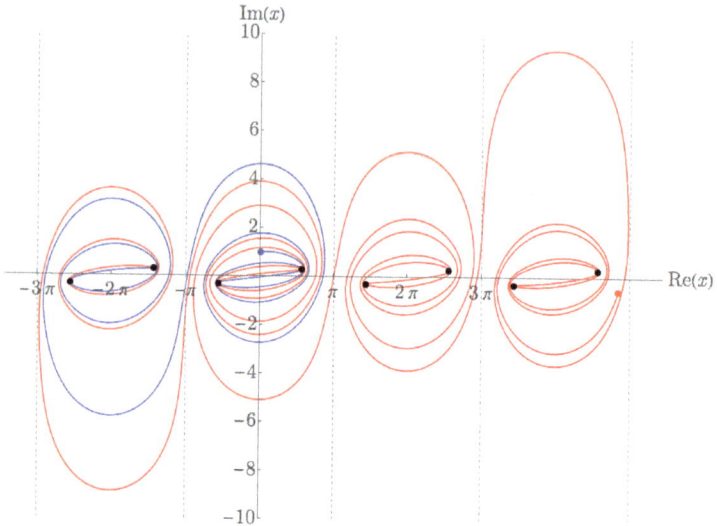

Fig. 4.16 Path of a classical particle in the potential $\cos x$. Instead of the energy being real, as in Fig. 4.15, the energy $E = \frac{1}{5} + \frac{1}{3}i$ here is complex. Consequently, the particle is no longer confined to just one strip and visits different strips in a deterministic random fashion. The trajectory $x(t)$ begins at the blue dot at $x(0) = i$ and moves along the blue path to the next strip to the left. The particle then goes rightward along the red path, revisits the strip where it began, and visits two more strips to the right. The plot stops at the red dot, but the trajectory continues its random but deterministic motion, going two strips to the left, then two to the right, and so on, eventually visiting all strips. This complex classical behavior is analogous to the random hopping of a quantum particle from site to site in a crystal.

4.3 Complex Probability

So far, this chapter has focused on the trajectories of complex-deformed classical-mechanical systems having $\mathcal{PT}$ symmetry. This section begins by considering another kind of classical system; namely, a probabilistic random walk. We show that it is possible to deform a random-walk process into the complex domain in such a way as to preserve the $\mathcal{PT}$ symmetry. This complex system can still have a meaningful probabilistic interpretation.

To construct a $\mathcal{PT}$-symmetric deformation of a random-walk process we begin with a conventional one-dimensional classical random walk and *deform the coin that is tossed to direct the random walk by giving it an imaginary bias.* For a conventional one-dimensional random walk, the walker visits sites on the real-x axis and there is a probability density of finding

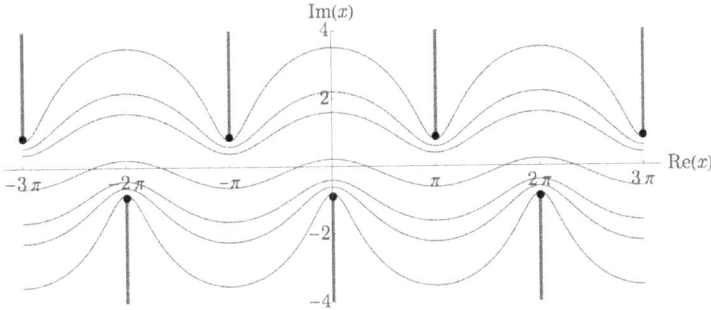

Fig. 4.17 Complex trajectories of classical energy $E = 1$ for the classical pendulum in an imaginary gravitational field $H = p^2 + i \sin x$. Seven trajectories drift through the crystal in a wave-like fashion and seven more trajectories run from the turning points upward and downward to imaginary infinity.

the random walker at any given site. However, for a PT-symmetric random walk the random walker can be thought of as moving along a contour in the complex-$z = x + iy$ plane. For this elementary problem we will see in Sec. 4.3.1 that the contour is merely a straight line parallel to the real-x axis. On this contour we can construct a local positive probability density.

Quantum mechanics is a fundamentally probabilistic theory. Since we discuss the probabilistic motion of classical particles in the complex plane in this chapter, we consider next the possibility of constructing a complex quantum-mechanical probability density that describes the motion of quantum particles in the complex plane.

4.3.1 *PT-symmetric classical random walk*

In the usual formulation of a one-dimensional random-walk problem the random walker visits the lattice sites $x_n = n\delta$, where $n = 0, \pm 1, \pm 2, \pm 3, \ldots$ on the x axis. These sites are separated by the characteristic length δ. At each time step the walker tosses a *fair coin* (a coin with an equal probability $1/2$ of getting heads or tails). If the result is heads, the walker then moves one step to the right; if the result is tails, the walker then moves one step to the left. To describe such a random walk we introduce the probability distribution $P(n, k)$, which represents the probability of finding the random walker at position x_n at time $t_k = k\tau$, where $k = 0, 1, 2, 3, \ldots$. The interval between these time steps is the characteristic time τ. The probability

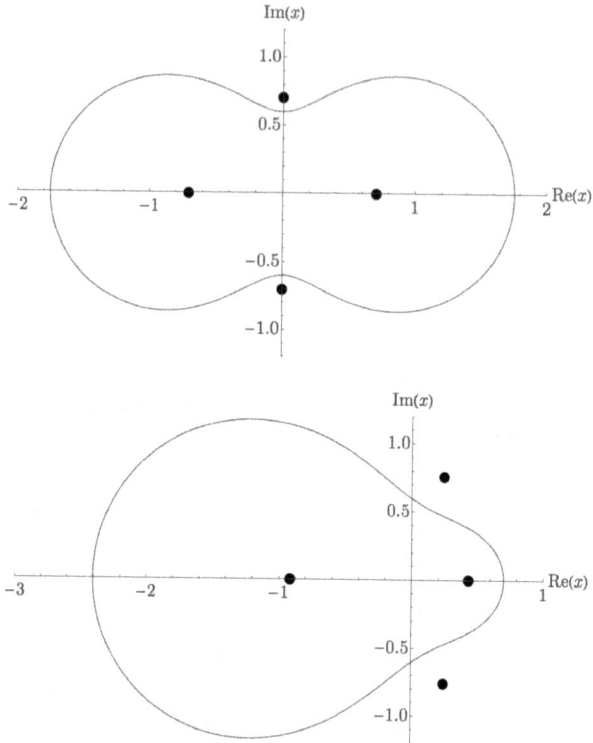

Fig. 4.18   Classical closed orbits for the Hamiltonian $h = p^2 + 4x^4 + 2\hbar x$, which has a quantum anomaly. The isospectral partner of h is the $\mathcal{PT}$-symmetric Hamiltonian $H = p^2 - x^4$ (see Sec. 2.3). In the upper panel we take the classical limit $\hbar = 0$ to eliminate the quantum anomaly and plot a classical orbit for the case $E = 1$. There are two real and two complex turning points. The period of this peanut-shaped orbit is $T = 1.854$, which agrees exactly with the classical period of an $E = 1$ orbit for the Hamiltonian H (see Fig. 1.15). In the lower panel we do not eliminate the quantum anomaly term and we plot an orbit for the Hamiltonian $h = p^2 + 4x^4 + 2x$. In this case we find that for $E = 1$ the classical period of the orbit changes to $T = 1.669$.

distribution $P(n, k)$ satisfies the well known partial difference equation

$$P_{n,k} = \tfrac{1}{2}P_{n+1,k-1} + \tfrac{1}{2}P_{n-1,k-1}. \tag{4.8}$$

We deform this random-walk problem into the complex domain by introducing a coin is not fair; we suppose that the coin has an *imaginary* bias. Thus, we assume that at each step the so-called "probability" of the random walker moving to the right is

$$\alpha = \tfrac{1}{2} + i\beta\delta \tag{4.9}$$

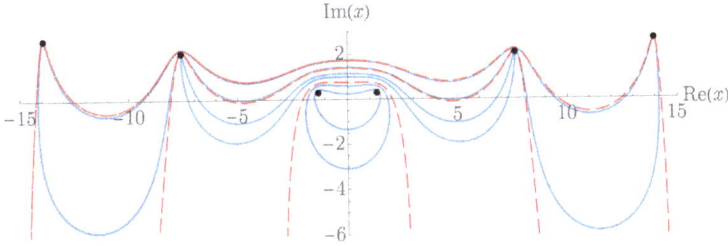

Fig. 4.19 Complex trajectories for the Hamiltonian $H = \frac{1}{2}p^2 - ixe^{ix}$ having classical energy $E = 1$. There are an infinite number of classes of orbits, where the nth class contains orbits that enclose n pairs of turning points. Four such orbits are shown (in blue), two in class $n = 1$, one in class $n = 2$, and one in class $n = 3$. The nth class is separated from the $n = 1$st class by separatrix curves (red dashed lines), which do not close but rather continue downward towards negative-imaginary infinity. Three such separatrix curves are shown.

and that the "probability" of moving to the left is

$$1 - \alpha = \tfrac{1}{2} - i\beta\delta = \alpha^*. \qquad (4.10)$$

The parameter β is a measure of the *imaginary* unfairness of the coin. Of course, a conventional probability must be a positive number between 0 and 1; this is why we use quotation marks in the term "probability."

Note that $\alpha + \alpha^* = 1$, so the total "probability" of a move is still unity. Consequently, (4.8) generalizes to

$$P(n, k) = \alpha P(n - 1, k - 1) + \alpha^* P(n + 1, k - 1). \qquad (4.11)$$

We would like to interpret $P(n, k)$ as a probability density, but the problem is that $P(n, k)$ obeys the *complex* equation (4.11) and thus is complex-valued. However, even though this equation is complex, it is $\mathcal{PT}$ symmetric because it is invariant under the combined operations of parity and time reversal. Under a parity reflection we interchange the "probabilities" (4.9) and (4.10) for rightward and leftward steps and this has the effect of interchanging the indices $n + 1$ and $n - 1$ in (4.11). Time reversal is realized by complex conjugation.

Let us choose the simple initial condition that the random walker stands at the origin $n = 0$ at time $k = 0$: $P(0, 0) = 1$. For this initial condition the exact solution to (4.11) is

$$P(n, k) = \frac{k!\alpha^{\frac{k+n}{2}}(\alpha^*)^{\frac{k-n}{2}}}{\left(\frac{k+n}{2}\right)!\left(\frac{k-n}{2}\right)!} = \frac{k!\alpha^{\frac{k+n}{2}}(1-\alpha)^{\frac{k-n}{2}}}{\left(\frac{k+n}{2}\right)!\left(\frac{k-n}{2}\right)!}.$$

This solution is manifestly $\mathcal{PT}$ symmetric because it is invariant under combined space reflection $n \to -n$ and time reversal (complex conjugation).

Let us find the continuum limit of this complex random walk. To do so we introduce the continuous variables t and x according to

$$t \equiv k\tau \quad \text{and} \quad x \equiv n\delta$$

subject to the requirement that the *diffusion constant* σ given by the ratio $\sigma \equiv \delta^2/\tau$ is held fixed. We then let $\delta \to 0$ and define $\rho(x,t)$ to be the probability density:

$$\rho(x,t) \equiv \lim_{\delta \to 0} \frac{P(n,k)}{\delta} = \frac{1}{\sqrt{2\pi\sigma t}} e^{-\frac{(x-2i\sigma\beta t)^2}{2\sigma t}}. \tag{4.12}$$

The function $\rho(x,t)$ solves the *complex* $\mathcal{PT}$-symmetric diffusion equation

$$\rho_t(x,t) = \sigma\rho_{xx}(x,t) - 2i\sigma\beta\rho_x(x,t) \tag{4.13}$$

and satisfies the delta-function initial condition $\lim_{t\to 0} \rho(x,t) = \delta(x)$, where x is *real*.

Interestingly, there is a contour in the complex-x plane on which the probability density $\rho(x,t)$ is *real*. For this simple problem the contour is merely the *straight horizontal line* $\operatorname{Im} x = 2\sigma\beta t$. At $t = 0$ this line lies on the real axis, but as time increases this line moves vertically at the constant velocity $2\sigma\beta$. Thus, although the "probability" density is a complex function of x and t, we have identified a contour in the complex-x plane on which the probability is real and positive and hence may be interpreted as a conventional probability density.

4.3.2 *Probability density for $\mathcal{PT}$ quantum mechanics*

In a $\mathcal{PT}$-symmetric classical dynamical system a typical trajectory of the classical particle lies in the complex plane. When we quantize such a system, we are faced with the difficult problem of constructing a local probabilistic description of the motion of such a quantum particle; that is, constructing a local probability current and probability density for the particle. While this is not an easy problem, we have shown above that there is a probabilistic interpretation for a complex random-walk process; that is, we have found a time-dependent path in the complex plane on which there is a positive probability density. This path is just a straight line.

The situation for complex quantum mechanics is not so simple because if there is a path in the complex plane on which we can define a probability density, the path is not just a simple straight line. Although there is much more work to be done, two papers have been published that make a first attempt at finding such a path in the complex plane [Bender *et al.* (2010b,c)]. Below is a brief summary of this work.

The time-dependent Schrödinger equation $i\psi_t = \hat{H}\psi$ in complex-coordinate space (z space) is

$$i\psi_t(z,t) = -\psi_{zz}(z,t) + V(z)\psi(z,t). \tag{4.14}$$

The coordinate z is complex, so the complex conjugate of (4.14) is

$$- i\psi_t^*(z^*,t) = -\psi_{z^*z^*}^*(z^*,t) + V^*(z^*)\psi^*(z^*,t). \tag{4.15}$$

Substituting $-z$ for z^* in (4.15), we obtain

$$- i\psi_t^*(-z,t) = -\psi_{zz}^*(-z,t) + V(z)\psi^*(-z,t), \tag{4.16}$$

where we have used the PT symmetry condition $V^*(-\hat{x}) = V(\hat{x})$ to replace $V^*(-z)$ with $V(z)$. To derive a local conservation law we first multiply (4.14) by $\psi^*(-z,t)$ and get

$$i\psi^*(-z,t)\psi_t(z,t) = -\psi^*(-z,t)\psi_{zz}(z,t) + V(z)\psi^*(-z,t)\psi(z,t). \tag{4.17}$$

Next, we multiply (4.16) by $-\psi(z,t)$ and obtain

$$i\psi(z,t)\psi_t^*(-z,t) = \psi(z,t)\psi_{zz}^*(-z,t) - V(z)\psi(z,t)\psi^*(-z,t). \tag{4.18}$$

We then add (4.17) to (4.18) and get

$$\frac{\partial}{\partial t}[\psi(z,t)\psi^*(-z,t)] + \frac{\partial}{\partial z}[i\psi(z,t)\psi_z^*(-z,t) - i\psi^*(-z,t)\psi_z(z,t)] = 0.$$

This equation has the generic form of a local continuity equation

$$\rho_t(z,t) + j_z(z,t) = 0 \tag{4.19}$$

with local density

$$\rho(z,t) \equiv \psi^*(-z,t)\psi(z,t) \tag{4.20}$$

and local current

$$j(z,t) \equiv i\psi_z^*(-z,t)\psi(z,t) - i\psi^*(-z,t)\psi_z(z,t). \tag{4.21}$$

Note that the local density $\rho(z,t)$ in (4.20) is *not* the absolute square of $\psi(z,t)$. Rather, because the PT symmetry is unbroken, $\psi^*(-z,t) = \psi_z(z,t)$ and thus $\rho(z,t) = [\psi(z,t)]^2$. This allows us to extend the density into the complex-z plane as an *analytic* function.

While (4.19) has the form of a local conservation law, the local density $\rho(z,t)$ in (4.20) is a complex-valued function. Thus, it is not clear whether $\rho(z,t)$ can serve as a probability density because for a locally conserved quantity to be interpretable as a probability density, it must be real and positive and its spatial integral must be normalized to unity. With this in mind, we attempt to identify a contour C in the complex-z plane on which $\rho(z,t)$ can be interpreted as being a probability density. We argue that on such a contour $\rho(z,t)$ must satisfy three conditions:

$$Condition\ I: \quad \mathrm{Im}\,[\rho(z)\,dz] = 0, \tag{4.22}$$

$$Condition\ II: \quad \mathrm{Re}\,[\rho(z)\,dz] > 0, \tag{4.23}$$

$$Condition\ III: \quad \int_C \rho(z)\,dz = 1. \tag{4.24}$$

Note that on such a contour the probability density $\rho(z,t)$ for finding a particle in the complex-z plane at time t is in general *not* real. Rather, it is the *product* $\rho\,dz$ (which represents the *local* contribution to the total probability) that should be real and positive. In the case of the complex random-walk problem the contour happens to be a horizontal line and thus both the infinitesimal line segment dx and the probability density ρ are individually real and positive.

A complex contour C that fulfills the above three requirements depends on the wave function $\psi(z,t)$ and thus it is time dependent. However, for simplicity, we restrict our attention to the wave functions $\psi(z,t) = e^{iE_n t}\psi_n(z)$, where E_n is an eigenvalue of the Hamiltonian and $\psi_n(z)$ is the corresponding eigenfunction of the time-independent Schrödinger equation. For this choice of $\psi(z,t)$ the local current $j(z,t)$ vanishes, and $\rho(z)$ and the contour C on which it is defined are both time independent.

Condition I enables us to construct a differential equation that defines the contour C, but this equation is difficult to solve. We illustrate the procedure for the harmonic-oscillator Hamiltonian whose eigenfunctions are just Gaussians multiplied by Hermite polynomials: $\psi_n(z) = e^{-z^2/2}\mathrm{He}_n(z)$. To find the probability density in the complex-z plane, we construct $\rho(z)$ according to (4.20) and then impose the three conditions in (4.22)-(4.24). The function $\rho(z)$ then has the general form

$$\rho(z,t) = e^{-x^2+y^2-2ixy}[S(x,y) + iT(x,y)],$$

where $z = x + iy$ and $S(x,y)$ and $T(x,y)$ are polynomials in x and y:

$$S(x,y) = \mathrm{Re}\left(\left[\mathrm{He}_n(z)\right]^2\right) \quad \text{and} \quad T(x,y) = \mathrm{Im}\left(\left[\mathrm{He}_n(z)\right]^2\right).$$

Thus, $\rho\,dz$ has the form

$$\rho\,dz = e^{-x^2+y^2}[\cos(2xy) - i\sin(2xy)][S(x,y) + iT(x,y)](dx + i\,dy). \tag{4.25}$$

Imposing Condition I in (4.22), we get a nonlinear differential equation for the contour $y(x)$ in the $z = x + iy$ plane on which the imaginary part of $\rho\,dz$ vanishes:

$$\frac{dy}{dx} = \frac{S(x,y)\sin(2xy) - T(x,y)\cos(2xy)}{S(x,y)\cos(2xy) + T(x,y)\sin(2xy)}. \tag{4.26}$$

This is a nontrivial differential equation, and even the differential equation for the special case $n = 0$,

$$\frac{dy}{dx} = \tan(2xy),$$

cannot be solved analytically in closed form. Nevertheless, we may proceed in principle. On this contour defined by (4.26) we impose Conditions II and III in (4.23) and (4.24). We calculate the real part of $\rho\,dz$:

$$\mathrm{Re}\,(\rho\,dz) = e^{-x^2+y^2}\,\{[S(x,y)\cos(2xy) + T(x,y)\sin(2xy)]dx$$
$$+ [S(x,y)\sin(2xy) - T(x,y)\cos(2xy)]dy\}.$$

Thus, using (4.26), we get

$$\mathrm{Re}\,(\rho\,dz) = e^{-x^2+y^2}\,\frac{[S(x,y)]^2 + [T(x,y)]^2}{S(x,y)\cos(2xy) + T(x,y)\sin(2xy)}\,dx \tag{4.27}$$

or, alternatively,

$$\mathrm{Re}\,(\rho\,dz) = e^{-x^2+y^2}\,\frac{[S(x,y)]^2 + [T(x,y)]^2}{S(x,y)\sin(2xy) - T(x,y)\cos(2xy)}\,dy. \tag{4.28}$$

Equations (4.27) and (4.28) suggest that there might be some problems with establishing the existence of a contour $y(x)$ in the complex plane on which ρ can be interpreted as a probability density. First, if the contour $y(x)$ should pass through a zero of either the numerator or the denominator of the right side of (4.26), then the sign of $\mathrm{Re}\,(\rho\,dz)$ will change, which would violate the requirement of positivity in (4.23). However, detailed analysis of these equations [Bender *et al.* (2010c)] shows that the sign of $\mathrm{Re}\,(\rho\,dz)$ actually does *not* change. Second, if the contour $y(x)$ passes through a zero of the numerator or the denominator of the right side of (4.26), one might expect ρ to be singular at this point. Indeed, it *is* singular, and one may worry that the integral in (4.24) would diverge and thus violate condition III that the total probability be normalized to unity. However, in [Bender *et al.* (2010c)] it is shown that the singularity in ρ is an *integrable* singularity and thus the probability is normalizable.

Of course, this initial work hardly begins to solve the difficult problem of constructing a local probability density in *PT*-symmetric quantum theory. This remains a rich and interesting area for future study.

4.4 $\mathcal{PT}$-Symmetric Classical Field Theory

The standard procedure for constructing a non-Hermitian $\mathcal{PT}$-symmetric Hamiltonian is to begin with a Hamiltonian that is both Hermitian and $\mathcal{PT}$ symmetric, and then to introduce a parameter ε that deforms the Hamiltonian into the complex domain while maintaining its $\mathcal{PT}$ symmetry. This is what was used to construct the class of Hamiltonians in (2.7). We can follow this same procedure for classical nonlinear wave equations because many of these wave equations are $\mathcal{PT}$ symmetric.

As an example, consider the Korteweg–de Vries (KdV) equation

$$u_t + uu_x + u_{xxx} = 0, \tag{4.29}$$

which describes shallow one-dimensional water waves. To show that this equation is $\mathcal{PT}$ symmetric, we define a classical parity reflection $\mathcal{P}$ to be the replacement $x \to -x$, and treating $u = u(x,t)$ as a velocity, the sign of u also changes under $\mathcal{P}$: $u \to -u$. We then define a classical time reversal $\mathcal{T}$ to be the replacement $t \to -t$, and again because u is a velocity, the sign of u also changes under $\mathcal{T}$: $u \to -u$. Note that while the KdV equation is symmetric under combined $\mathcal{PT}$ reflection, it is not symmetric under $\mathcal{P}$ or $\mathcal{T}$ separately. The KdV equation is a special case of the Camassa–Holm equation [Camassa and Holm (1993)], which is also $\mathcal{PT}$ symmetric. Other nonlinear wave equations such as the generalized KdV equation $u_t + u^k u_x + u_{xxx} = 0$, the Sine–Gordon equation $u_{tt} - u_{xx} + g\sin u = 0$, and the Boussinesq equation are $\mathcal{PT}$ symmetric as well.

The observation that there are many nonlinear wave equations possessing $\mathcal{PT}$ symmetry suggests that one can generate rich and interesting families of new complex nonlinear $\mathcal{PT}$-symmetric wave equations by following the same procedure that is used in $\mathcal{PT}$ quantum mechanics, and one can try to discover which properties (conservation laws, solitons, integrability, stochastic behavior) of the original wave equations are preserved and which are lost when performing a complex deformation. If we follow the quantum-mechanical formalism and also require that $i \to -i$ under time reversal, we immediately see that a possible way to generate new $\mathcal{PT}$-symmetric nonlinear wave equations from the KdV equation is to introduce the real parameter ε as follows:

$$u_t - iu(iu_x)^\varepsilon + u_{xxx} = 0. \tag{4.30}$$

Various members of this family of equations have been studied in [Bender *et al.* (2007a)]. Of course, there are other ways to deform the KdV equation into the complex domain while preserving $\mathcal{PT}$ symmetry; see Chap. 9 for more details.

Chapter 5

$\mathcal{PT}$-Symmetric Quantum Field Theory

"Most men would have given up...But not the crew of $\mathcal{PT}$-109."
—From the movie $\mathcal{PT}$-109

Two profound advances in the early 20th century transformed classical physics into what we view today as modern physics. The first was the discovery of quantum mechanics, a theory that describes the properties of matter. The second was the discovery of relativity, a theory that describes the geometry of space and time. When these two theories are combined, the resulting relativistic quantum theory is called *quantum field theory*. Quantum field theory describes the properties and interactions of elementary particles, which are the fundamental constituents of the natural world.

Quantum-mechanical theories that describe a nonrelativistic particle under the influence of a potential have only a finite number of degrees of freedom. Most of the $\mathcal{PT}$-symmetric quantum-mechanical models discussed in this book have just one degree of freedom, and the Hamiltonians for these theories are constructed from one pair of dynamical variables, $\hat{x}$ and $\hat{p}$. These dynamical variables obey the equal-time commutation relation

$$[\hat{x}(t), \hat{p}(t)] = i, \tag{5.1}$$

which is known as the *Heisenberg algebra*.

Quantum field theory is much more complicated than quantum mechanics. For a quantum field theory in $D + 1$-dimensional space-time, the operators $\hat{x}(t)$ and $\hat{p}(t)$ are generalized to the quantum fields $\varphi(\mathbf{x}, t)$ and $\pi(\mathbf{x}, t)$, which represent a *continuously infinite* number of degrees of freedom, one for each value of the spatial variable $\mathbf{x}$. These quantum fields obey the D-dimensional equal-time commutation relation

$$[\varphi(\mathbf{x}, t), \pi(\mathbf{y}, t)] = i\delta^{(D)}(\mathbf{x} - \mathbf{y}), \tag{5.2}$$

which is a D-dimensional generalization of (5.1).

We showed in Chap. 1 that complex mathematics provides deep insight into real mathematics. This suggested that it might be useful to deform conventional quantum mechanics into the complex domain, and in doing so we discovered new non-Hermitian $\mathcal{PT}$-symmetric quantum theories that are consistent with the basic principles of quantum mechanics. This chapter discusses the properties of quantum field theories that are obtained by $\mathcal{PT}$-symmetric deformation of conventional Hermitian quantum field theories.

5.1   Introduction to $\mathcal{PT}$-Symmetric Quantum Field Theory

Complex numbers are central in quantum mechanics, but we ordinarily consider space-time as being restricted to the real domain. A point in space-time is represented by the *real* four-dimensional vector (x, y, z, t). The *homogeneous Lorentz group* is defined as the six-parameter group of all *real* 4×4 matrices that perform rotations and boosts on the space-time vector (x, y, z, t) but that leave the numerical value of the Lorentz scalar $x^2 + y^2 + z^2 - t^2$ invariant. There are three spatial rotations; one can rotate about the x, y, and z axes. Repeating an experiment in a laboratory that has been rotated in space will yield the same result. One can also boost the velocity in three possible ways, along the x, y, or z axes. Again, repeating an experiment in a laboratory that is moving at a constant velocity relative to the original laboratory will yield the same result.

For many years it was thought that the Lorentz group (plus translations in space and time) constituted the complete geometrical symmetry group of the universe. However, along with the continuous transformations (rotations and boosts), the definition above of the homogeneous Lorentz group also allows for two discrete transformations that we now know are not fundamental symmetries of nature. The first of these, called *parity* $\mathcal{P}$, reverses the sign of the spatial part of a four-vector; that is, $\mathcal{P} : (x, y, z, t) \to (-x, -y, -z, t)$. (Note that $\mathcal{P}$ changes one's right hand into one's left hand; such a transformation cannot be achieved by a rotation.) The second, called *time reversal* $\mathcal{T}$, reverses the sign of the time component of a four-vector; that is, $\mathcal{T} : (x, y, z, t) \to (x, y, z, -t)$.

A Nobel prize was awarded to Lee and Yang in 1957 for demonstrating that parity is not a symmetry of nature and another was awarded to Cronin and Fitch in 1980 for demonstrating that time reversal is also not a symmetry of nature. (A left-handed laboratory can obtain different experimental results from a right-handed laboratory, and a laboratory traveling backward in time can obtain different results from a laboratory traveling forward in time.) After these advances, it was accepted that the correct geometrical symmetry group of nature must exclude $\mathcal{P}$ and $\mathcal{T}$. If we remove these symmetries from the homogeneous Lorentz group, we obtain a new smaller continuous symmetry group called the *proper orthochronous Lorentz group* (POLG).

The homogeneous Lorentz group consists of four disconnected parts: (i) the POLG, which is a subgroup of the homogeneous Lorentz group, (ii) all the elements of the POLG multiplied by parity $\mathcal{P}$, (iii) all the elements of the POLG multiplied by time reversal $\mathcal{T}$, and (iv) all the elements of the POLG multiplied by space-time reflection $\mathcal{PT}$. However, if we extend the homogeneous Lorentz group to the complex Lorentz group by including all *complex* 4×4 matrices that leave the numerical value of the Lorentz scalar $x^2 + y^2 + z^2 - t^2$ invariant, we find that the resulting group has only two disconnected parts: The POLG is now joined continuously to the POLG $\times$ $\mathcal{PT}$ and the POLG $\times$ $\mathcal{P}$ is joined continuously to the POLG $\times$ $\mathcal{T}$.

Thus, when we extend real geometry to complex geometry, a new discrete symmetry, namely, $\mathcal{PT}$ symmetry, emerges naturally. We see that $\mathcal{PT}$ symmetry means *combined* $\mathcal{P}$ and $\mathcal{T}$; a $\mathcal{PT}$ reflection changes the sign of all four components of a space-time vector $\mathcal{PT}$: $(x, y, z, t) \rightarrow (-x, -y, -z, -t)$. For uncharged particles that are their own antiparticles, this discrete symmetry is a fundamentally correct symmetry of nature.[1]

Quantum field theories are vastly more complicated than quantum-mechanical theories, but constructing quantum field theories that are formally non-Hermitian and $\mathcal{PT}$ invariant is not difficult. Indeed, cubic and quartic quantum-field-theoretic Hamiltonian densities that are analogous to the $\mathcal{PT}$-symmetric quantum-mechanical Hamiltonians $\hat{H} = \frac{1}{2}\hat{p}^2 + \frac{1}{2}\mu^2\hat{x}^2 + \frac{1}{3}i\hat{x}^3$ and $\hat{H} = \frac{1}{2}\hat{p}^2 + \frac{1}{2}\mu^2\hat{x}^2 - \frac{1}{4}\hat{x}^4$ are

$$\mathcal{H} = \tfrac{1}{2}\pi^2(\mathbf{x}, t) + \tfrac{1}{2}[\nabla_{\mathbf{x}}\varphi(\mathbf{x}, t)]^2 + \tfrac{1}{2}\mu^2\varphi^2(\mathbf{x}, t) + \tfrac{1}{3}ig\varphi^3(\mathbf{x}, t) \qquad (5.3)$$

and

$$\mathcal{H} = \tfrac{1}{2}\pi^2(\mathbf{x}, t) + \tfrac{1}{2}[\nabla_{\mathbf{x}}\varphi(\mathbf{x}, t)]^2 + \tfrac{1}{2}\mu^2\varphi^2(\mathbf{x}, t) - \tfrac{1}{4}g\varphi^4(\mathbf{x}, t). \qquad (5.4)$$

As in quantum mechanics, where the operators $\hat{x}$ and $\hat{p}$ change sign under parity $\mathcal{P}$, we assume here that the fields in these Hamiltonians are

---

[1]For fermions and charged particles, $\mathcal{PT}$ symmetry is augmented with an additional symmetry operator called *charge conjugation* C, which turns particles into antiparticles. This is the origin of the famous CPT theorem in particle physics [Streater and Wightman (2000)]. The charge-conjugation operator C in particle physics has similar mathematical properties, but is not the same as the C operator introduced in Chap. 3 for $\mathcal{PT}$-symmetric quantum mechanics.

pseudoscalars, so they also change sign under a parity reflection:

$$\mathcal{P}\varphi(\mathbf{x},t)\mathcal{P} = -\varphi(-\mathbf{x},t), \qquad \mathcal{P}\pi(\mathbf{x},t)\mathcal{P} = -\pi(-\mathbf{x},t). \qquad (5.5)$$

Non-Hermitian $\mathcal{PT}$-symmetric quantum field theories, such as those in (5.3) and (5.4), exhibit a rich variety of behaviors. Cubic field-theory models like that in (5.3) are of interest because they arise in the study of the Yang-Lee edge singularity [Fisher (1978); Cardy (1985); Cardy and Mussardo (1989)] and in Reggeon field theory [Brower *et al.* (1978); Harms *et al.* (1980)]. In these papers it was assumed that time evolution in an $i\varphi^3$ field theory is nonunitary. However, by constructing the $\mathcal{C}$ operator, we argue in Sec. 5.4 that such a quantum field theory is actually unitary with respect to $\mathcal{CPT}$ conjugation. We show in Sec. 5.6 that $\mathcal{PT}$ symmetry eliminates the ghosts in the Lee model, another cubic quantum field theory. The field theory described by (5.4) is striking because it is asymptotically free, as explained in Sec. 5.7. In that section we also examine a $\mathcal{PT}$-symmetric version of quantum electrodynamics, and the $\mathcal{PT}$-symmetric Thirring and Sine–Gordon models, and we briefly mention gravitational and cosmological theories, the Higgs model, and the double-scaling limit.

5.2   Perturbative and Nonperturbative Behavior

In courses on quantum field theory φ^3 and φ^4 field theories are used as pedagogical models to explain Feynman diagrams and perturbative renormalization even though the cubic model is not physically realistic because the energy density is unbounded below (the theory does not have a stable ground state) and the quartic model is not asymptotically free in four-dimensional space-time. We explain in the following subsections how the imaginary coupling in the cubic model in (5.3) repairs the problem of the energy density in a conventional φ^3 theory not being bounded below and why the negative coupling in the quartic model (5.4) does not lead to a spectrum that is unbounded below.

5.2.1   *Cubic* $\mathcal{PT}$-*symmetric quantum field theory*

The Feynman rules for the $g\varphi^3$ quantum field theory in (5.3) are simply $-6g$ for a three-point vertex amplitude and $1/(p^2+m^2)$ for a line amplitude in momentum space. To calculate the vacuum-state energy density we must evaluate the sum of all connected vacuum graphs. (The vacuum graphs of order g^2 and of order g^4 are shown in Fig. 5.1.) The resulting perturbation

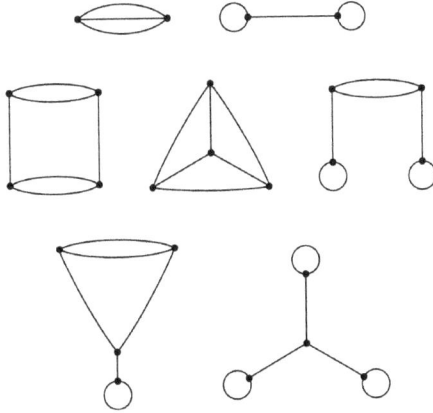

Fig. 5.1   Connected vacuum graphs of order g^2 (two graphs) and of order g^4 (five graphs) that contribute to the vacuum energy density in a $g\varphi^3$ quantum field theory. Note that for a cubic theory the vacuum graphs all have an even number of vertices and the amplitude for any graph is proportional to a power of g^2. Consequently, if g is real, the graphs all add in phase and the coefficients a_n in the Feynman perturbation series (5.6) all have the same sign. Thus, this perturbation series is not Borel summable. This nonsummability is the signal that the vacuum state of a conventional φ^3 theory is unstable. However, if g is pure imaginary, the series (5.6) is alternating and Borel summable. Thus, the ground-state energy density is real and the vacuum state is stable.

expansion contains only *even* powers of g:

$$\sum_{n=0}^{\infty} a_n g^{2n}. \tag{5.6}$$

The form of this perturbation series reveals a fundamental problem. Because the number of vacuum graphs grows factorially with the number of vertices [Bender and Wu (1973)], this Feynman perturbation series is *divergent* [Bender and Dunne (1999); Bender and Weniger (2001)].[2] Ordinarily, a divergent perturbation series is not a serious obstacle because the series is asymptotic and we can use a summation procedure, such as Padé or Borel summation, to obtain the sum of the series [Bender and Orszag (1999)]. However, in the case of a conventional $g\varphi^3$ field theory for which the coupling constant g is real, all the graphs contributing to the perturbation series have the *same sign*. The series (5.6) is not an alternating series because only even powers of g appear in the series. Thus, the divergent series is not Borel summable and there is a branch cut in the coupling-constant

---

[2]This divergence cannot be repaired by renormalization; renormalization is used to treat individual diagrams whose Feynman integrals are divergent. Here, the series itself diverges whether or not the perturbation coefficients a_n are finite [Bender *et al.* (1970)].

plane. The discontinuity across this cut leads to a complex value for the sum of the series even though the coefficients a_n are all real.

This complex result for the vacuum energy density is not a surprise. It merely confirms that the cubic potential is not bounded below, so the vacuum state becomes unstable when $g \neq 0$. The imaginary part of the vacuum energy density is a measure of the lifetime of the vacuum state.

The advantage of the 𝒫𝒯-symmetric Hamiltonian in (5.3) is that the coupling term is pure imaginary and therefore the square of the coupling constant is *negative*. Consequently, the divergent series (5.6) *alternates in sign*. The Borel sum of the alternating series is *real*, which implies that the vacuum state is stable. Of course, the cubic Hamiltonian (5.3) is not Hermitian. However, we show in Sec. 5.4 how to calculate the 𝒞 operator for the theory to first order in the coupling constant g. The calculation is difficult, but if this calculation can be extended to higher orders in powers of the coupling constant, then this would demonstrate that the 𝒞 operator exists. The existence of the 𝒞 operator implies that the Hamiltonian is self-adjoint with respect to 𝒞𝒫𝒯 conjugation, which in turn implies that the Hamiltonian density (5.3) defines a unitary theory.

5.2.2   *Quartic 𝒫𝒯-symmetric quantum field theory*

One might think that extending the graphical analysis in Subsec. 5.2.1 from the cubic Hamiltonian (5.3) to the quartic Hamiltonian (5.4) would lead to the conclusion that a negative coupling constant causes the vacuum state for this theory to become unstable. However, we have already proved in Sec. 2.3 that an upside-down quartic quantum-mechanical potential has a stable ground state. Here, we extend that argument to quantum field theory and show that even though the coupling constant in (5.4) is negative, the vacuum state is stable.

The Feynman rules for the quartic field theory in (5.4) are $6g$ for a four-point vertex amplitude and $1/(p^2 + m^2)$ for a line amplitude in momentum space. (The connected vacuum graphs of order g and of order g^2 are shown in Fig. 5.2.) The graphical perturbation expansion has the form

$$\sum_{n=0}^{\infty} b_n g^n. \qquad (5.7)$$

All integer powers of the coupling constant g appear in this expansion and not just even powers as in (5.6).

Since all integer powers of g appear in (5.7), in a conventional $g\varphi^4$ theory this graphical perturbation series alternates in sign and is therefore

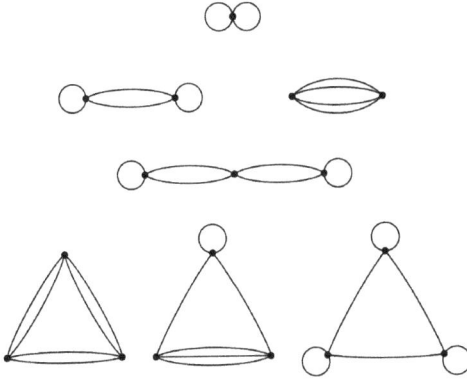

Fig. 5.2   Connected vacuum graphs of order g (two graphs) and of order g^2 (four graphs) that contribute to the vacuum energy density of a quartic quantum field theory. For a $g\varphi^4$ theory, vacuum graphs with all integer powers of g occur in the perturbation series (5.7). As in the case of a cubic quantum field theory, if g is real, the graphs all add in phase, so the coefficients b_n alternate in sign for a conventional $g\varphi^4$ theory but do *not* alternate in sign for a $\mathcal{PT}$-symmetric $-g\varphi^4$ theory. Thus, the graphical perturbation series for a $-g\varphi^4$ theory is not Borel summable and one might think that the energy of the vacuum state of a $-g\varphi^4$ theory is complex. However, the vacuum energy becomes real when one includes the nonperturbative (nongraphical) contributions to the vacuum energy. The result is that, like the vacuum state of a conventional $g\varphi^4$ theory, the vacuum state of a $-g\varphi^4$ theory is stable.

Borel summable. The Borel sum is real, and thus the energy of the vacuum state is real. Hence, the vacuum state of a conventional $g\varphi^4$ theory (which has a right-side-up potential) is stable. However, the coefficients b_n for a $\mathcal{PT}$-symmetric $-g\varphi^4$ theory do not alternate in sign. Therefore, like the perturbation series for the vacuum energy density of a conventional cubic quantum field theory, the Borel sum of the series is complex. There is a cut in the complex-g plane, and since the discontinuity across the cut is imaginary, the Borel sum of the series is complex.

It is quite remarkable that despite this result *the energy of the vacuum state of a $-g\varphi^4$ theory is still real!* This is because, in addition to the perturbative (graphical) contributions to the vacuum energy, the $\mathcal{PT}$-symmetric φ^4 theory also has *nonperturbative* contributions. The discontinuity across the cut in the Borel sum and these nonperturbative contributions are both pure imaginary and exponentially small. The amazing feature of quartic $\mathcal{PT}$-symmetric quantum field theory is that these two imaginary contributions exactly cancel. Thus, the vacuum energy for the upside-down quartic field theory is real, so the theory has a stable ground state. We explain the appearance of the nonperturbative terms for the simple case of a zero-dimensional quantum theory in the next subsection.

5.2.3 *Zero-dimensional PT-symmetric field theory*

To explain the origin of the nonperturbative contributions in PT-symmetric quartic quantum field theory, we consider the simple case of quantum field theory in zero-dimensional space-time. The partition function Z for such a theory is just a one-dimensional integral of the form $Z = \int d\varphi\, e^{-H}$, where H is the classical Hamiltonian. We examine first a *conventional* quartic field theory $H = \frac{1}{2}\varphi^2 + \frac{1}{4}g\varphi^4$ whose partition function is given by

$$Z = \int_{-\infty}^{\infty} d\varphi\, \exp\left(-\tfrac{1}{2}\varphi^2 - \tfrac{1}{4}g\varphi^4\right),$$

where the path of integration is the real-φ axis. To find the small-g perturbation expansion of Z, we expand $\exp\left(-\tfrac{1}{4}g\varphi^4\right)$ as a Taylor series in powers of g and then integrate term by term. The result is a divergent alternating series whose Borel sum is real. The coefficients in this expansion can also be obtained by calculating the vacuum graphs in Fig. 5.2.

A fancier way to obtain the weak-coupling expansion is to convert the integral to *Laplace form* [Bender and Orszag (1999)] by making the scaling transformation $\varphi = t/\sqrt{g}$. The resulting integral representation of the partition function is

$$Z = \frac{1}{\sqrt{g}} \int_{t=-\infty}^{\infty} dt\, \exp\left[-\frac{1}{g}f(t)\right], \tag{5.8}$$

where $f(t) = \frac{1}{2}t^2 + \frac{1}{4}t^4$. We then approximate this integral by using the *method of steepest descents* [Bender and Orszag (1999)]. We begin by locating the saddle points, which are the zeros of

$$f'(t) = t^3 + t.$$

There are three saddle points and these are located at $t = 0, i, -i$. (The saddle points are shown on Fig. 5.3.)

Next, we determine the constant-phase contours. The *phase* of the integral (5.8) [that is, the imaginary part of $f(t)$] vanishes at $t = 0$. Thus, if we let $t = x + iy$, we can express the equation for the curves of constant (and vanishing) phase as

$$xy + x^3y - xy^3 = 0.$$

Therefore, the curves of constant phase are the real-t axis $y = 0$, the imaginary-t axis $x = 0$, and the two hyperbolas $y = \pm\sqrt{x^2 + 1}$. These

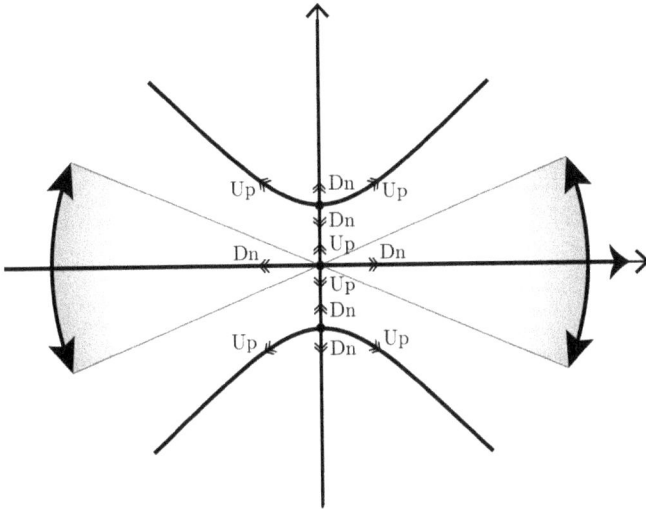

Fig. 5.3   Complex-t plane for the integral (5.8). Because the integrand is an analytic function of t, the integration contour, which begins at $t = -\infty$ and terminates at $t = \infty$, can be deformed into any path that originates in the left Stokes sector and terminates in the right Stokes sector. Since we treat g as a small positive parameter, we use the method of steepest descents to approximate the integral. There are three saddle points at $t = 0$, i, $-i$, which are shown as dots. The solid lines are paths of constant phase. The steepest-descent contour follows the real axis from $-\infty$ to $+\infty$, as indicated by the flattened arrow. This path passes through the saddle point at $t = 0$ and does not pass through the other two saddle points.

constant-phase paths are all shown as heavy solid lines on Fig. 5.3. In particular, there is a constant-phase contour in the complex-t plane that originates at $t = -\infty$, runs up the real-t axis, passes through the saddle point at $t = 0$, continues on up the real-t axis, and terminates at $t = +\infty$. Thus, the situation is extremely simple: The original path of integration in (5.8) is *already* a constant-phase contour and does not need to be deformed in order to follow a path of constant phase. This contour is a *steepest path* that goes *down* from the saddle point at $t = 0$ in two directions, namely, up and down the real-t axis.

Figure 5.3 shows that there are two other saddle points at $t = \pm i$. These saddle points lie on the network of constant-phase paths that include the saddle point at $t = 0$. However, these saddle points make no contribution to the asymptotic expansion of Z for small g because the path of steepest descent through $t = 0$ does not pass through these saddle points. Thus, these saddle points are called *irrelevant* saddle points. Only the *relevant* saddle point at $t = 0$ gives rise to the asymptotic series expansion of Z, and since this expansion is a series in powers of g, this saddle point is called a *perturbative* saddle point. The main conclusion from this analysis is that a conventional φ^4 quantum field theory (with a positive mass term and a positive self-interaction term) has no nonperturbative contributions; the perturbation series obtained by summing the Feynman graphs in Fig. 5.2 gives a complete description of the asymptotic behavior of the theory for small coupling constant g.

5.2.4 Saddle-point analysis of $\mathcal{PT}$-symmetric theories

Consider the $\mathcal{PT}$-symmetric class of zero-dimensional Hamiltonians

$$H = \tfrac{1}{2}\varphi^2 + \frac{1}{2+\varepsilon}g\varphi^2(i\varphi)^\varepsilon. \tag{5.9}$$

These Hamiltonians are $\mathcal{PT}$-deformed versions of the zero-dimensional quadratic Hamiltonian $H = \tfrac{1}{2}(1+g)\varphi^2$. The partition function Z for H in (5.9) is given by the integral $Z = \int d\varphi \, e^{-H}$, where the path of integration originates inside the left Stokes sector centered about the angle

$$\theta_{\text{center, left}} = -\pi + \frac{\pi\varepsilon}{2\varepsilon+4}$$

and terminates inside the right Stokes sector centered about the angle

$$\theta_{\text{center, right}} = -\frac{\pi\varepsilon}{2\varepsilon+4}.$$

These two Stokes sectors have the same opening angles $\Delta = \pi/(\varepsilon+2)$. For the undeformed theory $\varepsilon = 0$, the Stokes sectors are centered about the positive- and negative-real axes and have opening angles $\pi/2$. As ε increases, the Stokes sectors rotate downwards and become thinner.

Cubic zero-dimensional $\mathcal{PT}$-symmetric theory:
For the cubic case $\varepsilon = 1$ the Stokes sectors lie adjacent to and below the real axis and they have opening angles of $\pi/3$ (see Fig. 5.4). In this case we rescale the integration variable in (5.9) by $\varphi = t/g$ and obtain the following formula for the partition function:

$$Z = \frac{1}{g}\int_C dt \, \exp\left[-g^{-2}f(t)\right], \tag{5.10}$$

where $f(t) = \tfrac{1}{2}t^2 + \tfrac{1}{3}it^3$. The integral (5.10) converges if the path of integration C follows the real axis from $-\infty$ to ∞. This path lies at the upper edge of the Stokes sectors, so the cubic exponential in the integrand

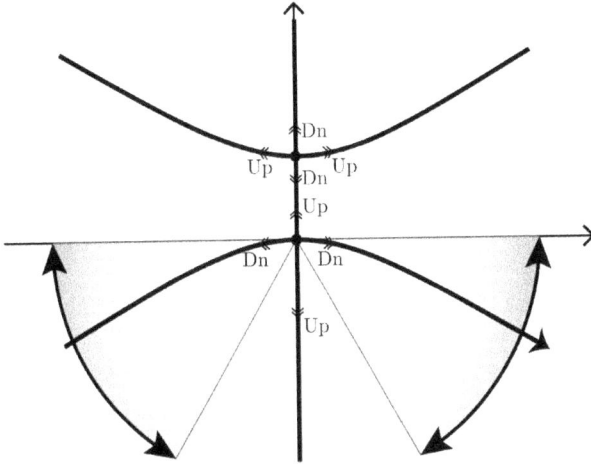

Fig. 5.4 Steepest-descent analysis of the partition function Z for the cubic zero-dimensional $\mathcal{PT}$-symmetric quantum field theory in (5.10). There are saddle points at $t = 0$ and $t = i$. Constant-phase curves are shown as heavy solid lines. The steepest-descent paths from the perturbative saddle point at $t = 0$ turn downwards and terminate in the centers of the Stokes sectors at 30° angles below the real axis. The saddle point at $t = i$ is irrelevant because it does not contribute to the asymptotic approximation of Z for small g.

is oscillatory, but the convergence is assured because of the negative-real quadratic term in the exponential. Note that if this theory were a conventional Hermitian cubic theory, the factor of i in the cubic exponential would be missing, and as a consequence, the integral would diverge. Thus, it is essential that this cubic quantum field theory be non-Hermitian and $\mathcal{PT}$ symmetric!

A quick way to find the small-g asymptotic behavior of Z in (5.10) is to expand the cubic exponential in a Taylor series and then to integrate term by term. The result is a divergent asymptotic series in powers of g^2. This is precisely the perturbation expansion that one obtains by summing the vacuum graphs in Fig. 5.1. However, let us use the method of steepest descents to examine this integral for small coupling constant. The equation for the saddle points is $f'(t) = t + it^2 = 0$. This equation reveals that there are two saddle points, which are located at $t = 0$ and $t = i$. These saddle points are shown in Fig. 5.4. The phase at the saddle point at $t = 0$ is 0, so we seek constant-phase paths along which the phase is identically 0. Setting $t = x + iy$, we find that there are three such paths, the y-axis and the two hyperbolas $x = \pm\sqrt{3y^2 - 3y}$. These are shown as heavy solid lines on Fig. 5.4. Only the lower hyperbola, which passes through the perturbative saddle point at $t = 0$, contributes to the asymptotic expansion of Z for small g; this steepest-descent path turns downward and terminates in the centers of the Stokes sectors. This steepest-descent contour avoids the saddle point at $t = i$, so this saddle point does not contribute to the asymptotic expansion of Z for small g and is an irrelevant saddle point.

Quartic zero-dimensional $\mathcal{PT}$-symmetric theory:

The quartic $\mathcal{PT}$-symmetric theory obtained by setting $\varepsilon = 2$ in (5.9) is especially interesting because the steepest-descent analysis of the integral reveals that Z has two relevant *nonperturbative* saddle points as well as a relevant perturbative saddle point. Thus, unlike the situation for a conventional quartic theory with a right-side-up potential, *Feynman graphs alone are not sufficient to obtain the behavior of this quartic field theory*. To treat the partition function for small g we introduce the scaling $\varphi = t/\sqrt{g}$ and obtain the integral representation

$$Z = \frac{1}{\sqrt{g}} \int_C dt \, \exp\left[-\frac{1}{g}f(t)\right], \qquad (5.11)$$

where $f(t) = \frac{1}{2}t^2 - \frac{1}{4}t^4$. For this integral to converge, the contour C must terminate in the Stokes sectors shown in Fig. 5.5.

Let us use the method of steepest descents to examine the integral (5.11). The equation for the saddle points, $f'(t) = t - t^3 = 0$, shows that there are three saddle points, which are located at $t = -1, 0, 1$. (These saddle points are shown in Fig. 5.5.) At the saddle point at $t = 0$ the phase is 0, so we seek constant-phase paths along which the phase is identically 0. Setting $t = x + iy$, we find that there are four such paths: the x axis $y = 0$, the y axis $x = 0$, and the two hyperbolas $x = \pm\sqrt{1 + y^2}$. These are shown as heavy solid lines on Fig. 5.5. This figure shows that the steepest-descent path that terminates in the Stokes sectors passes through *all three saddle points*, the perturbative saddle point at $t = 0$ and the two nonperturbative saddle points at $t = \pm 1$. The asymptotic expansion of Z for small g consists of two kinds of terms, a real perturbation expansion in powers of g, whose coefficients are precisely the sums of the vacuum graphs in Fig. 5.2, and a purely imaginary nonperturbative part that contains the exponentially small factor $ie^{1/g}$. The Borel sum of the graphical contributions contains an imaginary part that *exactly cancels* this nonperturbative part, leaving a *real* value for the partition function.

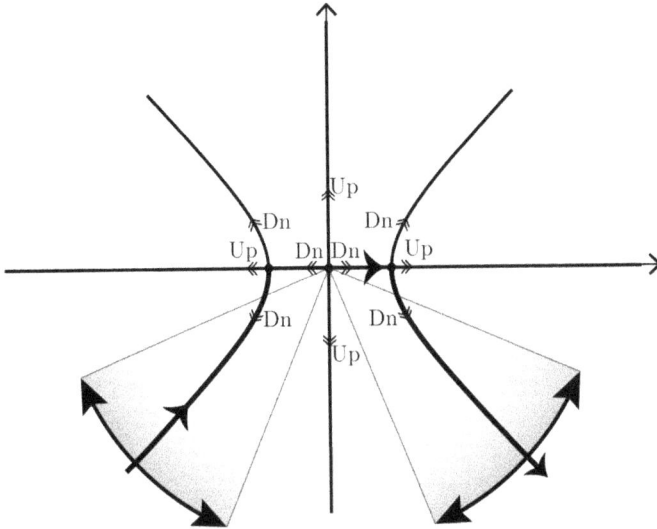

Fig. 5.5 Configuration in the complex-t plane for the steepest-descent analysis of the integral (5.11). There are three saddle points, all lying on the *real-t* axis (instead of on the imaginary-t axis, as in Fig. 5.3). The steepest-descent contour of integration passes through all three saddle points. This path originates in the left Stokes sector and rises up to the saddle point at $t = -1$. The path then makes a 90° right turn and runs along the real-t axis, where it passes through the perturbative saddle point at $t = 0$. Finally, the path makes another 90° right turn at the saddle point at $t = 1$, runs downward, and terminates in the center of the right Stokes sector.

Proving that this exact cancellation must occur is actually quite easy; all we need to show is that the integral in (5.11) is *real*, and this is so because the integral contour in Fig. 5.5 is $\mathcal{PT}$ symmetric (left-right symmetric). As a consequence, if the contribution to the integral along the left half of the contour has the form $A + iB$ (A and B real), then by symmetry the integral along the right half of the contour has the form $A - iB$. Thus, the full integral has the value $2A$, which is real.

This amazing and beautiful cancellation lies at the very heart of $\mathcal{PT}$-symmetric quantum field theory. This surgically precise cancellation of imaginary contributions occurs in *one-dimensional* quantum field theory (quantum mechanics) as well as in zero-dimensional quantum field theory; the proof that the energy is real in $D = 1$ is given in Chap. 6. It is not yet known at a rigorous level whether this exact cancellation persists in dimensions other than $D = 0$ and $D = 1$. A complete proof would be extremely difficult and so far it has only been verified perturbatively.

5.3 Nonvanishing One-Point Green's Function

A conventional φ^4 theory is parity invariant, and therefore the one-point Green's function G_1, which is the vacuum expectation value of the field, vanishes. However, the violation of parity symmetry in $\mathcal{PT}$-symmetric quantum field theories has a remarkable physical implication; namely, that the one-point Green's function G_1 does not vanish. To understand the nonvanishing of G_1 at a heuristic level, note that the classical trajectories of a $x^2(ix)^\varepsilon$ theory favor the *lower-half* complex plane (see, for example, Fig. 1.15). As a consequence, the one-point Green's function of a $\varphi^2(i\varphi)^\varepsilon$ theory is a negative-imaginary number.

This is easy to see in zero-dimensional quantum field theory. The one-point Green's function for the cubic theory in (5.3) is given explicitly by the ratio of integrals

$$G_1 = \int d\varphi\, \varphi e^{f(\varphi)} \Big/ \int d\varphi\, e^{f(\varphi)}, \qquad (5.12)$$

where $f(\varphi) = \frac{1}{2}\varphi^2 + \frac{1}{3}i\varphi^3$ and where the path of integration may be taken along the real-φ axis. By a simple symmetry argument (merely replace φ by $-\varphi$), we can see that G_1 is purely imaginary. This argument holds for any field theory with a self-interaction term of the form $\varphi^2(i\varphi)^\varepsilon$. For the cubic theory we can calculate G_1 as a graphical perturbation series (with no nonperturbative contributions); it is the sum of all Feynman diagrams having one external line.

The one-point Green's function for the quartic zero-dimensional $\mathcal{PT}$-symmetric field theory is given by the ratio of integrals in (5.12), where now $f(\varphi) = \frac{1}{2}\varphi^2 - \frac{1}{4}\varphi^4$ and the path of integration terminates in the Stokes sectors in Fig. 5.5. By the symmetry argument above, G_1 is nonvanishing and imaginary. However, unlike the case of a cubic theory, the nonvanishing of G_1 is a *purely nonperturbative effect*; G_1 cannot be calculated in terms of Feynman graphs because a quartic field theory has no graphs with one external line! Indeed, if we follow the steepest-descent path in Fig. 5.5, the contribution from the perturbative saddle point at $t = 0$ vanishes by oddness; the only nonzero contributions come from the nonperturbative saddle points. Thus, G_1 is exponentially small when $g > 0$ is small; it is of order $e^{-1/g}$.[3]

---

[3]A plausible conjectural consequence of the nonvanishing of G_1 is that the Higgs particle, which is a fundamental component of the Standard Model of particle physics, arises as a quantum parity anomaly (see Sec. 2.3).

The nonvanishing of the one-point Green's function in a quartic $-\varphi^4$ theory may seem surprising because the Hamiltonian density appears to be symmetric under the transformation $\varphi \to -\varphi$. The reason that $G_1 \neq 0$ is that the boundary conditions on the path-integral representation of the partition function are *not* symmetric under $\varphi \to -\varphi$ (see Figs. 5.5 and 5.6).

To present a clear picture of how a quartic PT-symmetric $g\varphi^4$ quantum field theory works, we calculate the one-point Green's function from the Dyson–Schwinger equations. We begin by giving an elementary derivation of these equations and then demonstrate that when we try to solve these equations by truncation, there is a parity-symmetric solution for the case of positive g, but a new parity-violating solution for negative g.

5.3.1 Derivation of the Dyson–Schwinger equations

The Dyson–Schwinger equations are an infinite set of coupled equations for the Green's functions of a quantum field theory. An easy way to derive these equations makes use of elementary formal functional methods. We begin with the Euclidean-space Lagrangian density in which we have coupled the field $\varphi(x)$ to an external c-number source $J(x)$:

$$\mathcal{L} = \tfrac{1}{2}(\partial\varphi)^2 + \tfrac{1}{2}m^2\varphi^2 - \frac{g}{2+\varepsilon}\varphi^2(i\varphi)^\varepsilon - J\varphi. \tag{5.13}$$

This Lagrangian density represents a self-interacting scalar quantum field theory in D-dimensional Euclidean space-time. If we vary the action (which is the integral of $\mathcal{L}$) with respect to $\varphi(x)$, we obtain the field equation

$$- \partial^2\varphi(x) + m^2\varphi(x) - ig[i\varphi(x)]^{\varepsilon+1} = J(x). \tag{5.14}$$

Next, leaving the source turned on, we take the expectation value of the field equation (5.14) in the vacuum state of the theory $|0\rangle$ and divide by the vacuum functional $Z[J] = \langle 0|0\rangle$:

$$- \partial^2 G_1^{(J)}(x) + m^2 G_1^{(J)}(x) - gi^{2+\varepsilon}\langle 0|\varphi^{1+\varepsilon}(x)|0\rangle/\langle 0|0\rangle = J(x). \tag{5.15}$$

Here, $G_1^{(J)}(x)$ is the one-point Green's function in the presence of the external source:

$$G_1^{(J)}(x) \equiv \langle 0|\varphi(x)|0\rangle/\langle 0|0\rangle. \tag{5.16}$$

The function $J(x)$ appears alone on the right side of (5.15) because it is a c-number and therefore can be factored out of matrix elements.

The objective is now to use (5.15) to calculate the Green's functions of the quantum field theory. The *connected* Green's functions in the presence of the source J are functional derivatives of the logarithm of the partition function $Z[J]$ with respect to the source $J(x)$:[4]

$$G_n^{(J)}(x_1, x_2, \ldots, x_n) \equiv \frac{\delta^n}{\delta J(x_1)\delta J(x_2)\cdots\delta J(x_n)} \ln(Z[J]). \qquad (5.17)$$

We now turn off the source; that is, we set $J(x) \equiv 0$:

$$G_n(x_1, x_2, \ldots, x_n) = G_n^{(J)}(x_1, x_2, \ldots, x_n)\Big|_{J\equiv0}. \qquad (5.18)$$

Turning off the source restores translation invariance. Therefore, the one-point Green's function G_1 is a constant independent of x; the two-point Green's function depends only on the difference $x - y$, $G_2(x, y) = G_2(x-y)$; $G_3(x, y, z) = G_3(x - y, x - z)$ depends on two differences; and so on.

To proceed, we must express the third term in (5.15) in terms of the connected Green's functions of the theory. To do so, we note that functional differentiation with respect to $J(x)$ is equivalent to inserting $\varphi(x)$ into the matrix elements [Bender *et al.* (1976)].

Let us consider the simple case $\varepsilon = 1$. In this case it is necessary to calculate the quantity $\langle 0|\varphi^2(x)|0\rangle$. We begin with (5.16) multiplied by $Z[J]$:

$$G_1^{(J)}(x)Z[J] = \langle 0|\varphi(x)|0\rangle. \qquad (5.19)$$

Taking the functional derivative of this equation with respect to $J(x)$ gives

$$[G_1^{(J)}(x)]^2 Z[J] + G_2^{(J)}(x, x)Z[J] = \langle 0|\varphi^2(x)|0\rangle. \qquad (5.20)$$

Hence, we can eliminate $\langle 0|\varphi^2(x)|0\rangle$ from (5.15) and get

$$-\partial^2 G_1^{(J)}(x) + m^2 G_1^{(J)}(x) + gi\left([G_1^{(J)}(x)]^2 + G_2^{(J)}(x, x)\right) = J(x). \qquad (5.21)$$

We now obtain the *first* of the Dyson–Schwinger equations by setting $J \equiv 0$ (that is, by turning off the source):

$$m^2 G_1 + gi\left[G_1^2 + G_2(0)\right] = 0.$$

Remember that by translation invariance G_1 is a constant, so its derivative vanishes. Also, $G_2(0) = G_2(x - x) = G_2(x, x)$.

To obtain the *second* Dyson–Schwinger equation for the case $\varepsilon = 1$, we take a functional derivative of (5.21) with respect to $J(y)$ and get

$$-\partial^2 G_2^{(J)}(x, y) + m^2 G_2^{(J)}(x, y) + gi\left[2G_1^{(J)}(x)G_2^{(J)}(x, y)\right.$$
$$\left. + G_3^{(J)}(x, x, y)\right] = \delta(x - y).$$

---

[4]The connected Green's functions of a quantum field theory are the connected part of the vacuum expectation value of the time-ordered product of the fields. In a perturbative expansion this is just the sum of the connected n-point Feynman diagrams.

We then set $J \equiv 0$ and obtain:

$$-\partial^2 G_2(x-y) + m^2 G_2(x-y)$$
$$+gi\left[2G_1 G_2(x-y) + G_3(0, x-y)\right] = \delta(x-y).$$

By continuing the process of functionally differentiating with respect to J and setting $J \equiv 0$, we obtain an infinite tower of coupled differential equations. This is the full set of Dyson–Schwinger equations. For example, the third in the sequence is

$$-\partial^2 G_3(x-y, x-z) + m^2 G_3(x-y, x-z) + gi[G_4(0, x-y, x-z)$$
$$+2G_1 G_3(x-y, x-z) + 2G_2(x-z)G_2(x-y)] = 0.$$

As another illustration, we derive the first four Dyson–Schwinger equations from (5.15) for the case $\varepsilon = 2$. Using the same approach as we did for the case $\varepsilon = 1$, we begin by re-expressing $\langle 0|\phi^3(x)|0\rangle$. We do this by taking the functional derivative of (5.20) with respect to $J(x)$ to obtain

$$[G_1^{(J)}(x)]^3 Z[J] + 3G_1^{(J)}(x)G_2^{(J)}(x,x)Z[J] + G_3^{(J)}(x,x,x)Z[J] = \langle 0|\phi^3(x)|0\rangle.$$

Substituting this result into the $\varepsilon = 2$ version of (5.15), we get

$$-\partial^2 G_1^{(J)}(x) + m^2 G_1^{(J)}(x) - g\left([G_1^{(J)}(x)]^3\right.$$
$$\left. +3G_1^{(J)}(x)G_2^{(J)}(x,x) + G_3^{(J)}(x,x,x)\right) = J(x). \qquad (5.22)$$

We obtain the first Dyson–Schwinger equation for $\varepsilon = 2$ by setting $J \equiv 0$:

$$m^2 G_1 - g\left[G_1^3 + 3G_1 G_2(0) + G_3(0,0)\right] = 0.$$

Following the same procedure as for $\varepsilon = 1$, we take a functional derivative of (5.22) with respect to $J(y)$ and get

$$-\partial^2 G_2^{(J)}(x,y) + m^2 G_2^{(J)}(x,y) - g\left(3[G_1^{(J)}(x)]^2 G_2^{(J)}(x,y)\right.$$
$$+3G_1^{(J)}(x)G_3^{(J)}(x,x,y) + 3G_2^{(J)}(x,x)G_2^{(J)}(x,y)$$
$$\left. +G_4^{(J)}(x,x,x,y)\right) = \delta(x-y). \qquad (5.23)$$

Thus, setting $J \equiv 0$ gives the second Dyson–Schwinger equation for $\varepsilon = 2$:

$$-\partial^2 G_2(x-y) + m^2 G_2(x-y) - g[3G_1^2 G_2(x-y) + G_4(0,0,x-y)$$
$$+3G_1 G_3(0, x-y) + 3G_2(0)G_2(x-y)] = \delta(x-y).$$

We repeat this process by functionally differentiating (5.23) with respect to $J(z)$ and set $J \equiv 0$ to get the third Dyson–Schwinger equation for $\varepsilon = 2$:

$$0 = -\partial^2 G_3(x-y, x-z) + m^2 G_3(x-y, x-z) - g[3G_1^2 G_3(x-y, x-z)$$
$$+6G_1 G_2(x-y)G_2(x-z) + 3G_1 G_4(0, x-y, x-z)$$
$$+3G_2(x-z)G_3(0, x-y) + 3G_2(0)G_3(x-y, x-z)$$
$$+3G_2(x-y)G_3(0, x-z) + G_5(0,0, x-y, x-z)].$$

The fourth Dyson–Schwinger equation is

$$
\begin{aligned}
0 = &-\partial^2 G_4(x-y, x-z, x-w) + m^2 G_4(x-y, x-z, x-w) \\
&- g[6G_1 G_2(x-z)G_3(x-y, x-w) + 6G_1 G_2(x-y)G_3(x-z, x-w) \\
&+ 6G_2(x-y)G_2(x-z)G_2(x-w) + 6G_1 G_2(x-w)G_3(x-y, x-z) \\
&+ 3G_1^2 G_4(x-y, x-z, x-w) + 3G_1 G_5(0, x-y, x-z, x-w) \\
&+ 3G_2(x-w)G_4(0, x-y, x-z) + 3G_2(x-z)G_4(0, x-y, x-w) \\
&+ 3G_3(0, x-y)G_3(x-z, x-w) + 3G_2(x-y)G_4(0, x-z, x-w) \\
&+ 3G_3(0, x-z)G_3(x-y, x-w) + 3G_3(0, x-w)G_3(x-y, x-z) \\
&+ 3G_2(0)G_4(x-y, x-z, x-w) + G_6(0, 0, x-y, x-z, x-w)].
\end{aligned}
$$

5.3.2   *Truncation of the Dyson–Schwinger equations*

If it is truncated, the system of coupled Dyson–Schwinger equations is incomplete because there are too many unknowns. The first equation contains G_1 and the higher Green's functions G_2, G_3, ...; the second contains G_1, G_2, G_3, G_4, and so on. Since each new equation contains new unknown Green's functions, the system never closes. To make progress we must *force* the system to close. A simple and effective way to do so is to set all higher Green's functions to zero. This systematic truncation procedure has been shown to give highly accurate results for the Green's functions and energy levels of conventional Hermitian quantum-mechanical problems [Bender *et al.* (1976)]. This truncation of the Dyson–Schwinger equations is a variational procedure; including more and more of the higher Green's functions is equivalent to enlarging the space of variational parameters.

Although this truncation procedure is useful, it leads to coupled nonlinear differential-equation systems. We illustrate the procedure here for the case of the cubic and quartic theories by considering only the simplest truncation in which we keep just the first two of the Dyson–Schwinger equations and set $G_n = 0$ for all $n > 2$. In this case, for integer values of ε we get an elementary set of equations to solve for the first two Green's functions. For $\varepsilon = 1$ the first two truncated Dyson–Schwinger equations are

$$
m^2 G_1 + g i G_1^2 + g i G_2(0) = 0,
$$

$$
-\partial^2 G_2(x-y) + \left[m^2 + 2g i G_1\right] G_2(x-y) = \delta(x-y), \qquad (5.24)
$$

and for $\varepsilon = 2$ the first two Dyson–Schwinger equations are

$$
m^2 G_1 - g G_1^3 - 3g G_1 G_2(0) = 0,
$$

$$
-\partial^2 G_2(x-y) + \left[m^2 - 3g G_1^2 + 3g G_2(0)\right] G_2(x-y) = \delta(x-y). \quad (5.25)
$$

In each of the two systems of equations (5.24) and (5.25) the expressions in square brackets represent a shift of the square of the unrenormalized mass m; we interpret this as a *mass renormalization*. Thus, in each case we replace the expression in square brackets by the term M^2, where M is the *renormalized mass*. The second member of each of the above two equations then has the generic form

$$- \partial^2 G_2(x - y) + M^2 G_2(x - y) = \delta(x - y). \tag{5.26}$$

Note that if we hypothesize that $G_1 = -iA$, where A is positive, then the mass shift is *real and positive*.

To solve (5.26) we take the Fourier transform and obtain

$$\tilde{G}_2(p) = \frac{1}{p^2 + M^2}.$$

In one-dimensional space-time the inverse Fourier transform is easy to evaluate, and we get

$$G_2(x) = \frac{1}{2M} e^{-M|x|}.$$

Thus, we see that $G_2(0) = \frac{1}{2M}$; substituting this result into the first equation in each of (5.24) and (5.25) gives an algebraic equation for the one-point Green's function. If we solve this equation, we verify our hypothesis that A is positive.

A conventional quartic field theory is parity invariant. This implies that $G_1 \equiv 0$, so the first equation of (5.25) simply disappears and we get a consistent solution to the truncated Dyson–Schwinger equations. However, for the PT-symmetric quartic theory, if we assume that $G_1 = 0$, we obtain a negative value for M^2, which is inconsistent. Thus, there are two completely different solutions for the quartic field theory depending on the sign of g; in the conventional Hermitian theory ($g > 0$) all of the odd Green's functions vanish, and in the unconventional PT-symmetric theory ($g < 0$), G_1 has a negative-imaginary value.

5.4 C Operator for Cubic PT-Symmetric Field Theory

To verify the unitarity of a PT-symmetric quantum theory, one must calculate the C operator. This is not easy to do in quantum field theory. However, we show in this section how to perform such a calculation at the perturbative level.

5.4.1 $i\varphi^3$ *quantum field theory*

In this subsection we perform a perturbative calculation of the $\mathcal{C}$ operator for the quantum-field-theoretic Hamiltonian (5.3) [Bender *et al.* (2004d)]. By calculating $\mathcal{C}$ perturbatively (treating $g = \epsilon$ as small), we obtain a fully acceptable Lorentz-invariant quantum field theory. This calculation shows that in principle it is possible to construct perturbatively the Hilbert space in which the Hamiltonian for this cubic field theory in $(D+1)$-dimensional Minkowski space-time is *self-adjoint*. Thus, the imaginary self-coupled cubic theory has a positive spectrum and exhibits unitary time evolution.

For this calculation we apply the powerful algebraic techniques explained in Subsecs. 3.8.1 and 3.8.2 for the calculation of the $\mathcal{C}$ operator in quantum mechanics. As in quantum mechanics, we express $\mathcal{C}$ in the form $\mathcal{C} = \exp\left(\epsilon Q_1 + \epsilon^3 Q_3 + \ldots\right)\mathcal{P}$, where now Q_{2n+1} ($n = 0, 1, 2, \ldots$) are real functionals of the fields $\varphi_{\mathbf{x}}$ and $\pi_{\mathbf{x}}$.

To find Q_n for $\mathcal{H}$ in (5.3) we must solve a system of operator equations. We begin by making an *ansatz* for Q_1 analogous to that in (3.33):

$$Q_1 = \iiint d\mathbf{x}\,d\mathbf{y}\,d\mathbf{z}\left(M_{(\mathbf{xyz})}\pi_{\mathbf{x}}\pi_{\mathbf{y}}\pi_{\mathbf{z}} + N_{\mathbf{x}(\mathbf{yz})}\varphi_{\mathbf{y}}\pi_{\mathbf{x}}\varphi_{\mathbf{z}}\right). \qquad (5.27)$$

Note that in quantum mechanics M and N are constants, but in field theory they are *functions*. The notation $M_{(\mathbf{xyz})}$ indicates that this function is totally symmetric in its three arguments, and the notation $N_{\mathbf{x}(\mathbf{yz})}$ indicates that this function is symmetric under the interchange of the second and third arguments. The unknown functions M and N are *form factors*; they describe the spatial distribution of three-point interactions of the field variables in Q_1. The nonlocal spatial interaction of the fields is an intrinsic property of $\mathcal{C}$. (Note that we have suppressed the time variable t in the fields, and we use subscripts to indicate the spatial dependence.)

To determine M and N we substitute Q_1 into the first equation in (3.32), namely $[H_0, Q_1] = -2H_1$, which now takes the form

$$\left[\int d\mathbf{x}\,\pi_{\mathbf{x}}^2 + \iint d\mathbf{x}\,d\mathbf{y}\,\varphi_{\mathbf{x}}G_{\mathbf{xy}}^{-1}\varphi_{\mathbf{y}}, Q_1\right] = -4i\int d\mathbf{x}\,\varphi_{\mathbf{x}}^3,$$

where the inverse Green's function is given by $G_{\mathbf{xy}}^{-1} \equiv (\mu^2 - \nabla_{\mathbf{x}}^2)\delta(\mathbf{x} - \mathbf{y})$. We obtain the following coupled system of partial differential equations:

$$(\mu^2 - \nabla_{\mathbf{x}}^2)N_{\mathbf{x}(\mathbf{yz})} + (\mu^2 - \nabla_{\mathbf{y}}^2)N_{\mathbf{y}(\mathbf{xz})} + (\mu^2 - \nabla_{\mathbf{z}}^2)N_{\mathbf{z}(\mathbf{xy})}$$
$$= -6\delta(\mathbf{x} - \mathbf{y})\delta(\mathbf{x} - \mathbf{z}),$$
$$N_{\mathbf{x}(\mathbf{yz})} + N_{\mathbf{y}(\mathbf{xz})} = 3(\mu^2 - \nabla_{\mathbf{z}}^2)M_{(\mathbf{wxy})}. \qquad (5.28)$$

We solve (5.28) by Fourier-transforming to momentum space. We get

$$M_{(\mathbf{xyz})} = -\frac{4}{(2\pi)^{2D}} \int\int d\mathbf{p}\, d\mathbf{q}\, \frac{e^{i(\mathbf{x}-\mathbf{y})\cdot\mathbf{p}+i(\mathbf{x}-\mathbf{z})\cdot\mathbf{q}}}{D(\mathbf{p},\mathbf{q})}, \tag{5.29}$$

where $D(\mathbf{p},\mathbf{q}) = 4[\mathbf{p}^2\mathbf{q}^2 - (\mathbf{p}\cdot\mathbf{q})^2] + 4\mu^2(\mathbf{p}^2 + \mathbf{p}\cdot\mathbf{q} + \mathbf{q}^2) + 3\mu^4 > 0$, and

$$N_{\mathbf{x}(\mathbf{yz})} = -3\left(\nabla_{\mathbf{y}}\cdot\nabla_{\mathbf{z}} + \tfrac{1}{2}\mu^2\right) M_{(\mathbf{xyz})}. \tag{5.30}$$

For the special case of a $(1 + 1)$-dimensional quantum field theory, the integral in (5.29) evaluates to $M_{(\mathbf{xyz})} = -K_0(\mu R)/\left(\sqrt{3}\pi\mu^2\right)$, where K_0 is the associated Bessel function and $R^2 = \tfrac{1}{2}[(\mathbf{x} - \mathbf{y})^2 + (\mathbf{y} - \mathbf{z})^2 + (\mathbf{z} - \mathbf{x})^2]$. Thus, we have calculated $\mathcal{C} = e^{\epsilon Q_1}\mathcal{P}$ to first order in perturbation theory.

The $\mathcal{C}$ operator for this cubic field theory transforms as a *scalar* under the homogeneous Lorentz group [Bender *et al.* (2005a)]. Because the Hamiltonian H_0 for the unperturbed theory ($g = 0$) commutes with the parity operator $\mathcal{P}$, the intrinsic parity operator $\mathcal{P}_I$ in the noninteracting theory transforms as a Lorentz scalar.[5] When the coupling constant g is nonzero, the parity symmetry of H is broken and $\mathcal{P}_I$ is no longer a scalar. However, $\mathcal{C}$ *is* a scalar. Since $\lim_{g\to 0}\mathcal{C} = \mathcal{P}_I$, one can interpret the $\mathcal{C}$ operator in quantum field theory as the complex extension of the intrinsic parity operator when the imaginary coupling constant is turned on. This means that $\mathcal{C}$ is frame invariant and it shows that the $\mathcal{C}$ operator plays a truly fundamental role in non-Hermitian quantum field theory.

5.4.2 *Other cubic quantum field theories*

We can repeat the calculation done in Subsec. 5.4.1 for cubic quantum field theories having several interacting scalar fields [Bender *et al.* (2004d,c)]. Consider first the case of *two* scalar fields $\varphi_{\mathbf{x}}^{(1)}$ and $\varphi_{\mathbf{x}}^{(2)}$ whose interaction is governed by

$$H = H_0^{(1)} + H_0^{(2)} + i\epsilon \int d\mathbf{x}\, \left[\varphi_{\mathbf{x}}^{(1)}\right]^2 \varphi_{\mathbf{x}}^{(2)},$$

which is the analog of the quantum-mechanical theory described by H in (3.35). Here,

$$H_0^{(j)} = \tfrac{1}{2} \int d\mathbf{x}\, \left[\pi_{\mathbf{x}}^{(j)}\right]^2 + \tfrac{1}{2} \int\int d\mathbf{x}\, d\mathbf{y}\, \left[G_{\mathbf{xy}}^{(j)}\right]^{-1} \varphi_{\mathbf{x}}^{(j)} \varphi_{\mathbf{y}}^{(j)}.$$

---

[5]The *intrinsic* parity operator $\mathcal{P}_I$ and the parity operator $\mathcal{P}$ have the same effect on the fields except that $\mathcal{P}_I$ does not reverse the sign of the spatial argument of the field. In both Hermitian and $\mathcal{PT}$-symmetric quantum mechanics $\mathcal{P}$ and $\mathcal{P}_I$ are indistinguishable.

To determine $\mathcal{C} = e^{\epsilon Q_1}\mathcal{P}$ to order ϵ we introduce the *ansatz*

$$Q_1 = \int\!\!\int\!\!\int dx\,dy\,dz \Big[N^{(1)}_{\mathbf{x(yz)}} \left(\pi^{(1)}_{\mathbf{z}}\varphi^{(1)}_{\mathbf{y}} + \varphi^{(1)}_{\mathbf{y}}\pi^{(1)}_{\mathbf{z}} \right) \varphi^{(2)}_{\mathbf{x}}$$

$$+ N^{(2)}_{\mathbf{x(yz)}} \pi^{(2)}_{\mathbf{x}}\varphi^{(1)}_{\mathbf{y}}\varphi^{(1)}_{\mathbf{z}} + M_{\mathbf{x(yz)}} \pi^{(2)}_{\mathbf{x}}\pi^{(1)}_{\mathbf{y}}\pi^{(1)}_{\mathbf{z}} \Big], \tag{5.31}$$

where $M_{\mathbf{x(yz)}}$, $N^{(1)}_{\mathbf{x(yz)}}$, and $N^{(2)}_{\mathbf{x(yz)}}$ are unknown functions and the parentheses indicate symmetrization. We find that

$$M_{\mathbf{x(yz)}} = -G_m(R_1)\,G_{\mu_2}(R_2),$$

$$N^{(1)}_{\mathbf{x(yz)}} = -\delta(2\mathbf{x} - \mathbf{y} - \mathbf{z})G_m(R_1),$$

$$N^{(2)}_{\mathbf{x(yz)}} = \tfrac{1}{2}\delta(2\mathbf{x} - \mathbf{y} - \mathbf{z})G_m(R_1) - \delta(\mathbf{y} - \mathbf{z})G_{\mu_2}(R_2).$$

Here, $G_\mu(r) = \frac{r}{\mu}\left(\frac{\mu}{2\pi r}\right)^{D/2} K_{-1+D/2}(\mu r)$, where $r = |\mathbf{r}|$, is the Green's function in D-dimensional space, $m^2 = \mu_1^2 - \tfrac{1}{4}\mu_2^2$, $R_1^2 = (\mathbf{y} - \mathbf{z})^2$, and $R_2^2 = \tfrac{1}{4}(2\mathbf{x} - \mathbf{y} - \mathbf{z})^2$.

Next, consider the case of *three* interacting scalar fields whose dynamics is governed by

$$H = H_0^{(1)} + H_0^{(2)} + H_0^{(3)} + i\epsilon \int dx\, \varphi^{(1)}_{\mathbf{x}}\varphi^{(2)}_{\mathbf{x}}\varphi^{(3)}_{\mathbf{x}},$$

which is the analog of H in (3.37). We make the *ansatz*

$$Q_1 = \int\!\!\int\!\!\int dx\,dy\,dz \Big[N^{(1)}_{\mathbf{xyz}}\pi^{(1)}_{\mathbf{x}}\varphi^{(2)}_{\mathbf{y}}\varphi^{(3)}_{\mathbf{z}} + N^{(2)}_{\mathbf{xyz}}\pi^{(2)}_{\mathbf{x}}\varphi^{(3)}_{\mathbf{y}}\varphi^{(1)}_{\mathbf{z}}$$

$$+ N^{(3)}_{\mathbf{xyz}}\pi^{(3)}_{\mathbf{x}}\varphi^{(1)}_{\mathbf{y}}\varphi^{(2)}_{\mathbf{z}} + M_{\mathbf{xyz}}\pi^{(1)}_{\mathbf{x}}\pi^{(2)}_{\mathbf{y}}\pi^{(3)}_{\mathbf{z}} \Big]. \tag{5.32}$$

The solutions for the unknown functions are as follows: $M_{\mathbf{xyz}}$ is given by the integral (5.29) where $D(\mathbf{p},\mathbf{q}) = 4[\mathbf{p}^2\mathbf{q}^2 - (\mathbf{p}\cdot\mathbf{q})^2] + 4[\mu_1^2(\mathbf{q}^2 + \mathbf{p}\cdot\mathbf{q}) + \mu_2^2(\mathbf{p}^2 + \mathbf{p}\cdot\mathbf{q}) - \mu_3^2\mathbf{p}\cdot\mathbf{q}] + \mu^4$ and $\mu^4 = 2\mu_1^2\mu_2^2 + 2\mu_1^2\mu_3^2 + 2\mu_2^2\mu_3^2 - \mu_1^4 - \mu_2^4 - \mu_3^4$. The N coefficients are expressed as derivatives acting on M:

$$N^{(1)}_{\mathbf{xyz}} = \left[4\nabla_{\mathbf{y}}\cdot\nabla_{\mathbf{z}} + 2(\mu_2^2 + \mu_3^2 - \mu_1^2) \right] M_{\mathbf{xyz}},$$

$$N^{(2)}_{\mathbf{xyz}} = \left[-4\nabla_{\mathbf{y}}\cdot\nabla_{\mathbf{z}} - 4\nabla_{\mathbf{z}}^2 + 2(\mu_1^2 + \mu_3^2 - \mu_2^2) \right] M_{\mathbf{xyz}},$$

$$N^{(3)}_{\mathbf{xyz}} = \left[-4\nabla_{\mathbf{y}}\cdot\nabla_{\mathbf{z}} - 4\nabla_{\mathbf{y}}^2 + 2(\mu_1^2 + \mu_2^2 - \mu_3^2) \right] M_{\mathbf{xyz}}.$$

Again, this calculation of $\mathcal{C}$ shows that these cubic $\mathcal{PT}$-symmetric field theories are fully consistent quantum theories.[6]

---

[6]In [Bender *et al.* (2005a,b)] it is shown that the $\mathcal{C}$ operator in quantum field theory has the form $\mathcal{C} = e^Q \mathcal{P}_{\mathrm{I}}$, where $\mathcal{P}_{\mathrm{I}}$ is the *intrinsic* parity operator. The difference between $\mathcal{P}$ and $\mathcal{P}_{\mathrm{I}}$ is that $\mathcal{P}_{\mathrm{I}}$ does not reflect the spatial arguments of the fields. For a cubic interaction Hamiltonian this distinction is technical. It does not affect the final results for the Q operator in (5.27), (5.31), and (5.32).

5.5   Bound States in a $\mathcal{PT}$-Symmetric Quartic Potential

In a Hermitian $g\varphi^4$ ($g > 0$) quantum field theory the force between two particles is repulsive. However, in a $\mathcal{PT}$-symmetric quartic field theory the coupling constant g is negative, so the force between two particles is attractive. This attractive force is a direct physical consequence of the parity anomaly discussed in Sec. 2.3. The immediate physical effect of this parity anomaly is the appearance of quantum-field-theoretic bound states.

To elucidate the connection between the parity anomaly and bound states, we consider the Hamiltonian

$$\hat{H} = \tfrac{1}{2m}\hat{p}^2 + \tfrac{1}{2}\mu^2\hat{x}^2 - g\hat{x}^4, \tag{5.33}$$

which is a scaled version of the Hamiltonian (2.41), and treat this Hamiltonian as defining a quantum field theory in one-dimensional space-time.[7]

In particle physics a *bound state* is a state having a negative binding energy. In the context of quantum mechanics we define a bound state as follows: Let the energy levels of the Hamiltonian be E_n ($n = 0, 1, 2, \ldots$). The *renormalized mass M* is the *mass gap* $M = E_1 - E_0$. Higher excitations are measured relative to the vacuum energy, $E_n - E_0$ ($n = 2, 3, 4, \ldots$), so the nth excitation is a bound state if the binding energy

$$B_n \equiv E_n - E_0 - nM \tag{5.34}$$

is negative. If B_n is positive, then the state is *unbound* because it can decay into n one-particle states of mass M in the presence of an external field.

The differential-equation analysis in Sec. 2.3 yields the equivalent Hermitian Hamiltonian [Buslaev and Grecchi (1993); Jones and Mateo (2006)]

$$\tilde{\hat{h}} = \frac{\hat{p}^2}{2m} - \hbar\sqrt{\frac{2g}{m}}\,\hat{x} + 4g\left(\hat{x}^2 - \frac{\mu^2}{8g}\right)^2. \tag{5.35}$$

The linear anomaly term $\hat{x}$ in this Hamiltonian gives rise to field-theoretic bound states, so the $\mathcal{PT}$-symmetric Hamiltonian (5.33) represents particles having attractive forces. Specifically, in [Bender *et al.* (2001a)] it was demonstrated numerically that for small positive values of g the first few states of H in (5.33) are bound. Surprisingly, as the coupling strength g increases, the binding effect gets *weaker* and the number of bound states *decreases* until when g/μ^3 exceeds the critical value 0.0465, there are no bound states at all.[8]

---

[7]It would be interesting to study these bound states in arbitrary dimension by using the Bethe-Salpeter equation. Such a study has not yet been done.

[8]In [Bender *et al.* (2001a)] a heuristic argument was given to explain why there is such a critical value. This argument is heuristic because the non-Hermitian Hamiltonian is evaluated for complex x. As explained in Chap. 1, when x is complex, one cannot use order relationships such as $>$ or $<$, which only apply to real numbers.

Because $\hat{H}$ in (5.33) has the same spectrum as the Hermitian Hamiltonian in (5.35), it is easy to explain the appearance of bound states and to show that the bound states are a direct consequence of the linear anomaly term. To probe the influence of the anomaly, we generalize (5.35) by inserting a dimensionless parameter ϵ to measure the strength of the anomaly:

$$\hat{H}(\epsilon) = \tfrac{1}{2}\hat{p}^2 - \epsilon\sqrt{2g}\,\hat{x} + 4g\left(\hat{x}^2 - \tfrac{1}{8g}\right)^2, \tag{5.36}$$

where for simplicity we have set $m = \mu = \hbar = 1$.

If we set $\epsilon = 0$ in (5.36), the anomaly term vanishes and the potential is a *symmetric* double well. The mass gap (the renormalized mass M) for a double well is exponentially small because it is the result of tunneling between the wells. Thus, B_n in (5.34) is positive and there are no bound states. In Fig. 5.6 we display the double-well potential and the first 18 states of the system when $g = 0.008$ and $\epsilon = 0$. All states are unbound.

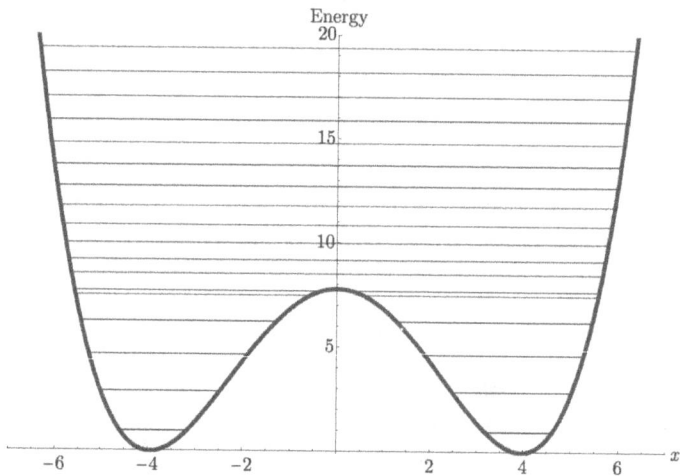

Fig. 5.6 Potential of the Hermitian Hamiltonian (5.36) plotted as a function of the real variable x for the case $\epsilon = 0$ and $g = 0.008$. The first 18 energy levels are indicated by horizontal lines. Because $\epsilon = 0$, there is no anomaly and the double-well potential is symmetric. Therefore, the mass gap is small and there are no bound states.

If the value of ϵ is increased to 1, the double-well potential becomes asymmetric and the two lowest states are not approximately degenerate. Therefore, bound states can occur near the bottom of the potential well. Higher-energy states eventually become unbound because, as we know from WKB theory, in a quartic well the nth energy level grows like $n^{4/3}$ for large

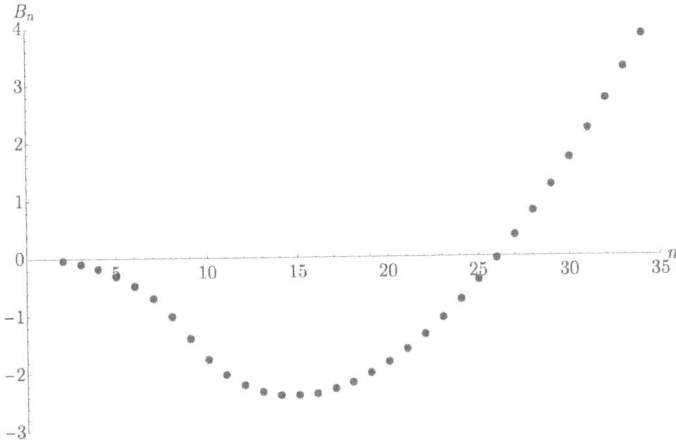

Fig. 5.7   Binding energies $B_n = E_n - E_0 - nM$ plotted as a function of n for $g = 0.008$ and $\epsilon = 1$ for the Hamiltonian (5.36). A negative value of B_n indicates a bound state. There are 25 bound states for these parameter values. Note that B_n varies smoothly as a function of n.

n. A nice way to display bound states is to plot the value of the binding energy B_n as a function of n. In Fig. 5.7 we display the bound states for $\epsilon = 1$ and for $g = 0.008$. For these values there are 25 bound states. Observe that the binding energy B_n is a smooth function of n.

As g becomes larger, the number of bound states becomes smaller because the depth of the double well decreases, and for large enough g there are no bound states at all. Figure 5.8 displays the potential for $\epsilon = 1$ and for $g = 0.046$. When we increase the value of g from 0.008 to $g = 0.046$, we find that there is just one bound state.

It is noteworthy that the bound-state spectrum depends so sensitively on the strength of the anomaly term in the Hamiltonian (5.36). If ϵ is slightly less than 1, the first few states become unbound. If ϵ is slightly greater than 1, the binding energy B_n is not a smooth function of n for small n. This demonstrates that the bound-state spectrum is highly sensitive to the value of ϵ in the anomaly term.

5.6   Lee Model

In this section we use the tools that we have developed to study non-Hermitian $\mathcal{PT}$-symmetric quantum theories to examine a famous early

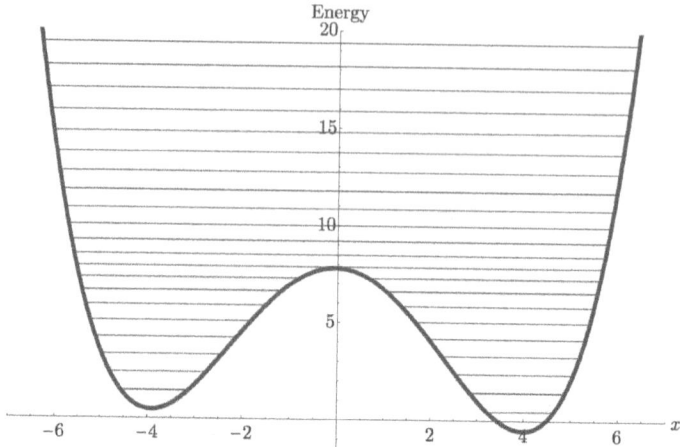

Fig. 5.8   Asymmetric potential well plotted as a function of the real variable x for the Hermitian Hamiltonian (5.36) with $\epsilon = 1$ and $g = 0.046$, which is larger than the choice $g = 0.008$ in Fig. 5.7. Energy levels are indicated by horizontal lines and 22 eigenvalues are shown; there is just one bound state.

quantum-field-theory model known as the *Lee model*. In 1954 the Lee model was proposed as a quantum field theory in which mass, wave function, and charge renormalization could be performed exactly and in closed form [Lee (1954); Källén and Pauli (1955); Schweber (1961); Barton (1963)]. In this model, when the renormalized coupling constant is taken to be larger than a critical value, the Hamiltonian becomes non-Hermitian and a *ghost state* (a negative-norm state) appears. This ghost state was assumed for more than fifty years to be a fundamental defect of the Lee model. However, we show here that the non-Hermitian Lee-model Hamiltonian is actually $\mathcal{PT}$ symmetric. When the states of this model are re-examined using the $\mathcal{C}$ operator, the ghost state is found to be an ordinary physical state having positive norm [Bender *et al.* (2005b); Jones (2008)].[9]

The Lee-model Hamiltonian has a cubic term that describes the interaction of three spinless particles called V, N, and θ [see (5.37)]. The V and N particles are treated as fermions and behave roughly like nucleons, and the θ particle is a boson and behaves roughly like a pion. In the model a V may emit a θ, and when it does so it becomes an N: $V \to N + \theta$. Also, an N may absorb a θ, and when it does so it becomes a V: $N + \theta \to V$.

---

[9]The original idea for studying the Lee model as an example of a $\mathcal{PT}$-symmetric field theory can be found in [Kleefeld (2004, 2005)].

The Lee model is solvable because it lacks crossing symmetry; that is, the N is forbidden to emit an anti-θ and become a V. Eliminating crossing symmetry makes the Lee model solvable because it introduces two conservation laws. First, *the number of N quanta plus the number of V quanta is fixed.* Second, *the number of N quanta minus the number of θ quanta is fixed.* These two highly constraining conservation laws decompose the Hilbert space of states into an infinite number of noninteracting sectors. The simplest sector is the vacuum sector. Because of the conservation laws, there are no vacuum graphs and the *bare* (unrenormalized) vacuum is the physical vacuum. The next two sectors are the one-θ-particle and the one-N-particle sectors. These sectors are also trivial because the conservation laws prevent any dynamical processes from occurring. As a result, the masses of the N particle and of the θ particle are not renormalized; that is, the physical masses of these particles are the same as their bare masses.

The first nontrivial sector is the $V/N\theta$ sector. The physical states in this sector of the Hilbert space are linear combinations of the bare V and the bare $N\theta$ states. The physical states are linear combinations of the one-physical-V-particle state and the physical $N\theta$ scattering states, and to find these states one looks for the poles and cuts of the Green's functions. Renormalization in the $V/N\theta$ sector is easy to perform and it can be done exactly and in closed form. First, one finds the renormalized (physical) mass of the V particle, which is different from the bare mass. Second, we define g^2, the square of the renormalized coupling constant, as the value of the $N\theta$ scattering amplitude at threshold, which in principle one could measure in a physical $N\theta$ scattering experiment. We then calculate the unrenormalized coupling constant g_0 in terms of the renormalized coupling constant g, and this allows us to use the scattering-experiment measurement to determine g_0.

The intriguing aspect of the Lee model is that in the $V/N\theta$ sector a strange new state may appear, which is called a *ghost state* because its conventional Hermitian norm is negative. This state arises as a consequence of coupling-constant renormalization. When we express g_0^2 as a function of g^2, we obtain the graph in Fig. 5.9. We emphasize that g^2 is a physical parameter whose numerical value is not something we can choose; it is determined by a laboratory experiment. If g^2 were to be measured to be near 0, then from Fig. 5.9 we see that g_0^2 is also small. However, we can see from this figure that if the experimental value of g^2 is larger than a critical value, then the square of the unrenormalized coupling constant is *negative*. Therefore, in this latter regime g_0 is *imaginary*, so the Hamiltonian is non-

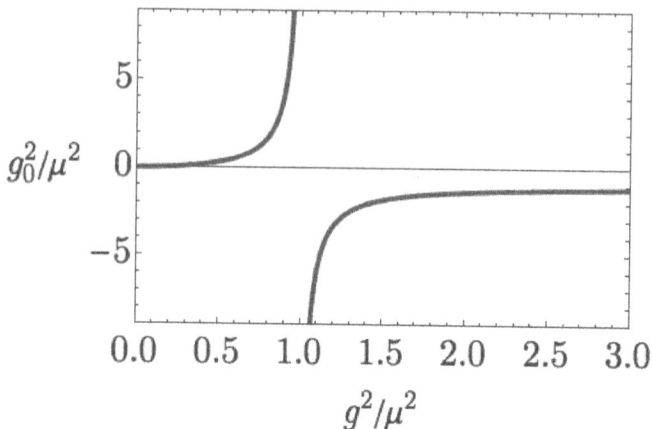

Fig. 5.9   Square of the unrenormalized coupling constant g_0^2 for the Lee model plotted as a function of the square of the renormalized coupling constant g^2. Note that $g^2 = 0$ when $g_0^2 = 0$ and that as g^2 increases from 0 so does g_0^2. However, as g^2 increases past a critical value, g_0^2 abruptly becomes negative. In this regime g_0 is imaginary, the Hamiltonian [see (5.37)] becomes non-Hermitian, and a ghost state appears.

Hermitian. The ghost state appears in the $V/N\theta$ sector in this range of the coupling constant. Ghost states are generally regarded as unacceptable in quantum theory because their presence signals a violation of unitarity and invalidates a probabilistic interpretation of the theory.

There have been many unsuccessful attempts in the context of Hermitian quantum theory to make sense of the Lee model as a physical quantum theory in the ghost regime [Barton (1963); Källén and Pauli (1955); Schweber (1961)]. However, in the context of $\mathcal{PT}$-symmetric quantum theory we can give a physical interpretation of the $V/N\theta$ sector of the Lee model when g_0 becomes imaginary and the Hamiltonian becomes non-Hermitian. The Lee model Hamiltonian has a cubic interaction term, and we showed in Sec. 5.4 how to make sense of a Hamiltonian in which there is a cubic interaction multiplied by an imaginary coupling constant. The procedure is to calculate the $\mathcal{C}$ operator and to use it to define a new positive-definite inner product.

For simplicity of presentation we focus here on the *quantum-mechanical* Lee model; the analysis of the field-theoretic Lee model [Bender *et al.* (2005b)] is more complicated but qualitatively identical. The Hamiltonian for the quantum-mechanical Lee model is

$$\hat{H} = \hat{H}_0 + g_0 \hat{H}_1$$
$$= m_{V_0} V^\dagger V + m_N N^\dagger N + m_\theta a^\dagger a + g_0 \left(V^\dagger N a + a^\dagger N^\dagger V \right). \quad (5.37)$$

The bare states are the eigenstates of $\hat{H}_0$ and the physical states are the eigenstates of the full Hamiltonian $\hat{H}$. The mass parameters m_N and m_θ represent the *physical* masses of the one-N-particle and one-θ-particle states because these states do not undergo mass renormalization. However, m_{V_0} is the *bare* mass of the V particle.

We treat the V, N, and θ particles as pseudoscalars; that is, we assume that the fields that create and destroy these three particles all change sign under $\mathcal{P}$ just as the creation and annihilation operators $a^\dagger$ and a, which combine to give the position operator $\hat{x} = (a + a^\dagger)/\sqrt{2}$ and the momentum operator $\hat{p} = i(a^\dagger - a)/\sqrt{2}$, change sign in quantum mechanics.[10] Thus, we assume that $\mathcal{P}V\mathcal{P} = -V$, $\mathcal{P}N\mathcal{P} = -N$, $\mathcal{P}a\mathcal{P} = -a$, $\mathcal{P}V^\dagger\mathcal{P} = -V^\dagger$, $\mathcal{P}N^\dagger\mathcal{P} = -N^\dagger$, and $\mathcal{P}a^\dagger\mathcal{P} = -a^\dagger$.

Under time reversal $\mathcal{T}$, $\hat{p}$ and i change sign but $\hat{x}$ does not: $\mathcal{T}\hat{p}\mathcal{T} = -\hat{p}$, $\mathcal{T}i\mathcal{T} = -i$, $\mathcal{T}\hat{x}\mathcal{T} = \hat{x}$. Thus, we assume that $\mathcal{T}V\mathcal{T} = V$, $\mathcal{T}N\mathcal{T} = N$, $\mathcal{T}a\mathcal{T} = a$, $\mathcal{T}V^\dagger\mathcal{T} = V^\dagger$, $\mathcal{T}N^\dagger\mathcal{T} = N^\dagger$, $\mathcal{T}a^\dagger\mathcal{T} = a^\dagger$. When the bare coupling constant g_0 is real, $\hat{H}$ in (5.37) is Hermitian, $\hat{H}^\dagger = H$, and when g_0 is imaginary, $g_0 = i|g_0|$, $\hat{H}$ is not Hermitian. However, by virtue of these transformation properties, $\hat{H}$ is $\mathcal{PT}$-symmetric: $\hat{H}^{\mathcal{PT}} = \hat{H}$.

Let us suppose first that g_0 is real so that $\hat{H}$ is Hermitian and examine the simplest nontrivial sector of the quantum-mechanical Lee model; namely, the $V/N\theta$ sector. We seek the eigenstates of $\hat{H}$ in the form of linear combinations of the bare one-V-particle and the bare one-N-one-θ-particle states. There are two eigenfunctions. We interpret the eigenfunction corresponding to the lower-energy eigenvalue as the physical one-V-particle state, and we interpret the eigenfunction corresponding to the higher-energy eigenvalue as the physical one-N-one-θ-particle state. (In the field-theoretic Lee model this higher-energy state corresponds to the continuum of physical N-θ scattering states.) We then make the *ansatz*

$$|V\rangle = c_{11}|1,0,0\rangle + c_{12}|0,1,1\rangle, \qquad |N\theta\rangle = c_{21}|1,0,0\rangle + c_{22}|0,1,1\rangle,$$

and demand that these states be eigenstates of $\hat{H}$ with eigenvalues m_V (the renormalized V-particle mass) and $E_{N\theta}$. The eigenvalue problem reduces to a pair of algebraic equations:

$$c_{11}m_{V_0} + c_{12}g_0 = c_{11}m_V, \qquad c_{21}g_0 + c_{22}\left(m_N + m_\theta\right) = c_{22}E_{N\theta}. \quad (5.38)$$

---

[10]Since $\hat{x}$ and $\hat{p}$ change sign under parity $\mathcal{P}$ ($\mathcal{P}\hat{x}\mathcal{P} = -\hat{x}$, $\mathcal{P}\hat{p}\mathcal{P} = -\hat{p}$), it follows that a and $a^\dagger$ must also change sign under the action of $\mathcal{P}$.

The solutions to (5.38) are

$$m_V = \frac{1}{2}\left(m_N + m_\theta + m_{V_0} - \sqrt{\mu_0^2 + 4g_0^2}\right),$$

$$E_{N\theta} = \frac{1}{2}\left(m_N + m_\theta + m_{V_0} + \sqrt{\mu_0^2 + 4g_0^2}\right), \qquad (5.39)$$

where $\mu_0 \equiv m_N + m_\theta - m_{V_0}$. Notice that m_V (the mass of the physical V particle) is different from m_{V_0} (the mass of the bare V particle) because the V particle undergoes mass renormalization.

Next, we perform wave-function renormalization. Following [Barton 1963], we define the wave-function renormalization constant Z_V by $\sqrt{Z_V} = \langle 0|V|V\rangle$. This gives

$$Z_V^{-1} = \frac{1}{2}g_0^{-2}\sqrt{\mu_0^2 + 4g_0^2}\left(\sqrt{\mu_0^2 + 4g_0^2} - \mu_0\right).$$

Finally, we perform coupling-constant renormalization. Again, following [Barton (1963)], we note that $\sqrt{Z_V}$ is the ratio between the renormalized coupling constant g and the bare coupling constant g_0. Thus, $g^2/g_0^2 = Z_V$. Elementary algebra then gives the bare coupling constant in terms of the renormalized mass and coupling constant:

$$g_0^2 = g^2 / \left(1 - g^2/\mu^2\right), \qquad (5.40)$$

where μ is defined as $\mu \equiv m_N + m_\theta - m_V$. We cannot freely choose g because g is, in principle, determined by an experimental measurement. Once g has been determined experimentally, we can use (5.40) to determine g_0. The relation in (5.40) is plotted in Fig. 5.9. This figure reveals the startling behavior of the Lee model: If g is larger than the critical value μ, then the square of g_0 is negative and g_0 is imaginary.

As g approaches the critical value from below, the two energy eigenvalues in (5.39) vary accordingly. These eigenvalues are the two zeros of the secular determinant $f(E)$ obtained from applying Cramer's rule to (5.38). As g (and g_0) increase, the energy of the physical $N\theta$ state increases and the energy of the $N\theta$ state becomes infinite as g reaches its critical value. As g increases past its critical value, the upper energy eigenvalue abruptly jumps from being large and positive to being large and negative. Then, as g continues to increase, this energy eigenvalue approaches the energy of the physical V particle from below.

Above the critical value of g, the Hamiltonian $\hat{H}$ in (5.37) becomes non-Hermitian, but its eigenvalues in the $V/N\theta$ sector remain real. (The eigenvalues remain real because $\hat{H}$ becomes PT symmetric. Indeed, all of the cubic PT-symmetric Hamiltonians that we have studied have real spectra.) However, in the PT-symmetric regime it is no longer appropriate to interpret the lower eigenvalue as the energy of the physical $N\theta$ state. Rather, it is the energy of a new ghost state designated as $|G\rangle$. As shown in [Källén and Pauli (1955); Schweber (1961); Barton (1963)], the Hermitian norm of this state is *negative*.

A physical interpretation of this ghost state emerges easily when we use the procedure developed in [Bender *et al.* (2005b)]. To begin we verify that in the PT-symmetric regime, where g_0 is imaginary, the states of $\hat{H}$ are eigenstates of the PT operator, and we choose the multiplicative phases of these states so that their PT eigenvalues are unity:

$$PT|G\rangle = |G\rangle, \qquad PT|V\rangle = |V\rangle.$$

It is then straightforward to verify that the PT norm of the V state is positive, while the PT norm of the ghost state is negative.

We now follow the procedures in Chap. 3 to calculate C. We express C as an exponential of a function Q multiplying the parity operator: $C = \exp\left[Q(V^\dagger, V; N^\dagger, N; a^\dagger, a)\right] P$. We then impose the operator equations $C^2 = 1$, $[C, PT] = 0$, and $[C, H] = 0$. The condition $C^2 = 1$ gives

$$Q(V^\dagger, V; N^\dagger, N; a^\dagger, a) = -Q(-V^\dagger, -V; -N^\dagger, -N; -a^\dagger, -a).$$

Thus, $Q(V^\dagger, V; N^\dagger, N; a^\dagger, a)$ is an odd function in total powers of $V^\dagger$, V, $N^\dagger$, N, $a^\dagger$, and a. Next, we impose the condition that $[C, PT] = 0$ and get

$$Q(V^\dagger, V; N^\dagger, N; a^\dagger, a) = Q^*(-V^\dagger, -V; -N^\dagger, -N; -a^\dagger, -a),$$

where $*$ denotes complex conjugation.

Last, we impose the condition that $C, \hat{H} = 0$, which translates to

$$\left[e^Q, \hat{H}_0\right] = g_0\left\{e^Q, H_1\right\}_+.$$

Although in Sec. 5.4 we were only able to find the C operator to leading order in perturbation theory, for the Lee model we can calculate C exactly and in closed form. To do so, we follow [Bender *et al.* (2005b)] and seek a solution for Q as a formal Taylor series in powers of g_0:

$$Q = \sum_{n=0}^{\infty} g_0^{2n+1} Q_{2n+1}. \tag{5.41}$$

Only odd powers of g_0 appear in this series and Q_{2n+1} are all anti-Hermitian: $Q_{2n+1}^\dagger = -Q_{2n+1}$. The exact formula for Q_{2n+1} is

$$Q_{2n+1} = (-1)^n \frac{2^{2n+1}}{(2n+1)\mu_0^{2n+1}} \left(V^\dagger N a n_\theta^n - n_\theta^n a^\dagger N^\dagger V\right),$$

where $n_\theta = a^\dagger a$ is the number operator for θ-particle quanta.

We then sum over all Q_{2n+1} and obtain the *exact* result that

$$Q = V^\dagger N a \frac{1}{\sqrt{n_\theta}} \tan^{-1}\left(\frac{2g_0\sqrt{n_\theta}}{\mu_0}\right) - \frac{1}{\sqrt{n_\theta}} \tan^{-1}\left(\frac{2g_0\sqrt{n_\theta}}{\mu_0}\right) a^\dagger N^\dagger V.$$

We exponentiate this result to obtain the C operator. The exponential of Q simplifies considerably because we are treating the V and N particles as fermions and therefore we can use the identity $n_{V,N}^2 = n_{V,N}$. The *exact* result for e^Q is

$$e^Q = \left[1 - \frac{2g_0\sqrt{n_\theta}a^\dagger N^\dagger V}{\sqrt{\mu_0^2 + 4g_0^2 n_\theta}} + \frac{\mu_0 n_N(1 - n_V)}{\sqrt{\mu_0^2 + 4g_0^2 n_\theta}} + \frac{\mu_0 n_V(1 - n_N)}{\sqrt{\mu_0^2 + 4g_0^2(n_\theta + 1)}}\right.$$
$$\left. + \frac{2g_0\sqrt{n_\theta}V^\dagger N a}{\sqrt{\mu_0^2 + 4g_0^2 n_\theta}} - n_V - n_N + n_V n_N\right].$$

We can also express the parity operator $\mathcal{P}$ in terms of number operators:

$$\mathcal{P} = e^{i\pi(n_V + n_N + n_\theta)} = (1 - 2n_V)(1 - 2n_N)e^{i\pi n_\theta}.$$

Combining e^Q and $\mathcal{P}$, we obtain the exact expression for C:

$$C = \left[1 - \frac{2g_0\sqrt{n_\theta}a^\dagger N^\dagger V}{\sqrt{\mu_0^2 + 4g_0^2 n_\theta}} + \frac{\mu_0 n_N(1 - n_V)}{\sqrt{\mu_0^2 + 4g_0^2 n_\theta}} + \frac{\mu_0 n_V(1 - n_N)}{\sqrt{\mu_0^2 + 4g_0^2(n_\theta + 1)}}\right.$$
$$\left. + \frac{2g_0\sqrt{n_\theta}V^\dagger N a}{\sqrt{\mu_0^2 + 4g_0^2 n_\theta}} - n_V - n_N + n_V n_N\right](1 - 2n_V)(1 - 2n_N)e^{i\pi n_\theta}.$$

Using this C operator to calculate the CPT norm of the V state and of the ghost state, we find that these norms are both positive. Furthermore, the time evolution is unitary. This establishes that if we use the CPT inner product, the quantum-mechanical Lee model is a physically acceptable and consistent quantum theory even in the ghost regime where the unrenormalized coupling constant is imaginary.

5.7 Other PT-Symmetric Quantum Field Theories

This section briefly mentions a variety of quantum-field-theory models and the role that PT symmetry may play in understanding these models. Sometimes one finds that, as in the case of the Lee model discussed in Sec. 5.6, renormalizing a formally Hermitian Hamiltonian makes the Hamiltonian non-Hermitian and gives rise to ghost states. The key insight, as we saw in Sec. 5.6, may be to recognize that the renormalized Hamiltonian is PT

symmetric. In such cases the procedure is to redefine the inner product and to show that the ghost state then becomes a physical state. This is a powerful technique that has been used to study other theories, such as the quantum-mechanical Pais–Uhlenbeck model [Bender and Mannheim (2008)] and more advanced field-theory problems [Curtright *et al.* (2007); Curtright and Veitia (2007); Ivanov and Smilga (2007)]. However, as we will see, there are other ways in which PT-symmetric structures may appear.

5.7.1 *Higgs sector of the Standard Model*

A crucially important field theory model is the Standard Model of particle physics. A serious difficulty arises, however, when one renormalizes this model: The renormalized Higgs vacuum state appears to be unstable. However, a PT-symmetric reinterpretaton of the renormalized model suggests that the vacuum state may actually be stable; that is, the ground-state energy appears to be *real* [Bender *et al.* (2016c)]. This model is particularly interesting because the PT symmetry is *broken*; all but a handful of the lowest-energy states have complex energy and are therefore unstable. This result is not surprising, however, because it is indeed true that almost all known higher-mass elementary particles are unstable.

Let us make some highly conjectural remarks: The distinguishing features of the PT-symmetric $-g\varphi^4$ quantum field theory in (5.4) are that (i) its spectrum is real and bounded below, (ii) it is perturbatively renormalizable, (iii) it has a dimensionless coupling constant in four-dimensional space-time, and (iv) it is *asymptotically free* [Bender *et al.* (2000b)]. (The property of asymptotic freedom was established long ago in [Symanzik (1971a,b, 1973)].) In short, while a $+g\varphi^4$ theory in four-dimensional space-time is trivial because it is not asymptotically free, Symanzik proposed a "precarious" theory with a negative quartic coupling constant as a candidate for an asymptotically free theory of strong interactions. Symanzik used the term "precarious" because the negative sign of the coupling constant suggests that this theory is energetically unstable. However, as is explained in Chap. 1, imposing PT-symmetric boundary conditions (in this case on the functional-integral representation of the quantum field theory) gives a spectrum that is bounded below. Thus, Symanzik's proposal to use a nontrivial $-g\varphi^4$ model as a basis for a theory of strong interactions may well be resurrected.

The $-g\varphi^4$ quantum field theory has another remarkable property. Although the theory seems to be parity invariant, the PT-symmetric boundary conditions violate parity invariance. Hence, as we saw in Sec. 5.3, the one-point Green's function (the expectation value of the field φ) does not vanish.[11] Thus, a nonzero vacuum expectation value can be achieved without having spontaneous symmetry breaking because parity symmetry is *permanently* broken.

These properties suggest that a $-g\varphi^4$ quantum field theory might be useful in describing the Higgs sector of the Standard Model. Perhaps the Higgs particle state is a consequence of the field-theoretic parity anomaly in the same way that the quantum-mechanical parity anomaly described in Sec. 5.5 gives rise to bound states. Early research in this area includes studies of transformations of functional integrals [Jones *et al.* (2006)] and large-N approximations and matrix models [Ogilvie and Meisinger (2007)], but much additional work must be done.

5.7.2 $\mathcal{PT}$-symmetric quantum electrodynamics

If the unrenormalized electric charge e in the Hamiltonian for quantum electrodynamics were imaginary, then the Hamiltonian would be non-Hermitian, but if one also specifies that the potential field A^μ in this theory transforms as an *axial* vector instead of as a vector, then the Hamiltonian becomes $\mathcal{PT}$ symmetric [Milton (2004)]. We thus assume that the four-vector potential and the electromagnetic fields transform under $\mathcal{P}$ like

$$\mathcal{P}: \quad \mathbf{E} \to \mathbf{E}, \quad \mathbf{B} \to -\mathbf{B}, \quad \mathbf{A} \to \mathbf{A}, \quad A^0 \to -A^0.$$

Under time reversal, the transformations are assumed to be conventional:

$$\mathcal{T}: \quad \mathbf{E} \to \mathbf{E}, \quad \mathbf{B} \to -\mathbf{B}, \quad \mathbf{A} \to -\mathbf{A}, \quad A^0 \to A^0.$$

The Lagrangian of the theory must have an imaginary coupling constant so that it will be invariant under the product of these two symmetries:

$$\mathcal{L} = -\tfrac{1}{4}F^{\mu\nu}F_{\mu\nu} + \tfrac{1}{2}\psi^\dagger\gamma^0\gamma^\mu\tfrac{1}{i}\partial_\mu\psi + \tfrac{1}{2}m\psi^\dagger\gamma^0\psi + ie\psi^\dagger\gamma^0\gamma^\mu\psi A_\mu.$$

The corresponding Hamiltonian density is then

$$\mathcal{H} = \tfrac{1}{2}(E^2 + B^2) + \psi^\dagger\left[\gamma^0\gamma^k\left(-i\nabla_k + ieA_k\right) + m\gamma^0\right]\psi.$$

The Lorentz transformation properties of the fermions are unchanged from the usual ones. Thus, the electric current appearing in the Lagrangian and Hamiltonian densities, $j^\mu = \psi^\dagger\gamma^0\gamma^\mu\psi$, transforms conventionally under both $\mathcal{P}$ and $\mathcal{T}$:

$$\mathcal{P}j^\mu(\mathbf{x},t)\mathcal{P} = \begin{pmatrix} j^0 \\ -\mathbf{j} \end{pmatrix}(-\mathbf{x},t), \qquad \mathcal{T}j^\mu(\mathbf{x},t)\mathcal{T} = \begin{pmatrix} j^0 \\ -\mathbf{j} \end{pmatrix}(\mathbf{x},-t).$$

---

[11]Techniques for calculating this expectation value are explained in [Bender *et al.* (2001c)].

Because its interaction is cubic, this non-Hermitian theory of "electrodynamics" is analogous to the spinless $i\varphi^3$ quantum field theory discussed in Sec. 5.4. Moreover, $\mathcal{PT}$-symmetric electrodynamics is asymptotically free (unlike ordinary electrodynamics). This non-Hermitian theory is especially interesting because the sign of the Casimir force is the opposite of that in ordinary electrodynamics [Milton (2004)]. The massless version of this theory is remarkable because finiteness conditions enable it to determine its own coupling constant [Bender and Milton (1999)].

The $\mathcal{C}$ operator for $\mathcal{PT}$-symmetric quantum electrodynamics has been constructed perturbatively to first order in e [Bender *et al.* (2005c)]. This construction is too technical to describe here, but it suggests that non-Hermitian quantum electrodynamics is a viable and consistent unitary quantum field theory. $\mathcal{PT}$-symmetric quantum electrodynamics is more elaborate than an $i\phi^3$ quantum field theory because it possesses some features of conventional quantum electrodynamics, such as Abelian gauge invariance. The only asymptotically free quantum field theories described by Hermitian Hamiltonians are those that possess a *non-Abelian* gauge invariance. Thus, $\mathcal{PT}$ symmetry allows for new kinds of asymptotically free theories, such as the $-\varphi^4$ theory discussed above, that do not possess a non-Abelian gauge invariance.

5.7.3 *Dual PT-symmetric quantum field theories*

Until now we have focused on bosonic $\mathcal{PT}$-symmetric Hamiltonians, but it is just as easy to construct fermionic $\mathcal{PT}$-symmetric Hamiltonians. We look first at free theories. The Lagrangian density for a conventional Hermitian free fermion field theory is

$$\mathcal{L}(\mathbf{x}, t) = \bar{\psi}(\mathbf{x}, t)(i\partial\!\!\!/ - m)\psi(\mathbf{x}, t), \tag{5.42}$$

and the corresponding Hamiltonian density is

$$\mathcal{H}(\mathbf{x}, t) = \bar{\psi}(\mathbf{x}, t)(-i\nabla\!\!\!\!/ + m)\psi(\mathbf{x}, t), \tag{5.43}$$

where $\bar{\psi}(\mathbf{x}, t) = \psi^\dagger(\mathbf{x}, t)\gamma_0$.

In $(1 + 1)$-dimensional space-time we adopt the following conventions: $\gamma_0 = \sigma_1$ and $\gamma_1 = i\sigma_2$, where σ are the Pauli matrices. With these definitions, we have $\gamma_0^2 = 1$ and $\gamma_1^2 = -1$. We also define $\gamma_5 = \gamma_0\gamma_1 = \sigma_3$, so that $\gamma_5^2 = 1$. The parity operator $\mathcal{P}$ has the effect

$$\mathcal{P}\psi(x, t)\mathcal{P} = \gamma_0\psi(-x, t), \qquad \mathcal{P}\bar{\psi}(x, t)\mathcal{P} = \bar{\psi}(-x, t)\gamma_0.$$

The time-reversal operator $\mathcal{T}$ has the effect

$$\mathcal{T}\psi(x, t)\mathcal{T} = \gamma_0\psi(x, -t), \qquad \mathcal{T}\bar{\psi}(x, t)\mathcal{T} = \bar{\psi}(x, -t)\gamma_0,$$

which is similar to the action of $\mathcal{P}$ except that $\mathcal{T}$ is antilinear and takes the complex conjugate of complex numbers. Using these definitions the Hamiltonian is $H = \int dx\,\mathcal{H}(x,t)$, where $\mathcal{H}$ in (5.43) is Hermitian, so $H = H^\dagger$. Note also that H is separately invariant under parity and under time reversal: $\mathcal{P}H\mathcal{P} = H$ and $\mathcal{T}H\mathcal{T} = H$.

To construct a non-Hermitian fermionic Hamiltonian we add a γ_5-dependent mass term to the Hamiltonian density (5.43):

$$\mathcal{H}(x,t) = \bar{\psi}(x,t)(-i\nabla\!\!\!\!/ + m_1 + m_2\gamma_5)\psi(x,t) \quad (m_2 \text{ real}). \tag{5.44}$$

The Hamiltonian $H = \int dx\,\mathcal{H}(x,t)$ associated with this Hamiltonian density (5.44) is not Hermitian because the m_2 term changes sign under Hermitian conjugation. This sign change occurs because γ_0 and γ_5 anticommute. Also, H is not invariant under $\mathcal{P}$ or under $\mathcal{T}$ separately because the m_2 term changes sign under each of these reflections. However, H *is* invariant under combined $\mathcal{P}$ and $\mathcal{T}$ reflection. Thus, H is $\mathcal{PT}$ symmetric.

To determine whether the $\mathcal{PT}$ symmetry of H is broken or unbroken, we check to see whether the spectrum of H is real. We do so by solving the field equation. The field equation associated with $\mathcal{H}$ in (5.44) is

$$\left(i\partial\!\!\!/ - m_1 - m_2\gamma_5\right)\psi(x,t) = 0.$$

If we iterate this equation and use $\partial\!\!\!/^2 = \partial^2$, we obtain

$$\left(\partial^2 + \mu^2\right)\psi(x,t) = 0. \tag{5.45}$$

This is the two-dimensional Klein–Gordon equation with $\mu^2 = m_1^2 - m_2^2$. The physical mass μ that propagates under this equation is real when

$$m_1^2 \geq m_2^2. \tag{5.46}$$

This inequality condition defines a *two-dimensional* parametric region of *unbroken* $\mathcal{PT}$ symmetry. When (5.46) is not satisfied, $\mathcal{PT}$ symmetry is *broken*. At the boundary between the regions of broken and unbroken $\mathcal{PT}$ symmetry (the line $m_2 = 0$), the Hamiltonian is Hermitian.[12]

The $\mathcal{C}$ operator associated with the $\mathcal{PT}$-symmetric Hamiltonian density $\mathcal{H}$ in (5.44) is given by $\mathcal{C} = e^Q \mathcal{P}$, where [Bender *et al.* (2005d)]

$$Q = -\tanh^{-1}\left[\varepsilon\int dx\,\bar{\psi}(x,t)\gamma_1\psi(x,t)\right]$$
$$= -\tanh^{-1}\left[\varepsilon\int dx\,\psi^\dagger(x,t)\gamma_5\psi(x,t)\right]. \tag{5.47}$$

---

[12]The same is true in quantum mechanics. For the Hamiltonian in (2.7) the region of broken (unbroken) $\mathcal{PT}$ symmetry is $\varepsilon < 0$ ($\varepsilon > 0$). At the boundary $\varepsilon = 0$ of these two regions the Hamiltonian is Hermitian.

The inverse hyperbolic tangent in this equation requires that $|\varepsilon| \leq 1$, or equivalently that $m_1^2 \geq m_2^2$, which corresponds to the unbroken region of $\mathcal{PT}$ symmetry. We use (5.47) to construct the equivalent Hermitian Hamiltonian h as in (3.43):

$$h = \exp\left[\tfrac{1}{2}\tanh^{-1}\varepsilon \int dx\, \psi^\dagger(x,t)\gamma_5\psi(x,t)\right] H$$
$$\times \exp\left[-\tfrac{1}{2}\tanh^{-1}\varepsilon \int dx\, \psi^\dagger(x,t)\gamma_5\psi(x,t)\right]. \qquad (5.48)$$

The commutation relations $[\gamma_5,\gamma_0] = -2\gamma_1$ and $[\gamma_5,\gamma_1] = -2\gamma_0$ simplify h in (5.48):

$$h = \int dx\, \bar{\psi}(x,t)(-i\slashed{\nabla} + \mu)\psi(x,t),$$

where $\mu^2 = m^2(1-\varepsilon^2) = m_1^2 - m_2^2$, in agreement with (5.45). Replacing H by h changes the γ_5-dependent mass term $m\bar{\psi}(1+\varepsilon\gamma_5)\psi$ to a conventional fermion mass term $\mu\bar{\psi}\psi$. Thus, the non-Hermitian $\mathcal{PT}$-symmetric Hamiltonian density in (5.44) is equivalent to the Hermitian Hamiltonian density in (5.43) with m replaced by μ.

If we introduce a four-point fermion interaction term in (5.42), we obtain the Lagrangian density for the massive Thirring model in $(1+1)$ dimensions:

$$\mathcal{L} = \bar{\psi}(i\slashed{\partial} - m)\psi + \tfrac{1}{2}g(\bar{\psi}\gamma^\mu\psi)(\bar{\psi}\gamma_\mu\psi), \qquad (5.49)$$

whose corresponding Hamiltonian density is

$$\mathcal{H} = \bar{\psi}(-i\slashed{\nabla} + m)\psi - \tfrac{1}{2}g(\bar{\psi}\gamma^\mu\psi)(\bar{\psi}\gamma_\mu\psi), \qquad (5.50)$$

We can then construct the $\mathcal{PT}$-symmetric Thirring model

$$\mathcal{H} = \bar{\psi}(-i\slashed{\nabla} + m + \varepsilon m\gamma_5)\psi - \tfrac{1}{2}g(\bar{\psi}\gamma^\mu\psi)(\bar{\psi}\gamma_\mu\psi) \qquad (5.51)$$

by introducing a γ_5-dependent mass term. The additional term is non-Hermitian but $\mathcal{PT}$ symmetric because it is odd under both parity reflection and time reversal. The Q operator for the interacting theory $g \neq 0$ is *identical* to the Q operator for $g = 0$ because in $(1+1)$-dimensional space, the interaction term $(\bar{\psi}\gamma^\mu\psi)(\bar{\psi}\gamma_\mu\psi)$ commutes with the Q in (2.4) [Bender *et al.* (2005d)]. Thus, the non-Hermitian $\mathcal{PT}$-symmetric Hamiltonian density in (5.51) is equivalent to the Hermitian Hamiltonian density in (5.50) with the mass m replaced by μ, where $\mu^2 = m^2(1-\varepsilon^2) = m_1^2 - m_2^2$. The same is true for the $(3+1)$-dimensional interacting Thirring model by virtue of the commutation relation $[\gamma_5, \gamma_0\gamma_\mu] = 0$, but because this higher-dimensional field theory is nonrenormalizable, the Q operator may only have a formal significance.

In $(1 + 1)$ dimensions the massive Thirring Model (5.49) is *dual* to the $(1 + 1)$-dimensional Sine–Gordon model [Abdalla *et al.* (2001)], whose Lagrangian density is $\mathcal{L} = \frac{1}{2}(\hat{p}_\mu \varphi)^2 + m^2 \lambda^{-2}(\cos \lambda \varphi - 1)$ and whose corresponding Hamiltonian density is

$$\mathcal{H} = \tfrac{1}{2}\pi^2 + \tfrac{1}{2}(\nabla \varphi)^2 + m^2 \lambda^{-2}(1 - \cos \lambda \varphi).$$

Here, $\pi(x, t) = \partial_0 \varphi(x, t)$ and in $(1 + 1)$-dimensional space $\nabla \varphi(x, t)$ is just $\hat{p}_1 \varphi(x, t)$. The duality between the Thirring model and the Sine–Gordon model is expressed as an algebraic relationship between the coupling constants g and λ: $\lambda^2/(4\pi) = 1/(1 - g/\pi)$. The free fermion theory ($g = 0$) is equivalent to the Sine–Gordon model with the special value for the coupling constant $\lambda^2 = 4\pi$.

The $\mathcal{PT}$-symmetric deformation (5.51) of the modified Thirring model is, by the same analysis, dual to a modified Sine–Gordon model with the Hamiltonian density

$$\mathcal{H} = \tfrac{1}{2}\pi^2 + \tfrac{1}{2}(\nabla \varphi)^2 + m^2 \lambda^{-2}(1 - \cos \lambda \varphi - i\varepsilon \sin \lambda \varphi), \tag{5.52}$$

which is $\mathcal{PT}$-symmetric and not Hermitian. The Q operator is

$$Q = \tfrac{2}{\lambda} \tanh^{-1} \varepsilon \!\int\! dx\, \pi(x, t)$$

for this Hamiltonian. Thus, the equivalent Hermitian Hamiltonian h is

$$h = \exp\left[-\tfrac{1}{\lambda} \tanh^{-1} \varepsilon \!\int\! dx\, \pi(x, t) \right] H \exp\left[\tfrac{1}{\lambda} \tanh^{-1} \varepsilon \!\int\! dx\, \pi(x, t) \right].$$

The operation that transforms H to h has the same effect as shifting the boson field φ by an imaginary constant:

$$\varphi \rightarrow \varphi + \tfrac{i}{\lambda} \tanh^{-1} \varepsilon. \tag{5.53}$$

Under this transformation the interaction term $m^2 \lambda^{-2}(1 - \cos \lambda \varphi - i\varepsilon \sin \lambda \varphi)$ in (5.52) becomes $-m^2 \lambda^{-2}(1 - \varepsilon^2) \cos \lambda \varphi$, apart from an additive constant. Therefore, h is the Hamiltonian for the Hermitian Sine–Gordon model but with mass μ given by $\mu^2 = m^2(1 - \varepsilon^2) = m_1^2 - m_2^2$. This change in the mass is the same as for the Thirring model. Being Hermitian, h is even in the parameter ε that breaks the Hermiticity of H.

The idea of generating a non-Hermitian but $\mathcal{PT}$-symmetric Hamiltonian from a Hermitian Hamiltonian by shifting the field operator as in (5.53), first introduced in the context of quantum mechanics in [Fernández *et al.* (2005)]. This suggests a way to construct solvable fermionic $\mathcal{PT}$-invariant models whenever there is a boson-fermion duality.

5.7.4 *PT-symmetric theories of gravity and cosmology*

In Subsec. 5.7.2 we showed that to construct a PT-symmetric model of quantum electrodynamics, one need only replace the electric charge e by ie and then replace the vector potential A^μ by an axial-vector potential. The result is a non-Hermitian but PT-symmetric Hamiltonian. Therefore, because $i^2 = -1$, the interesting classical feature of this model is that the sign of the Coulomb force, which is proportional to the square of the electric charge, is opposite to that of conventional electrodynamics. Thus, like charges feel an attractive force and unlike charges feel a repulsive force.

One can implement the same idea to construct a PT-symmetric model of massless spin-2 particles (gravitons). In a linearized theory of gravity, one simply replaces the gravitational coupling constant G by iG and then requires that the two-component tensor field behave like an axial tensor under parity. The result is a non-Hermitian PT-symmetric Hamiltonian, which at the classical level describes a *repulsive* gravitational force. It would be interesting to investigate the possible connections between such a model and the notion of dark energy and the recent observations that the expansion of the universe is accelerating.

There is also a plausible connection between PT symmetry and certain types of cosmological models. In the discussion preceeding (2.41) it is shown that the PT-symmetric $-x^4$ potential becomes reflectionless at its eigenenergies. That is, if a plane wave of energy E is incident on a $-x^4$ potential, there is no back-scattered wave (not even an exponentially small back-scattered wave) when E equals one of the (positive) eigenvalues of this potential. There have been many papers on anti-de Sitter cosmologies (see, for example, [Hertog and Horowitz (2005)]) and de Sitter cosmologies (see, for example, [Witten (2001); Bousso *et al.* (2002)]). In the AdS description the universe propagates *reflectionlessly* in the presence of a wrong-sign potential ($-x^6$, for example). In the dS case the usual Hermitian quantum mechanics must be abandoned and be replaced by a non-Hermitian one in which there are 'meta-observables'. The non-Hermitian inner product in the dS case is based on the traditional CPT theorem in the same way that the CPT inner product is used in PT-symmetric quantum theory [Bender *et al.* (2002)]. The reflectionless condition, which is equivalent to the requirement of PT symmetry, is what allows the wrong-sign potential to have a positive spectrum. Calculating the lowest energy level in this potential would be equivalent to determining the cosmological constant [Moffat (2005, 2006)].

5.7.5   Double-scaling limit

The double-scaling limit of a quantum field theory is a correlated limit in which the number N of species approaches infinity as the coupling constant g approaches a critical value. For the simplest case of an $O(N)$-symmetric quartic quantum field theory (a *vector* model), there is a serious problem with this limit because the critical coupling constant is *negative*. Thus, at the critical value of the coupling the Hamiltonian defines a quantum theory having an upside-down potential whose energy spectrum appears to be unbounded below. Furthermore, the integral representation of the partition function of the theory, if treated conventionally, appears not to exist. One can avoid all these difficulties if one approaches this correlated limit in a $\mathcal{PT}$-symmetric fashion [Bender *et al.* (2013b); Bender and Sarkar (2014)]. When this is done, one can calculate the partition function explicitly and exactly in low-dimensional field theories, and one finds that all of the apparent problems evaporate. This shows that a $\mathcal{PT}$-symmetric interpretation of a conventional Hermitian field theory may be required in instances other than renormalization. Much research still needs to be done on the correlated limits of more complicated theories, such a $O(N)$-symmetric matrix models.

5.7.6   Fundamental properties of fermionic theories

Fermionic theories are particularly interesting because for such theories the square of the time-reversal operator $\mathcal{T}^2$ is -1 rather than 1, as in the case of bosonic theories. Thus, constructing fundamental fermionic theories having $\mathcal{PT}$ symmetry can give rise to models having interesting new properties [Jones-Smith and Mathur (2014, 2016); Jones-Smith (2010); Ohlsson (2016)]. In these investigations, models of free fermion theories were constructed and the phenomenon of species oscillation was investigated. In conventional Hermitian theories it is accepted that neutrino oscillation implies that neutrinos have mass. However, it was shown that species oscillation can occur in non-Hermitian $\mathcal{PT}$-symmetric theories of *massless* neutrinos. This could have significant implications for cosmological models, and once again it is clear that much more work is required.

Part II
Advanced Topics in $\mathcal{PT}$ Symmetry

By contributing authors

"Reason is immortal, all else mortal."
—$\mathcal{Py}\mathcal{Thagoras}$

Chapter 6

Proof of Reality for Some Simple Examples

Contributed by Patrick E. Dorey, Clare Dunning, and Roberto Tateo

> "Build a man a fire, and he'll be warm for a day. Set a man on fire, and he'll be warm for the rest of his life."
>
> —$\mathcal{P}$ratchett, $\mathcal{T}$erry

The purpose of this chapter is to study the $\mathcal{PT}$-symmetric eigenvalue problem for the Hamiltonian $H = p^2 + x^2(ix)^\varepsilon$ at a rigorous level. We start with the $\mathcal{PT}$-symmetric cubic oscillator ($\varepsilon = 1$) that, as explained in the preface of this book, brought the issue of spectral reality to the attention of Bender and Boettcher [Bender and Boettcher (1998b)] and initiated the current upsurge of interest in the subject of $\mathcal{PT}$ symmetry. The Hamiltonian for this oscillator is

$$\hat{H} = \hat{p}^2 + i\hat{x}^3 \tag{6.1}$$

and the corresponding time-independent Schrödinger equation is

$$-\psi''(x) + ix^3\psi(x) = E\,\psi(x). \tag{6.2}$$

To set up a well-defined eigenvalue problem, this equation requires a boundary condition for $\psi(x)$. For now we take this to be that $\psi(x)$ be square-integrable on the real line, which is equivalent to the statement that the possibly-complex number E belongs to the spectrum of H if and only if (6.2) has a solution $\psi(x)$ that decays both as $x \to -\infty$ and as $x \to +\infty$.

Despite the non-Hermiticity of the problem, Bessis and Zinn-Justin conjectured that this spectrum is real and positive. In the following we describe a proof [Dorey *et al.* (2001a)] of this conjecture, which has a number of interesting ramifications and generalizations, some of which we also sketch. To do this we first give in Sec. 6.1 a brief introduction to the Stokes phenomenon and the variety of eigenvalue problems that can be associated with simple ordinary differential equations such as (6.2). Then in Sec. 6.2 we show how these apparently unrelated problems are linked via exact functional relations for their spectral determinants. The proof of the reality of the spectrum of (6.2) with the stated boundary condition is given in Sec. 6.3. In Sec. 6.4 we explain the extension of the proof to a more general class of problems that was found by Shin [Shin (2002)]. For more general problems such as these the spectrum is *not* always real. The following three sections (which readers only interested in reality proofs may skip) are devoted to further generalizations of the initial problem. The intricate structure of $\mathcal{PT}$-symmetry breaking is revealed and links to other topics of interest such as the complex square well and quasi-exact solvability are established. Finally, Sec. 6.8 discusses the connection between prior sections and the theory of quantum integrable models. This connection was the original motivation behind the work leading to the reality proof.

6.1   Stokes Phenomenon

The boundary conditions on (6.2) are imposed at $x = \pm\infty$, so we start by examining the behavior of solutions to the differential equation for large $|x|$. The WKB approximation

$$\psi(x) \sim \frac{1}{P^{1/4}(x)} \exp\left[\pm \int^x dt \sqrt{P(t)}\right] \qquad (|x| \to \infty), \qquad (6.3)$$

where $P(x) = (ix)^3 - E$, predicts two asymptotic behaviors as $x \to \pm\infty$:

$$\psi_\pm(x) \sim x^{-3/4} \exp\left[\pm \tfrac{2}{5}(ix)^{5/2}\right].$$

One of these approximations grows for large positive values of x, and is said to be *dominant* there, while the other decays and is said to be *subdominant*. The asymptotic solutions for large negative values of x also exhibit the same two behaviors. Because the equation is second order there are only two linearly independent solutions and the property of subdominance in a given (positive or negative) direction along the real axis selects a solution uniquely up to a multiplicative normalization. The constant E is an eigenvalue of (6.2) if and only if the solution that is subdominant as $x \to -\infty$ is also subdominant as $x \to +\infty$.

The WKB approximation (6.3) does not contain enough information to determine the value of E. While the WKB approximation is accurate on the disconnected portions of the x axis where $|x|$ is large, it fails in the intermediate region. In particular, the approximation breaks down near any *turning points* of $P(x)$, that is, those values of x for which $P(x) = 0$. This precludes the naive continuation of the approximate solution from large negative to large positive values of x.

The standard WKB approach to solving differential equations with turning points is to construct a chain of WKB approximations along a path that passes through the turning points and to match asymptotically these WKB approximations on either side of each turning point by using Airy functions. See [Bender and Orszag (1999)] for the general procedure and [Sorrell (2007)] for an application to $\mathcal{PT}$-symmetric Schrödinger equations.

An alternative approach, more in the spirit of complex analysis and $\mathcal{PT}$ symmetry, is to follow a path in the complex-x plane from large negative x to large positive x, keeping $|x|$ large throughout the process. The advantage of doing so is that the WKB approximation to the solutions of (6.2) is accurate when $|x|$ is large and such a path bypasses the regions where $P(x)$ is small and the WKB approximation is poor. However, this continuation into the complex plane brings with it the complications of the Stokes phenomenon, as we now explain. We set $x = -i\rho e^{i\theta}$ for $\rho, \theta \in \mathbb{R}$ so that ρ is the distance of x from the origin in the complex plane, and θ is the angle subtended from the negative-imaginary axis. While ρ remains large, the WKB approximation is valid and the possible behaviors of ψ are as before:

$$\psi_{\pm}(x) \sim \rho^{-3/4} \exp\left(\pm\tfrac{2}{5} e^{5i\theta/2} \rho^{5/2}\right).$$

At first sight, this is a puzzling formula: As θ increases from 0 to 2π, ψ_+ deforms into ψ_-, apparently contradicting the fact that all solutions to (6.2) are single valued. The resolution of this paradox is that the continuation of an asymptotic approximation is not always the same as the asymptotic approximation of a continuation. The mechanism responsible for this discrepancy, called the *Stokes phenomenon*, explains why the dominant component of an asymptotic solution, if nonzero, can obscure a discontinuous change in the size of its subdominant component.

To elucidate this mechanism, we consider how the phase $e^{5i\theta/2}$ controls the behavior of $\psi_{\pm}(x)$. If the phase has a nonzero real part, one of the solutions grows and the other decays for large ρ, so that the dominant and subdominant solutions can be distinguished. However, if

$$\mathrm{Re}(e^{5i\theta/2}) = 0, \tag{6.4}$$

then both solutions oscillate and neither solution dominates the other. As θ increases, the previously dominant solution becomes subdominant, and the subdominant solution becomes dominant. The exchange of dominance

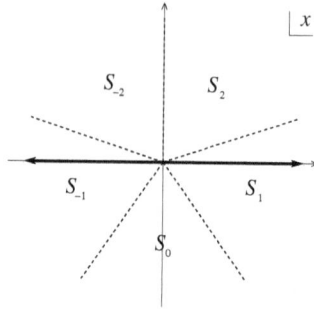

Fig. 6.1   Stokes sectors for the PT-symmetric cubic oscillator. The quantization contour (the double-arrowed line) runs along the real axis and connects S_{-1} and S_1.

and subdominance happens when (6.4) holds, that is, when

$$\theta = \tfrac{\pi}{5} \pm \tfrac{2\pi n}{5} \quad (n = 0, 1, 2). \tag{6.5}$$

The complex plane is thus split into five *Stokes sectors* separated by *anti-Stokes lines* at the angles (6.5) along which both WKB solutions oscillate. Figure 6.1 shows the five Stokes sectors $S_0, S_{\pm 1}, S_{\pm 2}$, defined by

$$S_k := \left\{ x \in \mathbb{C} : \left| \arg(ix) - \tfrac{2\pi k}{5} \right| < \tfrac{\pi}{5} \right\},$$

for the imaginary cubic oscillator (6.2).

In contrast to the behavior along the anti-Stokes lines, at the centers of the Stokes sectors, called *Stokes lines*, the dominant solution $\psi_d(x)$ is maximally dominant over the subdominant one $\psi_s(x)$: If a solution $\psi(x)$ in a Stokes sector S_k contains both dominant and subdominant parts,

$$\psi(x) = \alpha\, \psi_d(x) + \beta\, \psi_s(x),$$

in the vicinity of the Stokes line the asymptotic approximation to ψ is entirely determined by its dominant part $\alpha\, \psi_d$ provided that α is nonzero. As $\arg(ix) = \theta$ varies and x sweeps past the Stokes line, the coefficient β of the subdominant part can change discontinuously, as a function of θ, without contradicting any continuity properties of the full solution.[1] To summarize, while x remains within the same Stokes sector the change in the asymptotic

---

[1] A detailed analysis of the crossing of a Stokes line in [Berry (1989)] shows that the discontinuity is smoothed out on finer scales. This refinement is not required for the current discussion.

behavior is mild. However, if an anti-Stokes line is crossed, the dominant and subdominant solutions swap and the asymptotic behavior of the full solution is different from that of the continued asymptotic approximation. This is the *Stokes phenomenon*.

Suppose that the solution $\psi(x)$ is subdominant on the positive-real axis (in the sector S_1). To determine whether $\psi(x)$ is also subdominant on the negative-real axis (so that E would be an eigenvalue) we can continue $\psi(x)$ down into the lower-half plane, taking x though the sector S_0 and on to S_{-1}. Since $\psi(x)$ is subdominant in S_1, there is no Stokes phenomenon as x crosses the Stokes line there, but moving into S_0, $\psi(x)$ becomes dominant. Thus, as the Stokes line in S_0 is crossed the subdominant component can pick up a nonzero coefficient. Continuing through to S_{-1}, this subdominant component, if activated, becomes dominant, so the continued solution does not decay on the negative-real axis and, contrary to expectations based on a naive continuation of the asymptotic in S_1, E is not an eigenvalue. The only exceptions to this are if E is such that the coefficient of the subdominant component added to ψ when crossing the Stokes line in S_0 – sometimes called the *Stokes multiplier* – happens to be zero.

Thus, finding eigenvalues of a Schrödinger problem translates into a question about the zeros of a Stokes multiplier. We see in the next section that this is a useful reformulation because the Stokes multipliers for the cubic oscillator satisfy functional relations of a sort that had previously appeared in the study of integrable models of quantum field theory and statistical mechanics. In turn, this points the way to a proof of the surprising reality properties of the cubic-oscillator (6.1) spectrum.

We have discussed the eigenvalue problem associated with the specific pair of Stokes sectors S_1 and S_{-1} because the requirement of square integrability on the real line led us to look for solutions of (6.2) that decay simultaneously in these sectors. However, since we have allowed x to be complex, we may ask whether requiring that a solution of (6.2) decay simultaneously in *other* pairs of Stokes sectors leads to new eigenvalue problems having different discrete spectra. As follows quickly from [Sibuya (1975)], which is discussed in the next section, adjacent sectors define null eigenvalue problems, and the only other possibility seen from Fig. 6.1 is to consider pairs S_k and S_{k+2} for $k \neq -1$. A simple variable change rotates the eigenvalue problem with a quantization contour that begins and ends in sectors S_k and S_{k+2} onto the eigenvalue problem specified in S_{-1} and S_1 with the eigenvalues picking up a constant phase, and so there is nothing of interest here. In the terminology of the WKB method, quantization contours that begin and end in the neighborhood of $x = \infty$ are related to 'lateral' connection problems [Olver (1974)]. For higher-order anharmonic oscillators the Stokes sectors proliferate and the zoo of lateral connection problems does indeed grow [Bender and Klevansky (2010)], a phenomenon related to the so-called fusion hierarchies in integrable quantum field theory.

The origin provides another natural place where quantization contours may end, and contours that join $x = 0$ to $x = \infty$ lead to 'radial' (or central) connection problems. We may therefore seek solutions to the Schrödinger equation that decay for large x in sector S_k and satisfy either a Dirichlet $[\psi(0) = 0]$ or a Neumann $[\psi'(0) = 0]$ boundary condition at $x = 0$. The radial eigenvalue problem defined in the Stokes sector S_k may be mapped to the radial problem defined in S_j by rotating x and E.

In summary, there are three natural eigenvalue problems associated with the differential equation (6.1), each with its own set of eigenvalues and eigenfunctions. The first is the lateral $\mathcal{PT}$-symmetric eigenvalue problem wherein ψ is required to decay simultaneously in S_1 and S_{-1}. The other two are radial eigenvalue problems, defined with either a Dirichlet or a Neumann boundary condition at $x = 0$. The radial wave functions must all decay in one of the Stokes sectors, which we set to be S_0; this choice arises naturally as shown in the next section.

By studying Stokes multipliers we show that the three eigenvalue problems are linked via functional relations satisfied by their associated spectral determinants. The resulting constraints imply both the reality of the spectrum of the $\mathcal{PT}$-symmetric problem and a connection with integrable quantum field theory for all of the objects involved in the story so far.

6.2 Functional Relations

Following [Sibuya (1975)], we examine solutions to

$$- y''(x, E) + \left(ix^3 - E\right) y(x, E) = 0 \tag{6.6}$$

in the whole complex-x plane and treat E as a (complex) parameter. Sibuya's key result establishes the existence of an exact solution to (6.6) that is subdominant in S_0 and shares the WKB asymptotic approximation in (6.3) as $x \to \infty$ in any closed subsector $|\arg(ix)| \le \frac{3}{5}\pi - \delta$, $\delta > 0$, of $S_{-1} \cup S_0 \cup S_1$. We normalize this solution so that

$$y(x, E) \sim \frac{i}{\sqrt{2}}(ix)^{-3/4} \exp\left[-\tfrac{2}{5}(ix)^{5/2}\right] \tag{6.7}$$

in these sectors. In particular, $y(x, E)$ is subdominant in S_0, decaying as $|x| \to \infty$ along the negative imaginary axis. We can perform a simple variable change in $y(x, E)$, sometimes called *Symanzik (re)scaling*, to generate a collection of further solutions:

$$y_k(x, E) := \omega^{k/2} y(\omega^{-k} x, \omega^{2k} E). \tag{6.8}$$

We can check that $y_k(x, E)$ satisfies (6.6) for all integers k provided that $\omega = \exp(2i\pi/5)$. Furthermore, Sibuya's general results guarantee that

(**S1**) y_k exists and is an entire function of x and E;

(**S2**) y_k is subdominant in the Stokes sector S_k and is dominant in the neighboring sectors S_{k-1} and S_{k+1}.

Notice that (**S2**) justifies the claim made in the last section that eigenvalue problems for neighboring Stokes sectors are null: If $\psi(x)$ decays in S_k, then it must be proportional to y_k, and so it is dominant and does not decay in the two neighboring sectors $S_{k\pm1}$.

The Wronskian of y_k and y_{k+1}, evaluated as $|x| \to \infty$ in either S_k or S_{k+1}, where the asymptotic approximations to both solutions are fixed via (6.7), is

$$W[y_k, y_{k+1}] := y_k \, y'_{k+1} - y'_k \, y_{k+1} = 1. \tag{6.9}$$

Having a nonzero Wronskian, y_k and y_{k+1} must be independent, and so they form a basis of solutions to (6.6). All other solutions may be written as linear combinations of these two. In particular, we can expand y_{-1} in the basis of y_0 and y_1, yielding a *Stokes relation*

$$y_{-1}(x, E) = C(E) y_0(x, E) + \widetilde{C}(E) y_1(x, E), \tag{6.10}$$

where the *Stokes multipliers* $C(E)$ and $\widetilde{C}(E)$ depend on E but are independent of x.

Explicit expressions for the Stokes multipliers follow from taking the Wronskians of (6.10) first with y_1:

$$C(E) = \frac{W[y_{-1}, y_1]}{W[y_0, y_1]} = W[y_{-1}, y_1],$$

and then with y_0:

$$\widetilde{C}(E) = \frac{W[y_{-1}, y_0]}{W[y_1, y_0]} = -\frac{W[y_{-1}, y_0]}{W[y_0, y_1]} = -1,$$

where the final equality follows from (6.9) and (6.8). Since the y_k are entire functions of E, the same is true of $C(E)$. The Stokes relation (6.10) written in terms of $y_0(x, E) \equiv y(x, E)$ thus has the symmetrical form

$$C(E)y(x, E) = \omega^{-1/2} y(\omega x, \omega^{-2} E) + \omega^{1/2} y(\omega^{-2} x, \omega^2 E). \tag{6.11}$$

If we now set $x = 0$, then the three first arguments of the functions y all match. Defining $D(E) := y(0, E)$, we get

$$C(E)D(E) = \omega^{-1/2}D(\omega^{-2}E) + \omega^{1/2}D(\omega^2 E), \qquad (6.12)$$

where D, just like C, is an entire function of E. If we differentiate (6.11) with respect to x and then set $x = 0$, we obtain the second relation

$$C(E)\widetilde{D}(E) = \omega^{1/2}\widetilde{D}(\omega^{-2}E) + \omega^{-1/2}\widetilde{D}(\omega^2 E), \qquad (6.13)$$

where $\widetilde{D}(E) := y'(0, E)$ is another entire function of E.

The functional relations (6.12) and (6.13) are powerful because all functions involved are entire. In fact, they are well known in a different area of mathematics, namely, the study of integrable models (IMs) of quantum field theory and statistical mechanics. In that context (6.12) and (6.13) are known as *Baxter's TQ relations*. Relations of this form were first introduced by Baxter in the 1970s [Baxter (1971)] to solve the eight-vertex lattice model, a generalization of the square ice and six-vertex models solved in [Lieb (1967a–d)] and [Sutherland (1967)]. Baxter's TQ relations continue to play a key role in quantum integrable models and are of increasing interest in mathematics.

The relevance of the functions $C(E)$, $D(E)$, and $\widetilde{D}(E)$ to the eigenvalue problems associated with the differential equation (6.2) is that they can be interpreted as spectral determinants. A *spectral determinant* is a function that vanishes exactly at the eigenvalues of a given eigenvalue problem, just as the characteristic polynomial encodes the eigenvalues of a finite-dimensional matrix. The Stokes multiplier $C(E)$ is equal to the Wronskian of $y_{-1}(x)$ and $y_1(x)$ and it vanishes if and only if these two solutions are multiples of one another. This happens if and only if (6.2) has a solution decaying in the two sectors S_{-1} and S_1 simultaneously. This is precisely the quantization condition of the $\mathcal{PT}$-symmetric cubic-oscillator problem (6.2). Therefore, $C(E)$ is proportional to the spectral determinant of the oscillator problem (6.2) and its zeros are the corresponding eigenvalues.[2]

Similarly, let $y(x, E)$ be a solution of (6.2) that vanishes for large $|x|$ in S_0. Then $D(E) \equiv y(0, E)$ is zero if and only if the differential equation has a solution that vanishes at $x = 0$ and also as $|x| \to \infty$ in S_0. Therefore, the zeros of $D(E)$ are the eigenvalues of the radial problem with a (zero) Dirichlet boundary condition at the origin and decay along the negative

---

[2]Strictly speaking, we have only shown that $C(E)$ shares the same zeros as the spectral determinant. The ratio of the two could be an entire function with no zeros (for example, the exponential of a polynomial) but this option can be controlled by considering the asymptotic growth of the two functions, as touched on in the next section.

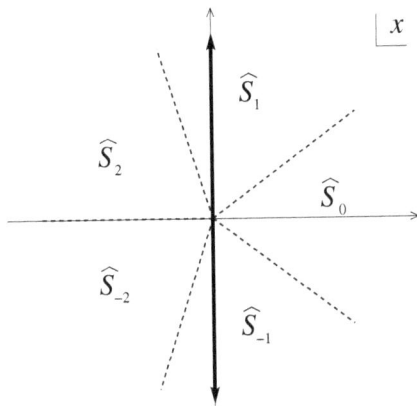

Fig. 6.2 Rotated Stokes sectors $\widehat{S}_k := iS_k$ for the real cubic oscillator (6.14). The quantization contour for the associated $\mathcal{PT}$-symmetric problem runs along the imaginary axis.

imaginary axis. An identical argument establishes that the spectral determinant $\widetilde{D}(E)$ encodes the spectrum of the radial problem with a Neumann boundary condition at the origin. We will see shortly that these two are Hermitian problems in disguise: The variable changes

$$x \to -ix, \quad \psi(-ix) = \phi(x)$$

remove the factor of i from the potential and negate E, mapping the differential equation to the real-cubic-oscillator problem

$$- \phi''(x) + x^3 \phi(x) = e\,\phi(x), \tag{6.14}$$

where $e = -E$, and rotating all Stokes sectors and quantization contours by 90° to those shown in Fig. 6.2.

The rotated Stokes sector $\widehat{S}_0$ includes the positive-real axis and so the negated zeros of $D(E)$, $\{e_j : D(-e_j) = 0\}_{j=0}^{\infty}$, are eigenvalues for solutions of (6.14) with boundary conditions

$$\phi(0) = 0 \qquad [\phi(x) \in L^2(\mathbb{R}^+)].$$

This is a Hermitian eigenvalue problem on the positive-real axis. Thus, the e_j are all real and (because the potential x^3 is positive on $\mathbb{R}^+$) all positive. Note that in the rotated picture, the quantization contour for the $\mathcal{PT}$-symmetric problem is the imaginary axis, linking $\widehat{S}_{-1}$ and $\widehat{S}_1$. Even though the differential equation is now real, the quantization contour is complex and the reality of the $\mathcal{PT}$-symmetric spectrum remains nontrivial.

The functional relations (6.12) and (6.13) each link the spectrum of the $\mathcal{PT}$-symmetric cubic oscillator with that of a real cubic oscillator, subject to either a Dirichlet or Neumann boundary condition at the origin. To see the relation between the eigenvalues explicitly, we write the spectral determinants in product form. For E large and away from the positive real axis it is safe to use the WKB approximation to find the asymptotic behavior of $y(0, E) \equiv D(E)$ [Dorey and Tateo (1999b)]:

$$\ln D(E) \sim \frac{\pi^2}{30} \frac{2^{11/3}}{\Gamma(2/3)^3} (-E)^{5/6} \quad (|\arg(-E)| < \pi).$$

The *order*[3] of $D(E)$ is $\frac{5}{6}$, which is strictly less than 1. By Hadamard's factorization theorem $D(E)$ can therefore be written as a simple infinite product over its zeros, which are the numbers $\{-e_i\}_{i=0}^{\infty}$:

$$D(E) = D(0) \prod_{i=0}^{\infty} \left(1 + \frac{E}{e_i} \right). \tag{6.15}$$

The precise (nonzero) value of the constant $D(0)$ given in [Dorey and Tateo (1999b)] is not needed here.

Consider (6.12) evaluated at $E = -e_k$, a zero of $D(E)$. Since $C(E)$ is entire, it is nonsingular at this point and the right side of (6.12) vanishes:

$$\omega^{-1/2} D(-\omega^{-2} e_k) + \omega^{1/2} D(-\omega^2 e_k) = 0 \quad (k = 0, 1, \dots).$$

Replacing $D(E)$ with its Hadamard product (6.15) and rearranging, we deduce that the eigenvalues of the Hermitian radial problem satisfy an infinite set of exact quantization conditions:

$$\prod_{i=0}^{\infty} \left(\frac{e_i - \omega^{-2} e_k}{e_i - \omega^2 e_k} \right) = -\omega \quad (k = 0, 1, \dots). \tag{6.16}$$

Such coupled equations are known in the study of integrable models as *Bethe-ansatz* equations.

Next we can evaluate (6.12) at a zero of $C(E)$, that is, at an eigenvalue E_k of the $\mathcal{PT}$-symmetric cubic oscillator (6.2). Reasoning as above, the right side of (6.12) is again zero, so

$$\omega^{-1/2} D(\omega^{-2} E_k) + \omega^{1/2} D(\omega^2 E_k) = 0 \quad (k = 0, 1, \dots).$$

The Hadamard product formula for $D(E)$ implies that the eigenvalues of the $\mathcal{PT}$-symmetric lateral problem are related to those of the Hermitian radial problem by the set of Bethe-ansatz-like equations

$$\prod_{i=0}^{\infty} \left(\frac{e_i + \omega^{-2} E_k}{e_i + \omega^2 E_k} \right) = -\omega \quad (k = 0, 1, \dots). \tag{6.17}$$

---

[3]The *order* of an entire function $f(z)$ is the lower bound of all positive numbers μ such that $|f(z)| = O[\exp(|z|^{\mu})]$ as $|z| \to \infty$ (see, for example, [Titchmarsh (1932)]).

To summarize, the spectrum of the radial Dirichlet problem, encoded in $D(E)$, is tangled both with itself via the Bethe-ansatz equations (6.16) and with the spectrum of the $\mathcal{PT}$-symmetric problem, encoded in $C(E)$, via the Bethe-ansatz-like equations (6.17). Similar considerations apply to the Neumann problem encoded in $\tilde{D}(E)$. These observations form the beginning of a connection between ordinary differential equations and integrable models now known as the *ODE/IM correspondence*. This was first observed in [Dorey and Tateo (1999a)], using results on the ODE side from [Voros (1982, 1983, 1999)]. It was elaborated further in [Bazhanov *et al.* (2001)], with the link with Baxter's TQ relation being uncovered in [Dorey and Tateo (1999b)]. Ideas stemming from the correspondence are key ingredients in the proof of the reality of the spectrum of the $\mathcal{PT}$-symmetric problem, which is the subject of the next section.

6.3 Proof of Reality

We now complete the reality proof of [Dorey *et al.* (2001a)]. For ease of exposition we specialize to the $\mathcal{PT}$-symmetric cubic oscillator (6.1). Our objective is to show that the eigenvalues $\{E_k\}$ of this oscillator are real. The proof uses the functional relation (6.12) and some standard facts about the eigenvalues $\{e_i\}$ of the radial problem with a Dirichlet boundary condition at the origin. (We could equivalently obtain the same result starting from the problem with a Neumann boundary condition at the origin.)

We begin by demonstrating that the eigenvalues e_i of the Dirichlet radial eigenvalue problem (6.14) are all real and positive. The eigenvalues e_i are real because the Schrödinger equation is Hermitian on $\mathcal{L}^2(\mathbb{R}^+)$ with the given boundary condition. Positivity follows from the positivity of x^3 for $x > 0$, as alluded to above. Taking $e = e_i$ to be an eigenvalue and ϕ_i to be the corresponding eigenfunction, we multiply each term of (6.14) by $\phi_i^*(x)$, integrate from zero to infinity, and simplify the first integral on the left side by using integration by parts and by imposing the boundary conditions:

$$\int_0^\infty dx \left[|\phi_i'(x)|^2 + x^3 |\phi_i(x)|^2 \right] = e_i \int_0^\infty dx \, |\phi_i(x)|^2.$$

Since all these integrals are positive, the eigenvalues e_i of the Dirichlet problem (6.14) are all real and positive.

Returning to the eigenvalues $\{E_k\}$, we take the modulus-squared of (6.17) and since ω is a root of unity, we get

$$\prod_{i=0}^\infty \left| \frac{e_i + \omega^{-2} E_k}{e_i + \omega^2 E_k} \right|^2 = 1 \quad (k = 0, 1, \dots). \tag{6.18}$$

Writing $E_k = \rho_k \exp(i\delta_k)$ with ρ_k and δ_k real and $\rho_k > 0$, and using the reality of the e_i, we obtain

$$\prod_{i=0}^{\infty} \frac{e_i^2 + \rho_k^2 + 2e_i\,\rho_k \cos(\frac{4\pi}{5} - \delta_k)}{e_i^2 + \rho_k^2 + 2\,e_i\,\rho_k \cos(\frac{4\pi}{5} + \delta_k)} = 1 \quad (k = 0, 1, \dots). \tag{6.19}$$

If $\cos\left(\frac{4\pi}{5} - \delta_k\right) = \cos\left(\frac{4\pi}{5} + \delta_k\right)$, then each term in the infinite product is equal to one, and (6.19) holds trivially. On the other hand, if

$$\cos\left(\tfrac{4\pi}{5} - \delta_k\right) > \cos\left(\tfrac{4\pi}{5} + \delta_k\right),$$

then since $e_i, e_i^2, \rho_k, \rho_k^2$ are all positive, every term in the product (6.19) is greater than one and so the equality cannot hold. Similarly, if

$$\cos\left(\tfrac{4\pi}{5} - \delta_k\right) < \cos\left(\tfrac{4\pi}{5} + \delta_k\right),$$

then every term in the product is less than one and (6.19) again fails. Therefore, $\cos\left(\frac{4\pi}{5} - \delta_k\right) = \cos\left(\frac{4\pi}{5} + \delta_k\right)$ and thus

$$\sin\left(\tfrac{4\pi}{5}\right)\sin\delta_k = 0.$$

Hence $\delta_k = 0 \mod \pi$, and E_k is real. This argument did not depend on a particular choice of eigenvalue E_k, so we have proved that the spectrum of the $\mathcal{PT}$-symmetric cubic oscillator (6.2) is entirely real.

Given reality, the positivity of the eigenvalues follows as for the radial problem. Suppose that E_k is an eigenvalue and that $\phi_k(x)$ is its eigenfunction. Multiplying (6.2) by $\phi_k^*(x)$, integrating both sides from $-\infty$ to $+\infty$ along the real axis, integrating the first term by parts, and finally taking the real part noting the reality of E_k, we get

$$\int_{-\infty}^{\infty} dx\,|\phi_k'(x)|^2 = E_k \int_{-\infty}^{\infty} dx\,|\phi_k(x)|^2,$$

from which we conclude that $E_k > 0$.

This proof extends almost unchanged to the general class of $\mathcal{PT}$-symmetric anharmonic oscillators

$$-\psi''(x) - (ix)^{2M}\psi(x) = E\,\psi(x) \qquad (M \in \mathbb{R}^+) \tag{6.20}$$

introduced in [Bender and Boettcher (1998b); Bender *et al.* (1999b)]. If $2M$ is not an integer, $(ix)^{2M}$ can be rendered single valued by running a branch cut along the positive-imaginary axis. The opening angle of the Stokes sectors changes from $\frac{2\pi}{5}$ to $\frac{\pi}{M+1}$, and the phase ω, which rotates solutions between these sectors, changes from $e^{2\pi i/5}$ to $e^{i\pi/(M+1)}$, as illustrated in Fig. 6.3. The boundary conditions are that $\psi(x)$ decays exponentially in both S_{-1} and S_1. For $M < 2$ we could equally impose $\phi \in L^2(\mathbb{R})$, but when

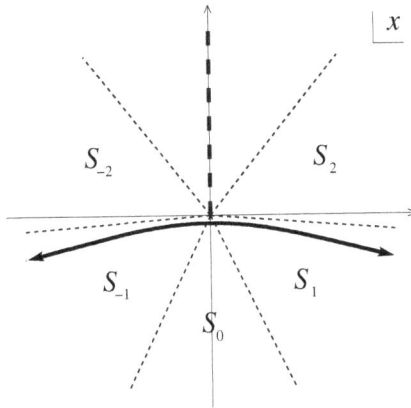

Fig. 6.3   Stokes sectors for the generalized $\mathcal{PT}$-symmetric oscillator (6.30) with $M = 2.1$. There is now a branch cut running up the positive-imaginary axis, shown by a dashed line, and the quantization contour has moved down from the real axis so as to continue to connect S_{-1} and S_1.

M passes 2, the correct analytic continuation of the eigenvalue problem is to deform the quantization contour downward so as to remain inside these two sectors, as shown in the figure. For $M > 1$ the reality proof given above then goes through essentially unchanged and establishes Bender and Boettcher's generalized reality conjecture. For $M \leq 1$ the order of the spectral determinant $D(E)$ turns out to be greater than or equal to one, so the simple Hadamard product (6.15) does not converge, and the proof fails. The boundary case of $M = 1$ is the standard simple harmonic oscillator so the spectrum is still real, while for $M < 1$ Bender and Boettcher had already found that full spectral reality is lost, so a breakdown of the proof was to be expected. This loss of reality as M decreases below 1 is notable in that infinitely-many eigenvalues immediately become complex, and this was analyzed further in [Dorey *et al.* (2005); Bender *et al.* (2017b)].

Further eigenvalue problems can be associated naturally with (6.20). If we do not deform the quantization contour downward as M increases past 2, and instead continue to demand that $\psi \in L^2(\mathbb{R})$, several cases arise. The case shown in Fig. 6.3 corresponds to the simultaneous decay of $\psi(x)$ in S_{-2} and S_2, which is *not* the analytic continuation of the $M < 2$ problem but rather a new problem belonging to the enlarged 'zoo' of lateral connection problems mentioned at the end of Sec. 6.1 with its own distinct spectrum. These problems were also discussed in [Bender *et al.* (1999b)], and found to have interesting reality properties. The relevant spectral determinants turn out to satisfy functional relations (related, as mentioned above, to the fusion hierarchies studied in the context of integrable models) [Dorey and Tateo (1999b)], which in suitable regimes again allow spectral reality to be established [Shin (2005b); Eremenko (2015)].

6.4   General Cubic Oscillators

The next sections discuss contrasting situations where only finitely many eigenvalues become complex for a given parametric value. One of the original motivations to look at the $\mathcal{PT}$-symmetric cubic oscillator was as a toy (zero-space-dimension) model of the Yang-Lee edge singularity [Fisher (1978); Cardy (1985); Cardy and Mussardo (1989)]. In this setting it is natural to consider more general eigenvalue problems of the form

$$- \psi''(x) + \left(igx^3 + \beta x^2 + i\gamma x \right) \psi(x) = E\psi(x) \quad [\psi \in L^2(\mathbb{R})]. \qquad (6.21)$$

This problem is $\mathcal{PT}$ symmetric for real parameters g, β, and γ and the $\mathcal{PT}$ symmetry is unbroken (the eigenvalues E are real) when g is large. For $\beta = \gamma = 0$ we recover the simple cubic oscillator discussed in the last section. For $g > 0$ the Stokes sectors are as before, where we require that the wave function decay in Stokes sectors S_{-1} and S_1.

The reality proof of the last section does not apply to this more general problem and indeed the spectrum is not always real. This is because the Symanzik scaling trick (6.8) no longer leaves the potential invariant. As a result, the $\mathcal{PT}$-symmetric problem finds itself linked not with a Hermitian radial problem but rather with a pair of radial problems with complex-conjugate potentials. Nevertheless, in 2002 Shin showed that for β, γ real, $g \neq 0$, and $g\gamma \geq 0$ the proof of [Dorey et al. (2001a)] could be extended to show that (6.21) does have an entirely real spectrum [Shin (2002)]. This section explains the extra step required, restricting for clarity to $\beta = 0$.

By rescaling x, γ, and E, we are free to set $g = 1$:

$$- \psi''(x) + \left(ix^3 + i\gamma x \right) \psi(x) = E\psi(x). \qquad (6.22)$$

Shin's result implies that whenever a solution of (6.22) simultaneously decays in $S_{\pm 1}$, the corresponding eigenvalue E is real not only for $\gamma = 0$ (the case covered in the last section) but also for all $\gamma > 0$. Spectral reality continues to hold for some negative values of γ; Fig. 6.4 shows the low-lying eigenvalues of (6.22) as a function of γ. Note that as γ becomes negative, the first merging of real eigenvalues to form a complex-conjugate pair occurs

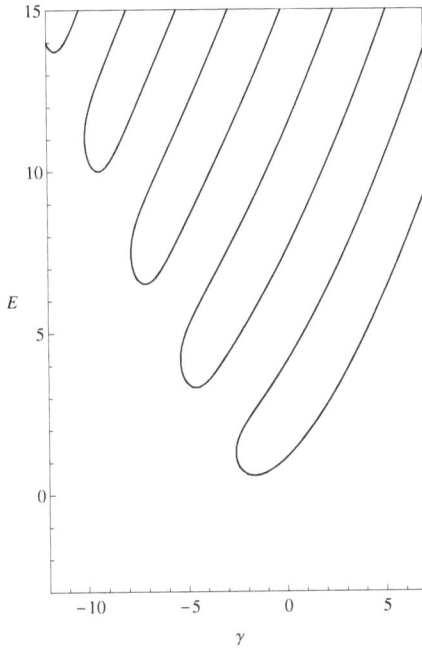

Fig. 6.4 Real eigenvalues of $p^2 + ix^3 + i\gamma x$ as a function of γ. All the eigenvalues are real for $\gamma \geq \gamma_c$, where $\gamma_c \approx -2.6118$. At $\gamma = \gamma_c$ the lowest two eigenvalues meet, forming a complex-conjugate pair for $\gamma < \gamma_c$. As γ decreases, more pairs of adjacent eigenvalues merge and become complex.

not at $\gamma = 0$ but rather at $\gamma = \gamma_c \approx -2.6118$. As γ becomes more negative, successive pairs higher in the spectrum merge and also become complex, as discussed in more detail in [Delabaere and Trinh (2000); Handy *et al.* (2001); Eremenko and Gabrielov (2011b)].

The reality proof starts just as in the last section. Sibuya's general results ensure the existence of an solution $y(x, E, \gamma)$ to (6.22), which is sub-dominant in S_0 and which possesses the WKB asymptotic approximation

$$y(x, E, \gamma) \sim \frac{i}{\sqrt{2}} (ix)^{-3/4} \exp\left[-\tfrac{2}{5}(ix)^{5/2} + \gamma(ix)^{1/2} \right] \qquad (6.23)$$

in closed subsectors of $S_{-1} \cup S_0 \cup S_1$. Using this basic solution, we define further solutions for integer k as

$$y_k(x, E, \gamma) = \omega^{k/2} \, y(\omega^{-k} x, \omega^{-2k}\gamma, \omega^{2k} E), \qquad (6.24)$$

so that y_k is the solution to (6.22) subdominant in the Stokes sector S_k, normalized such that $W[y_k, y_{k+1}] = 1$. Repeating the previous steps shows that $y(x, E, \gamma)$ satisfies the Stokes relation

$$C(E, \gamma)\, y(x, E, \gamma) = \omega^{-1/2} y(\omega x, \omega^{-2} E, \omega^2 \gamma) + \omega^{1/2} y(\omega^{-1} x, \omega^2 E, \omega^{-2} \gamma),$$

where we set $C(E, \gamma) := W[y_{-1}(x, E, \gamma), y_1(x, E, \gamma)]$. Defining $D(E, \gamma) := y(0, E, \gamma)$, we obtain

$$C(E, \gamma)\, D(E, \gamma) = \omega^{-1/2} D(\omega^{-2} E, \omega^2 \gamma) + \omega^{1/2} D(\omega^2 E, \omega^{-2} \gamma). \quad (6.25)$$

This result is a generalization of the functional relation (6.10) for the simple cubic oscillator and, as before, all functions involved are spectral determinants. However, because the Symanzik rescaling leading to (6.24) rotates γ into the complex plane, none of the spectral determinants correspond to Hermitian problems. Nevertheless, we can derive Bethe-ansatz-like equations. Let E_j be an eigenvalue of the $\mathcal{PT}$-symmetric problem, so that $C(E_j, \gamma) = 0$. Setting $E = E_j$ in (6.25) and using the entirety of the functions C and D, we find that

$$0 = \omega^{-1/2} D(\omega^{-2} E_j, \omega^2 \gamma) + \omega^{1/2} D(\omega^2 E_j, \omega^{-2} \gamma). \quad (6.26)$$

Now if $y(x, E, \gamma)$ is a solution of (6.22) normalized as in (6.23), then $y^*(-x^*, E^*, \gamma^*)$ is also a correctly normalized solution for all $\gamma, E \in \mathbb{C}$ [Shin (2002)] and consequently

$$D(E, \gamma) = D^*(E^*, \gamma^*).$$

Since we are taking γ to be real, this means that (6.26) implies that the zeros E_j of $C(E, \gamma)$ satisfy the coupled relations

$$\frac{D(\omega^{-2} E_j, \omega^2 \gamma)}{D^*(\omega^{-2} E_j^*, \omega^2 \gamma)} = -\omega \quad (j = 0, 1, \dots). \quad (6.27)$$

Letting $\{e_i(\gamma) : D(-e_i(\gamma), \gamma) = 0\}$ be the set of negated zeros of $D(E, \gamma)$, we again rewrite $D(E, \gamma)$ as a convergent Hadamard product:

$$D(E, \gamma) = D(0, \gamma) E^{n_0(\gamma)} \prod_{i=0}^{\infty} \left(1 + \frac{E}{e_i(\gamma)}\right),$$

where the factor $E^{n_0(\gamma)}$ allows for the theoretical possibility of there being a zero eigenvalue of multiplicity $n_0(\gamma)$. We then substitute into (6.27), cancel terms, take the modulus squared of both sides, and get

$$\prod_{i=0}^{\infty} \left| \frac{\omega^2 e_i(\omega^2 \gamma) + E_k}{\omega^2 e_i(\omega^2 \gamma) + E_k^*} \right|^2 = 1 \quad (k = 0, 1, \dots). \quad (6.28)$$

When $\gamma = 0$, the eigenvalues e_i are real and (6.28) is equivalent to (6.18).

Even though $D(E, \omega^2\gamma)$ is not the spectral determinant of a Hermitian problem for $\gamma \neq 0$, for $\gamma > 0$ we can still say enough about its zeros to complete the proof. The same variable change as before, namely

$$x \to -ix, \quad y(-ix) = \phi(x),$$

shows that the numbers $\{e_i(\omega^2\gamma)\}$ are the eigenvalues of

$$-\phi''(x) + \left(x^3 - \omega^2\gamma x\right)\phi(x) = e\,\phi(x)$$

with the eigenfunction $\phi(x)$ satisfying $\phi(0) = 0$ and subdominant in the rotated sector $\widehat{S}_0$, as shown in Fig. 6.2. For $\gamma \neq 0$ the complex factor $\omega^2 = e^{4\pi i/5}$ in the potential indicates that the problem is not Hermitian and its spectrum is not real. However, a weaker property holds that is enough for the proof: For each eigenvalue e, the quantity $\omega^2 e$ satisfies $\mathrm{Im}(\omega^2 e) > 0$ whenever $\gamma \geq 0$ [Shin (2002)].

To show this, consider the solution not on the positive-real axis, but along a more general ray lying in $\widehat{S}_0$, defined by setting $x = re^{i\theta}$ with r real and nonnegative and $|\theta| < \pi/5$. Define $f(r) = \phi(re^{i\theta})$. Then, the differential equation on this ray is

$$-e^{-2i\theta} f''(r) + (r^3 e^{3i\theta} - \omega^2\gamma re^{i\theta})f(r) = e\,f(r).$$

Next, multiply by $\omega^2 f^*(r)$ and integrate over all positive r:

$$\int_0^\infty dr\,\left[\omega^2 e^{-2i\theta}|f'(r)|^2 + \omega^2(r^3 e^{3i\theta} - \omega^2\gamma re^{i\theta})|f(r)|^2\right] = \omega^2 e \int_0^\infty dr\,|f(r)|^2,$$

where integration by parts and the boundary conditions have been used to simplify the first term on the left side. Take the imaginary part:

$$\int_0^\infty dr\,\left[\sin(\tfrac{4\pi}{5} - 2\theta)|f'(r)|^2 + \left(\sin(\tfrac{4\pi}{5} + 3\theta)r^3 - \sin(\tfrac{8\pi}{5} + \theta)\gamma r\right)|f(r)|^2\right]$$

$$= \mathrm{Im}\left(\omega^2 e\right)\int_0^\infty dr\,|f(r)|^2.$$

The choice $\theta = -\pi/10$ reduces the first term on the left to zero, leaving

$$\int_0^\infty dr\,|f(r)|^2(r^3 + \gamma r) = \mathrm{Im}(\omega^2 e)\int_0^\infty dr\,|f(r)|^2.$$

All integrals are positive if $\gamma \geq 0$, so $\mathrm{Im}(\omega^2 e) > 0$ for $\gamma \geq 0$.

The final step is to return to the Bethe-ansatz-like equations (6.28). Since $\mathrm{Im}(\omega^2 e_i) > 0$ for all i, if $\mathrm{Im}(E_k) > 0$, then

$$|\omega^2 e_i + E_k| > |\omega^2 e_i + E_k^*| \quad \forall i,$$

while if $\mathrm{Im}(E_k) < 0$,

$$|\omega^2 e_i + E_k| < |\omega^2 e_i + E_k^*| \quad \forall\, i.$$

Neither of these options is consistent with (6.28), so it must be that $\mathrm{Im}\,(E_k) = 0$. The argument works for all k, and hence the spectrum of the generalized PT-symmetric eigenproblem (6.22) is entirely real for $\gamma \geq 0$.

To finish this section, we introduce a further generalization of the PT-symmetric cubic potential, which provides some motivation for the content of the next section, and is perhaps of independent interest. We add a regular singularity at the origin and study

$$H = p^2 + ix^3 + i\gamma x + \left(\lambda^2 - \tfrac{1}{4}\right) x^{-2}, \tag{6.29}$$

where λ is a real nonnegative parameter. The contour along which the wave function is defined must be deformed to avoid the singularity at $x = 0$ if $\lambda \neq \tfrac{1}{2}$. The singularity induces a multivaluedness in the solutions: For $\lambda \neq \tfrac{1}{2}$ the problem is defined on a multisheeted Riemann surface and the number of Stokes sectors increases from five on the single sheet when $\lambda = \tfrac{1}{2}$, much as happens for the Bender–Boettcher generalization (6.20) of the cubic problem whenever $2M$ is not an integer. Znojil termed quantization contours that begin and end on different Riemann sheets as *toboggans* [Znojil (2005a)][4]. We investigate the λ dependence of the spectrum of (6.29) for the eigenvalue problem whose quantization contour lies on the first Riemann sheet joining S_{-1} to S_1 and passing below $x = 0$ when $\lambda \neq \tfrac{1}{2}$.

For $\lambda = \tfrac{1}{2}$ the problem reduces to the one studied earlier. Its real eigenvalues, as a function of γ, were shown in Fig. 6.4. The spectrum was proved above to be real for $\gamma \geq 0$, and it remains so until $\gamma = \gamma_c \approx -2.68$, at which point the lowest-lying pair of eigenvalues merge and become complex. In more-fancy language Fig. 6.4 depicts the *real spectral locus* $Z(E, \gamma)$ of (6.22), that is, those pairs $(E, \gamma) \in \mathbb{R}^2$ such that (6.22) has a solution satisfying the given boundary conditions. Equivalently, the spectral locus is the zero set of the spectral determinant $C(E, \gamma)$. In fact, the real spectral locus of (6.22) consists of a union of simple, disjoint curves $\Gamma_n \in \mathbb{R}^2$:

$$Z(E, \gamma) = \cup_{n=0}^{\infty} \Gamma_n,$$

with each curve Γ_n starting and ending at $\gamma \to \infty$ [Eremenko and Gabrielov (2011b)]. Figure 6.4 shows parts of the curves $\Gamma_0, \ldots, \Gamma_4$. At any fixed $\gamma < \gamma_c$, there are a finite number of pairs of complex eigenvalues and

---

[4]In some cases a variable change can be used to 'unwind' the toboggan [Dorey *et al.* (2008)], but we do not exploit this approach in this section. This unwinding technique is illustrated in (2.70).

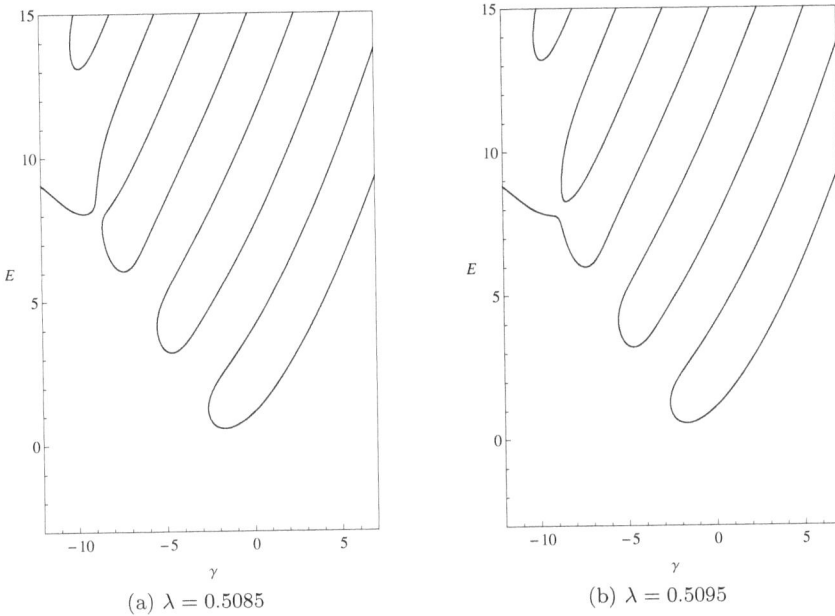

(a) $\lambda = 0.5085$

(b) $\lambda = 0.5095$

Fig. 6.5   Real eigenvalues of $p^2 + ix^3 + i\gamma x + (\lambda^2 - \frac{1}{4})x^{-2}$ as a function of γ for two values of λ. As λ increases, the connectivity of the eigenvalues is rearranged.

infinitely-many real eigenvalues, but as $\gamma \to -\infty$ every real level ultimately pairs with one of its nearest neighbors and becomes complex [Delabaere and Trinh (2000); Eremenko and Gabrielov (2011b)].

To explore the effect of varying λ away from $\lambda = \frac{1}{2}$ we solve the differential equation associated with (6.29) numerically along a straight contour just below the real axis. Figure 6.5 shows the low-lying real eigenvalues for two values of λ just larger than $\frac{1}{2}$. The new feature compared to the situation at $\lambda = \frac{1}{2}$ is that one level in the $\gamma \geq 0$ part of the real spectral locus is no longer connected to its nearest neighbor, resulting in a reversal of the pairing pattern of the level connectivity as this level is passed. In Fig. 6.5(a) this is the seventh $\gamma \geq 0$ level. For the slightly larger value of λ shown in Fig. 6.5(b), it is the fifth $\gamma \geq 0$ level, with the pairing of the sixth $\gamma \geq 0$ level switching from the fifth to the seventh level.

So far, this is similar to the reshuffling of levels in the Bender–Boettcher problem with an added x^{-2} singularity at the origin studied in [Dorey and Tateo (1999b); Dorey *et al.* (2005)]. Increasing λ further, as illustrated in Figs. 6.6 and 6.7(a), shows that as λ increases, the parity-flipping level moves further down the spectrum until it lies below all the rest of the real

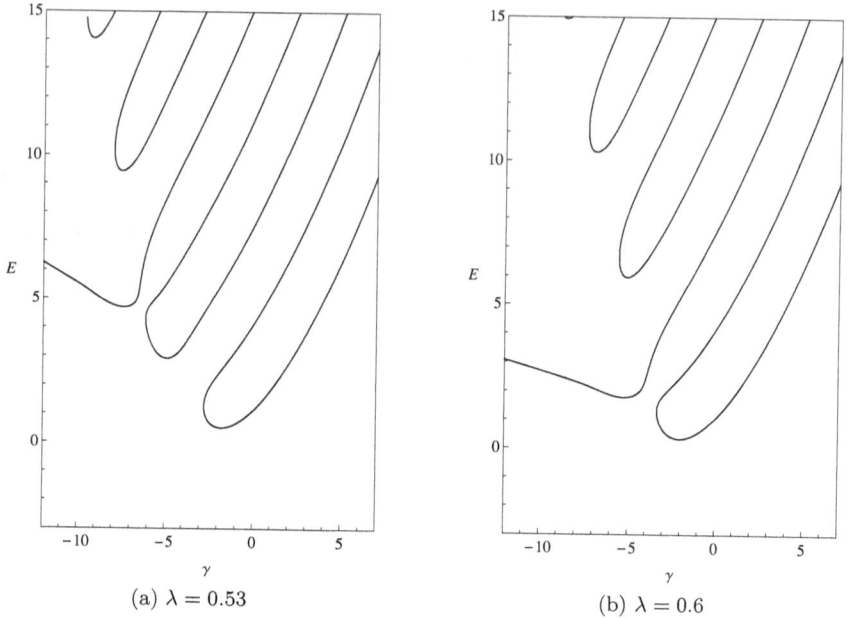

(a) $\lambda = 0.53$

(b) $\lambda = 0.6$

Fig. 6.6 Real eigenvalues of $p^2 + ix^3 + i\gamma x + (\lambda^2 - \frac{1}{4})x^{-2}$ as a function of γ for further values of λ.

spectral locus. However, when $\lambda = 3$ is reached, as depicted in Fig. 6.7(b), a new feature has emerged: A second parity-flipping level has moved through the spectrum, and a third one is about to approach the fourth and fifth levels. This pattern continues as λ grows, with an increasing number of parity-flipping levels traveling down through the spectrum. The numerics are delicate and it would be valuable to develop an analytic understanding of these parity-flipping levels to confirm these preliminary observations.

In PT-symmetric systems complex eigenvalues form via the coincidence of two or more previously real eigenvalues and the appearance of an *exceptional point* in the spectrum. The coincidence of the eigenvalues is accompanied by the simultaneous coincidence of the corresponding eigenfunctions. We have seen numerically that the real eigenvalues of (6.29) can coincide in two ways: Two previously-real eigenvalues can meet and form a complex-conjugate pair of eigenvalues, corresponding to a quadratic exceptional point; or, a complex-conjugate pair of eigenvalues can coincide with a real eigenvalue with all three eigenvalues momentarily real, corresponding to a cubic exceptional point.

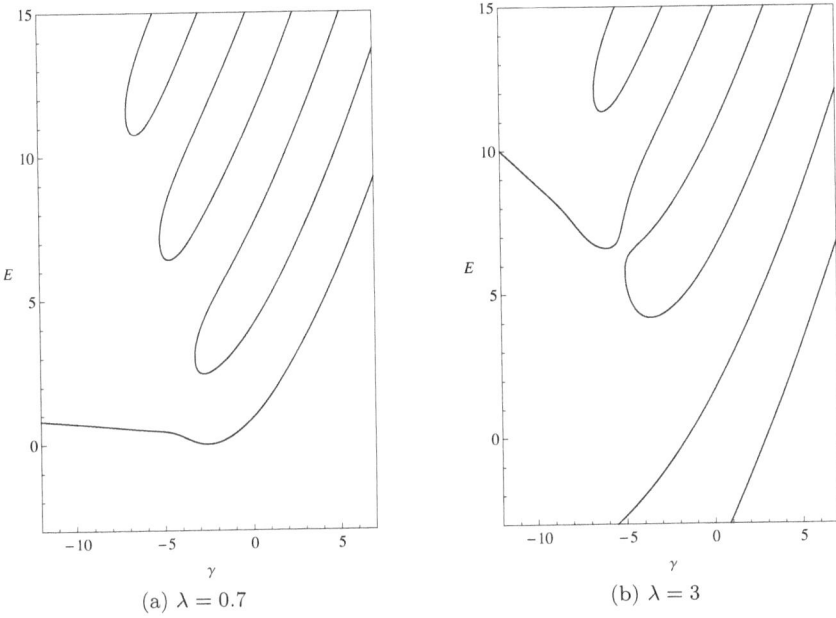

(a) $\lambda = 0.7$

(b) $\lambda = 3$

Fig. 6.7   Real eigenvalues of $p^2 + ix^3 + i\gamma x + (\lambda^2 - \frac{1}{4})x^{-2}$ as a function of γ for further values of λ.

To gain insight into the nature of these points we plot the real energy surface as a function of γ and λ. Figure 6.8 shows $E(\gamma, \lambda)$ for the cubic oscillator (6.29) in the vicinity of the cubic exceptional point at $(\gamma, \lambda) \approx (-8.737, 0.509)$, where $E \approx 8.029$. This exceptional point is located at the cusp where two lines of quadratic exceptional points meet, as is illustrated in blue in the (γ, λ)-plane above the real energy surface in the figure.

Extending the range of γ and λ allows one to anticipate an intricate pattern of quadratic and cubic exceptional points. In the next section we map such points for a slight modification of the current model. This modified model, a three-parameter family of $\mathcal{PT}$-symmetric problems generalizing the Bender–Boettcher Hamiltonian $H = p^2 - (ix)^{2M}$, is more analytically tractable. The lines of exceptional points along which complex-conjugate pairs of eigenvalues appear in the spectrum have interesting behaviors as functions of the three parameters. Many of these behaviors can be explained analytically [Dorey et al. (2009)].

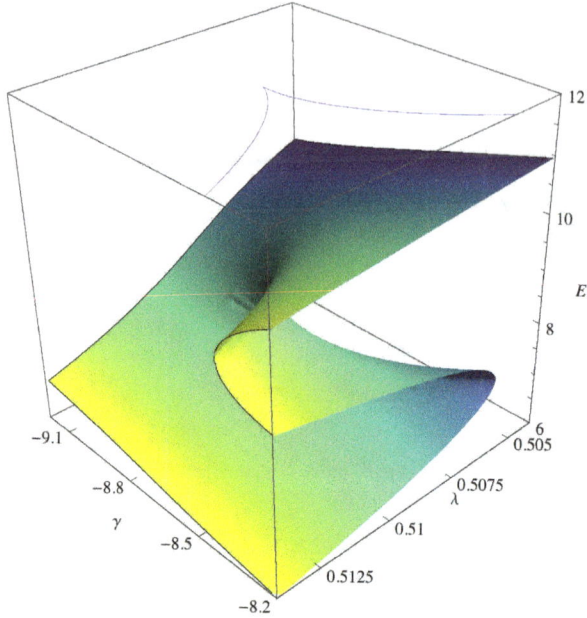

Fig. 6.8 Behavior of the real-energy-level surface $E(\gamma, \lambda)$ of (6.29) in the vicinity of a cubic exceptional point.

6.5 Generalized Bender–Boettcher Hamiltonian

The Bender–Boettcher problems (6.20) can be generalized to a three-parameter family that includes an interesting variety of special cases [Suzuki (2001); Dorey *et al.* (2001a)]. The Schrödinger equation is

$$\left[-\frac{d^2}{dx^2} - (ix)^{2M} - \alpha(ix)^{M-1} + \left(\lambda^2 - \tfrac{1}{4} \right) x^{-2} \right] \psi(x) = E\,\psi(x), \quad (6.30)$$

where $M > 0$, α, and λ are real parameters. For (6.20) a branch along the positive-imaginary axis renders the powers of ix single valued for noninteger values of $2M$. We take the boundary condition to be that the wave function $\psi(x)$ decays simultaneously in the Stokes sectors S_{-1} and S_1, where

$$S_k = \left\{ x \in \mathbb{C} : \left| \arg(ix) - \frac{2\pi k}{2M+2} \right| < \frac{\pi}{2M+2} \right\} \quad (k \in \mathbb{Z}).$$

This is the analytic continuation of the boundary condition for the $\mathcal{PT}$-symmetric cubic oscillator considered above. As before, for $M \geq 2$ the corresponding quantization contour cannot be the real axis, but rather must bend down into the lower-half plane to link S_{-1} and S_1 (see Fig. 6.3 for

$M = 2.1$). For all M the contour must pass below the origin if $\lambda^2 - \frac{1}{4} \neq 0$ to avoid the singularity there.

For both numerical and analytical work it is sometimes convenient to remove the bend in the contour. This can be achieved by a simple change of variables. One good choice for a change of variables is

$$ix = (i\kappa w)^\sigma, \quad \psi(x) = w^{(\sigma-1)/2}\phi(w),$$

with arbitrary σ and κ, which transforms (6.30) without inducing a first-derivative term. For $\sigma = 2/(M+1)$ and $\kappa = 1/\sqrt{\sigma}$, the dual equation is

$$\left[-\frac{d^2}{dw^2} + w^2 - \widetilde{\alpha} + \left(\widetilde{\lambda}^2 - \tfrac{1}{4} \right) w^{-2} - \widetilde{E}(iw)^{2\widetilde{M}} \right] \phi(w) = 0 \qquad (6.31)$$

with

$$\widetilde{M} = \tfrac{1-M}{M+1}, \quad \widetilde{E} = \left(\tfrac{2}{M+1} \right)^{2M/(M+1)} E, \quad \widetilde{\lambda} = \tfrac{2\lambda}{M+1}, \quad \widetilde{\alpha} = \tfrac{2\alpha}{M+1}.$$

The requirement that eigenfunctions $\psi(x)$ of the initial $\mathcal{PT}$-symmetric problem should decay in S_{-1} and S_1 translates into the simultaneous decay of $\phi(w)$ in the sectors $\widetilde{S}_{-1}$ and $\widetilde{S}_1$, with

$$\widetilde{S}_k = \{ w \in \mathbb{C} : |\arg(iw) - \pi k/2| < \pi/4 \}.$$

The angular size of the Stokes sectors is independent of M because w^2, the leading term in (6.31) for large $|w|$, is independent of M. However, even for integer values of M, the mapping introduces a branch cut running up the positive-imaginary axis (see Fig. 6.9).

Apart from the branch cut, the sectors match those of the simple harmonic oscillator and this makes (6.31) particularly useful for numerical work. Eigenvalues can be found by solving the ODE on a straight M-independent horizontal contour in the lower-half complex-w plane. An efficient approach uses WKB asymptotics at large $|w|$ as initial conditions for a pair of numerical solutions, $\phi_{-1}(w)$ and $\phi_1(w)$, decaying as $\text{Re}(w) \to -\infty$ and $\text{Re}(w) \to +\infty$ respectively. We find the eigenvalues by looking for zeros of the Wronskian $W[\phi_{-1}, \phi_1]$ evaluated in the neighborhood of the origin where both numerical solutions are reliable.

The transformed equation (6.31) allows for an easy analytical derivation of the leading behavior of the eigenvalues of (6.30) in the $M \to \infty$ limit, dubbed the *complex square well* in [Bender *et al.* (1999a)]. For simplicity we keep α and λ fixed as the limit is taken; then $\widetilde{M} \to -2$, $\widetilde{\alpha} \to 0$, $\widetilde{\lambda} \to 0$, and the dual problem reduces to

$$\left[-\frac{d^2}{dw^2} + w^2 + \left(\widetilde{E} - \tfrac{1}{4} \right) w^{-2} \right] \phi(w) = 0.$$

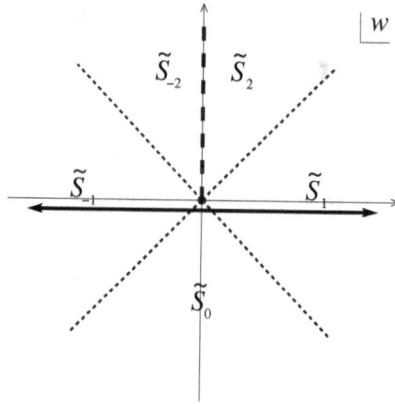

Fig. 6.9   Stokes sectors for the dual equation (6.31).

This equation is exactly solvable with (generalized) eigenvalues $\widetilde{E}_n = (2n + 1)^2$, $n = 0, 1, \ldots$. Translating back, we find that the leading behavior of the eigenvalues E_n of the original problem is

$$E_n \sim \tfrac{1}{4}(2n + 1)^2 M^2 \qquad (M \to \infty),$$

which reproduces a result that was found by a more involved route in [Bender *et al.* (1999a)].

6.6   Reality Domains for the Generalized Problem

A straightforward extension of the reality arguments from Sec. 6.3, given in [Dorey *et al.* (2001a)], proves that for $M > 1$ the spectrum of (6.30) is real provided that an associated Hermitian problem has a positive spectrum. This in turn holds provided that α and λ satisfy

$$\alpha < M + 1 + 2|\lambda|. \tag{6.32}$$

In this region the spectrum of (6.30) is positive if

$$\alpha < M + 1 - 2|\lambda|. \tag{6.33}$$

It is natural to examine the reality (or nonreality) of the spectrum in the region not covered by this result; that is, when $\alpha > M + 1 + 2|\lambda|$. Investigations begun in [Dorey *et al.* (2001b)] and continued in [Sorrell (2007); Dorey *et al.* (2009)] show that the region of complex eigenvalues has more structure than might be expected from the simple bound (6.32).

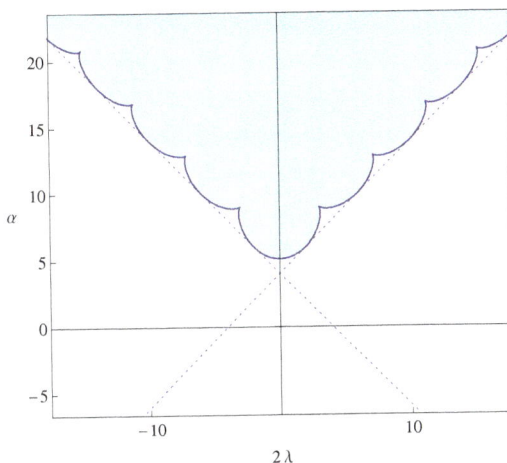

Fig. 6.10 The $(2\lambda, \alpha)$ plane for $M = 3$, showing the cusped line across which a pair of complex eigenvalues is formed. Below the cusped lines the spectrum is entirely real and below the dotted lines the spectrum is proved to be entirely real.

The shaded region of Fig. 6.10, adapted from [Dorey *et al.* (2001b)], shows the values of $(2\lambda, \alpha)$ for which the spectrum of the $M = 3$ model has at least one pair of complex-conjugate eigenvalues. To describe the situation more precisely, it is useful to work with alternative coordinates

$$\alpha_{\pm} = \frac{1}{2M+2}\left(\alpha - M - 1 \pm 2\lambda\right)$$

on the $(2\lambda, \alpha)$-plane. The dotted lines inclined at $45°$ to the 2λ and α axes in Fig. 6.10 correspond to $\alpha_+ = 0$ and $\alpha_- = 0$. These lines demarcate the *all-real* and *all-positive* regions given by (6.32) and (6.33). The cusped solid line shows the exceptional points in the $(2\lambda, \alpha)$-plane where a first pair of complex eigenvalues appears. This line has a regular pattern of cusps, and appears to touch the dotted boundary lines $\alpha_{\pm} = 0$ only at isolated points. We now discuss these two features, beginning with the second.

A special feature of (6.30) on the line $\alpha_- = 0$ is the existence of an exactly-zero eigenvalue [Dorey *et al.* (2001b)] with corresponding (unnormalized) eigenfunction

$$\psi(x) = (ix)^{1/2+\lambda} \exp\left(\tfrac{1}{M+1}(ix)^{M+1}\right). \tag{6.34}$$

The line $\alpha_+ = 0$ also has a protected zero eigenvalue and its eigenfunction is found from (6.34) by sending $\lambda \to -\lambda$. On the line $\alpha_- = 0$, $\alpha = M+1+2\lambda$, and the Schrödinger equation (6.30) factors as

$$\mathcal{Q}_+ \mathcal{Q}_- \psi(x) = E\psi(x), \tag{6.35}$$

where[5]

$$Q_\pm = \pm\frac{d}{dx} + i(ix)^M + (\lambda + \tfrac{1}{2})\frac{1}{x}.$$

The zero-energy eigenfunction (6.34) can then be recovered by solving

$$Q_-\psi(x) = 0. \tag{6.36}$$

The factorization is reminiscent of supersymmetric quantum mechanics. Let $\psi(x)$ be a normalizable solution (decaying in S_{-1} and S_1) to (6.35) with corresponding eigenvalue E. Set $\widetilde{\psi}(x) = Q_-\psi(x)$. Then, for $E \neq 0$ this is a nontrivial function, also decaying in S_{-1} and S_1. Applying the operator Q_- to both sides of (6.35) yields

$$Q_-Q_+\widetilde{\psi}(x) = EQ_-\psi(x) = E\widetilde{\psi}(x). \tag{6.37}$$

Thus, $\widetilde{\psi}(x)$ is an eigenfunction of the partner problem to (6.35) obtained by swapping the order of the operators Q_+ and Q_-:

$$Q_-Q_+ = -\frac{d^2}{dx^2} - (ix)^{2M} + (M - 1 - 2\lambda)(ix)^{M-1} + (\lambda + \tfrac{1}{2})(\lambda + \tfrac{3}{2})x^{-2}.$$

The same argument can be run in reverse to show that any nonzero eigenvalues for (6.37) are also eigenvalues for (6.35). Hence, the nonzero part of the spectrum at the point

$$(2\lambda, \alpha) = (2\lambda, M + 1 + 2\lambda), \quad (\alpha_+, \alpha_-) = \left(\tfrac{2\lambda}{M+1}, 0\right)$$

on the *supersymmetric* line $\alpha_- = 0$ is identical to that of its partner problem, which is at the point

$$(2\widehat{\lambda}, \widehat{\alpha}) = (2\lambda + 2, -M + 1 + 2\lambda), \quad (\widehat{\alpha}_+, \widehat{\alpha}_-) = \left(\alpha_+ - \tfrac{M-1}{M+1}, -1\right)$$

on the partner line $\widehat{\alpha}_- = -1$.

Solving the first-order equation (6.36) provides a normalizable zero-energy solution to (6.35) for all λ but solving the partner equation fails to provide a zero-energy solution to (6.37) because all nontrivial solutions $Q_+\widetilde{\psi}$ grow in both S_{-1} and S_1. Rather, any zero eigenvalues for the partner problem must arise from normalizable solutions to the full equation $Q_-Q_+\widetilde{\psi} = 0$, and as we see below these are only found for isolated values of λ.

The PT symmetry of the current problem means that a complex-conjugate pair of eigenvalues can only appear following the coincidence of two real eigenvalues. On the line $\alpha_- = 0$ these coincidences happen at

---

[5]This differs slightly from the formula in [Dorey *et al.* (2001b)] as that paper used the rotated version of the PT-symmetric problem.

$E = 0$. As just shown, at all points on this line the problem has a zero eigenvalue, so the points at which complex eigenvalues can form will be those where a *second* eigenvalue approaches zero. Zero eigenvalues have a special status here because, with the aid of some variable changes, the value of the spectral determinant $C(E, \alpha, \lambda)$ at $E = 0$ can be found exactly [Dorey and Tateo (1999b); Dorey *et al.* (2001b)]. In (α_+, α_-) coordinates the spectral determinant is given by

$$C(E, \alpha_+, \alpha_-)|_{E=0} = \left(\tfrac{M+1}{2}\right)^{1+\alpha_+ + \alpha_-} \frac{2\pi}{\Gamma(-\alpha_+)\Gamma(-\alpha_-)}. \qquad (6.38)$$

As expected, this vanishes for $\alpha_- = 0$, signaling the presence of a zero eigenvalue at all points on this line. Detecting the points where a second eigenvalue approaches zero is more delicate. The best tactic, following [Dorey *et al.* (2001b)], is to examine the zeros of $C(E, \widehat{\alpha}_+, \widehat{\alpha}_-)$ at $E = 0$, the spectral determinant of the partner problem on the partner line $\widehat{\alpha}_- = -1$. From (6.38) these occur at the isolated points $\widehat{\alpha}_+ = m - 1$ ($m = 1, \ldots$). Since $\widehat{\alpha}_+ = \alpha_+ - \frac{M-1}{M+1}$, this means that on the $\alpha_- = 0$ line a second eigenvalue merges with the already present $E = 0$ eigenvalue at the points

$$(\alpha_+, \alpha_-) = \left(m - \tfrac{2}{M+1}, 0\right) \quad (m = 1, 2, \ldots). \qquad (6.39)$$

We have thus identified the exact locations of a discrete subset of the exceptional points on the $(2\lambda, \alpha)$ phase diagram, evenly spaced on the line $\alpha = M + 1 + 2\lambda$.

Replacing λ with $-\lambda$ the above argument may be repeated to locate a second set of exceptional points, this time on the line $\alpha_+ = 0$:

$$(\alpha_+, \alpha_-) = \left(0, m - \tfrac{2}{M+1}\right) \quad (m = 1, 2, \ldots). \qquad (6.40)$$

On Fig. 6.10 the phase diagram for $M = 3$, the points (6.39) and (6.40) correspond, as expected, to the points where the cusped curve of exceptional points meets the boundary lines $\alpha = 4 + 2|\lambda|$. Examining the spectral surface, we see that these points are quadratic exceptional points where two real eigenvalues merge to form a complex-conjugate pair.

Along the lines $\alpha_\mp = n$ for $n \in \mathbb{Z}^+$, the eigenproblem also has an exact eigenfunction with a zero-energy eigenvalue:

$$\psi(x) = \frac{1}{\sqrt{2}} (ix)^{1/2 \pm \lambda} L_n^{\pm 2\lambda/(M+1)} \left(-\frac{2(ix)^{M+1}}{M+1}\right) e^{(ix)^{M+1}/(M+1)}, \qquad (6.41)$$

where $L_n^\rho(t)$ is the nth generalized Laguerre polynomial. It is natural to ask whether anything interesting happens to the spectrum on these lines.

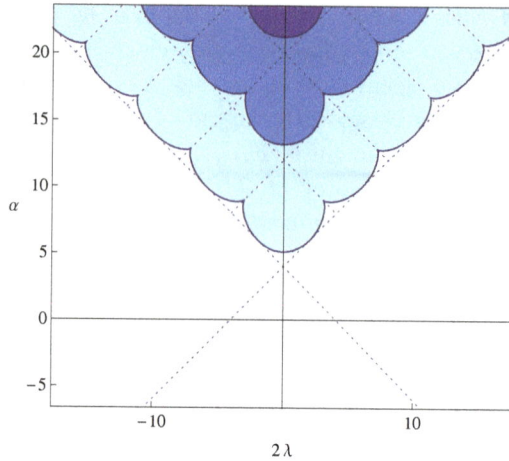

Fig. 6.11   Domain of unreality in the $(2\lambda, \alpha)$ plane for $M = 3$ showing in darker shades of blue the regions in which the spectrum has one, two, and three pairs of complex-conjugate eigenvalues, respectively. Portions of straight lines with a protected zero-energy level are also shown.

Figure 6.11, adapted from [Dorey *et al.* (2009)], adds to Fig. 6.10 the segments of the lines $\alpha_\pm = 1$, 2, 3, and 4 that lie in the region of possible unreality $\alpha > 4 + 2|\lambda|$. There are three shaded regions, each separated by cusped lines. There is one pair of complex-conjugate eigenvalues in the spectrum in the outer shaded region, two pairs of complex eigenvalues in the slightly-darker shaded region and three pairs of complex eigenvalues in the smallest, darkest inner region.

Not only is the pattern of cusps repeated but also the new cusped lines appear to touch the dotted lines $\alpha_\pm = 1, 2$ only at isolated points. The supersymmetry argument used to establish the existence of quadratic exceptional points on the lines $\alpha_\pm = 0$ does not work when $n > 0$ because the corresponding Hamiltonian no longer factors. A related argument using a higher-order supersymmetry may be used when $M = 3$ to locate the quadratic exceptional points for all $n \in \mathbb{Z}^+$ [Dorey *et al.* (2001a,b)]. However, this argument does not apply for other values of M.

Alternatively, the merging of two or more eigenvalues can be identified using *self-orthogonality* of the corresponding eigenfunction at the exceptional point, discussed in this context in [Dorey *et al.* (2009)]. An eigenfunction is self-orthogonal if

$$(\psi, \psi) = 0,$$

where $(\cdot, \cdot)$ is a suitable symmetric inner product. The inner product that is most useful to use here is

$$(f, g) \equiv \int_{\mathcal{C}} dx \, f(x) g(x),$$

where $\mathcal{C}$ is the contour along which the eigenfunction is defined.

The key idea is that a Hamiltonian is at an exceptional point if and only if there is a self-orthogonal eigenfunction at the relevant value of the parameters. This is easy to prove. First, suppose that the Hamiltonian H is tuned to an exceptional point where k eigenvalues coincide at some eigenvalue E. The Hamiltonian will then acquire a (partial) Jordan-block form, with corresponding Jordan chain $\{\psi^{(0)}, \ldots \psi^{(k-1)}\}$ such that

$$(H - E)\psi^{(0)}(x) = 0; \quad (H - E)\psi^{(j)}(x) = \psi^{(j-1)}(x) \quad (j = 1 \ldots k - 1).$$

Hence,

$$(\psi^{(0)}, \psi^{(0)}) = (\psi^{(0)}, H - E)\psi^{(1)}) = ((H - E)\psi^{(0)}\psi^{(1)}) = 0$$

by using the symmetry of the inner product. Therefore, $\psi^{(0)}(x)$ is a self-orthogonal eigenfunction.

The converse statement that the existence of a self-orthogonal eigenfunction implies that the spectrum has an exceptional point is also easy to prove. We use a result of [Trinh (2005)] that at an eigenvalue E_n the spectral determinant $C(E)$ satisfies

$$C'(E_n) = (\psi, \psi)|_{E_n},$$

where the differentiation is with respect to E. Therefore, a self-orthogonal eigenfunction $\psi(x)$ corresponding to an isolated eigenvalue $E = E_n$ is such that $C'(E_n) = 0$. Since by definition E_n is a zero of $C(E)$, it must be that $C(E)$, as a function of E, has a double zero at E_n. Therefore, two or more eigenvalues coincide and the eigenproblem is at an exceptional point for such E_n. This argument does not allow the multiplicity of the exceptional point to be determined.

We can use the exact eigenfunction (6.41) for $\alpha_{\pm} = n$ and self-orthogonality to locate exceptional points with eigenvalue $E = 0$ for general M. To calculate the inner product of the eigenfunction (6.41) with itself, we integrate along $-\gamma_{-1} + \gamma_1$, a piecewise linear contour connecting the Stokes sectors S_{-1} and S_1 where

$$\gamma_{\pm 1} := \{x = \tfrac{1}{i} e^{\pm i\pi/(M+1)} t, \, t \in [0, \infty)\},$$

and use the integral [Millican-Slater (2004)]

$$\int_0^\infty dt\, t^\alpha t^{(\gamma+\rho)/2} e^{-t} L_m^\rho(t) L_n^\gamma(t)$$

$$= \frac{(\frac{\gamma-\rho}{2} - \alpha)_n (\rho+1)_m}{n!\, m!} \Gamma\left(\frac{\gamma+\rho}{2} + 1 + \alpha\right) {}_3F_2\left(-m, \frac{\rho+\gamma}{2} + 1 + \alpha,\right.$$

$$\left.\frac{\rho-\gamma}{2} + 1 + \alpha; \rho+1, \frac{\rho-\gamma}{2} + 1 + \alpha - n; 1\right)$$

$$= \frac{\Gamma(\frac{1}{2}\gamma + \frac{1}{2}\rho + 1 + \alpha)}{m!\, n!} \sum_{k=0}^m \binom{m}{k} (\rho+1+k)_{m-k} (\tfrac{\rho+\gamma}{2} + 1 + \alpha)_k$$

$$\times (\tfrac{\rho-\gamma}{2} + 1 + \alpha)_k (\tfrac{\gamma-\rho}{2} - \alpha)_{n-k},$$

where $(a)_n = a(a+1)\ldots(a+n-1)$ is the *Pochhammer symbol* and ${}_3F_2$ is a generalized hypergeometric function.

The inner product on the line $\alpha_- = n$ [Dorey *et al.* (2009)] is

$$(\psi, \psi) = \frac{\pi}{2}\left(\frac{M+1}{2}\right)^{2n-1+(2+2\lambda)/(M+1)} Q_n(\lambda)/\Gamma(1 - \tfrac{2+2\lambda}{M+1}), \qquad (6.42)$$

where $Q_n(\lambda)$ is the polynomial of degree n in λ defined by

$$Q_n(\lambda) = \left(\tfrac{M-1}{M+1}\right)_n \left(1 + \tfrac{2\lambda}{M+1}\right)_n {}_3F_2\left(-n, \tfrac{2\lambda+2}{M+1}, \tfrac{2}{M+1}; \tfrac{M+1+2\lambda}{M+1}, -n + \tfrac{2}{M+1}; 1\right)$$

$$= \sum_{k=0}^n (-1)^k \binom{n}{k} \left(\tfrac{M-1}{M+1} - k\right)_n \left(\tfrac{2\lambda+2}{M+1}\right)_k \left(1 + \tfrac{2\lambda}{M+1} + k\right)_{n-k}. \qquad (6.43)$$

From (6.42), the wave function (6.41) is self-orthogonal if λ is such that

$$1/\Gamma(1 - \tfrac{2+2\lambda}{M+1}) = 0$$

or

$$Q_n(\lambda) = 0.$$

The infinitely many poles of the gamma term that arise when

$$2 + 2\lambda = m(M+1) \quad (m \in \mathbb{N})$$

predict exceptional points at

$$(\alpha_+, \alpha_-) = (n + m - \tfrac{2}{M+1}, n) \quad (m \in \mathbb{N},\, n \in \mathbb{Z}^+), \qquad (6.44)$$

and exchanging λ with $-\lambda$ gives the analogous result along the lines $\alpha_+ = n$:

$$(\alpha_+, \alpha_-) = (n, n + m - \tfrac{2}{M+1}) \quad (m \in \mathbb{N},\, n \in \mathbb{Z}^+). \qquad (6.45)$$

When $n = 0$, (6.44) and (6.45) reproduce the exceptional points (6.39) and (6.40) found using the supersymmetry of the problem on the reality boundary lines.

For the case $M = 3$ (6.44) and (6.45) match the locations of the isolated points where the cusped lines of Fig. 6.11 touch the lines $\alpha_+ = n$. These points are all quadratic exceptional points of the spectrum [Dorey *et al.* (2001a)], and from examining the spectral surface we expect that this statement extends to all $M > 1$.

What about the exceptional points predicted by the zeros of the polynomial $Q_n(\lambda)$? The first two polynomials are

$$Q_1(\lambda) = \frac{2\lambda}{M+1} + \frac{M^2+3}{(M+1)^2}$$

$$Q_2(\lambda) = \frac{8\lambda^2}{(M+1)^2} + \frac{(12M^2+8\,M+28)\lambda}{(M+1)^3} + \frac{4M^4+4M^3+16M^2+20M+20}{(M+1)^4}.$$

Setting $M = 3$, the single zero of $Q_1(\lambda)$ is located at

$$(2\lambda, \alpha) = (-3, 9),$$

which matches to high numerical accuracy the position of the single cusp shown in Fig. (6.11) on the line $\alpha_- = 1$. The zeros of $Q_2(\lambda)$ match the positions of the two cusps found on the line $\alpha_- = 2$ at

$$(2\lambda, \alpha) = \left(-5 - \frac{3\sqrt{2}}{2}, 15 - \frac{3\sqrt{2}}{2}\right), \quad (2\lambda, \alpha) = \left(-5 + \frac{3\sqrt{2}}{2}, 15 + \frac{3\sqrt{2}}{2}\right).$$

The condition $Q_n(\pm\lambda) = 0$ at $M = 3$ captures the positions of all of the cusps [Dorey *et al.* (2001a)].

For values of M other than 3, Figs. 6.12 and 6.13 show how the reality phase diagrams deform for smaller and larger values of M. The data for these figures was obtained using the mapping between (6.30) and (6.31) that was explained in the last section.

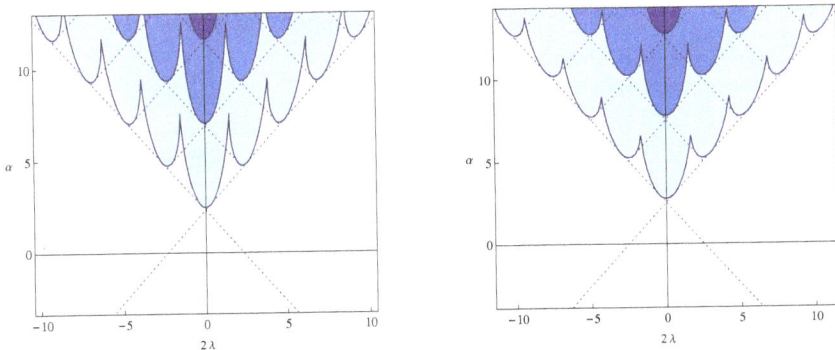

Fig. 6.12   Exceptional lines for $M = 1.3$ (left) and $M = 1.5$ (right).

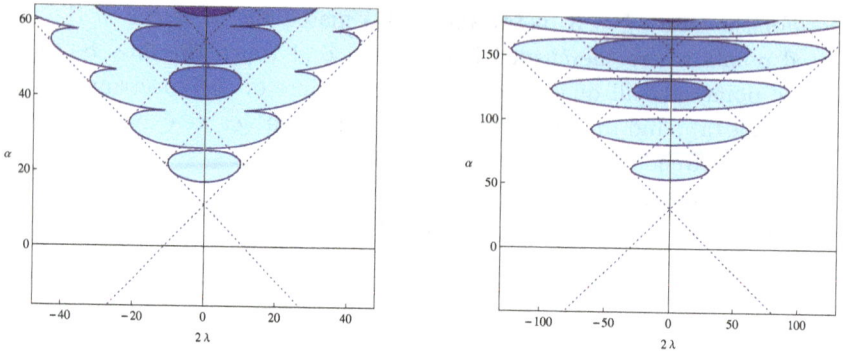

Fig. 6.13   Exceptional lines for $M = 10$ (left) and $M = 30$ (right).

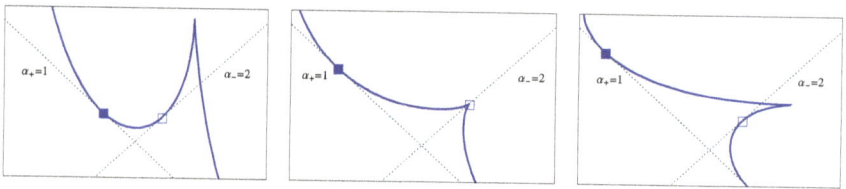

Fig. 6.14   Movement of a cusp as M moves away from 3. Empty boxes show exceptional points arising from zeros of $Q_2(\lambda)$, while filled boxes mark exceptional points arising from poles of the gamma factor of (6.42).

For values of M close to three the pattern of cusps persists, though the cusps no longer lie on the lines $\alpha_\pm \in \mathbb{N}$ as they did when $M = 3$. For $M < 3$ the cusps move upwards, while for $M > 3$ they move sideways, ultimately merging in pairs to leave *islands* where the spectrum is complex, entirely surrounded by regions of reality.

Figure 6.14 zooms in on the movement of one particular cusp, for $M = 1.5$, 3, and 6, lying near the intersection of the lines $\alpha_+ = 1$ and $\alpha_- = 2$. The boxes indicate the exceptional points predicted by self-orthogonality of the zero-energy eigenfunction. The empty boxes mark the points arising from zeros of the polynomial $Q_2(\lambda)$ and the filled boxes mark the points (6.44) arising from the gamma factor of (6.42). The zeros of $Q_n(\lambda)$ correspond to the position of the cubic exceptional point at the cusp when $M = 3$, whereas for all other M the zeros of Q_n correspond to quadratic exceptional points.

The movement of the cusps away from the lines $\alpha_\pm = n$ for $M \neq 3$ was explored in more detail in [Dorey *et al.* (2009)] by using the fact that the extreme cases of $M = 1$ and $M = \infty$ can be solved exactly. The former is the $\mathcal{PT}$-symmetric simple harmonic oscillator [Dorey and Tateo (1999b); Znojil (1999a)], and its spectrum is entirely real. In the large-M limit, (6.30) becomes an inhomogeneous complex square well [Bender *et al.* (1999a); Dorey *et al.* (2009)] as discussed above, which also has an entirely real spectrum. Near the exactly-solvable $M = 1$ and $M = \infty$ limits, perturbation theory can be used to explore the phase diagrams and to provide an analytical understanding of Figs. 6.12 and 6.13, confirming that the cusps move vertically as $M \to 1$ and horizontally as $M \to \infty$. As M increases, the cusps merge and the lines of exceptional points rearrange themselves to form the isolated islands of unreality mentioned above, as shown in Fig. 6.13.

For general M the domains of different degrees of unreality on the phase diagram are demarcated by cusped curves of exceptional points consisting of smooth segments of quadratic exceptional points connecting cubic exceptional points at the cusps. For $M = 3$ all of these cusps lie on the lines $\alpha_{\pm} = n$ along which there is an exactly-zero-energy eigenfunction. We explore the reasons behind this curiosity in the next section.

6.7 Quasi-Exactly-Solvable Models

The values $M = 1$ and $M = \infty$ are the only cases for which *all* eigenvalues and eigenfunctions of (6.30) can be found explicitly for all α and λ. However, if $M = 3$ and

$$\alpha = 4J + 2\lambda \quad \text{or} \quad \alpha = 4J - 2\lambda$$

with J a positive integer, then a *finite* subset can be found algebraically. Such cases are said to be *quasi-exactly solvable* (QES).[6]

The Hermitian sextic oscillator with $\alpha = -(4J + 2\lambda)$ and boundary conditions imposed at the origin and infinity is one of the most studied examples of quasi-exact solvability. Here, following [Dorey *et al.* (2001a, 2009)], we look at the $\mathcal{PT}$-symmetric version of this problem. We eliminate factors of i by setting $z = ix$ and $\Psi(z) = \psi(x/i)$ and get

$$\left[-\frac{d^2}{dz^2} + z^6 + \alpha z^2 + (\lambda^2 - \tfrac{1}{4})z^{-2} \right] \Psi(z) = -E\,\Psi(z), \tag{6.46}$$

where $\Psi(z) \in \mathcal{L}^2(i\mathcal{C})$ and $i\mathcal{C}$ is a quantization contour linking the rotated Stokes sectors $\widehat{S}_{-1}$ and $\widehat{S}_1$ shown in Fig. 6.15.

---

[6]For a comprehensive review of quasi-exact solvability see [Turbiner (2016)].

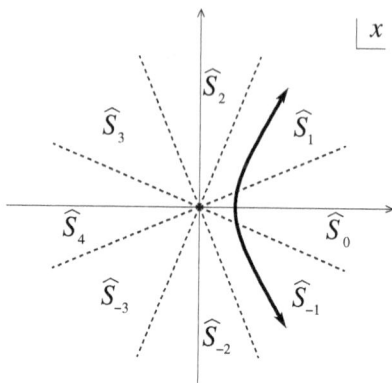

Fig. 6.15   Rotated Stokes sectors for the sextic $\mathcal{PT}$-symmetric oscillator (6.46). The quantization contour $i\mathcal{C}$ connects $\widehat{S}_{-1}$ and $\widehat{S}_1$.

A convenient approach to find the exact solutions, adapted from Bender and Dunne's *ansatz* for the radial problem [Bender and Dunne (1996)], is to write the sought-after eigenfunctions as an exponential term multiplying a power series in x and E. An appropriate *ansatz* for the $\mathcal{PT}$-symmetric eigenproblem (6.46) with the given boundary conditions is

$$\Psi(z) = e^{z^4/4} z^{\lambda+1/2} \sum_{k=0}^{\infty} a_k(\lambda) p_k(E, \alpha, \lambda) z^{2k} \qquad (6.47)$$

where

$$a_k(\lambda) = \left(-\tfrac{1}{4}\right)^k \frac{1}{k!\,\Gamma(k+\lambda+1)}$$

and the functions $p_k(E, \alpha, \lambda)$ are to be determined. If λ is not a negative integer, $\Psi(z)$ is an everywhere-convergent series solution to (6.46). If λ is a negative integer, then $a_{-\lambda}(\lambda) = 0$ and the *ansatz* should be modified. We return to this point shortly but for now we assume that λ is not a negative integer.

We substitute (6.47) into (6.46) and collect coefficients, and find that for $k \geq 1$ the functions $p_k(E) \equiv p_k(E, \alpha, \lambda)$ satisfy the three-term relation

$$p_k(E) = -E p_{k-1} + 4(\alpha - 2\lambda + 4 - 4k)(k-1)(k-1+\lambda) p_{k-2}. \qquad (6.48)$$

Setting $p_0(E) = 1$ fixes the normalization of the solution and $p_1(E) = -E$ then follows from (6.48); $p_k(E)$ is a polynomial of degree k in E with leading term $(-E)^k$. Moreover, $p_k(E)$ is a function of E^2 for even k, and E times a function of E^2 for odd k. Something special happens if

$$J := (\alpha - 2\lambda)/4 \qquad (6.49)$$

is a positive integer: Equation (6.48) implies that

$$p_{J+1}(E) = -E\,p_J(E),$$

and therefore

$$p_{J+k}(E) \propto p_J(E) \quad (k = 1, \dots).$$

All higher coefficients are proportional to $p_J(E)$ and vanish if $p_J(E)$ vanishes. Thus, the power series in (6.47) truncates to a $(2J-2)$-degree polynomial in z. The $e^{z^4/4}$ factor then causes $\Psi(z)$ to decay in the sectors $\widehat{S}_{\pm 1}$. Thus, E is an eigenvalue of the $\mathcal{PT}$-symmetric problem. With all factors included the full asymptotic behavior of $\Psi(z)$ in such a case is

$$\Psi(x) \sim a_{J-1}(\lambda) p_{J-1}(E)\, e^{z^4/4} z^{\lambda+1/2} z^{2J-2}.$$

Given (6.49), we see that $\Psi(x)$ agrees with the subdominant WKB asymptotic approximation in the sectors $\widehat{S}_{\pm 1}$, which follows from (6.3). If E is *not* a zero of $p_J(E)$, then the power series does not terminate and, in general, sums to an exponentially growing function that overwhelms the $e^{z^4/4}$ term in the sectors $\widehat{S}_{\pm 1}$, rendering Ψ non-normalizable. The J solutions of $p_J(E) = 0$ are the quasi-exactly-solvable eigenvalues. At these values of E, (6.47) gives the corresponding normalizable eigenfunctions.

The lines $\alpha = 4J + 2\lambda$ correspond to

$$(\alpha_+, \alpha_-) = [\tfrac{1}{2}(J - 1 + \lambda), \tfrac{1}{2}(J - 1)].$$

Negating λ through all of the above, we see that there is a second set of quasi-exactly-solvable lines at $\alpha = 4J - 2\lambda$, $J \in \mathbb{N}$, corresponding to

$$(\alpha_+, \alpha_-) = [\tfrac{1}{2}(J - 1), \tfrac{1}{2}(J - 1 - \lambda)].$$

For J odd, E is a factor of $p_J(E)$, so one of the QES eigenvalues is zero. Indeed, for J odd the QES lines are precisely the lines of exactly-zero eigenvalues discussed in the last section.

When λ is a negative integer, say $-K$, the *ansatz* (6.47) must be revisited because the factor $a_K(\lambda)$ is zero. If $K > J$, this is not a problem because the power-series part of (6.47) truncates at $k = J - 1$. But when $K \leq J$ the recurrence relation (6.48) can no longer be inferred at $k = K$.

To see what has happened, we can start from a simpler form of the *ansatz* as in [Dorey *et al.* (2012)] and set

$$\Psi(z) = e^{z^4/4} z^{-K+1/2} \sum_{k=0}^{\infty} \left(-\tfrac{1}{4}\right)^k \tfrac{1}{k!} P_k(E, K, J) z^{2k} \tag{6.50}$$

with $\alpha = 4J - 2K$. The functions $P_k \equiv P_k(E, K, J)$ then satisfy

$$(k - K)P_k = -EP_{k-1} + 16(J + 1 - k)(k - 1)P_{k-2} \quad (k \geq 1). \qquad (6.51)$$

The recurrence relation (6.51) determines $P_k(E)$ in terms of J, K, and the initial condition $P_0(E)$ for $k = 1 \ldots K - 1$. When $k = K$, (6.51) takes the form

$$0 = P_0(E)\, R_K(E) \qquad (6.52)$$

with $R_K(E)$ being a degree-K polynomial in E that depends on J. For (6.52) to hold, either $P_0(E) = 0$ with E (for the moment) arbitrary or E must be one of the K zeros of $R_K(E)$ with $P_0(E)$ arbitrary.

If $P_0(E) = 0$, it follows from (6.51) that

$$P_1 = P_2 = \cdots = P_{K-1} = 0,$$

while $P_K(E)$ is arbitrary and $P_k(E)$ for $k \geq K$ are degree-$(k - K)$ polynomials in E, each multiplied by $P_K(E)$. As before, when $k = J + 1$ we get

$$P_{J+1}(E) = -EP_J(E). \qquad (6.53)$$

Consequently,

$$P_{k+J} \propto P_J \quad (k = 1, \ldots). \qquad (6.54)$$

Thus, if E is such that $P_J(E) = 0$, the power-series part of (6.50) truncates. Since $P_J(E)$ is a degree-$J - K$ polynomial in E multiplied by $P_K(E)$, there are $J - K$ solutions to $P_J(E) = 0$, and for each solution E_n the corresponding QES eigenfunction $\Psi(z)$ has normalization constant $P_K(E_n)$.

If E is one of the K zeros of $R_K(E)$, $P_0(E)$ is initially arbitrary. The recurrence relation (6.51) then determines $P_k(E)$ for $k = 1 \ldots K - 1$ to be a polynomial in E of degree k multiplied by P_0. For (6.51) to hold at $k = K$, E must be equal to E_m, one of the K zeros of $R_K(E)$. The value of $P_K(E)$ is unconstrained, and for $k > K$ it is then fixed by (6.51) to be one polynomial in E_m times $P_0(E_m)$ plus another polynomial in E_m times $P_K(E_m)$. As before, the recurrence relation implies (6.53) and (6.54). Requiring that

$$P_J(E_m) = 0,$$

the power-series term of the wave function truncates and fixes $P_K(E_m)$. Thus, the second solution of (6.52) yields K more QES eigenvalues and corresponding normalizable eigenfunctions.

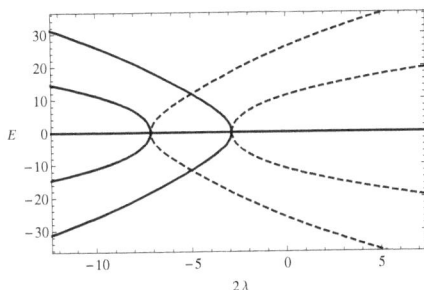

Fig. 6.16   Real (solid lines) and imaginary (dashed lines) parts of the QES eigenvalues for $J = 5$ as functions of 2λ.

What lies behind this splitting of the QES spectrum at $\lambda = -K$ for $K = 1 \ldots J - 2$ on the QES line $\alpha = 4J + 2\lambda$? The answer is that such points lie on *two* QES lines: $\alpha = 4J + 2\lambda$ and $\alpha = 4(J - K) - 2\lambda$. Thus, from one perspective, the problem is QES with $J - K$ QES eigenvalues, while from another perspective the problem is QES with J QES eigenvalues.

To explore the merging of eigenvalues at $E = 0$ we use the quasi-exact solvability on the lines $\alpha_\pm = n$, where $J = 2n + 1$. Consider first $J = 3$ and the line $\alpha = 12 + 2\lambda$, or equivalently $(\alpha_+, \alpha_-) = (1 + \lambda/2), 1)$. Solving

$$p_3(E) = -E^3 - 32(2\lambda + 3)E = 0,$$

we find that the QES eigenvalues are

$$E = 0, \pm 4\sqrt{2}\sqrt{-2\lambda - 3}.$$

The paired eigenvalues are real for $2\lambda < -3$, vanish at $2\lambda = -3$, and are imaginary for $2\lambda > -3$. Thus, the cubic exceptional point at $(2\lambda, \alpha) = (-3, 9)$ on the line $\alpha_- = 1$ of Fig. 6.11 arises from the coincidence of the three QES eigenvalues at $E = 0$.

Figure 6.16 shows the real and imaginary parts of the QES eigenvalues as functions of λ when $J = 5$. All eigenvalues are real for $2\lambda < -7$, there are two values of λ at which three eigenvalues coincide at $E = 0$, and for sufficiently large λ all of the nonzero QES eigenvalues are imaginary. Thus, on the line $\alpha_- = 2$ the two cubic exceptional points arise in the QES sector of the problem.

These examples suggest that the n cusps on each of the lines $\alpha_\pm = n$ arise from the merging of three eigenvalues in the QES sector of the spectrum, which we explain following [Dorey *et al.* (2009)]. The idea is to track the QES eigenvalues as functions of λ along the line $\alpha_- = n$. We prove that the QES eigenvalues are all real for $\lambda > 1 - J$ and all imaginary or zero for $\lambda < -1$. By exploring the intermediate-λ range we deduce that there are precisely n cubic exceptional points.

The first step is to prove a result of independent interest, namely that the levels that become complex always lie in the QES portion of the spectrum. This was established in [Dorey *et al.* (2001b)] for sextic oscillators using two spectral equivalences from [Dorey *et al.* (2001a)]. A similar conjecture for QES PT-symmetric quartic oscillators [Bender and Boettcher (1998a)] was proved [Eremenko and Gabrielov (2011a)].

In Sec. 6.6 we found a pair of eigenproblems with the same eigenvalues except for a single zero-energy eigenvalue present in one of the pairs. The eigenproblem with the additional eigenvalue corresponds to the $J = 1$ QES case $(2\lambda, \alpha_J) = (2\lambda, 4J + 2\lambda)$. Here, we find a partner to the QES eigenproblem for $J > 1$ with the same eigenvalues except for the J QES eigenvalues.

Consider the polynomials $p_k(E)$ when $k > J$. They can be written as

$$p_{k+J}(E) = p_J(E)\,\widehat{p}_k(E),$$

where the degree-k polynomials in E, $\widehat{p}_k(E, \alpha_J, \lambda) \equiv \widehat{p}_k(E)$ satisfy

$$\widehat{p}_k(E) = -E\widehat{p}_{k-1} + 16(1-k)(k+J-1)(n+J-1+\lambda)\widehat{p}_{k-2}\,(k \geq 1) \quad (6.55)$$

with initial conditions $\widehat{p}_0 = 1, \widehat{p}_1(E) = -E$. Comparing (6.55) with (6.48), we see that

$$\widehat{p}_k(E, \alpha_J, \lambda) = p_k(E, \widehat{\alpha}, \widehat{\lambda}),$$

where

$$(2\widehat{\lambda}, \widehat{\alpha}) = (2\lambda + 2J, 2\lambda + 2J). \quad (6.56)$$

This implies a relation between the eigenproblem at the points (6.56) and

$$(2\lambda, \alpha_J) = (2\lambda, 4J + 2\lambda). \quad (6.57)$$

To make this precise, we rewrite the series expansion (6.47) as

$$\Psi(z, E, \alpha_J, \lambda) = e^{z^4/4} z^{\lambda+1/2} \left(\sum_{k=0}^{J-1} a_k(\lambda) p_k(E, \alpha, \lambda) z^{2k} \right.$$
$$\left. + \sum_{k=0}^{\infty} a_{k+J}(\lambda) p_J(E, \alpha_J, \lambda)\, \widehat{p}_k(E, \alpha_J, \lambda) z^{2(k+J)} \right)$$

and compare it with

$$\Psi(z, E, \widehat{\alpha}_J, \widehat{\lambda}_J) = e^{z^4/4} z^{\lambda+J+1/2} \sum_{k=0}^{\infty} a_k(\lambda + J) P_k(E, \alpha_J, \lambda) z^{2k}.$$

The differential operator

$$Q_J(\lambda) = z^J \left[\frac{1}{z}\frac{d}{dz} - z^2 - \frac{\lambda + \frac{1}{2}}{z^2} \right]^J$$

acts as

$$Q_J(\lambda)\, e^{z^4/4} z^{\lambda + 1/2}\, z^{2k} = 0 \quad (k = 0, \ldots, J - 1),$$

and thus annihilates the first J terms of $\Psi(z, E, \alpha_J, \lambda)$. Therefore, $\Psi(z, E, \alpha_J, \lambda)$ is mapped by $Q_J(\lambda)$ onto a function proportional to $\Psi(z, E, \hat{a}, \hat{\lambda})$. Moreover, if $\Psi(z)$ decays in the Stokes sectors $\widehat{S}_{\pm 1}$, then $Q_J(\lambda)\Psi(z)$ also decays in these sectors.

We conclude that $Q_J(\lambda)$ maps the QES eigenfunctions of the problem at the point $(2\lambda, \alpha_J)$ to zero and the remaining eigenfunctions are mapped to those of the problem at the point (6.56). The operator $Q_J(\lambda)$ is the generalization of the supersymmetry operator $Q_+ \equiv Q_1$, discussed in Sec. 6.5, to QES problems with $J > 1$. This proves that the spectrum at the QES points (6.57) consists of the eigenvalues of (6.56) plus the J QES eigenvalues defined by $p_j(E) = 0$.

The reality result of Sec. 6.5 applied to the partner problem (6.56) establishes that its spectrum is real for all λ. Combining this with the spectral equivalence established above, we conclude that the eigenvalues of $(2\lambda, \alpha_J)$ that are *not* in the QES sector are real for all λ. Applying the reality result directly to the points $(2\lambda, \alpha_J)$, we see that all the eigenvalues are real if

$$\alpha_J < 4 + 2|\lambda| \quad \Rightarrow \quad \lambda < -J + 1.$$

Therefore, any complex eigenvalues at the points $(2\lambda, \alpha_J)$ must appear when $\lambda \geq -J + 1$, and since they lie in the QES sector, we locate the exceptional points by studying the zeros of the polynomials $p_J(E, \alpha_J, \lambda)$.

The zeros of $p_J(\alpha_J, \lambda)$ are all real when $\lambda < -J+1$ and are also distinct [Dorey *et al.* (2001a, 2009)]. The solutions of $p_J(-E, \alpha_J, \lambda) = 0$ are the QES eigenvalues of the radial sextic eigenproblem at $\alpha = -(4J+2\lambda)$, which is a Hermitian problem for this range of λ. The *reflection* symmetry [Dorey *et al.* (2009)]

$$p_J(E, \alpha_J, \lambda) = (-i)^J p_J(iE, \alpha_J, -J - \lambda) \tag{6.58}$$

implies that the zeros of $p_J(E, \alpha_J, \lambda)$ are distinct and imaginary (or zero) when $\lambda > -1$. Since p_J is an even or an odd function of E, it has $2\lfloor J/2 \rfloor$ complex zeros for $\lambda > -1$ in addition to the zero eigenvalue present for all odd J.

To treat the interval $-J + 1 \leq \lambda \leq -1$ recall that the points $\lambda = -J + n$, $n = 2, \ldots J - 1$ lie on the QES lines $\alpha = 4J + 2\lambda$ and on the QES lines $\alpha = 4n - 2\lambda$. Swapping λ to $-\lambda$, the results above applied to $\alpha = 4n - 2\lambda$ imply that $2\lfloor n/2 \rfloor$ of the J QES eigenvalues are complex when $\lambda > 1$. Therefore, between $-J + 2m - 1 \leq \lambda \leq -J + 2m$ on the line $\alpha = 4J + 2\lambda$ for $m = 1, 2, \ldots \lfloor (J + 1)/2 \rfloor - 1$, the number of nonreal eigenvalues changes from $2m - 2$ to $2m$. Since nonreal eigenvalues in $\mathcal{PT}$ problems occur in complex-conjugate pairs and the QES sector is invariant under $E \to -E$, any nonreal eigenvalues created or destroyed away from $E = 0$ must appear in quartets. So, if the number of nonreal eigenvalues changes by two, then at least one complex-conjugate pair has been created or destroyed at $E = 0$, and this subinterval must contain at least one exceptional point with eigenvalue zero.

Recall that the real zeros of the polynomial $Q_n(\lambda)$ (6.43) on the line $\alpha = 4J + 2\lambda$ ($\lambda \in \mathbb{R}$) correspond to an exceptional point in the spectrum with eigenvalue zero. Therefore, when $J = 2n + 1$, $Q_n(\lambda)$ must have at least one real zero in each interval $[2m - 1, 2m]$ ($m = 1 \ldots n$). However, $Q_n(\lambda)$ is a polynomial of degree n, so the zeros of $Q_n(\lambda)$ must all be simple. Thus, on the lines $\alpha = 4J + 2\lambda$ with $J = 2n + 1$ there are n odd-order exceptional points with eigenvalue zero. Making the mild assumption that such exceptional points have degree three,[7] the n exceptional points found as the zeros of $Q_n(\lambda)$ account for all n of the complex QES levels that are created on the line $\alpha = 4J + 2\lambda$ with $J = 2n + 1$. Hence, on the lines $\alpha = 4J + 2\lambda$ with $J = 2n + 1$ there are precisely n cubic exceptional points arising from three QES eigenvalues merging at $E = 0$. Swapping $\lambda \to -\lambda$ establishes the same result on the QES line $\alpha = 4J - 2\lambda$.

We now use the Bender–Dunne polynomials $p_J(E)$ to locate the cusps on the lines $\alpha = 4J + 2\lambda$ with $J = 2n + 1$ and $n = 0, 1 \ldots$. Since $p_{2n+1}(E, \lambda)$ factors as E times a polynomial in E^2, the triply-degenerate zero-energy levels can be identified by solving

$$\frac{d}{dE} p_{2n+1}(E, \lambda) \bigg|_{E=0} = 0.$$

We consider the polynomials $p_{2m+1}(E, \lambda)$ ($m = 0 \ldots n$) and remove some trivial factors by setting

$$q_k(\lambda, n) = \frac{-1}{2^7 k!} \frac{d}{dE} p_{2k+1}(E, \lambda) \bigg|_{E=0}.$$

---

[7]See [Dorey *et al.* (2009)] for a justification of the statement that there are no exceptional points of higher degree when $\lambda \in \mathbb{R}$.

Examining the recurrence relation (6.48) and its derivative with respect to E at $E = 0$, we deduce that $q_k(\lambda, n)$ satisfies a solvable first-order recurrence relation [Dorey *et al.* (2009)]

$$q_k(\lambda, n) = (n - k + \tfrac{1}{2})(k + \tfrac{1}{2}\lambda)q_{k-1} + \binom{n}{k}\prod_{j=1}^{k}(j - \tfrac{1}{2})(j - \tfrac{1}{2} + \tfrac{1}{2}\lambda)$$

with $q_0(\lambda, n) = 1$. When $k = n$, we find that

$$q_n(\lambda) \equiv q_n(\lambda, n) = (1 + \tfrac{1}{2}\lambda)_n(\tfrac{1}{2})_n \, {}_3F_2(-n, \tfrac{1}{2}, \tfrac{1}{2} + \tfrac{1}{2}\lambda; 1 + \tfrac{1}{2}\lambda, \tfrac{1}{2} - n; 1)$$

$$= \sum_{j=0}^{n}(-1)^j \binom{n}{j}(\tfrac{1}{2} - j)_n(\tfrac{1}{2} + \tfrac{1}{2}\lambda)_j(1 + \tfrac{1}{2}\lambda + j)_{n-j},$$

which is a polynomial of degree n in λ. Note that $q_n(\lambda)$ is precisely the polynomial $Q_n(\lambda)$ (6.43) evaluated at $M = 3$, as expected, although obtained here by an entirely different route.

From the study of the $M = 3$ model, we have proved that the cusps shown in Fig. 6.11 correspond to cubic exceptional points, that there are n such points on the lines $\alpha_\pm = n$, and that the locations are identified with the n solutions of $q_n(\lambda) = 0$. Moreover, the siting of the cusps on the lines $\alpha_\pm = n$, special to $M = 3$, is now understood to follow from the quasi-exact solvability of the model.

We finish the exploration of the sextic $\mathcal{PT}$-symmetric oscillator with a discussion of its spectral locus. (This discussion is motivated by similar considerations of the real QES spectral locus of the $\mathcal{PT}$-symmetric quartic oscillator [Eremenko and Gabrielov (2009a,b, 2011a,b, 2012a,b); Shapiro and Tater (2014)] and the QES spectral locus of the Hermitian sextic oscillator without an angular-momentum term [Eremenko and Gabrielov (2009b); Alexandersson and Gabrielov (2012); Alexandersson (2012); Shapiro and Tater (2017)].) In the simplest case $\lambda^2 = 1/4$, the $1/x^2$ term is absent from (6.46) and the complex spectral locus is an analytic hypersurface in $\mathbb{C}^2$ [Alexandersson and Gabrielov (2012); Alexandersson (2012)].

We first elucidate the role that the polynomials $p_J(E, \lambda)$ and $q_n(\lambda)$ play in the QES spectral locus. We set $\alpha = 4J + 2\lambda$ with J a fixed positive integer. The QES spectral locus $Z_J(E, \lambda)$ is defined as the set $(E, \lambda) \in \mathbb{C}^2$ such that (6.46) has a nonzero QES eigenfunction. Therefore, $Z_J(E, \lambda)$ is described explicitly by the algebraic curve in $\mathbb{C}^2$ defined by $p_J(E, \lambda) = 0$. The QES spectral locus for $\alpha = 4J - 2\lambda$ is defined by $p_J(E, -\lambda) = 0$. Figure 6.17 shows the *real* QES spectral locus $Z_J \cap \mathbb{R}^2$ when $\alpha = 4J + 2\lambda$ for $J = 20$ and $J = 21$. A consideration of such plots for other values of

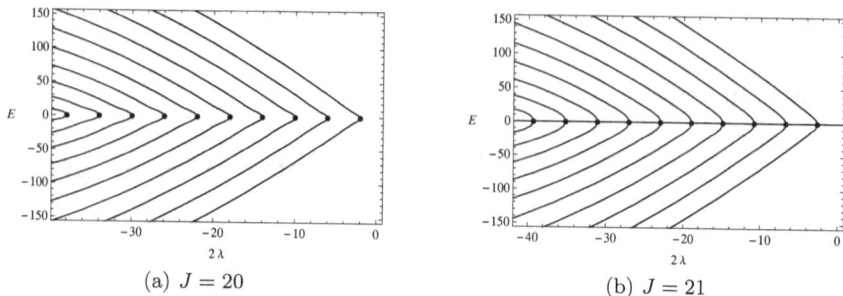

(a) $J = 20$ (b) $J = 21$

Fig. 6.17   Real QES spectral locus for $J = 20$ and $J = 21$. The dots mark the positions of the exceptional points in the $(2\lambda, E)$-plane.

J leads to the conjecture that $Z_J \cap \mathbb{R}^2$ consists of $\lfloor J/2 \rfloor$ mutually disjoint analytic curves plus the line $E = 0$ when J is odd.

The exceptional points in the QES spectrum, marked by dots in Fig. 6.17, can be located explicitly. For J odd the dots mark eigenvalue-zero cubic exceptional points with λ values fixed by the zeros of $q_n(\lambda)$. For J even we seek the values of λ such that $p_J(E)$ has two zeros at $E = 0$. We fix n to be a positive integer and set $J = 2n$. Since $p_{2n}(E, \lambda)$ is an even polynomial in E, $p'_{2n}(0, \lambda) = 0$ holds automatically for all λ and there are two zero eigenvalues in the QES sector whenever $p_{2n}(0, \lambda) = 0$. Let

$$r_k(\lambda, n) = \frac{(-1)^k}{2^{7k}\Gamma(k + \frac{1}{2})} p_{2k}(0, \lambda).$$

From (6.48) we deduce that r_k satisfies the first-order recurrence relation

$$r_k = (k - \tfrac{1}{2} + \tfrac{1}{2}\lambda)(n - k + \tfrac{1}{2})r_{k-1}$$

with initial condition $r_0 = \sqrt{\pi}$. The solution is

$$r_k(\lambda, n) = (\tfrac{1}{2}\lambda + \tfrac{1}{2})_k(\tfrac{1}{2} - n)_k,$$

so

$$r_n(\lambda, n) = (-1)^n \frac{\Gamma(n + \frac{1}{2})}{2^n \sqrt{\pi}} \prod_{j=1}^{n} (\lambda + 2j - 1),$$

from which we conclude that there are quadratic exceptional points at $\lambda = 1 - 2j$, $j = 1 \ldots J/2$.

Although we cannot plot $Z_J(E, \lambda)$ when $(E, \lambda) \in \mathbb{C}^2$, we can explore some of its structure by locating the positions of the branch points of $E(\lambda)$ by using the polynomials $p_J(E)$. The branch points occur at the multiple zeros of $p_J(E, \lambda)$ and we must find $(E, \lambda) \in \mathbb{C}^2$ such that

$$p_J(E, \lambda) = 0 \quad \text{and} \quad p'_J(E, \lambda) = 0,$$

where $'$ denotes differentiation with respect to E.

To locate the common zeros of p_J and p'_J we calculate the $2J-1 \times 2J-1$ determinant

$$R_J \equiv R(p_J, p'_J)(\lambda) = \begin{vmatrix} c_0 & c_1 & \cdots & 0 & 0 \\ 0 & c_0 & \cdots & 0 & 0 \\ \vdots & \vdots & \ddots & \vdots & \vdots \\ 0 & 0 & \cdots & c_{J-1} & c_J \\ Jc_0 & (J-1)c_1 & \cdots & 0 & 0 \\ 0 & Jc_0 & \cdots & 0 & 0 \\ \vdots & \vdots & \ddots & \vdots & \vdots \\ 0 & 0 & \cdots & c_{J-1} & 0 \end{vmatrix}$$

in which $p_J(E, \lambda) = c_0(\lambda)E^J + c_1(\lambda)E^{J-1} + \cdots + c_J(\lambda)$. The resultant R_J is a polynomial in λ with leading term

$$(-1)^J 8^{J(J-1)} \left(\prod_{j=0}^{J-1} j! \right)^2 \lambda^{J(J-1)/2},$$

and the first few nontrivial examples are

$$R_2 = 2^6(\lambda + 1),$$
$$R_3 = -2^{17}(2\lambda + 3)^3,$$
$$R_4 = 2^{32}3^2(\lambda + 3)(\lambda + 1)(16\lambda^2 + 64\lambda + 73)^2,$$
$$R_5 = -2^{57}3^4(8\lambda^2 + 40\lambda + 41)^3(4\lambda^2 + 20\lambda + 33)^2,$$
$$R_6 = 2^{82}3^6 5^2(\lambda + 5)(\lambda + 3)(\lambda + 1)(4096\lambda^6 + 73728\lambda^5 + 564992\lambda^4$$
$$+ 2356224\lambda^3 + 5610816\lambda^2 + 7178112\lambda + 3879657)^2.$$

Figure 6.18 shows the zeros of $R_{15}(\lambda)$ and $R_{20}(\lambda)$, that is, the positions of the branch points of $E(\lambda)$ for Z_{15} and Z_{20}.

The symmetry of the plots about $\lambda = -J/2$ is a consequence of the reflection symmetry (6.58). The zeros of $p_J(E, \lambda)$ are constrained by the $\mathcal{PT}$ symmetry of the eigenproblem to appear in quartets $E = a \pm ib$ and $E = -a \pm ib$ or, if real, as quadratic or cubic zeros at $E = 0$. From our earlier work the branch points of $E(\lambda)$ at real E, and consequently real λ, can all be found exactly: They are the zeros of $q_n(\lambda)$ when $J = 2n + 1$ or are the zeros of $r_n(\lambda)$ when $J = 2n$. Therefore, a factor of the resultant is

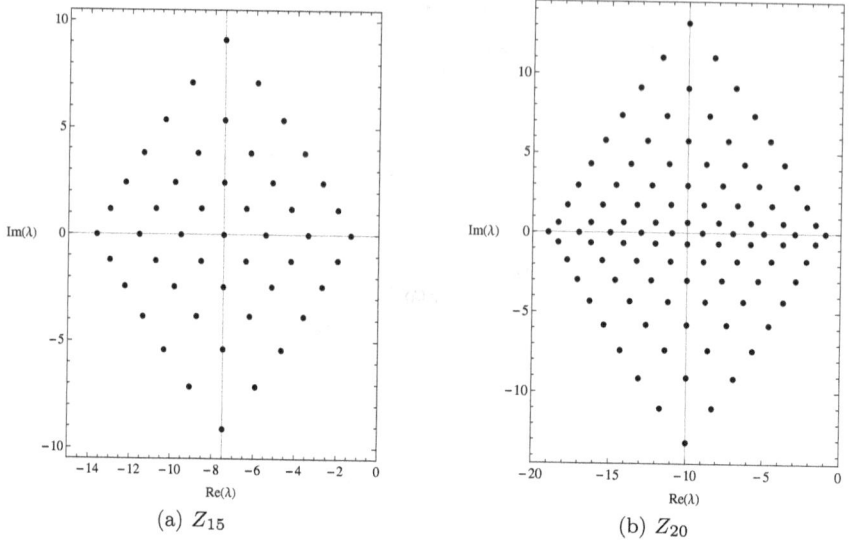

(a) Z_{15} (b) Z_{20}

Fig. 6.18   Branch points of $E(\lambda)$ of the QES spectral locus Z_{15} and Z_{20}.

either $q_n(\lambda)$ or $r_n(\lambda)$. For example,

$$R_2 = r_1(\lambda),$$
$$R_3 = -2^2 3\, [q_1(\lambda)]^3,$$
$$R_4 = r_2(\lambda)\, [16(\lambda + 2)^2 + 9]^2,$$
$$R_5 = -2^6 93^4\, [q_2(\lambda)]^3\, [4(\lambda + \tfrac{5}{2})^2 + 8]^2,$$
$$R_6 = r_3(\lambda)\, [4096(\lambda + 3)^6 + 12032(\lambda + 3)^4 - 15552(\lambda + 3)^2 + 59049]^2.$$

For $J \geq 4$ there are also branch points at complex E and λ, encoded in the other polynomial factor in the resultants. It is harder to find a generating function for these polynomials. Moreover, the pattern of zeros is curiously reminiscent of zero plots [Clarkson (2003)] of the Okamoto polynomials of type II [Okamoto (1986)]. The branch points of $E(b)$ of the QES spectral locus of the Hermitian sextic potential

$$\left[-\frac{d^2}{dx^2} + x^6 + 2bx^4 + b^2 - (4J - 1)x^2 \right] \psi(x) = E\, \psi(x) \quad [\psi(x) \in L^2(\mathbb{R})]$$

also form rhombi [Eremenko and Gabrielov (2009b); Shapiro and Tater (2017)], though these models have only complex branch points. Figure 6.19 shows that the branch points of $\lambda(E)$ of the $\mathcal{PT}$-symmetric sextic oscillator plotted as a function of complex E also form an interesting pattern.

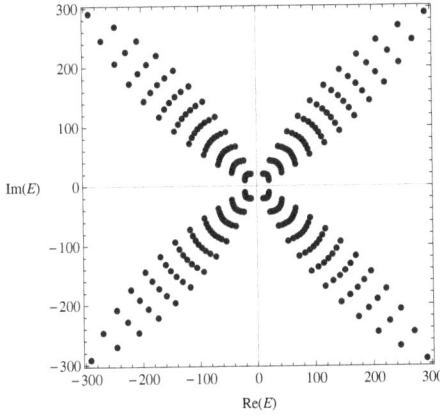

Fig. 6.19   Branch points of $\lambda(E)$ of the QES spectral locus Z_{26}.

6.8   Concluding Remarks

We have discussed various aspects of $\mathcal{PT}$-symmetric quantum-mechanical models that include and generalize the one-dimensional Schrödinger problem with a cubic potential, the spectral reality of which could be proved using ideas inspired by a link with integrable systems now called the ODE/IM correspondence. Extending the scope to include more general $\mathcal{PT}$-symmetric problems reveals an intricate structure of $\mathcal{PT}$-symmetry breaking, which was explored in detail for the sextic oscillator, where a variety of methods could be applied to obtain exact results.

The link with integrable systems stems from the observation that the spectral determinants, functional relations, and Bethe-ansatz equations introduced in Sec. 6.2 coincide with fundamental objects studied in the field of integrable models. The precise models in this story involve an extension of Baxter's approach to integrable lattice models to integrable massless (conformal) quantum field theories [Bazhanov *et al.* (1996, 1997a,b)]. The cubic oscillator (6.2) and the Bender–Boettcher problems (6.20) with general $M \in \mathbb{R}^+$ are related to a family of conformal field theories with conformal central charge $c = 1 - M^2/(M+1)$ [Dorey and Tateo (1999a); Bazhanov *et al.* (2001); Dorey and Tateo (1999b)], while the generalized problems (6.30) are related to continuum versions of the Perk–Schultz integrable lattice models [Suzuki (2001); Dorey *et al.* (2001a); Fateev and Lukyanov (2006)]. The Stokes multiplier $C(E)$ in (6.12) and (6.13) takes the role of the ground-state eigenvalue of the transfer matrix of the integrable model, while $D(E)$ and $\widetilde{D}(E)$ correspond to the ground-state eigenvalue of the auxiliary Q operator. More elaborate ordinary differential equations, with added regular singularities constrained by a zero-monodromy condition, control the corresponding eigenvalues for the excited states in the quantum models [Bazhanov *et al.* (2003b)].

The precise match between sets of functional equations appearing in the two fields extends to an exact match between many other objects defined on both sides of the correspondence. This observation, together with the discovery of generalizations involving classes of multiparameter higher-order differential equations [Dorey and Tateo (2000); Suzuki (2000); Dorey *et al.* (2000, 2007a); Bazhanov *et al.* (2002, 2003a)], has led to the beginnings of a unifying program for differential equations and integrable models, though it should be stressed that there are many more complicated $\mathcal{PT}$-symmetric potentials with more involved connection formulas for which links with the theory of integrable models have yet to be found. More recently, inspired by developments in gauge and string theory [Gaiotto *et al.* (2013); Alday *et al.* (2011)], the correspondence has been extended to a link between *massive* integrable quantum field theories and connection problems for exactly-solvable nonlinear classical field theories [Lukyanov and Zamolodchikov (2010); Dorey *et al.* (2013); Lukyanov (2013); Ito and Locke (2014); Adamopoulou and Dunning (2014); Negro (2017); Ito and Locke (2015); Ito and Shu (2017)]. This is a promising area for future research with connections to many currently active topics in pure mathematics. It is remarkable that it should be linked through a relatively simple sequence of steps to the $\mathcal{PT}$-symmetric cubic oscillator, where our discussions began.

Chapter 7

Exactly Solvable $\mathcal{PT}$-Symmetric Models

Contributed by Géza Lévai

"Inspiration is a guest that does not willingly visit the lazy."
—*Pyotr Tchaikovsky*

In attempting to understand an emergent area of physics, it is extremely helpful to study elementary solvable models. $\mathcal{PT}$-symmetric quantum mechanics is a complex generalization of conventional quantum mechanics, and some of its features are unexpected and nonintuitive. By examining the solutions to exactly solvable $\mathcal{PT}$-symmetric quantum-mechanical models, we can develop an understanding of the nature of $\mathcal{PT}$-symmetric quantum mechanics. Exactly solvable $\mathcal{PT}$-symmetric quantum-mechanical models are usually constructed as generalizations of exactly solvable models in conventional quantum mechanics. This section explores the properties of some carefully selected $\mathcal{PT}$-symmetric potentials.

7.1 Exactly Solvable Potentials

Since the discovery of quantum mechanics local potentials have been constructed to model the behavior of the subatomic world. For example, the Coulomb potential accurately describes some attributes of atomic systems and the harmonic-oscillator potential gives a good approximation to bound quantum systems. Physical forces may be attractive or repulsive, short-range or long-range, and to model such forces one must choose an appropriate form of potential together with a relevant set of parameters.

Some quantum-mechanical potentials are exactly solvable. This means that the energy eigenvalues, the bound-state eigenfunctions, and the scattering matrix can be determined in closed analytical form. In recent years the number of known exactly solvable potentials has expanded considerably due to the emergence of symmetry-based approaches. The concept of solvability has also been extended beyond the classically exactly solvable potentials to conditionally exactly solvable and quasi-exactly solvable potentials. The range of exactly solvable potential shapes has increased and their solutions can be obtained both analytically and numerically.

In addition to their role in describing realistic physical systems, exactly solvable quantum-mechanical potentials are mathematically elegant because of their connection with symmetry. Symmetry and invariance are important characteristic features of any physical system and often afford a deep insight into the physics of the system. The structure of the energy spectrum often reveals the underlying symmetry of the potential.

Exactly solvable potentials are particularly useful in exploring the nature of $\mathcal{PT}$-symmetric quantum mechanics. Physical systems defined by complex potentials are harder to understand intuitively than those governed by real potentials. Fortunately, the methods developed for the study of real potentials can be adapted to the $\mathcal{PT}$-symmetric versions of these potentials. These methods enable us to study the subtle properties of $\mathcal{PT}$-symmetric systems, such as the breaking of $\mathcal{PT}$ symmetry.

7.2   Generating Real Exactly Solvable Potentials

This section reviews two methods for constructing potentials for which the associated one-dimensional Schrödinger equation is exactly solvable. These potentials may be defined on the real axis, the positive-real axis (radial potentials), or on a finite domain.

The first method constructs potentials whose bound-state solutions have the form $\psi(x) = f(x)F[z(x)]$, where $F(z)$ is usually a classical orthogonal polynomial, but can also be some other special function. Typical choices for orthogonal polynomials are the Jacobi or the generalized Laguerre polynomials, and in this case one obtains the Natanzon and the Natanzon confluent potential classes. These classes contain the most well-known exactly solvable potentials as special cases.

The second method uses supersymmetric quantum mechanics (SUSYQM) to generate new solvable potentials from already known solvable potentials. The bound-state energy spectrum of the new potential is essentially the same as that of the original potential. The energy levels of the two potentials are related by a first-order differential operator.

7.2.1 Method 1: Variable transformations

For some potentials the Schrödinger equation can be solved by transforming it into the differential equation of some special function. The procedure was first used to derive some simple potentials [Bhattacharjie and Sudarshan (1962)] and was later developed by Natanzon, who applied it systematically to transform the Schrödinger equation into the hypergeometric and the confluent hypergeometric differential equation [Natanzon (1971, 1979)]. Let us transform the Schrödinger equation with units $2m = \hbar = 1$

$$\psi''(x) + [E - V(x)]\psi(x) = 0 \tag{7.1}$$

into a second-order differential equation for the special function $F(z)$,

$$F''(z) + Q(z)F'(z) + R(z)F(z) = 0. \tag{7.2}$$

To do this, we seek solutions of the form [Lévai (1989, 1994b)]

$$\psi(x) = f(x)F[z(x)]. \tag{7.3}$$

At this stage we do not specify the domain of x. We return to this point in our discussion of PT-symmetric potentials in Sec. 7.4.

Substituting (7.3) into (7.1), comparing it with (7.2), and eliminating $f(x)$ gives

$$E - V(x) = \tfrac{1}{2}z'''(x)/z'(x) - \tfrac{3}{4}[z''(x)/z'(x)]^2$$
$$+[z'(x)]^2\left(R[z(x)] - \tfrac{1}{2}Q'[z(x)] - \tfrac{1}{4}Q^2[z(x)]\right), \tag{7.4}$$

where the first two terms on the right side are called the *Schwartzian derivative*. While $Q(z)$ and $R(z)$ in (7.2) define the special function $F(z)$, (7.4) does not depend on $F(z)$ and defines the variable transformation $z(x)$. The eigenfunction $\psi(x)$ is given by

$$\psi(x) = [z'(x)]^{-1/2}F[z(x)]\exp\left[\tfrac{1}{2}\int^{z(x)} \mathrm{d}s\, Q(s)\right]. \tag{7.5}$$

Our task now is to find the function $z(x)$ that transforms (7.4) to an exactly solvable Schrödinger equation. Given a particular potential $V(x)$ and energy E it is virtually impossible to find an analytic expression for $z(x)$. However, it is possible to construct pairs of functions $V(x)$ and $z(x)$ that give rise to exactly solvable Schrödinger equations [Bhattacharjie and Sudarshan (1962)]. Since the left side of (7.4) contains the constant E, we construct a corresponding constant C on the right side. This observation can be used to compile systematically a list of solvable potentials [Lévai

(1989)] that are associated with a first-order differential equation for z of the form

$$[z'(x)]^2 \Phi(z) = C, \tag{7.6}$$

where $\Phi(z)$ depends on $R(z)$ and $Q(z)$.

The solution to (7.6),

$$\int^z \mathrm{d}s\, \Phi^{1/2}(s) = C^{1/2} x + x_0, \tag{7.7}$$

defines an implicit function $x(z)$, and in some cases one can obtain the explicit function $z(x)$ [Lévai (1989)]. If only $x(z)$ is available, one obtains the potential in implicit form. [Even if $z(x)$ is not known explicitly, integrations and derivatives can be performed.] We usually choose $x_0 = 0$ and $z(0) = 0$. Choosing $x_0 \neq 0$ corresponds to a coordinate shift and thus it is a trivial and rarely useful transformation for potentials defined on the real-x axis. However, such a transformation is important for the PT-symmetric potentials discussed in Sec. 7.5.

7.2.2   *Method 2: Supersymmetric quantum mechanics*

Let us consider two one-dimensional Hamiltonians, $H_- = A^\dagger A$ and $H_+ = AA^\dagger$, defined in terms of the first-order differential operators

$$A = \frac{\mathrm{d}}{\mathrm{d}x} + W(x) \quad \text{and} \quad A^\dagger = -\frac{\mathrm{d}}{\mathrm{d}x} + W(x). \tag{7.8}$$

The Schrödinger equations for these Hamiltonians are

$$H_\pm \psi^{(\pm)}(x) = \left[-\frac{\mathrm{d}^2}{\mathrm{d}x^2} + V_\pm(x) \right] \psi^{(\pm)}(x) = E^{(\pm)} \psi^{(\pm)}(x), \tag{7.9}$$

where the potentials are expressed in terms of the superpotential $W(x)$ as

$$V_\pm(x) = W^2(x) \pm \frac{\mathrm{d}}{\mathrm{d}x} W(x). \tag{7.10}$$

The energy of the ground state of H_- is set to 0: $H_- \psi_0^{(-)}(x) = 0$.

Apart from the ground state of H_+, the energy levels of the two Hamiltonians are the same

$$H_+ \left[A\psi^{(-)}(x) \right] = AA^\dagger \left[A\psi^{(-)}(x) \right] = AH_- \psi^{(-)}(x) = E^{(-)} A\psi^{(-)}(x)$$

and the eigenfunctions are related by $\psi^{(+)}(x) = \left[E^{(-)} \right]^{-1/2} A\psi^{(-)}(x)$. However, for the ground state of H_-, we have $E_0^{(-)} = \langle \psi_0^{(-)} | A^\dagger A | \psi_0^{(-)} \rangle = 0$, so

$A\psi_0^{(-)} = 0$. Thus, the corresponding energy is missing from the spectrum of H_+. The eigenvalues of the two Hamiltonians are related by

$$E_{n+1}^{(-)} = E_n^{(+)} \quad (n = 0, 1, 2, \ldots) \quad \text{with } E_0^{(-)} = 0. \tag{7.11}$$

The $A\psi_0^{(-)}(x) = 0$ relation also connects the ground-state eigenfunction of H_- with the $W(x)$ superpotential via $W(x) = -\frac{d}{dx} \ln \psi_0^{(-)}(x)$. We conclude that if the solutions of a one-dimensional potential $V_-(x)$ are known, the solutions of its supersymmetric partner potential

$$V_+(x) = V_-(x) - 2\frac{d^2}{dx^2} \ln \psi_0^{(-)}(x) \tag{7.12}$$

can also be obtained and the energy spectra of the two potentials are related by (7.11). Given V_-, the potential V_+ can be constructed either analytically or numerically. In fact, a sequence of isospectral solvable potentials can be constructed by consecutive application of this procedure. Adjacent members of this hierarchy of potentials are supersymmetric partners, and each potential has one less bound state than the previous one.

Equation (7.4) may be reinterpreted in the context of SUSYQM. That is, whenever $R(z) = 0$ for the ground state, we have $E - V(x) = -W^2(x) + W'(x)$, where the superpotential is

$$W(x) = -\frac{d}{dx} \ln f(x) = -\frac{1}{2}Q[z(x)]z'(x) + \frac{1}{2}z''(x)/z'(x).$$

If $F(z)$ is an orthogonal polynomial, the condition $R_{n=0}(z) = 0$ always holds and thus orthogonal polynomials are an ideal subject of SUSYQM.

SUSYQM was introduced as an attempt to understand supersymmetry breaking in quantum field theory [Witten (1981)], but it soon became a useful tool for analyzing quantum-mechanical potentials. In a widely known supersymmetric quantum-mechanical model referred to as $N = 2$ supersymmetry [Cooper *et al.* (1995); Junker (1996)], the supersymmetric Hamiltonian and charge operators satisfy the sl(1|1) superalgebra:

$$\{Q, Q^\dagger\} = \mathcal{H}, \quad Q^2 = (Q^\dagger)^2 = 0, \quad [Q, \mathcal{H}] = [Q^\dagger, \mathcal{H}] = 0.$$

Here, the operators are 2×2 matrices: $\mathcal{H}$ contains the H_- (bosonic) and H_+ (fermionic) Hamiltonians from (7.9) on the diagonal; Q and $Q^\dagger$ contain A and $A^\dagger$ in off-diagonal positions. These operators swap between the $\psi^{(-)}$ *bosonic* and $\psi^{(+)}$ *fermionic* sectors. There are many further SUSYQM models beyond the $N = 2$ case discussed here with alternative realizations of the operators [Cooper *et al.* (1995); Junker (1996)].

If the ground-state energy of a potential $V_-(x)$ is eliminated using a superpotential $W(x)$ constructed from the ground state $\psi_0^{(-)}(x)$ of $V_-(x)$, this case is referred to as *unbroken* SUSY. The formalism can be generalized to the case of *broken* SUSY, for which the energy spectrum is left unchanged, or a new ground state is inserted. To accomplish this the formalism must be modified by introducing a *factorization* energy

$$H_- = A^\dagger(\epsilon)A(\epsilon) + \epsilon, \tag{7.13}$$

where,

$$A(\epsilon) = [A^\dagger(\epsilon)]^\dagger = -\frac{\mathrm{d}}{\mathrm{d}x} + \frac{\mathrm{d}}{\mathrm{d}x}\ln\chi(x) \equiv -\frac{\mathrm{d}}{\mathrm{d}x} - W(x). \tag{7.14}$$

In this equation the function χ is a solution to $H_-\chi = \epsilon\chi$; $\chi(x)$ must be nodeless or else the resulting potential would be singular at the nodes. We define the SUSY partner of $V_-(x)$ as

$$V_+(x) = V_-(x) - 2\frac{\mathrm{d}^2}{\mathrm{d}x^2}\ln\chi(x). \tag{7.15}$$

For $\epsilon = 0$, $\chi(x) = \psi_0^{(-)}(x)$ is recovered. For $\epsilon < 0$ nodeless $\chi(x)$ solutions can be chosen and, depending on their boundary conditions, it is possible to define operations that either insert a new ground state below $E = 0$ or leave the spectrum unchanged. These techniques may be applied to problems defined for all x, positive x, or a finite interval of x. The boundary conditions on $\chi(x)$ determine the change of the phase shifts and, in the case of radial potentials, the coefficient of the x^{-2} singularity. The $\chi(x)$ function is sometimes written as the product of the ground-state eigenfunction and a nodeless function:

$$\chi(x) = \psi_0^{(-)}(x)\xi(x). \tag{7.16}$$

This implies that

$$W(x) = -\frac{\mathrm{d}}{\mathrm{d}x}\ln\chi(x) = -\frac{\mathrm{d}}{\mathrm{d}x}\left[\ln\psi_0^{(-)}(x) + \ln\xi(x)\right]. \tag{7.17}$$

For completeness, we note that SUSYQM can be considered as a reformulation of the factorization method originating in [Schrödinger (1940a,b, 1941)] and developed further in [Infeld and Hull (1951)]. This approach is also related to Darboux's technique for solving second-order differential equations [Darboux (1882)].

7.3   Types of Exactly Solvable Potentials

This section summarizes the common types of exactly solvable Hermitian potentials and describes how these can be generated by applying the methods in Sec. 7.2.

7.3.1 *Natanzon and Natanzon-confluent potentials*

An extensive class of exactly solvable potentials can be generated from classical orthogonal polynomials by using the method discussed in Subsec. 7.2.1. For example, for the Jacobi polynomials [Abramowitz and Stegun (1965)] we choose $Q(z) = [\beta - \alpha - (\alpha + \beta + 2)z]/(1 - z^2)$ and $R(z) = n(n + \alpha + \beta + 1)/(1 - z^2)$ in (7.4), leading to [Lévai (1989)]

$$E_n - V(x) = \frac{z'''(x)}{2z'(x)} - \frac{3}{4}\left[\frac{z''(x)}{z'(x)}\right]^2 - \frac{z(x)[z'(x)]^2}{2[1 - z^2(x)]^2}\left(\alpha^2 - \beta^2\right)$$

$$+ \frac{[z'(x)]^2}{1 - z^2(x)}\left[n + \tfrac{1}{2}(\alpha + \beta)\right]\left[n + \tfrac{1}{2}(\alpha + \beta) + 1\right]$$

$$+ \frac{[z'(x)]^2}{[1 - z^2(x)]^2}\left[1 - \tfrac{1}{4}(\alpha + \beta)^2 - \tfrac{1}{4}(\alpha - \beta)^2\right]. \quad (7.18)$$

Then, (7.6) becomes

$$[z'(x)]^2\Phi(z) \equiv \frac{[z'(x)]^2\phi(z)}{[1 - z^2(x)]^2} \equiv [z'(x)]^2\frac{p_\mathrm{I}(1 - z^2) + p_\mathrm{II} + p_\mathrm{III}z}{[1 - z^2(x)]^2} = C. \quad (7.19)$$

Thus, the most general differential equation defining the $z(x)$ variable transformation can be obtained from the combination of three terms. Equation (7.5) becomes $\psi(x) = (\phi[z(x)])^{1/4}[1 + z(x)]^{\beta/2}[1 - z(x)]^{\alpha/2}P_n^{(\alpha,\beta)}[z(x)]$.

The reparametrization $\omega = \tfrac{1}{2}(\alpha + \beta)$ and $\rho = \tfrac{1}{2}(\alpha - \beta)$ gives

$$E_n - V(x) = \frac{z'''(x)}{2z'(x)} - \frac{3}{4}\left[\frac{z''(x)}{z'(x)}\right]^2$$

$$+ \frac{C}{\phi(z)}\left\{\left[1 - z^2(x)\right]\left[(n + \tfrac{1}{2} + \omega)^2 - \tfrac{1}{4}\right]\right.$$

$$\left. + \left(1 - \omega^2 - \rho^2\right) - 2\omega\rho z(x)\right\}. \quad (7.20)$$

The potential $V(x)$ in (7.20) has the general form

$$V(x) = -\frac{z'''(x)}{2z'(x)} + \frac{3}{4}\left[\frac{z''(x)}{z'(x)}\right]^2 + C\frac{s_I - s_I z^2(x) + s_{II} + s_{III}z(x)}{\phi(z)}. \quad (7.21)$$

If we combine (7.21) and (7.20) and multiply by $\phi(z)$, we see that the three terms must each vanish independently, giving the equations

$$\left(n + \tfrac{1}{2} + \omega\right)^2 - \tfrac{1}{4} + s_I - p_I E_n/C = 0, \quad (7.22)$$

$$1 - \omega^2 - \rho^2 + s_{II} - p_{II}E_n/C = 0, \quad (7.23)$$

$$-2\omega\rho + s_{III} - p_{III}E_n/C = 0. \quad (7.24)$$

Solving the potential problem requires expressing ω and ρ in terms of n. This gives a (possibly implicit) determination of E_n.

Let us analyze the origin and the role of the parameters appearing in $V(x)$ and E_n. The first group consists of the p_i parameters, which appear in $z(x)$ via (7.6) and (7.7), where $\phi(z)$ is defined by (7.19). The latter equation shows that C is an unimportant parameter that scales the energy (or the coordinate) and can be absorbed into the p_i parameters, so there are three independent parameters of this type. The constant of integration x_0 in (7.7) is a coordinate shift, which is unimportant. However, for $\mathcal{PT}$-symmetric potentials this can be an *imaginary* coordinate shift, which redefines the boundary conditions. Hence, this shift may have a significant effect. The parameters assigned to the $z(x)$ function appear both in $V(x)$ and E_n.

The other set of parameters s_i (or their reparametrized versions) originate from the $Q(z)$ and $R(z)$ functions in (7.2) defining the special function (the Jacobi polynomial in the present case). These parameters appear in E_n and the main terms of $V(x)$ in (7.21), but not in the potential terms originating from the Schwartzian derivative. This will be relevant to our discussion of $\mathcal{PT}$-symmetric potentials.

The formalism used here differs from that used originally by Natanzon [Natanzon (1971, 1979)] in which $F(z)$ is the hypergeometric function $_2F_1(a, b; c; z)$. However, since the Jacobi polynomials can be obtained from the hypergeometric function (by setting a or b to a nonpositive integer [Abramowitz and Stegun (1965)]), the two approaches are closely related. We use the present framework because it is easier to describe the transformation properties of $z(x)$ under $\mathcal{PT}$ reflection. Thus, this formalism extends to the $\mathcal{PT}$-symmetric setting. The relation between the present approach and the original one is discussed in [Lévai (2015)].

For the Natanzon confluent potentials the procedure must be applied to the generalized Laguerre polynomials $L_n^{(\alpha)}(z)$ [Abramowitz and Stegun (1965)] with $Q(z) = -1 + (\alpha + 1)/z$ and $R(z) = n/z$. The function $z(x)$ is determined from

$$[z'(x)]^2 \, \Phi(z) \equiv [z'(x)]^2 \, \frac{\phi(z)}{z^2(x)} \equiv [z'(x)]^2 \, \frac{p_{\mathrm{I}}z + p_{\mathrm{II}}z^2 + p_{\mathrm{III}}}{z^2(x)} = C, \quad (7.25)$$

where the sequence $\{p_i\}$ is chosen to conform with already established notation. The most general potential can be written as

$$V(x) = -\frac{z'''(x)}{2z'(x)} + \frac{3}{4}\left[\frac{z''(x)}{z'(x)}\right]^2 + \frac{C}{\phi(z)}\left[s_I z(x) + s_{II} z^2(x) + s_{III}\right], \quad (7.26)$$

and the analogs of (7.22), (7.23), and (7.24) are

$$n + \tfrac{1}{2}(\alpha + 1) + \omega + s_I - p_I E_n/C = 0,$$
$$-\tfrac{1}{4} + s_{II} - p_{II} E_n/C = 0,$$
$$\tfrac{1}{4}(1 - \alpha^2) + s_{III} - p_{III} E_n/C = 0. \quad (7.27)$$

The Natanzon confluent potentials were introduced [Natanzon (1979, 1971)] using the transition from hypergeometric to confluent hypergeometric functions [Abramowitz and Stegun (1965)] but they are now referred to under the present name [Cordero and Salamo (1991)]. (For a list of references on Natanzon and Natanzon confluent potentials see [Lévai (2015)].)

7.3.2 *Shape-invariant potentials*

The Natanzon and Natanzon-confluent potential classes contain a remarkable subset of potentials called *shape-invariant potentials*, which are the most well-known exactly solvable problems. These potentials are obtained by choosing a form of $\phi(x)$ in (7.19) and (7.25) in which only a single nonzero p_i parameter is used [Gendenshtein (1983)]. This leads to simple $z(x)$ functions. The PI and PII subclasses are obtained from (7.19) for $p_I \neq 0$ and $p_{II} \neq 0$ and contain three potentials each with different $z(x)$ solutions. The choice $p_{III} \neq 0$ does not lead to a shape-invariant potential [Lévai (1991)]. The LI, LII, and LIII subclasses are generated from (7.25) similarly, leading to a single $z(x)$ solution, that is, to a single potential in each case. Shape-invariant potentials are listed in Table 7.1 together with the names they are usually referred to in the literature [Cooper *et al.* (1995)]. The one-dimensional harmonic oscillator (HI) is obtained from the standard procedure applied to the Hermite polynomials. It can also be obtained from the radial harmonic oscillator (LI) by setting $\alpha = \pm 1$ and extending the domain of definition to $x < 0$, so it is considered to be a separate shape-invariant potential for historical reasons. This limiting case can also be obtained by utilizing the relation between the generalized Laguerre and Hermite polynomials [Abramowitz and Stegun (1965)]. Two more potentials, the Pöschl–Teller I and II potentials, can be obtained from the Scarf I and the generalized Pöschl–Teller potentials by the transformation $x \to 2x$ [Lévai (1989); Cooper *et al.* (1995)].

The PI and LI classes are special; choosing $p_I \neq 0$ implies that E_n is determined from (7.22) and the first line of (7.27), which contain n explicitly. This explains the quadratic and linear dependence of E_n on n in the two cases and also why the parameters appearing in the remaining equations are independent of n. This also implies that $f(x)$ in (7.3) is the ground state $\psi_0(x)$.

A common feature of shape-invariant potentials is that the Schwartzian derivative terms in (7.21) and (7.26) blend into the remaining (s_i-dependent) terms, reducing the independent potential terms to two (apart

Table 7.1   Shape-invariant potentials in their real form. The notation of [Lévai (1989); Cooper *et al.* (1995)] is followed except that the constants setting $E_0 = 0$ are omitted and the nonradial potentials are shifted to a position symmetric about the origin.

$(z')^2/C =$ (Class)	C	$z(x)$	$V(x)$	$x \in$	Name
$(1 - z^2)$ (PI)	$-a^2$	$i \sinh(ax)$	$\frac{B^2 - A^2 - A}{\cosh^2(ax)}$ $+ \frac{B(2A+1)\sinh(ax)}{\cosh^2(ax)}$	$(-\infty, \infty)$	Scarf II
	$-a^2$	$\cosh(ax)$	$\frac{B^2 + A^2 + A}{\sinh^2(ax)}$ $- \frac{B(2A+1)\cosh(ax)}{\sinh^2(ax)}$	$[0, \infty)$	generalized Pöschl–Teller
	a^2	$\sin(ax)$	$\frac{B^2 + A^2 - A}{\cos^2(ax)}$ $- \frac{B(2A-1)\sin ax}{\cos^2(ax)}$	$[-\frac{\pi}{2a}, \frac{\pi}{2a}]$	Scarf I
$(1 - z^2)^2$ (PII)	a^2	$\tanh(ax)$	$-\frac{A(A+1)}{\cosh^2(ax)}$ $+ 2B\tanh(ax)$	$(-\infty, \infty)$	Rosen–Morse II
	a^2	$\coth(ax)$	$\frac{A(A-1)\cosh(ax)}{\sinh^2(ax)}$ $- 2B\coth(ax)$	$[0, \infty)$	Eckart
	$-a^2$	$i\tan(ax)$	$\frac{A(A+1)}{\cos^2(ax)}$ $+ 2B\tan(ax)$	$[-\frac{\pi}{2a}, \frac{\pi}{2a}]$	Rosen–Morse I
z (LI)	2ω	$\frac{1}{2}\omega x^2$	$\frac{\omega^2}{4}x^2 + \frac{l(l+1)}{x^2}$	$[0, \infty)$	3D harmonic oscillator
1 (LII)	$\frac{e^4}{(n+l+1)^2}$	$C^{1/2}x$	$-\frac{e^2}{x} + \frac{l(l+1)}{x^2}$	$[0, \infty)$	Coulomb
z^2 (LIII)	a^2	$\frac{2B}{a}e^{-ax}$	$-B(2A+1)e^{-ax}$ $+ B^2 e^{-2ax}$	$(-\infty, \infty)$	Morse
1 (HI)	$\frac{\omega}{2}$	$C^{1/2}x$	$\frac{\omega^2}{4}x^2$	$(-\infty, \infty)$	1D harmonic oscillator

from a constant). This also explains why the PIII potential (obtained from $p_{III} \neq 0$) cannot be shape invariant [Lévai (1991)].

Shape-invariant potentials can also be discussed in terms of SUSYQM, which also explains the origin of their name. Considering the relation of the supersymmetric partner potentials as in (7.12), one might expect that their mathematical structure would differ. However, in some cases $V_-(x)$ and $V_+(x)$ have the same functional dependence on the coordinate and differ only by some parameters that set their depth and shape. This is the case with the shape-invariant potentials [Gendenshtein (1983)]

$$V_+(x; a_0) - V_-(x; a_1) \equiv W^2(x; a_0) + W'(x; a_0) - W^2(x; a_1) + W'(x; a_1)$$

$$= \mathcal{R}(a_1), \tag{7.28}$$

where a_0 and a_1 are parameters of the supersymmetric partner potentials and $\mathcal{R}(a_1)$ is a constant. The two sets of parameters a_0 and a_1 are connected by the formal expressions $a_1 = \mathcal{F}(a_0)$, where $\mathcal{F}$ is a simple addition $a_{i+1} = a_i + \text{constant}$ for shape-invariant potentials. Equation (7.28) shows that for shape-invariant potentials the consecutive application of a SUSY transformation (7.12) and a change of the potential parameters recovers the original potential, apart from an energy shift. The discrete energy spectrum of $V_-(x; a_0)$ is $E_n^{(-)} = \sum_{k=1}^{n} \mathcal{R}(a_k)$ [Gendenshtein (1983)], where a_k is generated by the repeated application of $\mathcal{F}$. The eigenfunctions can be generated by consecutive application of the SUSYQM ladder operators [Dutt *et al.* (1986)] $A^\dagger(x; a_k)$ as

$$\psi_n^{(-)}(x; a_0) = N_0 A^\dagger(x; a_0) A^\dagger(x; a_1)...A^\dagger(x; a_{n-1}) \psi_0^{(-)}(x; a_n).$$

Almost all of the potentials in the shape-invariant class were found using the factorization method [Infeld and Hull (1951)] and their classification (almost identical to the present one) was based on the Lie theory of special functions [Miller Jr. (1968)]. However, a classification scheme inspired by SUSYQM [Cooper *et al.* (1987)] differs from the present one. The relation of shape-invariance and the potential algebra approach [Alhassid *et al.* (1983)] is discussed in [Lévai (1994a)].

7.3.3 *Beyond Natanzon class: More general $\psi(x)$*

The bound-state eigenfunctions cannot always be expressed in terms of a single special function as in (7.3). In such cases $\psi(x)$ must be expressed in terms of two or more special functions of the same type, as we see below.

In general, the supersymmetric partners of Natanzon and Natanzon confluent potentials do not belong to the same potential class. This is because the action of the first-order differential operators (7.8) on the eigenfunctions (7.3) results in a linear combination of two terms, one containing the orthogonal polynomial $F_n(z)$ and another containing its derivative $F_n'(z)$. The derivative of an orthogonal polynomial of degree n can be expressed as a linear combination of orthogonal polynomials of degree n and degree $n-1$ [Abramowitz and Stegun (1965)], so the structure of the bound-state eigenfunction of the supersymmetric partner potential has a structure different from that characterizing Natanzon confluent potentials. Shape-invariant potentials are an exception. These are defined by (7.28) using the SUSYQM transformation that eliminates the ground-state level of the $V_-(x)$ potential, which corresponds to unbroken SUSY. In this case the two orthogonal polynomials can be rewritten as a single orthogonal polynomial by using a recurrence relation [Abramowitz and Stegun (1965)].

Thus, while the shape-invariant potential class is closed under SUSYQM transformations that are based on unbroken SUSY, this is not always true for the Natanzon (confluent) class. When the SUSY partner of any Natanzon or Natanzon confluent potential is constructed for broken SUSY [that is, using a more general solution $\chi(x)$ (7.16)], then the eigenfunctions of the SUSY partner will necessarily possess a two-term structure.

Another situation in which the bound-state eigenfunctions of a potential are expressed in terms of two special functions is when the boundary conditions require the presence of both independent solutions of the Schrödinger equation (7.1) and thus that of (7.2). This happens when a radial potential is generated by cutting a potential defined on the whole x axis. This is the case for the Woods–Saxon potential [Bencze (1966)], which can be recognized as the Rosen–Morse II potential (see Table 7.1) with $x = (r - R)/(2a)$ defined for $r \geq 0$, where the bound-state eigenfunctions are expressed in terms of two linearly independent solutions of the hypergeometric differential equation. One can approximate the bound-state eigenfunctions of a simplified one-term version of this potential in terms of a single hypergeometric function if we assume that $R \gg a$; that is, if the radius R is much larger than the thickness parameter a [Flügge (1971)]. In analogy with the Woods–Saxon potential, the radial version of the Scarf II potential has also been constructed [Lévai *et al.* (2017)], and its application has been proposed in nuclear physics.

7.3.4 *Beyond Natanzon class: Other functions $F(z)$*

A new class of potentials can be generated by using *exceptional orthogonal polynomials* [Gomez-Ullate *et al.* (2009, 2010b)] as special functions $F(z)$. These polynomials differ from the classical polynomials [Abramowitz and Stegun (1965)] in that the degree of the lowest polynomial in the sequence is larger than zero. Nevertheless, these polynomials form a square-integrable orthogonal basis with an appropriate weight function. After the X_1 Laguerre and Jacobi polynomials were found, more general X_m versions were also discovered [Gomez-Ullate *et al.* (2010a, 2012)]. These exceptional polynomials may be used to extend the radial harmonic oscillator and the Scarf I [Quesne (2008)], as well as the generalized Pöschl–Teller [Bagchi *et al.* (2009)] and the Scarf II [Midya and Roy (2013)] potentials. These problems cannot be transformed continuously into their conventional correspondents. Rather, the discrete parameter m is chosen to be zero, and the exceptional orthogonal polynomials reduce to their classical analogs, which also recovers the classical potentials.

The approach of Subsec. 7.2.1 can be used with the appropriate $Q(z)$ and $R(z)$ functions, while the $z(x)$ functions describing the variable transformations are the same as those in the PI and LI classes (see Table 7.1). The eigenvalues also have the same quadratic (PI) and linear (LI) dependence on n, as explained in Subsec. 7.3.2. These potentials satisfy the requirement of shape invariance; that is, the exceptional orthogonal polynomials have the same property as the classical ones in that the bound-state eigenfunctions maintain the same structure after a SUSYQM transformation.

The potentials solved by the X_m orthogonal polynomials are the SUSYQM partners of PI and LI type shape-invariant potentials generated with broken supersymmetry [Bagchi *et al.* (2009); Quesne (2009)]. The discussion in Subsec. 7.2.2 must be applied with superpotentials of the type (7.17) generated from nonphysical solutions (7.16) of the corresponding LI and PI shape-invariant potentials. The $\xi(x)$ function appearing there must be chosen as an appropriate rational expression of the $z(x)$ function characterizing the given potential. The eigenfunctions of the extended potential contain the linear combination of two classical orthogonal polynomials $L_n^{(\alpha)}[z(x)]$ or $P_n^{(\alpha,\beta)}[z(x)]$. These special linear combinations are the X_m exceptional orthogonal polynomials.

Some potentials may be solved in terms of Bessel functions $J_{\pm\nu}(z)$. The discussion of Subsec. 7.2.1 applies: We take $Q(z) = 1/z$ and $R(z) = 1 - \nu^2/z^2$ in (7.2). Two $z(x)$ functions originate from (7.6) if we substitute $\Phi(z) = 1$ and $\Phi(z) = z^{-2}$. Each gives a solvable problem. These choices correspond to the same $z(x)$ functions as in the shape-invariant LII and LIII cases (see Table 7.1) with $z(x) = C^{1/2}x + x_0$ and $z(x) = A\exp(C^{1/2}x)$; these describe a particle confined to a sphere of radius R [Flügge (1971)] and in an exponential potential [ter Haar (1975)]. Only one of the independent solutions of the Bessel differential equation is used to construct the bound-state eigenfunctions. In both cases the eigenvalues are expressed in terms of the zeros of Bessel functions and are determined implicitly.

Potentials can also be generated by transforming some versions of the Heun differential equation into the Schrödinger equation [Batic *et al.* (2013); Ishkhanyan (2015, 2016); Ishkhanyan and Ishkhanyan (2017)]. However, since the mathematical properties of the Heun functions are less known than those of the hypergeometric and confluent hypergeometric functions, in practical calculations they are usually expanded in terms of the latter and related functions. These potentials could thus be included among the examples presented in Subsec. 7.3.3. Similar potentials have also been found using SUSYQM techniques [López-Ortega (2015)].

7.3.5   *Further types of solvable potentials*

There is a special class of potentials, called *quasi-exactly solvable* (QES) potentials, for which it is possible to obtain the lowest few (a finite number) eigenfunctions analytically while the remaining eigenfunctions cannot be expressed in closed form [Ushveridze (1994)]. The lowest eigenfunctions resemble (7.3) with a polynomial playing the role of $F(z)$. The coefficients of this polynomial satisfy a recurrence relation, which must terminate. However, with some choices of the parameters the sequence can be forced to terminate, leading to finite-degree polynomials that yield the lowest few eigenfunctions. The eigenvalues and various matrix elements can then be calculated exactly for this limited set of eigenfunctions. The simplest example of a QES potential is the sextic potential, which contains quadratic and quartic, and in the case of a radial problem, centrifugal terms. However, the coefficients are correlated and are not arbitrary.

Conditionally exactly solvable (CES) potentials are potentials for which the Schrödinger equation can be solved only if the parameters in the potential are chosen to have some specific values. An early example of such a potential, the Dutt–Khare–Varshni (DKV) potential, has three terms and two free parameters [Dutt *et al.* (1995)]. This potential belongs to the Natanzon class [Roychoudhury *et al.* (2001)], and the potential term with a fixed coupling coefficient originates from the Schwartzian derivative in (7.21), the terms of which do not contain any free parameters.

Other types of CES potentials were obtained from SUSYQM partners of shape-invariant potentials constructed by inserting a new ground state below that of the original one [Junker and Roy (1997)]. This construction, based on a nonphysical solution $\chi(x)$ as in (7.15), can be used only for certain energies. This potential belongs to the class described in Subsec. 7.3.3, which includes potentials whose bound-state solutions are expressed in terms of two special functions of the same type.

Finally, we remark that there are piecewise-continuous potentials, and one can solve the associated Schrödinger equation by patching solutions and their derivatives at the boundaries of individual regions. One such potential is the finite square well [Flügge (1971)]. In this case the solutions do not involve special functions and are expressed in terms of trigonometric and/or exponential functions. The eigenvalues are obtained as roots of transcendental equations and cannot expressed in closed form.

7.4 Aspects of $\mathcal{PT}$-Symmetric Potentials

$\mathcal{PT}$ symmetry requires that $V^*(-x) = V(x)$. If x is real this means that the real/imaginary potential component must be an even/odd function of x. Here, we extend the discussion of Sec. 7.2 to $\mathcal{PT}$-symmetric potentials and point out the main differences with respect to real potentials.

7.4.1 *Constructing $\mathcal{PT}$-symmetric potentials*

In order to satisfy the requirement of normalizability, Schrödinger eigenvalue problems for $\mathcal{PT}$-symmetric potentials are defined with boundary conditions imposed in Stokes wedges in the complex plane [Bender and Boettcher (1998b)]. Unlike their Hermitian counterparts, the Stokes wedges for $\mathcal{PT}$-symmetric potentials may not include the real axis. The operator $\mathcal{P}$ usually performs space reflection about the origin in the complex plane (for complex x, $\mathcal{P} : x \to -x$), so $\mathcal{PT}$-symmetric eigenvalue problems are typically defined on domains that are symmetric with respect to $x = 0$.

For radial potentials $\mathcal{PT}$-symmetric eigenvalue problems are an exception. Generalizing a Hermitian problem in the radial variable $r > 0$ to a $\mathcal{PT}$-symmetric setting requires that we analytically continue the potential into the complex domain and pose the eigenvalue problem on a $\mathcal{PT}$-symmetric contour that avoids the singularity at the origin. To construct such a contour we can introduce an imaginary coordinate shift in which we replace $r > 0$ by $x - \mathrm{i}c$ (in which we take x and c real). Taking x to be the coordinate operator and c to be a numerical constant, we see that the $\mathcal{PT}$ operator transforms $x - \mathrm{i}c$ into $-(x - \mathrm{i}c)$. Thus, $x - \mathrm{i}c$ is a $\mathcal{PT}$-symmetric trajectory. Using the approach in Subsec. 7.2.1 this imaginary coordinate shift can be considered to be the $x_0 = -\mathrm{i}c$ constant of integration in (7.7). Although $c \neq 0$ may change the potential and the eigenfunctions substantially, it has no effect on the energy eigenvalues. (This is a generalization of the fact that finite real coordinate shifts do not change the energy eigenvalues of real potentials.) However, an imaginary coordinate shift changes the boundary conditions of a Schrödinger eigenvalue problem, and this affects the eigenfunctions and thus the physics of the problem.

An imaginary coordinate shift can also be employed to extend nonradial Hermitian eigenvalue problems to the corresponding $\mathcal{PT}$-symmetric eigenvalue problems. This imaginary coordinate shift is a linear automorphism (similarity transformation) using the operator e^{cp}, where p is the momentum operator. Under this automophism the potential energy operator $V(x)$

transforms as $\mathrm{e}^{cp}V(x)\mathrm{e}^{-cp} = V(x - \mathrm{i}c)$ [Ahmed (2001b)] and the action of e^{cp} on the eigenfunctions is $\mathrm{e}^{cp}\psi(x) = \psi(x - \mathrm{i}c)$.

Note that an imaginary coordinate shift is not sufficient to define the PT-symmetric version of some potentials. For example, more general PT-symmetric trajectories must be used for the Coulomb [Znojil and Lévai (2000)] and the Morse potentials [Znojil (1999b)]. Also, the P operator can be defined as a reflection about a point other than the origin. This is the case for the Khare–Mandal potential [Khare and Mandal (2000)].

In summary, we are generalizing Hermitian exactly solvable potentials to PT-symmetric exactly solvable potentials. For shape-invariant potentials most $z(x)$ functions [see (7.2)] are either PT even or PT odd; that is, they satisfy $PT z(x) = \pm z(x)$ [Lévai and Znojil (2000)]. The $\phi(z)$ functions are trivial in this case, so implementing PT symmetry requires only the appropriate choice of the s_i parameters in (7.21) and (7.26). The situation is more difficult for Natanzon and Natanzon confluent potentials because the function $\phi(z)$ is more complicated, so the p_i parameters must be chosen with care. This has been done for the Natanzon potentials [Lévai (2015)]. [For these potentials it is not always possible to construct a $z(x)$ function with definite PT-parity.] The situation is even more difficult for potentials outside the Natanzon confluent potential classes.

7.4.2   *Energy spectrum and breaking of PT symmetry*

Perhaps the most striking finding of PT-symmetric quantum mechanics is that the discrete energy eigenvalues of these complex potentials may be partly or even completely real (see Chap. 2). From a mathematical point of view this was found to be the consequence of pseudo-Hermiticity [Mostafazadeh (2002b)]; PT-symmetric quantum systems are P-pseudo-Hermitian. It was also demonstrated in many cases that with increasing non-Hermiticity, that is, applying stronger coupling of the imaginary potential component, the eigenvalues merge pairwise and continue as complex-conjugate pairs. This is the phenomenon of PT-symmetry breaking; the eigenstates cease to be PT symmetric after this transition even though the potential remains PT symmetric. The merging levels carry the quasiparity quantum number $q = \pm 1$ [Bagchi *et al.* (2001)] and the same n quantum number. In the domain with real energy eigenvalues the bound-state eigenfunctions are eigenfunctions of the PT operator, while in the domain of complex energy eigenvalues the PT operator transforms the two functions into one another: $PT\psi_n^{(q)}(x) = \psi_n^{(-q)}(x)$.

There are two reasons for this doubling of the energy spectrum (which does not occur for Hermitian (real) potentials). In the case of $\mathcal{PT}$-symmetric potentials generated by an imaginary coordinate shift from radial potentials it appears as a result of the more relaxed boundary conditions, while in the case of potentials defined on the full x axis the second set of solutions corresponds to resonances of the Hermitian potential. We may view q as a sign change of some parameter that leaves the potential intact but gives a different normalizable eigenfunction. Note that q does not appear for all $\mathcal{PT}$-symmetric potentials.

There are two kinds of breakdown of $\mathcal{PT}$ symmetry, sudden and gradual. In the former case all the energy eigenvalues become complex at the same value of the control parameter, while in the latter case pairs of energy eigenvalues become complex one-by-one as the control parameter is varied. The sudden mechanism appears to characterize all shape-invariant potentials and some others as well; the gradual mechanism occurs in the case of more complicated exactly solvable potentials. Finally, there are $\mathcal{PT}$-symmetric potentials for which there is no $\mathcal{PT}$ transition; some energy eigenvalues remain real over the entire parameter domain.

7.4.3 *Inner product, pseudonorm, and the $\mathcal{C}$ operator*

In general, for complex potentials the eigenfunctions are not orthogonal. However, $\mathcal{PT}$-symmetric potentials are $\mathcal{P}$-pseudo-Hermitian, and orthogonality (but not orthonormality) of the eigenstates can be restored by introducing the modified inner product $\langle \psi_i | \mathcal{P} | \psi_j \rangle$. The pseudonorm defined in this way has indefinite sign.

For exactly solvable potentials the transition to a modified inner product does not require much modification. The $\mathcal{PT}$ analog of the equation used to prove the reality of the eigenvalues of Hermitian systems is

$$(E_m - E_n^*) \int_{-a}^{a} \psi_m(x)\psi_n^*(-x)\mathrm{d}x = 0. \tag{7.29}$$

If $\mathcal{PT}$ symmetry is unbroken, E_m and E_n are real and unequal. Consequently the integral in (7.29) vanishes. In this case one also obtains

$$\int_{-a}^{a} \psi_n(x)\psi_n^*(-x)\mathrm{d}x = \pi_n \int_{-a}^{a} \psi_n^2(x)\mathrm{d}x$$

because $\psi_n^*(-x) = \mathcal{PT}\psi_n(x) = \pi_n\psi_n(x)$, where π_n is the $\mathcal{PT}$ parity of $\psi_n(x)$ state. The sign of the pseudonorm is thus determined by the π_n $\mathcal{PT}$ parity of the state and usually (but not always) behaves like $(-1)^n$.

The norm becomes positive if the inner product is modified by introducing the operator $\mathcal{C}$ [Bender *et al.* (2002)]. This operator is defined in terms of the energy eigenfunctions but calculating $\mathcal{C}$ in closed form is nontrivial, even for solvable potentials; $\mathcal{C}$ is known explicitly only in special cases.

7.4.4   *SUSYQM and $\mathcal{PT}$ symmetry*

The requirement of $\mathcal{PT}$ symmetry separates the even and odd components of the potential; the real part is even and the imaginary part is odd. In SUSYQM the parity properties of the superpotential $W(x)$ influence those of the supersymmetric partner potentials (7.10). Separating the components of $W(x)$ into real/imaginary as well as even/odd parts $W(x) = W_{\mathrm{Re}}(x) + W_{\mathrm{Ro}}(x) + iW_{\mathrm{Ie}}(x) + iW_{\mathrm{Io}}(x)$ and allowing the factorization energy ϵ in (7.13) to be complex leads to an inhomogeneous system of linear first-order differential equations for $W_{\mathrm{Re}}(x)$ and $W_{\mathrm{Io}}(x)$ [Lévai (2004, 2006)], where the inhomogeneity is represented by the constant $\mathrm{Im}(\epsilon)$. By examining all the possibilities of nonvanishing or vanishing $W_{\mathrm{Re}}(x)$ and $W_{\mathrm{Io}}(x)$ one can show that the supersymmetric partner $V_+(x)$ of $V_-(x)$ can be $\mathcal{PT}$ symmetric only when ϵ is real [Lévai (2004, 2006)].

Another unusual feature of $\mathcal{PT}$-symmetric potentials is that the presence of the quasiparity quantum number q (see Subsec. 7.4.2) allows one to construct *two* rather than one supersymmetric partner of $V_-(x)$, originating from the ground states $\psi_0^{(\pm)}(x)$ [Lévai and Znojil (2002); Lévai *et al.* (2003)]. Whenever the ground-state energy is real (or zero, after a real shift of the energy scale), both supersymmetric partner potentials $V_+^{(\pm)}(x)$ are $\mathcal{PT}$ symmetric. However, if the ground states assigned to $q = \pm$ are $\mathcal{PT}$ transforms of one another [that is, if the $\mathcal{PT}$ symmetry of $V_-(x)$ is broken], then the SUSY partner potentials $V_+^{(\pm)}(x)$ cease to be $\mathcal{PT}$ symmetric. This finding also holds in the more general case for which the factorization energy ϵ has an imaginary component.

The doubling of states according to the quasiparity has implications for the potential algebras associated with some shape-invariant $\mathcal{PT}$-symmetric potentials. These are also doubled and appear as $SO(2,2) \sim SO(2,1) \otimes SO(2,1)$ or $SO(4) \sim SO(3) \otimes SO(3)$, depending on whether the potentials have a finite or an infinite number of real energy levels [Lévai *et al.* (2002a)].

Finally, we note that for $\mathcal{PT}$-symmetric potentials the operators A and $A^\dagger$ in Subsec. 7.2.2 cease to be Hermitian adjoints. Nevertheless, the dagger notation is used throughout this chapter.

7.4.5 Scattering in PT-symmetric potentials

The delicate balance of emissive and absorptive potential components manifests itself in the scattering properties of PT-symmetric potentials. The reflection and transmission coefficients $R(k)$ and $T(k)$ can be calculated not only for potentials defined on the real axis but also for those defined on general PT-symmetric trajectories in the complex plane. As in the Hermitian case the amplitudes $R(k)$ and $T(k)$ are determined from the asymptotic behaviors of the solutions to the Schrödinger equation.

In the non-Hermitian PT-symmetric case unitarity is violated and thus $|T(k)|^2 + |R(k)|^2 \neq 1$. Additionally, PT-symmetric potentials are characterized by *chirality* (handedness) in the sense that $R(-k) \neq R(k)$; that is, the reflection coefficient depends on the direction of the incident wave. Nevertheless, the transition coefficient behaves in the same way as for real potentials, that is, $T(-k) = T(k)$. These results show that PT-symmetric potentials lie between real and general complex potentials; some of their features mimic the properties of real potentials, while others are manifestly different. For PT-symmetric potentials unitarity is replaced by pseudo-unitarity conditions concerning $T(k)$, $R(k)$, and $R(-k)$ [Ge *et al.* (2012a)]. Further unusual properties have also been identified, such as unidirectional invisibility, where $R(k) = 0$ and $T(k) = 1$ from one direction for some k [Lin *et al.* (2011)]. Another special feature of PT-symmetric potentials is their ability to exhibit spectral singularities [Mostafazadeh (2009)]. These are infinite peaks in $T(k)$ and $R(k)$ at positive energies (real values of k) that behave like zero-width resonances. These results have inspired work on complex non-Hermitian scattering potentials that can be used to model active optical systems [Mostafazadeh (2014b)].

7.5 Examples of Solvable PT-Symmetric Potentials

This section presents examples of the solvable potentials discussed in Sec. 7.3 and shows how they can be made PT symmetric. The special properties outlined in Sec. 7.4 are also demonstrated.

7.5.1 Shape-invariant potentials

Table 7.2 displays the main features of the PT-symmetric versions of the potentials listed in Table 7.1. (A real shift of $\pi/2a$ gives rise to an independent version of the Scarf I and Rosen–Morse I potentials when $c \neq 0$. These are omitted in Table 7.1 for brevity.) In the PT-symmetrization of most shape-invariant potentials one identifies the parameter ranges where the bound-state energy eigenvalues are real or complex.

The PT-symmetric generalization of the *Scarf II* potential

$$V(x) = -V_R \operatorname{sech}^2 x + iV_I \tanh x \operatorname{sech} x \tag{7.30}$$

(see Table 7.1) is exactly solvable. In terms of α and β the parameters V_R and V_I are $V_R = \frac{1}{4}(2\alpha^2 + 2\beta^2 - 1)$ and $V_I = \frac{1}{2}(\beta^2 - \alpha^2)$. We express α and β in terms of V_R and V_I as

$$\alpha = -\sqrt{V_R + \frac{1}{4} - V_I}, \qquad \beta = -\sqrt{V_R + \frac{1}{4} + V_I}. \tag{7.31}$$

Both α and β are real and negative when $|V_I| \le V_R + \frac{1}{4}$. The potential $V(x)$ is plotted in Fig. 7.1.

From the parametrization in Table 7.1, $A/a = s = -(\alpha + \beta + 1)/2$ and $B/a = \lambda = i(\alpha - \beta)/2$, and for simplicity $a = 1$. The Scarf II potential can also be obtained (see Subsec. 7.3.1) by choosing $p_I = 1$, $p_{II} = p_{III} = 0$

Table 7.2 PT symmetrization of shape-invariant potentials [Lévai and Znojil (2000, 2001)]. Reference to specific results for the individual potentials are indicated in the footnotes.

Name	$x - ic$	Condition for PT symmetry	Breaking of PT symmetry	q	Pseudo-norm	$T(k)$, $R(k)$
Scarf II	yes	$\alpha^* = \pm\alpha$, $\beta^* = \pm\beta$	sudden	yes	yes[a]	yes[b]
Generalized Pöschl–Teller	yes, $c \ne 0$	$\alpha^* = \pm\alpha$, $\beta^* = \pm\beta$	sudden	yes		yes[c]
Scarf I[d]	yes, $c \ne 0$	$\alpha^* = \pm\beta$	no	no	yes[e]	n.a.
Rosen–Morse II	yes	$s(s+1)$, λ real	no	no	yes[f]	yes[f]
Eckart	yes, $c \ne 0$	$s(s+1)$, λ real	sudden	yes		
Rosen–Morse I	yes	$s(s+1)$, λ real	no	no	yes[g]	n.a.
Harmonic oscillator	yes, $c \ne 0$	$\omega^* = \omega$, $\alpha^* = \pm\alpha$	sudden	yes		n.a.
Coulomb	n.a.[h]	$e^2 = iZ$	sudden[i]	yes	yes	yes[j]
Morse	n.a.[k]	$A^* = A$, $B^* = B$	no	yes	yes	
1D harm. osc.	yes	$\omega^* = \omega$	no	no		n.a.

[a][Lévai *et al.* (2002b)], [b][Lévai *et al.* (2001)], [c][Lévai *et al.* (2002a)], [d]closed formula for the C operator available [Roychoudhury and Roy (2007)], [e][Lévai (2006)], [f][Lévai and Magyari (2009)], [g][Lévai (2008a)], [h][Znojil and Lévai (2000); Znojil *et al.* (2009)], [i][Lévai (2009)], [j][Lévai *et al.* (2009)], [k][Znojil (1999b)].

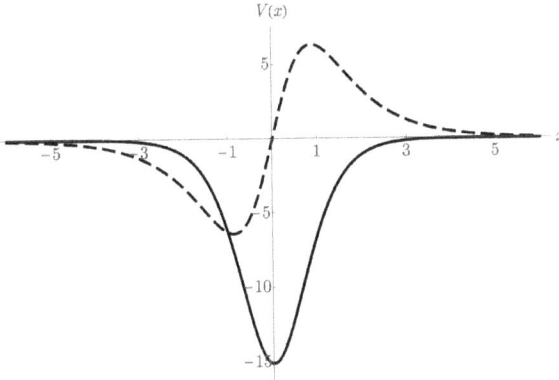

Fig. 7.1   Real part (solid line) and imaginary part (dashed line) of the Scarf II potential $V(x)$ in (7.30) with $V_R = 15.1$, $V_I = 12.7$ ($\alpha = -1.628$, $\beta = -5.296$); $\mathrm{Re}\,V(x)$ is even in x and $\mathrm{Im}\,V(x)$ is odd in x. Interchanging α and β corresponds to $x \to -x$.

and $s_I = 1/4$, solving for s_{II} and s_{III}, and substituting into (7.21). For this choice of parameters $\lim_{x \to \pm \infty} V(x) = 0$. Interchanging α and β corresponds to $V(x) \leftrightarrow V(-x)$. If both α and β are pure real or pure imaginary, $V(x)$ is $\mathcal{PT}$ symmetric. The potential $V(x)$ is invariant under $\alpha \leftrightarrow -\alpha$ but the bound-state eigenfunctions and eigenvalues are not invariant under this transformation. Finally, we introduce the quasiparity quantum number q discussed in Subsec. 7.4.2 via $q\alpha \equiv \pm\alpha$. In terms of these parameters the eigenfunctions and eigenvalues are

$$\psi_n^{(q)}(x) = N_n^{(q)}(1 - \mathrm{i}\sinh x)^{\frac{q\alpha}{2} + \frac{1}{4}}(1 + \mathrm{i}\sinh x)^{\frac{\beta}{2} + \frac{1}{4}} P_n^{(q\alpha, \beta)}(\mathrm{i}\sinh x), \quad (7.32)$$

$$E_n^{(q)} = -\left[n + \tfrac{1}{2}(q\alpha + \beta + 1)\right]^2. \quad (7.33)$$

The normalizability of $\psi_n^{(q)}(x)$ in (7.32) requires that

$$n < -\tfrac{1}{2}\mathrm{Re}(q\alpha + \beta + 1). \quad (7.34)$$

For the Scarf II potential the second set of bound states corresponds to resonances appearing for the Hermitian version of this potential, which is obtained from (7.30) for $\alpha^* = \beta$ [Lévai *et al.* (2001)].

The $\mathcal{PT}$ symmetry is broken when pairs of real eigenvalues (7.33) with the same n and opposite q merge and become complex. This happens as α reaches zero through real values and then becomes pure imaginary. From (7.33) we see that this transition occurs for *all* n, so the breakdown of $\mathcal{PT}$ symmetry occurs via the *sudden* mechanism described in Subsec. 7.4.2. This transition is illustrated in Fig. 7.2.

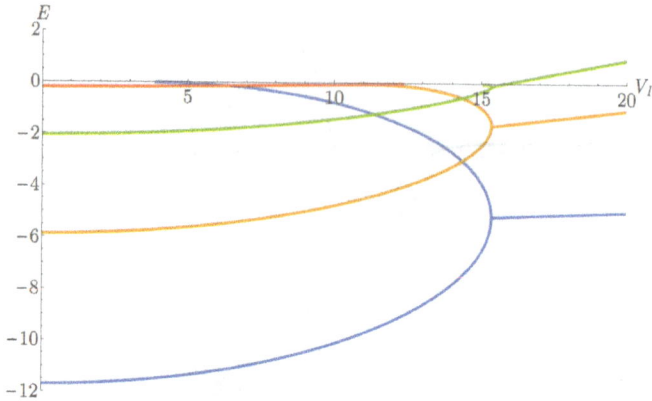

Fig. 7.2   Energy eigenvalues $E_n^{(q)}$ in (7.33) plotted as functions of V_I with $V_R = 15.1$ kept constant [see (7.31)]. The energies for $n = 0$ (blue), $n = 1$ (yellow), $n = 2$ (green), and $n = 3$ (red) for $q = \pm 1$ are shown. The energies become complex at $V_I = 15.35$ and are real below this point. In general, the energies become complex when $|V_I| > V_R + \frac{1}{4}$ [Ahmed (2001c)], which corresponds to $\alpha = 0$ and real β [Lévai and Znojil (2001)]. The energies are plotted for values of V_I that satisfy the normalizability condition (7.34).

Figure 7.2 illustrates a key feature of the $\mathcal{PT}$-symmetric Scarf II potential: Pairs of energy levels with opposite quasiparity q (and different n) cross at specific values of the potential parameters. According to (7.33) $E_n^{(+)} = E_m^{(-)}$ when $\alpha = m - n \equiv -j < 0$. As a consequence of the property $P_n^{(-j,\beta)}(z) = C(1-z)^j P_{n-j}^{(j,\beta)}(z)$ $(j = 0, 1, \ldots n)$ of Jacobi polynomials, the eigenfunctions in (7.32) become linearly dependent at the crossing points [Ahmed *et al.* (2015b)]. The crossing points in Fig. 7.2 occur at $v_I = 14.35$ $(j = 1)$ for $n = 0$ and $m = 1$, $v_I = 11.35$ $(j = 2)$ for $n = 0$ and $m = 2$ and at $v_I = 6.35$ $(j = 3)$ for $n = 0$ and $m = 3$.

The quasiparity doubling of states implies that there are *two* SUSYQM partners of the $\mathcal{PT}$-symmetric Scarf II potential depending on which node-less $n = 0$ state is used in the procedure outlined in Subsecs. 7.4.4 and 7.2.2 [see (7.12)]. Both supersymmetric partner potentials are $\mathcal{PT}$ symmetric if $\mathcal{PT}$ symmetry is unbroken (real α), while they cease to be $\mathcal{PT}$ symmetric in the case of broken $\mathcal{PT}$ symmetry (imaginary α) [Lévai and Znojil (2002)].

The eigenvalue problem for the $\mathcal{PT}$-symmetric Scarf II potential exhibits the symmetry of the $\mathrm{SO}(2,2) \sim \mathrm{SO}(2,1) \oplus \mathrm{SO}(2,1)$ Lie algebra. The ladder operators $J_\pm$ of both $\mathrm{SO}(2,1)$ algebras are first-order differential operators related to the two sets of A and $A^\dagger$ supersymmetric operators, and the Casimir operator is related to the Hamiltonian [Lévai *et al.* (2002a)]. The

$J_\pm$ commute with H, so the algebra is a potential algebra [Alhassid *et al.* (1983)] and the ladder operators connect states having the same energy eigenvalue (and quasiparity) associated with *different* coupling coefficients V_R and V_I. [Earlier algebraic studies of the $\mathcal{PT}$-symmetric Scarf II potential considered the SL(2, C) [Bagchi and Quesne (2000)] and SU(1, 1) $\sim$ SO(2, 1) [Lévai *et al.* (2001)] algebras, corresponding to only one of the two possible values of q.]

The modified inner product and the pseudonorm discussed in Subsec. 7.4.3 was calculated in [Lévai *et al.* (2002b)]. The quantity

$$I_{nl}^{(\alpha,\beta,\delta)} = \int_{-\infty}^{\infty} \psi_n^{(\alpha,\beta)}(x) \left[\psi_l^{(\delta,\beta)}(-x)\right]^* dx$$

contains all the necessary inner products for the $\mathcal{PT}$-symmetric Scarf II potential, where the notation $\psi_n^{(\alpha,\beta)}(x) = \psi_n^{(q)}(x)$ is used. Here β is real, while α and δ are either real or imaginary. There are four possibilities:

(i) $\delta = \alpha$ real: This corresponds to unbroken $\mathcal{PT}$ symmetry and states with the same quasiparity. In this case

$$I_{nl}^{(\alpha,\beta,\alpha)} = \delta_{nl} \left|N_n^{(\alpha,\beta)}\right|^2 \frac{(-1)^n \pi 2^{\alpha+\beta+2}\Gamma(-\alpha-\beta-n)}{(-\alpha-\beta-2n-1)n!\Gamma(-\alpha-n)\Gamma(-\beta-n)}. \quad (7.35)$$

Because of (7.34) each term in (7.35) is positive except for $(-1)^n$ and $[\Gamma(-\alpha-n)\Gamma(-\beta-n)]^{-1}$, but its sign depends on the relative magnitude of α, β, and n. Except for extreme values of α and β the argument of the two gamma functions is positive for the first few n, so then the alternating $(-1)^n$ factor determines the sign of the pseudonorm. However, as n reaches $-\alpha$ and/or $-\beta$, this regular pattern changes.

(ii) $\delta = -\alpha$ real: States with opposite quasiparity are considered with unbroken $\mathcal{PT}$ symmetry leading to $I_{nl}^{(\alpha,\beta,-\alpha)} = 0$ [Lévai *et al.* (2002b)].

(iii) $\delta = \alpha$ imaginary: The case of broken $\mathcal{PT}$ symmetry and states with the same quasiparity, resulting in $I_{nl}^{(\alpha,\beta,\alpha)} = 0$ [Lévai *et al.* (2002b)].

(iv) $\delta = -\alpha$ imaginary: Here, $\mathcal{PT}$ symmetry is broken and the states have different quasiparity. The $I_{nl}^{(\alpha,\beta,-\alpha)}$ overlap has the same form as (7.35) except that $\left|N_n^{(\alpha,\beta)}\right|^2$ must be replaced with $N_n^{(\alpha,\beta)}\left[N_l^{(-\alpha,\beta)}\right]^*$. This nonorthogonality of two different states is a typical feature of $\mathcal{PT}$-symmetric problems with broken $\mathcal{PT}$. The findings specific to the Scarf II potential can also be interpreted qualitatively by (7.29).

Scattering states have also been analyzed for the Scarf II potential with the $x \to x - ic$ imaginary coordinate shift. The transmission and reflection coefficients are [Lévai *et al.* (2001)]

$$T(k, \alpha, \beta) = \frac{\Gamma[\frac{1}{2}(\alpha + \beta + 1) - ik]\Gamma[-\frac{1}{2}(\alpha + \beta - 1) - ik]}{\Gamma(-ik)\Gamma(1 - ik)\Gamma^2(\frac{1}{2} - ik)}$$
$$\times \Gamma[\frac{1}{2}(\beta - \alpha + 1) - ik]\Gamma[\frac{1}{2}(\alpha - \beta + 1) - ik], \quad (7.36)$$

$$R(k, \alpha\beta) = i\exp(-2ck) \left(\frac{\cos[\frac{\pi}{2}(\alpha + \beta + 1)]\sin[\frac{\pi}{2}(\alpha - \beta)]}{\cosh(\pi k)} \right.$$
$$\left. - \frac{\sin[\frac{\pi}{2}(\alpha + \beta + 1)]\cos[\frac{\pi}{2}(\alpha - \beta)]}{\sinh(\pi k)} \right) T(k, \alpha, \beta). \quad (7.37)$$

These equations contain the Hermitian case with $c = 0$ and $\alpha = \beta^* = -s - \frac{1}{2} - i\lambda$ [Lévai *et al.* (2001); Dabrowska *et al.* (1988)], the PT-symmetric case with unbroken symmetry (α real) and the PT-symmetric case with broken symmetry (α imaginary and β real). The imaginary coordinate shift affects *only* the reflection coefficient (7.37).

In general, $T(k, \alpha, \beta) = T(-k, \alpha, \beta)$, but $R(k, \alpha, \beta) \neq R(-k, \alpha, \beta)$, so the PT-symmetric Scarf II potential exhibits chirality [Ahmed (2004)] (see Subsec. 7.4.5). For specific values of the parameters $|T(k, \alpha, \beta)|^2$ and $|R(k, \alpha, \beta)|^2$ can be zero or unity, giving rise to unidirectional invisibility [Ahmed (2013)]. The poles of (7.36) also determine the bound-state energies of (7.30). These poles occur where the argument of the gamma functions in the numerator take on nonpositive integer values. The two cases that are compatible with the normalizability condition (7.34) give the energy eigenvalues (7.33).

Some PT-symmetric potentials possess spectral singularities (see Subsec. 7.4.5), which correspond to zero-width resonances at positive energies. The analysis of (7.36) proves that they also exist for the potential (7.30) [Ahmed (2009)]. This phenomenon occurs in the case of broken PT symmetry; that is, when α is imaginary and β is real. To have spectral singularities the argument of the gamma functions must be a nonpositive integer $-N$. This condition determines the possible value of β, while k must have the value $\pm\frac{1}{2}i\alpha$. This gives a single spectral singularity at $E = -\alpha^2/4 > 0$. Spectral singularities only occur for the specific combination of the coupling coefficients $V_R + V_I + 1/4 = \beta^2 = (2N + 1)^2$ [Ahmed (2009)].

The PT-symmetric version of the *Rosen–Morse II* potential has the same even component as the Scarf II potential, but its odd component approaches finite values for large x:

$$V(x) = -s(s + 1)\text{sech}^2(x) + 2i\lambda\tanh(x), \quad (7.38)$$

where $s(s+1)$ and λ are real (see Fig. 7.3). (An imaginary coordinate shift $-ic$ can also be used but it is not discussed here.) We obtain this potential from Table 7.1 by taking $z(x) = \tanh(ax)$ and $C = a^2$ and choosing $A/a = s$, $B/a^2 = i\lambda$, and for simplicity $a = 1$. The α and β parameters of the Jacobi polynomial depend on n, and they are related to s and λ via $\alpha = s - n + \lambda/(s-n)$ and $\beta = s - n - \lambda/(s-n)$. Another way to obtain (7.38) is described in Subsec. 7.3.1: Choose $p_{II} = 1$ and $p_I = p_{III} = 0$ in (7.19) and (7.22 – 7.24), and substitute $s_{II} = -1$ in (7.21).

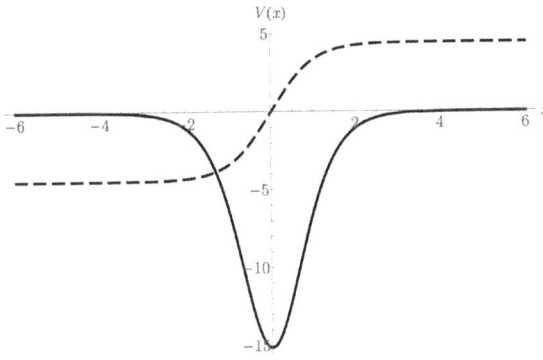

Fig. 7.3 Real part (solid line) and imaginary part (dashed line) of the Rosen–Morse II potential $V(x)$ in (7.38) with $s = 3.418$ and $2\lambda = 4.420$. The solid line is an even function of x and the dashed line is an odd function of x.

For the potential (7.38) (and its real counterpart) the general solutions involve hypergeometric functions [Lévai and Magyari (2009)]:

$$\psi^{(\pm)}(x) \sim (1 - \tanh x)^{\pm\alpha/2}(1 + \tanh x)^{\beta/2}$$
$$\times\, {}_2F_1[-s + \tfrac{1}{2}(\beta \pm \alpha), s + 1 + \tfrac{1}{2}(\beta \pm \alpha); 1 \pm \alpha; \tfrac{1}{2}(1 - \tanh x)].$$

Note that $\psi^{(\pm)}(x)$ are related by $\alpha \leftrightarrow -\alpha$, so only one of these solutions can be normalized. For this potential the doubling of bound states does not occur; the quasiparity quantum number does not appear. The bound-state solutions are

$$\psi_n(x) = N_n(1 - \tanh x)^{\alpha/2}(1 + \tanh x)^{\beta/2} P_n^{(\alpha,\beta)}(\tanh x),$$

the corresponding energy eigenvalues are

$$E_n = -(s-n)^2 + \lambda^2(s-n)^{-2}, \tag{7.39}$$

and normalizability requires that $s > n$. The normalization constant is

$$N_n = \frac{i^n 2^{n-s}\sqrt{n!}}{|\Gamma[s+1+i\lambda/(s-n)]|}\left(\frac{\Gamma(2s-n+1)[(s-n)^2 + \lambda^2(s-n)^{-2}]}{s-n}\right)^{1/2}$$

and the pseudonorm follows the regular $(-1)^n$ pattern with the quantum number n [Lévai and Magyari (2009)].

For this potential the $\mathcal{PT}$ symmetry cannot be broken, so the energy spectrum cannot become complex. This is because complex energy eigenvalues can occur only if s is complex, but $\mathcal{PT}$ symmetry requires that $s(s+1)$ is real. In principle, $s = -1/2 + i\sigma$ is compatible with this requirement, but the real part of the potential becomes repulsive in this case and no normalizable states can occur. The nonbreaking of $\mathcal{PT}$ symmetry has been proved to hold [Lévai (2015)] for the $\mathcal{PT}$-symmetric versions of all PII type shape-invariant potentials (the Rosen–Morse I, II and Eckart potentials) in Table 7.1.

We emphasize that complex eigenvalues do not appear even if the imaginary part of the potential is increased. (In the case of the Scarf II potential, which has the same real potential component, increasing the imaginary part of the potential does lead to the breakdown of $\mathcal{PT}$ symmetry via the sudden mechanism.) Figure 7.4 shows that increasing the imaginary part of the potential merely increases the energies. The energy levels E_n start to be positive for $s - |\lambda|^{1/2} \leq n < s$, and eventually all the energies are positive for $|\lambda| > s^2$. The $\mathcal{PT}$-symmetric Rosen–Morse II potential cannot support bound states if its real component vanishes (for $s = 0$). This represents an interesting difference between the Scarf II and the Rosen–Morse II potentials [Lévai (2011)]. The scattering amplitudes have also been determined [Lévai and Magyari (2009)] and were found to exhibit the chirality effects discussed in Subsec. 7.4.5.

Because the *Coulomb potential* is singular at the origin its $\mathcal{PT}$ symmetrization requires a domain of definition that avoids $x = 0$. However, the usual imaginary coordinate shift $x - ic$ does not solve the problem because the eigenfunctions are not normalizable due to their asymptotic behaviors [Lévai and Znojil (2000)]. In the first attempt to find such an integration path compatible with the requirements of $\mathcal{PT}$ symmetry, the standard Coulomb-harmonic oscillator mapping was considered [Znojil and Lévai (2000)]. Applying it to the imaginary-shifted coordinate of the $\mathcal{PT}$-symmetric harmonic oscillator, it produced an upward parabolic curve in the complex x plane of the Coulomb problem, circumventing the origin from below. Considering the x^{-1} character of the potential, this shape was especially interesting compared to the downward trajectories associated with the Bender–Boettcher potentials $x^2(ix)^\epsilon$ for $\epsilon > 2$. It was also found that the energy spectrum was inverted for positive values of the mass m.

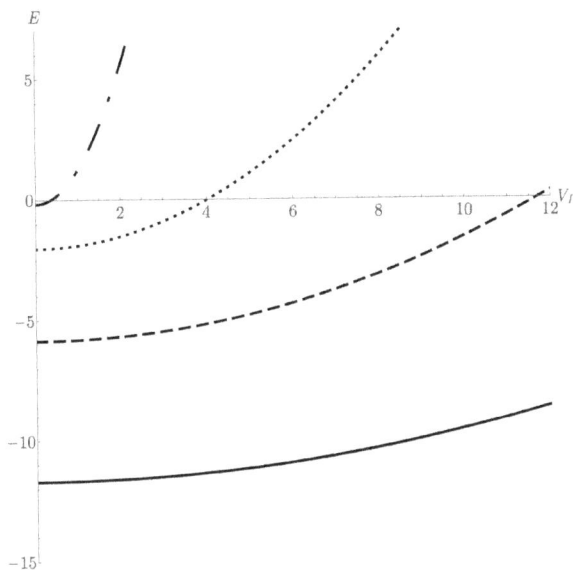

Fig. 7.4   Energy eigenvalues E_n in (7.39) plotted as functions of V_I with $V_R = 15.1$ held fixed. These parameters are $V_R = s(s+1)$ and $V_I = 2\lambda$ [see (7.38)]. For $V_I = 0$ the energy spectrum coincides with that in Fig. 7.2. $n = 0$ (solid line), $n = 1$ (dashed line), $n = 2$ (dotted line), $n = 3$ (dot-dashed line).

In a systematic study an upward U-shaped trajectory was considered [Znojil *et al.* (2009)]:

$$t(x) = \begin{cases} -\mathrm{i}(x + \frac{\pi}{2}c) - c, & x \in (-\infty, -\frac{\pi}{2}c), \\ ce^{\mathrm{i}(x/c + 3\pi/2)}, & x \in (-\frac{\pi}{2}c, \frac{\pi}{2}c), \\ \mathrm{i}(x - \frac{\pi}{2}c) + c, & x \in (\frac{\pi}{2}c, \infty), \end{cases} \tag{7.40}$$

with $c > 0$. With this PT-odd trajectory $t^*(-x) = -t(x)$, the potential

$$V_C(x) = \frac{L(L+1)}{t^2(x)} + \frac{\mathrm{i}Z}{t(x)},$$

is PT symmetric and the eigenfunctions are normalizable. Like other PT-symmetric potentials, there are two sets of such solutions characterized by the quasiparity quantum number q:

$$\Psi^{(q)}(x) = N^{(q)} e^{-kt(x)} [t(x)]^{\omega(q,L)} {}_1F_1[\omega(q,L) + \mathrm{i}Z/(2k), 2\omega(q,L), 2kt(x)], \tag{7.41}$$

where $\omega(q,L) = \frac{1}{2} + q\left(L + \frac{1}{2}\right)$. Bound states occur when the first argument of the confluent hypergeometric function is a nonnegative integer n. In this case the energy eigenvalues are

$$E_n^{(q)} = -\frac{\hbar^2 k^2}{2m} = \frac{\hbar^2 Z^2}{8m[n + \omega(q, L)]^2}.$$

These energy eigenvalues are positive for $m < 0$, which is the appropriate choice considering that (7.40) implies the presence of an effective mass in the theory [Znojil *et al.* (2009)].

Breakdown of $\mathcal{PT}$ symmetry occurs for $L = -\frac{1}{2} + i\lambda$ [Lévai (2009)]. In this case $(E_n^{(q)})^* = E_n^{(-q)}$ holds. Furthermore, $\mathcal{PT}\Psi^{(q)}(x) = \Psi^{(-q)}(x)$ applies for appropriate choice of $N^{(q)}$ in (7.41). By analyzing the scattering solutions one can recover the bound-state energy eigenvalues as the poles of the transmission coefficient [Lévai *et al.* (2009)]. The $\mathcal{PT}$-symmetric Coulomb problem is similar to other $\mathcal{PT}$-symmetric scattering potentials in that it exhibits the chirality effect described in Subsec. 7.4.5 [Lévai (2011)].

7.5.2   *Natanzon potentials — An example*

The Natanzon potentials discussed in Subsec. 7.3.1 contain shape-invariant potentials as special cases. This is also true for the $\mathcal{PT}$-symmetric version of the Natanzon potentials. The example discussed here contains all the PI- and PII-type shape-invariant potentials [Lévai (2012)]. It is defined by the parameters $p_I = 1$, $p_{II} = \delta$, and $p_{III} = 0$ in the $\phi(z)$ function appearing in (7.19). Table 7.1 shows that for $\delta = 0$ one obtains the PI-type shape-invariant potentials, while for $\delta \to \infty$ PII-type potentials can be generated, provided that $C/\delta \equiv \widetilde{C}$ is kept constant. The sign of C and δ determines which one of the six PI- and PII-type potentials is obtained. In general, solving (7.19) yields an implicit function $x(z)$.

Substituting the above p_i parameters in (7.22)–(7.24) and eliminating E_n and ρ, we obtain a quartic algebraic equation for ω. Choosing the s_i parameters in (7.21) as $s_I = 1/4$, $s_{II} = \Sigma - 3/4$ and $s_{III} = 2\Lambda$, we recover the general form of the potential in [Lévai (2012)]

$$V(x) = \frac{3C\delta(3\delta + 2)}{4[\delta + 1 - z^2(x)]^2} - \frac{5C\delta^2(\delta + 1)}{4[\delta + 1 - z^2(x)]^3}$$
$$+ \frac{C\Sigma}{\delta + 1 - z^2(x)} + \frac{2C\Lambda z(x)}{\delta + 1 - z^2(x)} \tag{7.42}$$

together with the energy eigenvalues

$$E_n = C\left[n + \tfrac{1}{2}(\alpha + \beta + 1)\right]^2 \tag{7.43}$$

and the corresponding eigenfunctions

$$\Psi_n(x) = N_n[\delta + 1 - z^2(x)]^{1/4}[1 - z(x)]^{\alpha/2}[1 + z(x)]^{\beta/2} P_n^{(\alpha,\beta)}[z(x)], \tag{7.44}$$

where $\alpha = \omega + \Lambda/\omega$ and $\beta = \omega - \Lambda/\omega$. The potential depends on four parameters: C and δ determine $z(x)$, and Σ and Λ act as coupling coefficients of two of the potential terms. Two further potential terms in (7.42) depend only on C and δ. As discussed in Subsec. 7.3.1, these originate from the Schwartzian derivative term in (7.21). The energy eigenvalues (7.43) are determined by the roots of the quartic algebraic equation mentioned above:

$$\Pi(\omega) = (\delta+1)\omega^4 + \delta(2n+1)\omega^3 + \left[\tfrac{1}{4}\delta(2n+1)^2 - \delta - \Sigma - \tfrac{1}{4}\right]\omega^2 + \Lambda^2 = 0. \tag{7.45}$$

Figure 7.5 displays the shape-invariant potentials that are obtained as limiting cases for the actual sign of C and δ.

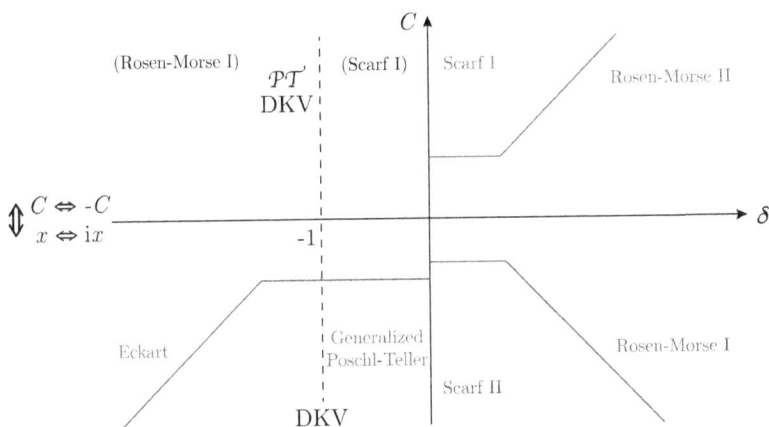

Fig. 7.5   Potentials obtained from (7.42) as special limits. PI-type potentials appear on either side of the vertical axis ($\delta = 0$), while PII-type potentials correspond to directions defined by $C/\delta \equiv \tilde{C}$. The six shape-invariant potentials are arranged in pairs, corresponding to some quadrant of the map. These pairs can be transformed into each other continuously along the broken lines inserted for illustration. Changing the sign of C implies changing x into ix, which transforms the trigonometric and hyperbolic versions of the potentials into each other. The DKV (Dutt–Khare–Varshni) potential [Dutt *et al.* (1995); Roychoudhury *et al.* (2001)] is also obtained as a special case for $\delta = -1$. For $\delta < 0$, $C > 0$ (Scarf I) and (Rosen–Morse I) indicate the $\pi/2$-shifted versions of the named potentials, that is, interchanging the $\sin x$ and $\cos x$ functions.

Of the possible choices $C = -a^2$ and $\delta \geq 0$ is considered here in detail. In this case the solution of (7.19) is

$$C^{1/2}x = \arctan[z(\delta+1-z^2)^{-1/2}] + \delta^{1/2}\text{arctanh}[\delta^{1/2}z(\delta+1-z^2)^{-1/2}].$$

This implicit solution becomes explicit in the two shape-invariant limits $z(x) = i\sinh(ax)$ for $\delta \to 0$ and $z(x) = i\tan(ax)$ for $\delta \to \infty$ with $\tilde{C} = C/\delta$, recovering the Scarf II and Rosen–Morse II potentials, respectively. The implicit nature of $z(x)$ does not hinder calculations involving the eigenfunctions. The N_n in (7.44) can be calculated in closed form [Lévai (2012)] both in the real and the $\mathcal{PT}$-symmetric case.

The potential (7.42) can be made $\mathcal{PT}$ symmetric by choosing real values for Σ and Λ [Lévai (2012)]. The energy eigenvalues are determined by the roots of (7.45) and both real and complex-conjugate roots can appear. Complex-conjugate ω values imply complex-conjugate energy eigenvalues in (7.43), so the $\mathcal{PT}$ symmetry can be broken. Normalizable solutions require that $-n - 1/2 > \mathrm{Re}(\omega)$ [Lévai (2012)]. Figure 7.6 shows that the energy eigenvalues turn complex as the strength of the imaginary potential component is varied by changing Λ. The critical points correspond to the minima of the quartic curves (7.45) at $\Pi(\omega_{\mathrm{crit.}})$ for ω values compatible with the normalizability condition above.

The simple structure of $\Pi(\omega)$ allows the analysis of the extrema to be carried out analytically. The breakdown of $\mathcal{PT}$ symmetry occurs here via the *gradual* mechanism, which is different from most exactly solvable potentials and similar to some semi-analytically [Znojil and Lévai (2001)] or numerically [Bender and Boettcher (1998b)] solvable examples. A major difference in the latter case is that for the Bender–Boettcher type potentials the complexification of energy eigenvalues starts from the higher end of the spectrum, while in the present case it starts at the lower end at $n = 0$. The accidental crossing of energy levels has been discussed for this potential too; in fact, it has been proved to be a generic feature of $\mathcal{PT}$-symmetric Natanzon-class potentials [Lévai (2017)].

The potential (7.42) is similar to the Scarf II potential for moderate values of δ, so for illustration a situation close to the Rosen–Morse I limit is presented in Fig. 7.7 [Lévai (2012)]. Note that this potential resembles both the $\mathcal{PT}$-symmetric Rosen–Morse I potential [Lévai (2008a)] and the finite $\mathcal{PT}$-symmetric square-well potential [Znojil and Lévai (2001)].

Besides the shape-invariant limits, (7.42) contains two non-shape-invariant Natanzon potentials. The first is the real Ginocchio potential [Ginocchio (1984)], which is obtained for $\Lambda = 0$, $\delta = 1/(\lambda - 1)$ and $C = \lambda^2/(\lambda^2 - 1)$. Here, $\Pi(\omega)$ in (7.45) reduces to a quadratic form, so the energy eigenvalues can be obtained in closed form. The second is the Dutt–Khare–Varshni (DKV) potential [Dutt *et al.* (1995); Roychoudhury *et al.* (2001)], which arises when $\delta = -1$. Here, $\Pi(\omega)$ reduces to a cubic form, so the ω roots and thus the energy eigenvalues can be obtained in closed form. Also, $z(x)$ can be found explicitly from (7.19) as $z(x) = 1 + c\exp(-2x)$.

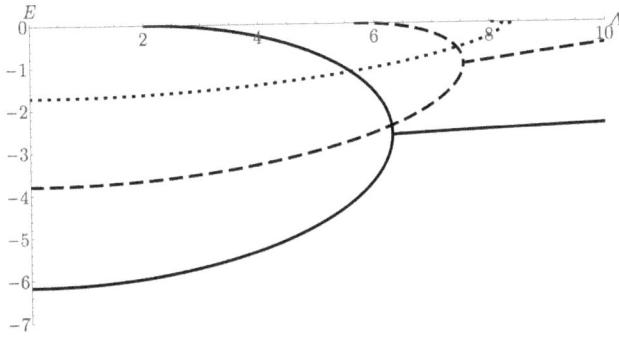

Fig. 7.6   Real parts of the eigenvalues (7.43) as functions of Λ for $C = -1$, $\delta = 1.25$, and $\Sigma = 15.1$. [Expressing ω in terms of Λ requires that (7.45) be solved numerically.] Complexification occurs at $\Lambda_{\text{crit 0}} = 6.298$, $\Lambda_{\text{crit 1}} = 7.542$, and $\Lambda_{\text{crit 2}} = 8.234$. For $\delta = 0$ the potential (7.42) reduces to that in Fig. 7.1 with $v_R = \Sigma = 15.1$, $v_I = -2\Lambda$, and $z(x) = \mathrm{i} \sinh x$.

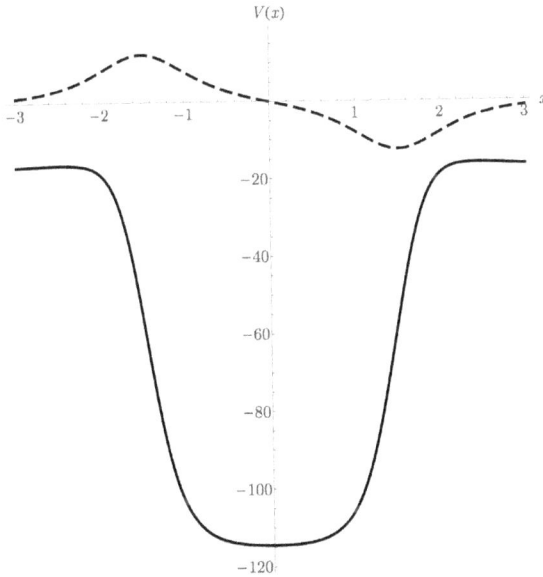

Fig. 7.7   Potential (7.42) for $\Sigma = 15.1$, $\Lambda = 1.25$, $\delta = 100$, and $C = -100$ (that is, $\tilde{C} = -1$) adapted to the energy scale convenient near the Rosen–Morse I limit.

The $\mathcal{PT}$-symmetric version of this potential is obtained from the Hermitian version by the transformation $x \to ix$ (that is, $C \to -C$) [Znojil *et al.* (2001)].

7.5.3 *Potentials generated by SUSY transformations*

The examples discussed here are generated by SUSYQM transformations in which the eigenfunctions are written as the linear combination of two special functions. According to Subsec. 7.3.3, there are two situations in which wave functions with this structure are obtained in a natural way.

The first occurs when the SUSY partners of general (non-shape-invariant) Natanzon or Natanzon confluent potentials are constructed by eliminating the ground-state energy of the original potential. An example of this type is the SUSYQM partner of the $\mathcal{PT}$-symmetric generalized Ginocchio potential, which we discuss here without citing explicit expressions for the potential, eigenfunctions, and energy spectrum [Lévai *et al.* (2003)]. The generalized Ginocchio potential [Ginocchio (1985)] is an implicit Natanzon potential; the function $x(z)$ can be obtained from the differential equation (7.19) with $p_I = \frac{1}{4}(\gamma^2 - 1)\lambda^{-4}$, $p_{II} = p_{III} = \frac{1}{2}\lambda^{-2}$. The potential depends on s and λ, in addition to γ. Two terms in the potential are determined via $s(s+1)$ and $\lambda(\lambda - 1)$. The other two terms originate from the Schwartzian derivative and depend only on γ. One of the terms in the potential has a $\lambda(\lambda - 1)x^{-2}$ singularity at $x = 0$, so the generalized Ginocchio potential is a radial potential. While $z(x)$ is implicit, the bound-state energies E_n are explicit.

The generalized Ginocchio potential can be $\mathcal{PT}$-symmetrized by introducing the familiar $x - ic$ imaginary coordinate shift while keeping γ, $s(s+1)$, and $\lambda(\lambda - 1)$ real [Lévai *et al.* (2003)]. This gives two sets of normalizable eigenfunctions distinguished by the quasiparity q. The breakdown of $\mathcal{PT}$ symmetry occurs via the sudden mechanism (see Subsec. 7.4.2) when the originally real λ parameter reaches $1/2$ and continues as $1/2 + il$, leading to pairs of complex-conjugate energy eigenvalues belonging to the same quantum number n and opposite values of q.

There are two sets of normalizable states, so the $\mathcal{PT}$-symmetric generalized Ginocchio potential has two SUSY partners, which remain $\mathcal{PT}$ symmetric if the parameter λ and the ground-state energy $E_{0,-}^{(q)}$ stay real. For complex λ the SUSY partner ceases to be $\mathcal{PT}$ symmetric [Lévai *et al.* (2003)]. This also happens for the $\mathcal{PT}$-symmetric Scarf II potential, where the eigenfunctions $\psi_{n,+}^{(q)}(x)$ are a linear combination of two Jacobi polynomials, $P_n^{(\alpha,\beta)}[z(x)]$ and $P_{n-1}^{(\alpha,\beta)}[z(x)]$ [Lévai and Znojil (2002)].

A second example is generated from the PT-symmetric version of the radial harmonic oscillator by a SUSY transformation corresponding to broken SUSY. Again, an imaginary coordinate shift $x-ic$ avoids the singularity at the origin. In this case the function $z(x)$ appearing in Table 7.1 must be modified as $z(x) = \frac{1}{2}\omega(x-ic)^2$, and the potential is

$$V(x) = \frac{1}{4}\omega^2(x-ic)^2 + (\alpha^2 - \frac{1}{4})(x-ic)^{-2}. \tag{7.46}$$

Because this avoids the boundary condition at the origin, solutions with both α and $-\alpha$ are allowed and labeled by the quasiparity quantum number q. Hence, there are two sets of normalizable eigenfunctions [Znojil (1999b); Lévai and Znojil (2000)]

$$\psi_n^{(q)}(x) = N_n^{(q)} \exp\left[-\frac{1}{4}\omega(x-ic)^2\right](x-ic)^{q\alpha-1/2}L_n^{(q\alpha)}\left[\frac{1}{2}\omega(x-ic)^2\right] \tag{7.47}$$

and energy eigenvalues

$$E_n^{(q)} = \omega(2n+1+q\alpha). \tag{7.48}$$

As can be seen from Table 7.2, PT symmetry requires that $\omega^* = \omega$ and $\alpha^* = \pm\alpha$, so ω is real, while α can be either real or imaginary. The conventional radial harmonic oscillator is obtained from (7.47) and (7.48) by taking $c = 0$, $q = 1$, and $\alpha = l + \frac{1}{2}$. When $q = -1$ the eigenfunction (7.47) is unphysical because it has a pole of order l. The PT symmetry is broken and the eigenvalues (7.48) are complex when α is imaginary. Pairs of eigenfunctions (7.47) with opposite q are PT reflections.

Because (7.46) is a shape-invariant potential, the SUSY transformation eliminating the ground state with $n = 0$ leads to the same potential with $q\alpha$ shifted by one. However, a more general SUSY transformation with a factorization energy different from $E_0^{(q)}$ gives nontrivial SUSY partners of (7.46). This can be done by adding a rational function of $z(x)$ to the usual superpotential [Sinha *et al.* (2004)]:

$$W^{(q)}(x) = \frac{\omega}{2}(x-ic) + \frac{q\alpha + 1/2}{x-ic} + \frac{2g(x-ic)}{1+g(x-ic)^2}. \tag{7.49}$$

The supersymmetric partners $V_-(x)$ and $V_+(x)$ are constructed according to (7.15). A suitable choice of the g parameter must be made to eliminate terms containing $[1+g(x-ic)^2]^{-1}$; a natural choice is $g = \omega/(2q\alpha+2)$. As discussed in Subsec. 7.3.5, potentials generated in this way are sometimes called *conditionally exactly solvable* (CES) because a solution can only be obtained for specific values of g [Junker and Roy (1997)]. This procedure corresponds to extending the superpotential in the spirit of (7.17).

The potential $V_-(x)$ is q-dependent, as are $W^{(q)}(x)$ and the factorization energy $q\alpha\omega$. One finds that

$$v_+(x) \equiv V_+^{(q)}(x) - q\alpha\omega = \frac{\omega^2}{4}(x - \mathrm{i}c)^2 + \frac{\alpha^2 - 1/4}{(x - \mathrm{i}c)^2} + 3\omega,$$

which is $V(x)$ shifted in energy by 3ω and independent of q. The q-dependent SUSY partners are

$$v_-^{(q)}(x) \equiv V_-^{(q)}(x) - q\alpha\omega = \frac{\omega^2}{4}(x - \mathrm{i}c)^2 + \frac{(q\alpha + 1/2)(q\alpha + 3/2)}{(x - \mathrm{i}c)^2} + 2\omega$$

$$+ \frac{4\omega}{2q\alpha + 2 + \omega(x - \mathrm{i}c)^2} - \frac{8\omega(2q\alpha + 2)}{(2q\alpha + 2 + \omega(x - \mathrm{i}c)^2)^2}.$$

The eigenvalues of $v_+(x)$ and $v_-(x)$ are

$$E_{n,+}^{(q)} = \omega(2n + 4 + q\alpha) \quad \text{and} \quad E_{n,-}^{(q)} = \omega(2n + 2 + q\alpha).$$

When α is imaginary the PT symmetry of $v_+(x)$ is broken and its SUSY partners $v_-^{(q)}(x)$ cease to be PT symmetric. This is qualitatively similar to the behavior of the PT-symmetric Scarf II [Lévai and Znojil (2002)] and generalized Ginocchio [Lévai *et al.* (2003)] potentials and is consistent with the discussion in Subsec. 7.4.4.

The eigenfunctions of $v_+(x)$ are insensitive to the energy shift, so they are equivalent to those in (7.47). The $(n + 1)$st eigenfunction of $v_-^{(q)}(x)$ can be obtained by acting on the (7.47) functions with the SUSY ladder operator $-\frac{\mathrm{d}}{\mathrm{d}x} + W^{(q)}(x)$, where $W^{(q)}(x)$ is given in (7.49):

$$\psi_{n,-}^{(q)}(x) = N_{n,-}^{(q)} e^{-\omega(x - \mathrm{i}c)^2/4}(x - \mathrm{i}c)^{q\alpha - \frac{1}{2}} \left\{ -2(n+1)L_{n+1}^{(q\alpha)}\left[\tfrac{1}{2}\omega(x - \mathrm{i}c)^2\right] \right.$$

$$\left. + \left[2n + 2q\alpha + 4 - \frac{4(q\alpha + 1)}{2q\alpha + 2 + \omega(x - \mathrm{i}c)^2}\right] L_n^{(q\alpha)}\left[\tfrac{1}{2}\omega(x - \mathrm{i}c)^2\right] \right\},$$

and the ground-state eigenfunction is

$$\psi_{0,-}^{(q)}(x) = N_{0,-}^{(q)} \exp\left[-\int W^{(q)}(x)\mathrm{d}x\right]$$

$$= N_{0,-}^{(q)} e^{-\omega(x - \mathrm{i}c)^2/4}(x - \mathrm{i}c)^{-q\alpha - \frac{1}{2}}\left[2q\alpha + 2 + \omega(x - \mathrm{i}c)^2\right]^{-1}.$$

7.5.4   *Potentials solved by other functions*

Applying the method in Subsec. 7.2.1 to the X_1 Jacobi polynomials $\hat{P}^{(\alpha,\beta)}(z)$ amounts to choosing [Gomez-Ullate *et al.* (2009, 2010b)]

$$Q(z) = \frac{(\beta - \alpha) - (\beta + \alpha + 2)z}{1 - z^2} - \frac{2(\beta - \alpha)}{(\beta - \alpha)z - (\beta + \alpha)},$$

$$R(z) = \frac{(n - 1)(n + \beta + \alpha) - (\beta + \alpha)z}{1 - z^2} - \frac{(\beta - \alpha)^2}{(\beta - \alpha)z - (\beta + \alpha)} \quad (7.50)$$

in (7.2). In contrast with the conventional Jacobi polynomials, the polynomials $\hat{P}^{(\alpha,\beta)}(z)$ are defined only for $\alpha \neq \beta$ [Abramowitz and Stegun (1965)]. As discussed in Subsec. 7.3.4, this family of exceptional orthogonal polynomials starts with the $n = 1$ polynomial but it possesses all the features of classical orthogonal polynomials. The analog of (7.4) contains more terms than (7.18), so there are more possibilities to select an appropriate $z(x)$ function via (7.6). However, the selection corresponding to the PI class with $(z')^2/(1 - z^2) = C$ is special because the the n-dependent term originating from $R(z)$ in (7.50) generates an expression for the energy E_n. The rationally extended versions of the Scarf I, Scarf II, and generalized Pöschl–Teller potentials can be derived by using the $z(x)$ functions for these potentials (see Table 7.1) [Quesne (2008); Bagchi et al. (2009)].

To illustrate, the $\mathcal{PT}$-symmetric rationally extended Scarf II potential is obtained by choosing $z(x) = i \sinh x$. It contains the usual terms of the $\mathcal{PT}$-symmetric Scarf II potential and two more terms

$$V(x) = -\frac{B^2 + A(A+1)}{\cosh^2 x} + \frac{iB(2A+1)\sinh x}{\cosh^2 x}$$
$$-\frac{2(2A+1)}{2A+1-2iB\sinh x} + \frac{2[(2A+1)^2 - 4B^2]}{(2A+1-2iB\sinh x)^2}, \quad (7.51)$$

where $A = (\alpha + \beta - 1)/2$ and $B = (\beta - \alpha)/2$ (see Figs. 7.8 and 7.9). The bound-state eigenfunctions are

$$\psi_n = N_n[1 - i\sinh(x)]^{\frac{\alpha}{2}-\frac{1}{4}}[1 + i\sinh(x)]^{\frac{\beta}{2}-\frac{1}{4}}$$
$$\times [\alpha + \beta - i(\beta - \alpha)\sinh x]^{-1}\hat{P}_{n+1}^{(\alpha,\beta)}[i\sinh(x)] \quad (7.52)$$

and the energy eigenvalues are

$$E_n = -\left[n + \tfrac{1}{2}(\alpha + \beta - 1)\right]^2.$$

We recognize (7.51) as the supersymmetric partner of the Scarf II potential generated by a SUSY transformation corresponding to broken SUSY [Bagchi et al. (2009); Bagchi and Quesne (2010)]. The corresponding superpotential is $W(x) = a \tanh x + ib\,\text{sech}\,x - i\cosh x/(c + i\sinh x)$, which is generated via (7.14) from the factorization function

$$\chi(x) = [1 - i\sinh(x)]^{-a/2}[1 + i\sinh(x)]^{-b/2}(c + i\sinh x). \quad (7.53)$$

From $V_-(x) = W^2(x) - W'(x) + \epsilon$ we recover the $\mathcal{PT}$-symmetric Scarf II potential (7.30) for $2a + 1 = \beta + q\alpha$ and $2b = \beta - q\alpha$, keeping in mind the invariance of that potential under $q \leftrightarrow -q$. Thus, c and the factorization energy also depend on q as $c = c^{(q)} = -(\beta - q\alpha)/(\beta + q\alpha - 2)$

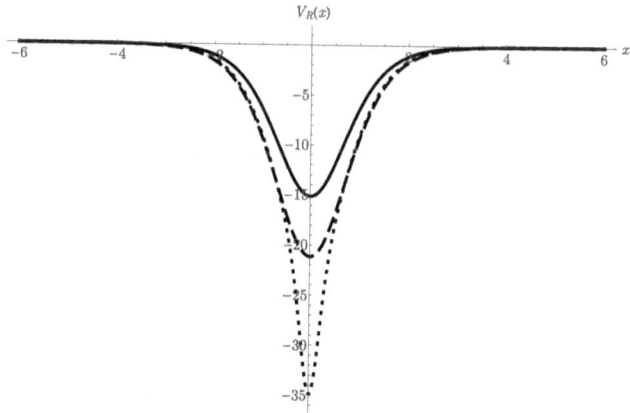

Fig. 7.8   Real component of the $\mathcal{PT}$-symmetric Scarf II potential in (7.30) with $V_R = 15.1$, $V_I = 12.7$ ($\alpha = -1.628$, $\beta = -5.296$), together with its supersymmetric partners obtained from (7.51) constructed with $A = \frac{1}{2}(\beta - q\alpha - 1)$ and $B = \frac{1}{2}(\beta + q\alpha - 2)$. The latter two potentials possess a new ground state at $E = -24.621$ and $E = -11.116$.

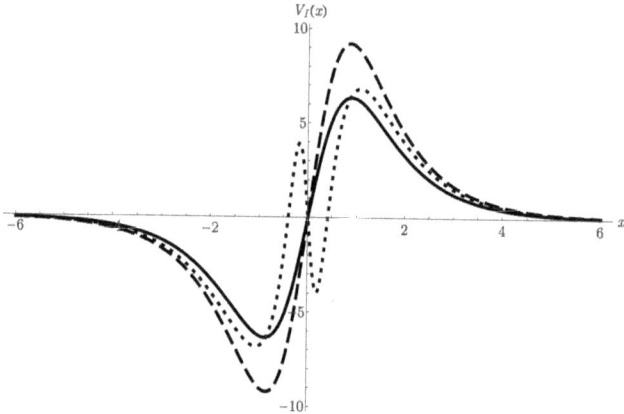

Fig. 7.9   Same as Fig. 7.8 for the imaginary potential components.

and $\epsilon = \epsilon^{(q)} = -(\beta + q\alpha - 3)^2/4$. The prescription for c cancels a potential term proportional to $(c + \mathrm{i}\sinh x)^{-1}$. The supersymmetric partner $V_+(x) = W^2(x) + W'(x) + \epsilon$ then becomes (7.51) with $A = (\beta - q\alpha - 1)/2$ and $B = (\beta + q\alpha - 2)/2$. Since (7.30) is invariant under the $\alpha \to -\alpha$ transformation, α can be replaced by $q\alpha$. In parallel with the discussion of the $\mathcal{PT}$-symmetric Scarf II potential in Subsec. 7.5.1, it can be established that the present SUSY transformation generates *two* SUSY partners distinguished by the quasiparity quantum number q. This duality of the SUSY partners is discussed in [Bagchi *et al.* (2009)] in terms of the $B \leftrightarrow A + 1/2$ permutation, while in [Bagchi and Quesne (2010)] certain signs are applied in combination with another parametrization of the potential parameters. Note also that starting with the α and β parameters of (7.30) the SUSY transformation recovers (7.51) with α and β shifted by one.

The structure of the eigenfunctions (7.52) also hints at the SUSY transformation: The X_1-type exceptional Jacobi polynomials can be expressed in terms of two Jacobi polynomials [Gomez-Ullate *et al.* (2009, 2010b)] as

$$\hat{P}_{n+1}^{(\alpha,\beta)}(z) = \left[-\frac{z}{2} + \frac{\beta+\alpha}{\beta-\alpha}\left(\frac{1}{2} + \frac{1}{\alpha+\beta+2n}\right)\right] P_n^{(\alpha,\beta)}(z) - \frac{P_{n-1}^{(\alpha,\beta)}(z)}{\alpha+\beta+2n}.$$

This structure arises naturally when the linear differential operators A and $A^\dagger$ act on the eigenfunctions (7.32). The SUSY transformation inserts a new ground state at the factorization energy ϵ with the wave function (7.53) as $\psi_0(x) \sim \chi^{-1}(x)$.

Like other $\mathcal{PT}$-symmetric models discussed in Subsec. 7.5.3, the energy levels may become complex but the $\mathcal{PT}$ symmetry of the SUSY $V_+(x)$ partner potential will be lost. As we have already seen, $\mathcal{PT}$ symmetry is broken if α is imaginary. The spectral singularity appearing in the $\mathcal{PT}$-symmetric Scarf II potential has an analog in the spectrum of rationally extended partner potentials [Bagchi and Quesne (2010)].

The procedure above for the X_1 case can be generalized to the rationally extended potentials based on the X_m Jacobi exceptional polynomials and has also been carried out in the $\mathcal{PT}$-symmetric context. In this case the $Q(z)$ and $R(z)$ functions in (7.2) and the superpotential $W(x)$ contain mth-order polynomial expressions written in terms of Jacobi polynomials.

The rationally extended version of the harmonic oscillator (LI shape-invariant class) has also been discussed in relation to the Laguerre X_1 exceptional orthogonal polynomials [Quesne (2008)]. The approach is similar to that presented in Subsec. 7.5.3 with a conditionally exactly solvable model, so $\mathcal{PT}$ symmetrization is possible by implementing the $x - ic$ imaginary coordinate shift [Sinha *et al.* (2004)].

The Bessel differential equation in Subsec. 7.3.4 can be used to construct $\mathcal{PT}$-symmetric potentials that do not belong to the Natanzon class. This procedure is based on the solution $z(x) = A\exp\left(C^{1/2}x\right)$ obtained from (7.6) with $\Phi(z) = z^{-2}$, which yields the exponential potential [ter Haar

(1975)]. This is a *radial* potential, so $\mathcal{PT}$ symmetrization requires extending to the full x axis so that the potential $V(x)$ contains an odd imaginary component. Accordingly, we define [Ahmed *et al.* (2015a)]

$$V(x) = \begin{cases} ig[1 - \exp(-2x/a)] \text{ for } x \leq 0, \\ ig[\exp(2x/a) - 1] \quad \text{ for } x \geq 0. \end{cases} \qquad (7.54)$$

The shape of this purely imaginary potential is qualitatively similar to that of the imaginary cubic oscillator with the difference that it tends to infinity exponentially. The eigenfunction is

$$\psi(x) = \begin{cases} H_{ipa}^{(1)}[sa \exp(-x/a)]/H_{ipa}^{(1)}(sa) \text{ for } x \leq 0, \\ K_{iqa}[sa \exp(x/a)]/K_{iqa}(sa) \quad \text{ for } x \geq 0, \end{cases}$$

where $p = (E - ig)^{1/2}$, $q = (E + ig)^{1/2}$, and $s = (1 + i)(g/2)^{1/2}$ [Ahmed *et al.* (2015a)]. We obtain the energy eigenvalues by patching $\psi'(x)$ at $x = 0$. The poles of the reflection amplitude give the energy eigenvalues [Ahmed *et al.* (2015a)].

The discrete spectrum of this potential contains a finite number of real energy eigenvalues at lower energies, followed by complex-conjugate eigenvalues. When g is increased, the real eigenvalues merge pairwise and become complex-conjugate pairs. However, the lowest eigenvalue remains real. [As $a \to \infty$ this level is eliminated because (7.54) tends to the linear imaginary potential.] There is no parametric separation of unbroken and broken $\mathcal{PT}$ symmetry [Ahmed *et al.* (2015a)].

7.5.5 *Further solvable potentials and generalizations*

The eigenfunctions of some $\mathcal{PT}$-symmetric potentials cannot be expressed in the form (7.3) either because there is no special function $F(z)$ or because not all solutions can be obtained explicitly. An example of the former case is the $\mathcal{PT}$-symmetric square well potential confined to $x \in [-1, 1]$, where $V(x) = iZ\text{sgn}(x)$ [Znojil and Lévai (2001)]. The eigenfunctions are hyperbolic (or exponential) functions and the eigenvalues can be obtained numerically. The energy spectrum of this potential is real for small Z but gradually becomes complex as Z increases. The complexification of the energy spectrum starts from below with the ground state. Such piecewise constant potentials are common in the Hermitian setting and their $\mathcal{PT}$-symmetric versions can be useful in the construction of higher-dimensional $\mathcal{PT}$-symmetric potentials.

Another type of $\mathcal{PT}$-symmetric potential not meeting the usual criteria of solvability belongs to the quasi-exactly solvable class [Ushveridze (1994)] mentioned briefly in Subsec. 7.3.5. An example of this is the Khare–Mandal potential [Khare and Mandal (2000)].

One-dimensional potentials are the most frequently occurring examples for PT-symmetric systems. However, this symmetry has been generalized to further physical situations. The main features of several such systems are illustrated below.

We can also define PT-symmetric potentials in higher spatial dimensions. Considering noncentral potentials in spherical coordinates in two and three dimensions, the separation of the radial and angular variables becomes possible. The resulting second-order differential equations will be essentially the radial Schrödinger equation (in the ρ and the r variable), and Schrödinger-like equations confined to a finite domain of the angular variables with the important difference that the boundary conditions are different from the usual ones. In particular, the azimuthal equation must be solved in $\varphi \in [0, 2\pi]$ with 2π-periodic boundary conditions, while the polar equation must be solved in $\theta \in [0, \pi]$ without requiring the vanishing of the eigenfunction at the boundaries. The expressions appearing in the kinetic energy term determine the general structure of the potential for which exact solutions can be obtained:

$$V(\rho, \varphi) = V_0(\rho) + \frac{1}{\rho^2} K(\varphi) \quad \text{(two dimensions)},$$

$$V(r, \theta, \varphi) = V_0(r) + \frac{P(\theta)}{r^2} + \frac{1}{r^2 \sin^2 \theta} \left[K(\varphi) - k + \tfrac{1}{4} \right] \quad \text{(three dimensions)}.$$

The PT-symmetry requirement $V(\rho, \varphi) = V^*(\rho, \varphi+\pi)$ and $V(r, \theta, \varphi) = V^*(r, \pi-\theta, \varphi+\pi)$ [Lévai (2007)] constrains the expressions appearing in the Schrödinger(-like) equations in each variable resulting in $V_0^*(r) = V_0(r)$ (or $V_0^*(\rho) = V_0(\rho)$ in two dimensions), $P^*(\pi-\theta) = P(\theta)$ and $K^*(\varphi+\pi) = K(\varphi)$ [Lévai (2007, 2008b)].

To find an exact solution in three dimensions we first solve the azimuthal equation with $K(\varphi)$ as above, and then solve the polar equation, which contains the k eigenvalue of the azimuthal equation along with $P(\theta)$. The function $P(\theta)$ must contain a term with $\sin^{-2} \theta$; that is, it must be either the Scarf I or the Rosen–Morse I potential (see Table 7.1). Finally, we solve the radial equation with $V_0(r)$ and a centrifugal equation that depends on the p eigenvalue of the polar equation. The potential $V_0(r)$ must contain an r^{-2} term, so it can be the radial harmonic oscillator or the Coulomb potential with arbitrary (noninteger) values of the angular momentum [Lévai (2007)]. Breaking of PT symmetry is possible if the azimuthal equation has complex eigenvalues. This complex eigenvalue is associated with the coupling of the

$\sin^{-2}\theta$ term of the polar equation and leads to an overall complex eigenvalue of the full system of equations. Two-dimensional $\mathcal{PT}$-symmetric potentials have also been considered in Cartesian coordinates [Tichy and Skála (2010)].

Eigenvalue problems with periodic potentials have also been studied [Berry and O'Dell (1998); Bender *et al.* (1999d); Jones (1999); Ahmed (2001a)]. For example, the $\mathcal{PT}$-symmetric Kronig–Penney model has a $\mathcal{PT}$-symmetric arrangement of delta functions in a unit cell. The band structure for this model is qualitatively similar to that of the corresponding Hermitian model in a portion of the parameter domain [Bender *et al.* (1999d)]. However, in other parts of the parameter domain there are marked differences [Ahmed (2001a)]. A generalization containing more delta functions in the unit cell was considered in [Cerveró and Rodríguez (2004)].

$\mathcal{PT}$ symmetry can also be extended to quantum-mechanical problems involving a position-dependent effective mass. In this case the Schrödinger equation includes an expression containing a mass function $m(x)$. However, for such theories the complicated nature of the kinetic energy term in the Hamiltonian poses serious difficulties [von Roos (1983)] so the variable transformation method outlined in Subsec. 7.2.1 must be reformulated. For special choices of $m(x)$ it is possible to transform the effective-mass Schrödinger equation into an exactly solvable potential problem. For example, the solutions of the $\mathcal{PT}$-symmetric Scarf II potential are used to describe scattering of a particle with position-dependent mass in a $\mathcal{PT}$-symmetric heterojunction [Sinha (2012)]. $\mathcal{PT}$ symmetry has also been implemented for solvable coupled-channel[Znojil (2006)], many-body [Znojil and Tater (2001)], and relativistic quantum-mechanical problems [Şimşek and Egrifes (2004); Egrifes and Sever (2005); Bender and Mannheim (2011)].

Acknowledgment

GL acknowledges support from the OTKA grant number K112962.

Chapter 8

Krein-Space Theory and PTQM

Contributed by Sergii Kuzhel

"My life is one long curve, full of turning points."
— *Pierre Trudeau*

The key objective of PTQM is the development of modern physics using Hamiltonians that are not Hermitian, but have certain characteristics of symmetry (like $\mathcal{PT}$ symmetry). Usually, $\mathcal{PT}$-symmetric Hamiltonians can be naturally considered as Hermitian operators with respect to an indefinite inner product ($\mathcal{PT}$ inner product). This opens up a promising way to reach deeper insights into the structural subtleties of the $\mathcal{PT}$-symmetric quantum theory with the use of the Krein-space methods.

We begin with an elementary presentation of the Krein-space theory focusing on those aspects that may be more appealing to PTQM-related studies (the mathematical background of $\mathcal{PT}$ and $\mathcal{C}$-symmetries, the meaning of the $\mathcal{CPT}$ inner product). The final part of the chapter deals with more advanced results and open problems.

8.1 Introduction

$\mathcal{PT}$-symmetric Hamiltonians are symmetric with respect to some indefinite inner product (indefinite metric). For instance, the Hamiltonian for the $\mathcal{PT}$-symmetric cubic oscillator $H = -d^2/dx^2 + ix^3$ is symmetric with respect to the $\mathcal{PT}$ inner product

$$(Hf, g)_{\mathcal{PT}} = (f, Hg)_{\mathcal{PT}},$$

where

$$(f, g)_{\mathcal{PT}} = \int_{\mathbb{R}} dx [\mathcal{PT} f(x)] g(x), \qquad f, g \in L_2(\mathbb{R}). \tag{8.1}$$

The main difference between the conventional inner product

$$(f,g) = \int_{\mathbb{R}} dx [\mathcal{T} f(x)] g(x), \qquad f,g \in L_2(\mathbb{R}), \qquad (8.2)$$

and the $\mathcal{PT}$ inner product $(\cdot,\cdot)_{\mathcal{PT}}$ is that the quadratic form $(f,f)_{\mathcal{PT}}$ is indefinite; that is, there exist nonzero functions $f \in L_2(\mathbb{R})$ such that $(f,f)_{\mathcal{PT}} < 0$. The set of functions $L_2(\mathbb{R})$ endowed with the $\mathcal{PT}$ inner product (8.1) is a Kreĭn space and the cubic oscillator Hamiltonian H is Hermitian with respect to the indefinite inner product $(\cdot,\cdot)_{\mathcal{PT}}$.

The interpretation of $\mathcal{PT}$-symmetric Hamiltonians as Hermitian operators in Kreĭn space allows one to use the well-developed methods of Kreĭn-space theory in the study of PTQM. Such an approach is not merely a reformulation of the conventional language of $\mathcal{PT}$ symmetry into the language of Kreĭn-space theory. An important benefit of using Kreĭn-space theory is the possibility to understand and to discover structural subtleties of PTQM that are manifestly new in comparison with conventional QM. This shows that PTQM is mathematically distinct from conventional QM and that is the principal point of this chapter.

First, we note that the Hermiticity of a $\mathcal{PT}$-symmetric Hamiltonian H with respect to the indefinite inner product $(\cdot,\cdot)_{\mathcal{PT}}$ is not completely satisfactory (and H cannot have any obvious relevance to physics) until it is shown that the dynamics generated by H is unitary with respect to an inner product whose associated norm is positive definite. This problem can be overcome by finding a new (hidden) symmetry represented by a linear operator $\mathcal{C}$, which commutes with both the Hamiltonian H and the $\mathcal{PT}$ operator. In terms of $\mathcal{C}$ one must construct a $\mathcal{CPT}$ inner product

$$(f,g)_{\mathcal{CPT}} \equiv (\mathcal{C}f,g)_{\mathcal{PT}} = \int_{\mathbb{R}} dx [\mathcal{CPT} f(x)] g(x), \qquad (8.3)$$

whose associated norm is positive definite and one must show that H is Hermitian with respect to $(\cdot,\cdot)_{\mathcal{CPT}}$.

The definition of $\mathcal{C}$-symmetry for H can be expressed formally as follows: $\mathcal{C}$ is a linear operator that obeys the following equations:

$$\mathcal{C}^2 = I, \qquad \mathcal{PTC} = \mathcal{CPT}, \qquad H\mathcal{C} = \mathcal{C}H. \qquad (8.4)$$

Requiring that the $\mathcal{CPT}$ inner product (8.3) determine a positive-definite norm implies that $\mathcal{CP}$ is a positive Hermitian operator in $L_2(\mathbb{R})$:

$$\mathcal{CP} > 0, \qquad (\mathcal{CP})^\dagger = \mathcal{CP}, \qquad (8.5)$$

where $\dagger$ represents the Dirac adjoint in $L_2(\mathbb{R})$.

The first two equations in (8.4) and (8.5) do not depend on the choice of H and they are equivalent to the following representation of $\mathcal{C}$:

$$\mathcal{C} = e^{\mathcal{Q}}\mathcal{P}, \tag{8.6}$$

where $\mathcal{Q}$ is Hermitian in $L_2(\mathbb{R})$ and $\mathcal{P}\mathcal{Q} = -\mathcal{Q}\mathcal{P}$ and $\mathcal{T}\mathcal{Q} = -\mathcal{Q}\mathcal{T}$.

Let $\mathcal{C}$ be defined as in (8.6). Then, the subspaces

$$\mathfrak{L}_+ = \tfrac{1}{2}(I+\mathcal{C})L_2(\mathbb{R}) \quad \text{and} \quad \mathfrak{L}_- = \tfrac{1}{2}(I-\mathcal{C})L_2(\mathbb{R}) \tag{8.7}$$

are maximal positive and maximal negative subspaces of the Kreĭn space $(L_2(\mathbb{R}), (\cdot, \cdot)_{\mathcal{P}\mathcal{T}})$ (see Example 8.2), which are mutually orthogonal with respect to the $\mathcal{P}\mathcal{T}$ inner product $(\cdot, \cdot)_{\mathcal{P}\mathcal{T}}$.

The subspaces $\mathfrak{L}_\pm$ form the direct sum $\mathcal{D}(\mathcal{C}) = \mathfrak{L}_+[\dotplus]\mathfrak{L}_-$, which is a dense set in $L_2(\mathbb{R})$. Moreover (see Theorem 8.7, or Theorem 8.1 for the more general case), there exists a one-to-one correspondence between the set of operators $\mathcal{C} = e^{\mathcal{Q}}\mathcal{P}$ and the set of all possible $\mathcal{P}\mathcal{T}$-orthogonal direct sums $\mathfrak{L}_+[\dotplus]\mathfrak{L}_-$, where the $\mathcal{P}\mathcal{T}$-invariant subspaces $\mathfrak{L}_\pm$ are maximal positive/negative in the Kreĭn space $(L_2(\mathbb{R}), (\cdot, \cdot)_{\mathcal{P}\mathcal{T}})$. If the direct sum is given, then the corresponding operator $\mathcal{C}$ acts as the identity operator on $\mathfrak{L}_+$ and the negative identity operator on $\mathfrak{L}_-$.

This crucial relationship allows one to reduce the investigation of $\mathcal{C}$ to the analysis of pairs of $\mathcal{P}\mathcal{T}$-orthogonal maximal definite subspaces $\mathfrak{L}_\pm$ in a Kreĭn space. In particular, the boundedness of $\mathcal{C}$ in $L_2(\mathbb{R})$ is equivalent to the property of $\mathfrak{L}_\pm$ to be maximal uniformly definite in the Kreĭn space $(L_2(\mathbb{R}), (\cdot, \cdot)_{\mathcal{P}\mathcal{T}})$. The latter means that the direct sum $\mathfrak{L}_+[\dotplus]\mathfrak{L}_-$ coincides with $L_2(\mathbb{R})$ and that the $\mathcal{C}\mathcal{P}\mathcal{T}$ inner product (8.3) is equivalent to the conventional inner product (8.2) (see Subsecs. 8.3.1 and 8.3.2).

In summary, if a $\mathcal{P}\mathcal{T}$-symmetric Hamiltonian H has the property of *bounded* $\mathcal{C}$-symmetry, then it is Hermitian in $L_2(\mathbb{R})$ with respect to the $\mathcal{C}\mathcal{P}\mathcal{T}$ inner product. These Hamiltonians are isospectral and are related by similarity transformation (Subsec. 8.3.6).

Bounded $\mathcal{C}$ operators appear naturally in matrix models. Other examples include Hamiltonians whose eigenfunctions correspond to real eigenvalues and form Riesz bases (Subsec. 8.4.2). Sometimes, these examples lead to the incorrect assertion that PTQM is the result of renormalization of conventional QM by a similarity transformation between $\mathcal{P}\mathcal{T}$-symmetric Hamiltonians with unbroken $\mathcal{P}\mathcal{T}$-symmetry and Hermitian ones.

More interesting is the case of $\mathcal{P}\mathcal{T}$-symmetric Hamiltonians having unbroken $\mathcal{P}\mathcal{T}$-symmetry and the property of unbounded $\mathcal{C}$ symmetry. If the operator $\mathcal{C}$ is unbounded in $L_2(\mathbb{R})$, then the maximal definite subspaces $\mathfrak{L}_\pm$

in (8.7) cannot be uniformly definite in the Kreĭn space $(L_2(\mathbb{R}), (\cdot, \cdot)_{\mathcal{PT}})$. Also, the direct sum $\mathfrak{L}_+[\dotplus]\mathfrak{L}_-$ does not coincide with $L_2(\mathbb{R})$ and the $\mathcal{CPT}$ inner product $(\cdot, \cdot)_{\mathcal{CPT}}$ is not equivalent to the conventional inner product $(\cdot, \cdot)$ in $L_2(\mathbb{R})$. Therefore, the completion of $\mathfrak{L}_+[\dotplus]\mathfrak{L}_-$ with respect to $(\cdot, \cdot)_{\mathcal{CPT}}$ gives rise to a new Hilbert space $\mathfrak{H}_C$ that *does not coincide with $L_2(\mathbb{R})$ as a linear set* (Subsec. 8.3.3). For this reason, while there is a formal equivalence between a $\mathcal{PT}$-symmetric Hamiltonian in $L_2(\mathbb{R})$ and a conventionally Hermitian Hamiltonian in $\mathfrak{H}_C$ in the sense that the two Hamiltonians are isospectral, the Hilbert spaces are inequivalent.

It should be emphasized that there is a *deeper reason* why PTQM is mathematically distinct from conventional QM. This can be explained naturally in the Kreĭn-space framework; Subsecs. 8.4.1 and 8.4.4 contain the relevant discussion. We present the principal arguments for a $\mathcal{PT}$-symmetric Hamiltonian H in $L_2(\mathbb{R})$ assuming that H has a complete set of eigenvectors corresponding to real simple eigenvalues. Let $\{f_n\}$ be the set of eigenfunctions of H. Then $(f_n, f_n)_{\mathcal{PT}} \neq 0$ and separating the sequence $\{f_n\}$ by the sign of the $\mathcal{PT}$ inner product, we obtain two subsequences of positive $\{f_n^+\}$ and negative $\{f_n^-\}$ elements with respect to $(\cdot, \cdot)_{\mathcal{PT}}$.

Let $\mathfrak{L}_+^0$ and $\mathfrak{L}_-^0$ be the closure [with respect to the conventional inner product (8.2)] of the linear spans of $\{f_n^+\}$ and $\{f_n^-\}$, respectively. By construction, $\mathfrak{L}_\pm^0$ are $\mathcal{PT}$-orthogonal definite subspaces in the Kreĭn space $(L_2(\mathbb{R}), (\cdot, \cdot)_{\mathcal{PT}})$ and the direct sum $\mathfrak{L}_+^0[\dotplus]\mathfrak{L}_-^0$ is a dense set in $L_2(\mathbb{R})$. Let $\mathcal{C}_0$ be an operator defined on $\mathfrak{L}_+^0[\dotplus]\mathfrak{L}_-^0$ such that $\mathcal{C}_0$ acts as the identity operator and negative identity operator on $\mathfrak{L}_+^0$ and $\mathfrak{L}_-^0$. The operator $\mathcal{C}_0$ is densely defined in $L_2(\mathbb{R})$ and $\mathcal{C}_0^2 f = f$ for any $f \in \mathcal{D}(\mathcal{C}_0) = \mathfrak{L}_+^0[\dotplus]\mathfrak{L}_-^0$.

The operator $\mathcal{C}_0$ cannot always be an operator of $\mathcal{C}$-symmetry. The issue is that the definite subspaces $\mathfrak{L}_\pm^0$ may be *proper subspaces of maximal definite subspaces* in the Kreĭn space $(L_2(\mathbb{R}), (\cdot, \cdot)_{\mathcal{PT}})$. If at least one of the subspaces $\mathfrak{L}_\pm^0$ loses the property of maximality in the classes of definite subspaces (positive or negative), then the sum $\mathfrak{L}_+^0[\dotplus]\mathfrak{L}_-^0$ it *cannot be the domain of definition* of an operator of $\mathcal{C}$-symmetry with properties (8.4) and (8.5). To be more precise, $\mathcal{C}_0$ may satisfy (8.4) but it cannot satisfy the second condition in (8.5); that is, the operator $\mathcal{C}_0$ cannot have the form $e^{\mathcal{Q}}\mathcal{P}$, where $e^{\mathcal{Q}}$ is *Hermitian* in $L_2(\mathbb{R})$. This problem arises only in the case of unbounded operators $\mathcal{C}$ and does not occur in matrix models.

In order to construct a proper operator $\mathcal{C}$ associated with $\mathfrak{L}_+^0[\dotplus]\mathfrak{L}_-^0$, we must extend the $\mathcal{PT}$-orthogonal direct sum $\mathfrak{L}_+^0[\dotplus]\mathfrak{L}_-^0$ to a $\mathcal{PT}$-orthogonal direct sum $\mathfrak{L}_+[\dotplus]\mathfrak{L}_-$, where $\mathfrak{L}_\pm \supset \mathfrak{L}_\pm^0$ are *maximal* definite subspaces (positive and negative). The operator $\mathcal{C}$ associated with the latter direct sum is an extension of C_0. In general, the extension $\mathfrak{L}_+^0[\dot{+}]\mathfrak{L}_-^0 \to \mathfrak{L}_+[\dot{+}]\mathfrak{L}_-$ is not uniquely determined. This leads to the *nonuniqueness* of unbounded operators of C-symmetry $C \supset C_0$ associated with $\mathfrak{L}_+^0[\dot{+}]\mathfrak{L}_-^0$.

Let C be an operator of C-symmetry associated with $\mathfrak{L}_+^0[\dot{+}]\mathfrak{L}_-^0$ and let $\mathfrak{H}_C$ be the corresponding Hilbert space obtained by the completion of $\mathfrak{L}_+[\dot{+}]\mathfrak{L}_-$ with respect to the CPT inner product. As a result of this construction, the original direct sum $\mathcal{D}_0 = \mathfrak{L}_+^0[\dot{+}]\mathfrak{L}_-^0$ is a linear manifold in the Hilbert space $(\mathfrak{H}_C, (\cdot, \cdot)_{CPT})$. Recent mathematical investigations [Kamuda *et al.* (2018); Kuzhel and Sudilovskaja (2017)] show that this manifold may lose the density property in $\mathfrak{H}_C$. Precisely, the following cases are possible: (a) the operator C is uniquely defined and $\mathcal{D}_0$ is dense in $\mathfrak{H}_C$; (b) there are infinitely many operators $C \supset C_0$ associated with $\mathcal{D}_0$ such that the linear manifold $\mathcal{D}_0$ remains dense in the corresponding Hilbert spaces $\mathfrak{H}_C$. Simultaneously, there are infinitely many operators $C \supset C_0$ which generate Hilbert spaces $\mathfrak{H}_C$ containing the closure of $\mathcal{D}_0$ in the CPT norm as a proper subspace; (c) $\mathcal{D}_0$ loses the density property in $\mathfrak{H}_C$ for any choice of C-symmetry associated with $\mathcal{D}_0$. The total amount of possible operators $C \supset C_0$ is not specified. (A unique operator C is possible as well as infinitely many.)

In the case (a) the eigenfunctions $\{f_n\}$ of H form an orthonormal basis of $\mathfrak{H}_C$ and there is a formal equivalence between the PT-symmetric operator in $L_2(\mathbb{R})$ and the Hermitian Hamiltonian in $\mathfrak{H}_C$ in the sense that these Hamiltonians are isospectral but the Hilbert spaces are inequivalent. The eigenfunctions of the shifted harmonic oscillator give an example of (a).

In the case (b) everything depends on the choice of C-symmetry. If $\mathcal{D}_0$ remains dense in $\mathfrak{H}_C$, then we obtain an isospectral realization of H in $\mathfrak{H}_C$. On the other hand, if $\mathcal{D}_0$ loses the density property in $\mathfrak{H}_C$, then H is a symmetric nondensely defined operator with respect to the CPT inner product. Its extension to a Hermitian operator in $\mathfrak{H}_C$ leads to the appearance of new spectral points. The constructed Hermitian operator cannot be isospectral to the original PT-symmetric operator. The eigenfunctions of the imaginary cubic oscillator are suspected to be an example of (b).

With regard to (c), an example of eigenfunctions $\{f_n\}$ for which there exists a unique unbounded operator of C-symmetry but where $\{f_n\}$ is not a complete set in $\mathfrak{H}_C$ can be found in [Kamuda *et al.* (2018)]. In my opinion, this is a mathematical construction without any relation to physics.

The mathematical setting discussed above is a good framework for PTQM. Indeed, the existence of different unbounded operators C associated with $\mathfrak{L}_+^0 [\dot{+}] \mathfrak{L}_-^0$ in the case (b) ensures a flexibility in the construction of model Hermitian operators associated with a $\mathcal{PT}$-symmetric operator H. One can choose a $\mathcal{C}$-symmetry that gives rise to the orthonormal basis $\{f_n\}$ of eigenfunctions and the isospectral realization of H in $\mathfrak{H}_\mathcal{C}$. Simultaneously, the choice of $\mathcal{C}$ that leads to the orthonormal system $\{f_n\}$ and the nondensely defined symmetric operator H in $\mathfrak{H}_\mathcal{C}$ is also possible. In the latter case, the appearance of a free-parameter subspace $\mathfrak{H}_\mathcal{C} \ominus_{\mathcal{CPT}} \mathcal{D}_0$ provides an excellent possibility for the additional modeling of structural subtleties of $\mathcal{PT}$-symmetric Hamiltonians.

The results of the Kreĭn-space theory presented in this chapter reflect the author's preferences in PTQM-related studies. The monographs [Azizov and Iokhvidov (1998); Bognar (1974); Nesemann (2011)] and the papers [Albeverio and Kuzhel (2015); Langer (1982); Tanaka (2006)] are recommended as complementary reading.

This chapter is organized as follows. Section 8.2 contains basic results on operator theory in Hilbert spaces. We emphasize the importance of the proper choice of domains of definition for unbounded operators and we review the different approaches in the literature to Hermitian operators. An introduction to Kreĭn-space theory is given in Sec. 8.3. The focus is on (equivalent) definitions of $\mathcal{C}$ operators and their relationship to $\mathcal{P}$-orthogonal direct sums of maximal definite subspaces. The method of construction of $\mathcal{C}$ operators for $\mathcal{P}$-symmetric operators with complete sets of eigenvalues is presented in Sec. 8.4. In addition to the usual discussion of Riesz and Schauder bases, the case of quasi bases is considered and the problem of the best choice of $\mathcal{C}$ operator is discussed. Section 8.5 deals with the theory of $\mathcal{PT}$-symmetric operators. The $\mathcal{PT}$-symmetry property is defined in an abstract setting assuming that $\mathcal{P}$ is the operator of fundamental symmetry and $\mathcal{T}$ is the conjugation operator acting in a Hilbert space $\mathfrak{H}$.

8.2 Terminology and Notation

In what follows $\mathfrak{H}$ denotes a complex Hilbert space with inner product $(\cdot, \cdot)$ linear in the second argument. It is sometimes useful to specify the inner product associated with $\mathfrak{H}$ and in this case we use the notation $(\mathfrak{H}, (\cdot, \cdot))$. All topological notions refer to the Hilbert-space-norm topology. For instance, a subspace of $\mathfrak{H}$ is a linear manifold in $\mathfrak{H}$, which is closed with respect to the norm $\| \cdot \| = \sqrt{(\cdot, \cdot)}$. The symbol $\mathcal{D}(H)$ represents the *domain of definition* and the symbol $\ker H$ denotes the kernel space of a linear operator H. The *range* of H is the image $\mathcal{R}(H) \equiv H\mathcal{D}(H)$.

A nonzero vector $f \in \mathcal{D}(H)$ is called a *root vector* of an operator H corresponding to the eigenvalue λ if $(H - \lambda I)^n f = 0$ for some $n \in \mathbb{N}$, where I is the identity operator in $\mathfrak{H}$. The set of all root vectors corresponding to a given eigenvalue λ, together with the zero vector, form a linear subspace $\mathcal{L}_\lambda(H)$, which is called the *root subspace* of H. A root vector $f \in \mathcal{L}_\lambda(H)$ is an *eigenvector* if f belongs to the kernel subspace $\ker(H - \lambda I)$. The algebraic and geometric multiplicities of an eigenvalue λ are denoted by m_λ^a and m_λ^g and these are defined by

$$m_\lambda^a = \dim \mathcal{L}_\lambda(H) \quad \text{and} \quad m_\lambda^g = \dim \ker(H - \lambda I),$$

where $m_\lambda^a \geq m_\lambda^g$. An eigenvalue λ is *semi-simple* if its algebraic and geometric multiplicities coincide: $m_\lambda^a = m_\lambda^g$. A semi-simple eigenvalue λ becomes *simple* if $\dim \ker(H - \lambda I) = 1$.

The set $\sigma_p(H)$ of all eigenvalues $\lambda \in \mathbb{C}$ of H is called the *point spectrum*. The *residual spectrum* $\sigma_r(H)$ of H contains all $\lambda \notin \sigma_p(H)$ for which $\mathcal{R}(H - \lambda I)$ is not dense in $\mathfrak{H}$; that is, the inverse operator $(H - \lambda I)^{-1}$ exists but is nondensely defined. The *continuous spectrum* $\sigma_c(H)$ of H is composed of all λ for which the inverse $(H - \lambda I)^{-1}$ is a densely defined unbounded operator. The three components of the spectrum of H defined above are pairwise disjoint and the *spectrum* of H is defined as union $\sigma(H) = \sigma_p(H) \cup \sigma_r(H) \cup \sigma_c(H)$. The *resolvent set* $\rho(H)$ of the operator H is the *complement* of $\sigma(H)$; that is, $\rho(H) \equiv \mathbb{C} \setminus \sigma(H)$. This means that $\lambda \in \rho(H)$ (here, λ is *not* an eigenvalue) $\iff$ the inverse $(H - \lambda I)^{-1}$ is a densely defined bounded operator in $\mathfrak{H}$.

The symbol $H \upharpoonright_\mathcal{D}$ indicates the restriction of an operator H onto a set $\mathcal{D}$. We say that an operator H_1 is an *extension* of H and denote it as $H \subset H_1$ if the domain of definition $\mathcal{D}(H)$ is a subspace of $\mathcal{D}(H_1)$ and

$$Hf = H_1 f, \qquad f \in \mathcal{D}(H).$$

Thus, the symbol $H \subset H_1$ means that $H_1 \upharpoonright_{\mathcal{D}(H)} = H$.

The adjoint of H with respect to the natural inner product $(\cdot, \cdot)$ in $\mathfrak{H}$ is denoted by $H^\dagger$. The operator $H^\dagger$ is defined on the set of all $g \in \mathfrak{H}$ for which there are $\gamma \in \mathfrak{H}$ such that

$$(Hf, g) = (f, \gamma), \qquad f \in \mathcal{D}(H). \tag{8.8}$$

The element γ in (8.8) is determined uniquely for given $g \in \mathcal{D}(H^\dagger)$ [because $\mathcal{D}(H)$ is a dense set in $\mathfrak{H}$] and $H^\dagger g = \gamma$. Therefore,

$$(Hf, g) = (f, H^\dagger g), \qquad f \in \mathcal{D}(H), \quad g \in \mathcal{D}(H^\dagger).$$

The domain $\mathcal{D}(H^\dagger)$ is the set of vectors g for which the functional $F(f) = (Hf, g)$ is continuous. If an operator H is *bounded*, that is,

$$\|Hf\| \leq c\|f\|, \qquad f \in \mathcal{D}(H),\ c \in \mathbb{R}^+,$$

then its domain of definition can be extended by continuity to all elements in $\mathfrak{H}$ and the adjoint $H^\dagger$ is defined on the whole $\mathfrak{H}$.

An operator H is *Hermitian* if

$$H = H^\dagger. \tag{8.9}$$

In matrix theory this definition of Hermiticity means that the (matrix) H is invariant under Hermitian conjugation †. Matrix operators are bounded and for bounded operators the definition (8.9) is equivalent to

$$(Hf, g) = (f, Hg) \qquad f, g \in \mathfrak{H}, \tag{8.10}$$

which can be used as a definition of a bounded Hermitian operator.

Most observables appearing in QM are *unbounded* and their domains of definition are dense sets in $\mathfrak{H}$. For unbounded operators (8.9) and (8.10) are inequivalent. The weaker condition (8.10) restricted to $f, g \in \mathcal{D}(H)$ means that $H^\dagger$ is an extension of H; that is, $H \subset H^\dagger$. An operator H satisfying $H \subset H^\dagger$ is called *symmetric*. For an unbounded operator H the property of Hermiticity means that H is symmetric [(8.10) holds for all $f, g \in \mathcal{D}(H)$] and its domain $\mathcal{D}(H)$ coincides with the domain $\mathcal{D}(H^\dagger)$.

The term "Hermitian operator" is used in a variety of ways in physics and mathematics. In mathematical literature the term "self-adjoint operator" is well defined and can be synonymous with the term "Hermitian operator" [Takhtajan (2008); Faddeev and Yakubovskii (2009)]. In the theory of finite-dimensional spaces endowed with an inner product, the concept of a "Hermitian operator" is unambiguous. However, in (infinite-dimensional) Hilbert space theory, the term "Hermitian operator" is used instead to denote a *nondensely defined* symmetric operator [Kuzhel and Kuzhel (1998)]. In the classic monograph [von Neumann (1996)] the term "Hermitian operator" is what we call a "densely defined symmetric operator," and what von Neumann referred to as a "hypermaximal Hermitian operator" we simply call a "Hermitian operator." Summing up:

> *In this chapter, "Hermiticity" is a synonym of "self-adjointness." A Hermitian operator is symmetric, but a symmetric operator is not in general Hermitian.*

Every Hermitian operator H is *closed*. This means that $\mathcal{D}(H)$ is a Hilbert space with respect to the *graph inner product*

$$(f,g)_H \equiv (f,g) + (Hf, Hg), \qquad f, g \in \mathcal{D}(H).$$

An operator H is called *closable* if there is a closed operator H_1 such that $H \subset H_1$. The minimal closed extension H_1 of H is called the *closure* of H and it is denoted by $\overline{H}$. If H is closed, then $H = \overline{H}$.

A subspace $\mathcal{D}$ of $\mathcal{D}(H)$ is a *core* of a closed operator H if $\overline{H \!\restriction_\mathcal{D}} = H$. A subspace $\mathcal{D} \subset \mathcal{D}(H)$ is a core for H if and only if $\mathcal{D}$ is a dense set in the Hilbert space $(\mathcal{D}(H), (\cdot, \cdot)_H)$. Every symmetric operator H is closable and its closure $\overline{H}$ is the result of consecutive calculations of adjoint operators: $\overline{H} = H^{\dagger\dagger}$, where $H^{\dagger\dagger} \equiv (H^\dagger)^\dagger$. Moreover, $\overline{H}^\dagger = H^\dagger$.

A symmetric operator H is *essentially Hermitian* (same as *essentially self-adjoint*) if its closure $\overline{H}$ is a Hermitian operator, that is, $\overline{H} = \overline{H}^\dagger$.

Lemma 8.1. *For a symmetric operator H the following are equivalent:*

(i) *H is essentially Hermitian;*
(ii) $\ker(H^\dagger + iI) = \ker(H^\dagger - iI) = \{0\}$;
(iii) *$\mathcal{R}(H + iI)$ and $\mathcal{R}(H - iI)$ are dense sets in $\mathfrak{H}$.*

The formal concept of a Hermitian operator is theoretically convenient, but in practice the domain of a Hermitian operator H may be too difficult to be described. Instead of the whole domain $\mathcal{D}(H)$ it is natural to consider a core $\mathcal{D} \subset \mathcal{D}(H)$ of H and to work with the essentially Hermitian operator $H \!\restriction_\mathcal{D}$. The advantage of this approach is the possibility of choosing the most convenient domain $\mathcal{D}$ among all possible cores.

Example 8.1. Let a Hermitian operator H have a *complete set of eigenvectors* $\{f_n\}$ in $\mathfrak{H}$. Here, complete set means that the linear span of eigenvectors $\{f_n\}$, that is, the set of all possible *finite* linear combinations

$$\mathrm{span}\{f_n\} \equiv \left\{ \sum_{n=1}^d c_n f_n : \forall d \in \mathbb{N}, \forall c_n \in \mathbb{C} \right\},$$

is a dense set in $\mathfrak{H}$.

Usually, the explicit description of $\mathcal{D}(H)$ is complicated. So, instead of $\mathcal{D}(H)$ we may consider $\mathcal{D} \equiv \mathrm{span}\{f_n\}$. Then, $H \!\restriction_\mathcal{D}$ is symmetric and the sets $\mathcal{R}(H \!\restriction_\mathcal{D} \pm iI)$ contain $\mathcal{D}$. Therefore, $\mathcal{R}(H \!\restriction_\mathcal{D} \pm iI)$ are dense in $\mathfrak{H}$. From Lemma 8.1 (iii) the operator $H \!\restriction_\mathcal{D}$ is essentially Hermitian. The subspace $\mathcal{D} = \mathrm{span}\{f_n\}$ is a core of the Hermitian operator H.

8.3   Elements of Kreĭn-Space Theory

We present Kreĭn-space theory in an abstract setting. This is natural because various Hamiltonians of PTQM are Hermitian with respect to different $\mathcal{PT}$ inner products.

8.3.1   *Definition and elementary properties*

An operator $\mathcal{P}$ is a *fundamental symmetry* in a Hilbert space $(\mathfrak{H}, (\cdot, \cdot))$ if $\mathcal{P}$ is Hermitian and a reflection operator. Equivalently, an operator $\mathcal{P}$ is a fundamental symmetry if

$$(i)\ \mathcal{P}^2 = I, \qquad (ii)\ (\mathcal{P}f, \mathcal{P}g) = (f, g) \quad \forall\ f, g \in \mathfrak{H}. \tag{8.11}$$

If $\mathcal{P}$ is a nontrivial fundamental symmetry ($\mathcal{P} \neq \pm I$), then the Hilbert space $\mathfrak{H}$ endowed with the indefinite inner product (indefinite metric)

$$[f, g] \equiv (\mathcal{P}f, g), \qquad f, g \in \mathfrak{H} \tag{8.12}$$

is called a *Kreĭn space* $(\mathfrak{H}, [\cdot, \cdot])$. The fundamental symmetry $\mathcal{P}$ determines the *fundamental decomposition* of the Kreĭn space $(\mathfrak{H}, [\cdot, \cdot])$:

$$\mathfrak{H} = \mathfrak{H}_+ \dotplus \mathfrak{H}_-, \qquad \mathfrak{H}_\pm = P_\pm \mathfrak{H}, \tag{8.13}$$

where $P_\pm = \frac{1}{2}(I \pm \mathcal{P})$ are orthogonal projection operators in the Hilbert space $(\mathfrak{H}, (\cdot, \cdot))$. We can see that

$$\mathcal{P}P_\pm = \pm P_\pm, \quad P_+ - P_- = \mathcal{P}, \quad P_- P_+ = P_+ P_- = 0. \tag{8.14}$$

The subspaces $\mathfrak{H}_\pm$ are mutually orthogonal with respect to the initial inner product $(\cdot, \cdot)$ *and* with respect to the indefinite metric $[\cdot, \cdot]$. The restriction of $\mathcal{P}$ onto $\mathfrak{H}_\pm$ coincides with $\pm I$.

Remark 8.1. The fundamental decomposition (8.13) is often used for an equivalent definition of Kreĭn space. To be precise, let $\mathfrak{H}$ be a complex linear space with a Hermitian sesquilinear form $[\cdot, \cdot]$ (that is, a mapping $[\cdot, \cdot] : \mathfrak{H} \times \mathfrak{H} \to \mathbb{C}$ such that $[g, \alpha_1 f_1 + \alpha_2 f_2] = \alpha_1 [g, f_1] + \alpha_2 [g, f_2]$ and $[g, f] = [f, g]^*$ for all $f_1, f_2, f, g \in \mathfrak{H}$, $\alpha_1, \alpha_2 \in \mathbb{C}$). Then $(\mathfrak{H}, [\cdot, \cdot])$ is called a Kreĭn space if $\mathfrak{H}$ has a decomposition (8.13), where $\mathfrak{H}_\pm$ are orthogonal with respect to $[\cdot, \cdot]$ and the linear manifolds $\mathfrak{L}_+$ and $\mathfrak{L}_-$ endowed with sesquilinear forms $[\cdot, \cdot]$ and $-[\cdot, \cdot]$ are Hilbert spaces.

In contrast to the conventional inner product, the indefinite inner product $[f, f]$ may have nonpositive real values. A nonzero vector $f \in \mathfrak{H}$ is *neutral, negative,* or *positive* if

$$[f, f] = 0, \qquad [f, f] < 0, \qquad \text{or}\quad [f, f] > 0.$$

A subspace $\mathfrak{L}$ (that is closed with respect to the norm $\| \cdot \| = \sqrt{(\cdot, \cdot)}$) of the Kreĭn space $(\mathfrak{H}, [\cdot, \cdot])$ is *neutral, nonnegative,* or *positive* if

$$[f, f] = 0, \qquad [f, f] \geq 0, \qquad \text{or} \quad [f, f] > 0$$

for all nonzero $f \in \mathfrak{L}$. *Nonpositive* and *negative* subspaces are introduced similarly. Additionally, $\mathfrak{L}$ is *uniformly positive* or *uniformly negative* if

$$[f, f] \geq \alpha(f, f) \quad \text{or} \quad - [f, f] \geq \alpha(f, f), \quad \alpha > 0, \quad f \in \mathfrak{L}.$$

A subspace $\mathfrak{L}$ of $\mathfrak{H}$ is *definite* if it is either positive or negative, and *indefinite* if it is neither positive nor negative. The term *uniformly definite* is defined accordingly.

For each of the above classes we can define maximal subspaces. For instance, a closed positive subspace $\mathfrak{L}$ is *maximal positive* if $\mathfrak{L}$ is not a proper subspace of a positive subspace in $\mathfrak{H}$. The concept of maximality for other classes is defined similarly. The subspaces $\mathfrak{H}_\pm$ in (8.13) are maximal uniformly definite. Indeed, any $f_+ \in \mathfrak{H}_+$ has the form $f_+ = P_+ f$, $f \in \mathfrak{H}$ and

$$[f_+, f_+] = (\mathcal{P} P_+ f, P_+ f) = (P_+ f, P_+ f) = \|f_+\|^2$$

from the first identity in (8.14). Therefore, $\mathfrak{H}_+$ is a uniformly positive subspace. Similarly, $\mathfrak{H}_-$ is a uniformly negative subspace since $[f_-, f_-] = -\|f_-\|^2$ for all $f_- \in \mathfrak{H}_-$. The maximality of $\mathfrak{H}_+$ ($\mathfrak{H}_-$) in the class of uniformly positive (negative) subspaces follows from the decomposition (8.13).

Vectors f, g of the Kreĭn space $(\mathfrak{H}, [\cdot, \cdot])$ are said to be $\mathcal{P}$-*orthogonal* and are denoted by $f[\perp]g$ if $[f, g] = 0$. Similarly, subspaces $\mathfrak{L}_1, \mathfrak{L}_2$ of $\mathfrak{H}$ are $\mathcal{P}$-*orthogonal* and are denoted by $\mathfrak{L}_1[\perp]\mathfrak{L}_2$ if $[f, g] = 0$ for all $f \in \mathfrak{L}_1$ and $g \in \mathfrak{L}_2$. The $\mathcal{P}$-orthogonal complement of a subspace $\mathfrak{L}$ of $\mathfrak{H}$ is defined as

$$\mathfrak{L}^{[\perp]} = \{g \in \mathfrak{H} : [f, g] = 0, \ \forall f \in \mathfrak{L}\}$$

and it is a closed subspace of $\mathfrak{H}$.

The *isotropic part* $\mathfrak{L}_0$ of a closed subspace $\mathfrak{L}$ is defined by $\mathfrak{L}_0 = \mathfrak{L} \cap \mathfrak{L}^{[\perp]}$ and its (nonzero) elements are called *isotropic vectors of* $\mathfrak{L}$. The existence of isotropic vectors depends on the choice of $\mathfrak{L}$. In particular, any vector of a neutral subspace $\mathfrak{L}$ is isotropic (since $\mathfrak{L} \subset \mathfrak{L}^{[\perp]}$ for neutral subspaces). On the other hand, there are no isotropic elements of the whole space $\mathfrak{H}$. Indeed, if $f \in \mathfrak{H}$, then $[f, g] = 0$ for all $g \in \mathfrak{H}$. Setting $g = \mathcal{P}f$, we obtain

$$0 = [f, \mathcal{P}f] = (\mathcal{P}f, \mathcal{P}f) = (\mathcal{P}^2 f, f) = (f, f) = \|f\|^2.$$

Therefore, $f = 0$.

Each isotropic vector f of $\mathfrak{L}$ is neutral (since f belongs to $\mathfrak{L}$ and, simultaneously, f is $\mathcal{P}$-orthogonal to all elements of $\mathfrak{L}$). The inverse implication

$$\text{neutral vector of } \mathfrak{L} \quad \Rightarrow \quad \text{isotropic vector of } \mathfrak{L}$$

is not true in general. However, the subsets of neutral and isotropic vectors of $\mathfrak{L}$ coincide when $\mathfrak{L}$ is a nonnegative (nonpositive) subspace. This fact follows from the Cauchy-Schwartz inequality

$$|[f,g]|^2 \le [f,f] \cdot [g,g] \qquad (f, g \in \mathfrak{L}),$$

which holds for nonnegative (nonpositive) subspaces.

A closed subspace $\mathfrak{L}$ is said to be *nondegenerate* if its isotropic part is trivial: $\mathfrak{L}_0 = \{0\}$. Otherwise, it is called *degenerate*. Each neutral subspace is degenerate, while definite subspaces are nondegenerate. If $\mathfrak{L}$ is nondegenerate, then $\mathfrak{L} \cap \mathfrak{L}^{[\perp]} = \{0\}$ and therefore we may consider the direct sum $\mathfrak{L} \dotplus \mathfrak{L}^{[\perp]}$ of these subspaces.

Lemma 8.2. *A $\mathcal{P}$-orthogonal sum*[1] *$\mathfrak{L}[\dotplus]\mathfrak{L}^{[\perp]}$ is a dense set in $\mathfrak{H}$ if and only if the subspace $\mathfrak{L}$ is nondegenerate.*

Proof. Let a vector f be orthogonal to $\mathfrak{L}[\dotplus]\mathfrak{L}^{[\perp]}$ in the Hilbert space $(\mathfrak{H}, (\cdot, \cdot))$. Then the vector $\mathcal{P}f$ is $\mathcal{P}$-orthogonal to $\mathfrak{L}$ and $\mathfrak{L}^{[\perp]}$. Therefore, $\mathcal{P}f$ belongs to $\mathfrak{L} \cap \mathfrak{L}^{[\perp]}$ (since $(\mathfrak{L}^{[\perp]})^{[\perp]} = \mathfrak{L}$). This means that $\mathcal{P}f$ is an isotropic vector for $\mathfrak{L}$ and hence $\mathcal{P}f = 0$ due to the nondegeneracy of $\mathfrak{L}$. The proof is completed.

Let $\mathfrak{L}_+$ be a maximal positive subspace. Then $\mathfrak{L}_+$ is nondegenerate, and by Lemma 8.2, the direct $\mathcal{P}$-orthogonal sum

$$\mathcal{D} = \mathfrak{L}_+[\dotplus]\mathfrak{L}_- \qquad \left(\mathfrak{L}_- = \mathfrak{L}_+^{[\perp]}\right) \tag{8.15}$$

is a dense set in $\mathfrak{H}$. The subspace $\mathfrak{L}_-$ in (8.15) is maximal negative. Indeed, if $g \in \mathfrak{L}_-$ were positive, we could construct the positive subspace $\mathfrak{L}_+[\dotplus]\,\text{span}\{g\}$, which would be bigger than $\mathfrak{L}_+$, which is impossible due to the maximality of $\mathfrak{L}_+$. Hence, $\mathfrak{L}_-$ is nonpositive and any (hypothetical) nonzero neutral vector $g \in \mathfrak{L}_-$ is isotropic for $\mathfrak{L}_-$. The latter is in contradiction with the density of $\mathcal{D}$ in (8.15). Therefore, $\mathfrak{L}_-$ is a negative subspace and its maximality can be deduced from the relation $\mathfrak{L}_-^{[\perp]} = \mathfrak{L}_+$.

The particular case where $\mathfrak{L}_\pm$ in (8.15) are maximal *uniformly* definite subspaces is of great importance because only in that case does their direct sum $\mathcal{D}$ coincide with the whole space $\mathfrak{H}$. To be precise, *a maximal definite subspace $\mathfrak{L}$ is maximal uniformly definite if and only if the space $\mathfrak{H}$ can be*

---

[1]The symbol $[\dotplus]$ emphasizes the mutual $\mathcal{P}$-orthogonality of subspaces.

decomposed into a P-orthogonal direct sum $\mathfrak{H} = \mathfrak{L}[\dot{+}]\mathfrak{L}^{[\perp]}$. The subspace $\mathfrak{L}^{[\perp]}$ in this decomposition is also maximal uniformly definite (that is, if $\mathfrak{L}$ is positive, then $\mathfrak{L}^{[\perp]}$ is negative and vice versa).

The subspaces $\mathfrak{L}_\pm$ in (8.15) are maximal definite and their 'deviation' from the subspaces $\mathfrak{H}_\pm$ of the fundamental decomposition (8.13) can be characterized by the use of a Hermitian strong contraction[2] T, which anticommutes with $\mathcal{P}$. To be precise,

$$\mathfrak{L}_+ = (I + T)\mathfrak{H}_+, \qquad \mathfrak{L}_- = (I + T)\mathfrak{H}_-. \tag{8.16}$$

The operator T is an *operator of transition* from the fundamental decomposition (8.13) to the direct sum (8.15). The collection of transition operators admits a simple description: A Hermitian operator T in $\mathfrak{H}$ is an operator of transition if and only if

$$\|Tf\| < \|f\| \quad (\forall f \in \mathfrak{H}, \ f \neq 0), \qquad \mathcal{P}T = -T\mathcal{P}.$$

The important particular case of maximal uniformly definite subspaces $\mathfrak{L}_\pm$ in (8.16) is completely characterized by the condition $\|T\| < 1$ (the case of uniformly strong contraction).

It can be seen that the subspaces $\mathcal{P}\mathfrak{L}_\pm$ are also maximal definite and their $\mathcal{P}$-orthogonal direct sum

$$\mathcal{P}\mathcal{D} = \mathcal{P}\mathfrak{L}_+[\dot{+}]\mathcal{P}\mathfrak{L}_- \tag{8.17}$$

is called *dual* to (8.15). The transition operator corresponding to the dual decomposition (8.17) coincides with $-T$. Denote by $P_{\mathfrak{L}_\pm} : \mathcal{D} \to \mathfrak{L}_\pm$ the projection operators onto $\mathfrak{L}_\pm$ with respect to the decomposition (8.15). Then the operators $P_{\mathfrak{L}_\pm}$ are defined on $\mathcal{D} = \mathfrak{L}_+[\dot{+}]\mathfrak{L}_-$ and

$$P_{\mathfrak{L}_\pm}f = P_{\mathfrak{L}_\pm}(f_{\mathfrak{L}_+} + f_{\mathfrak{L}_-}) = f_{\mathfrak{L}_\pm}, \qquad f = f_{\mathfrak{L}_+} + f_{\mathfrak{L}_-} \in \mathcal{D}, f_{\mathfrak{L}_\pm} \in \mathfrak{L}_\pm.$$

It has been shown that [Kuzhel (2009)]

$$P_{\mathfrak{L}_+} = (I - T)^{-1}(P_+ - TP_-), \qquad P_{\mathfrak{L}_-} = (I - T)^{-1}(P_- - TP_+), \tag{8.18}$$

where $P_\pm = \frac{1}{2}(I \pm \mathcal{P})$ are orthogonal projection operators on $\mathfrak{H}_\pm$.

A subspace $\mathfrak{L}$ of $\mathfrak{H}$ is called *hypermaximal neutral* if $\mathfrak{L}$ coincides with its $\mathcal{P}$-orthogonal complement $\mathfrak{L}^{[\perp]}$; that is, $\mathfrak{L} = \mathfrak{L}^{[\perp]}$. Let $\mathfrak{L}$ be a hypermaximal neutral subspace of the Kreĭn space $(\mathfrak{H}, [\cdot, \cdot])$. Then its dual subspace $\mathcal{P}\mathfrak{L}$ is also hypermaximal neutral. The subspaces $\mathfrak{L}$ and $\mathcal{P}\mathfrak{L}$ are orthogonal with respect to the initial inner product $(\cdot, \cdot)$ (since $(\mathcal{P}f, g) = [f, g] = 0$ for any $f, g \in \mathfrak{L}$) and the space $\mathfrak{H}$ can be decomposed:[3]

$$\mathfrak{H} = \mathfrak{L} \oplus \mathcal{P}\mathfrak{L}. \tag{8.19}$$

---

[2]'Strong contraction' means that $\|Tf\| < \|f\|$, $f \neq 0$.
[3]The symbol $\oplus$ means the orthogonality in the Hilbert space $(\mathfrak{H}, (\cdot, \cdot))$.

The decomposition (8.19) is completely determined by the operator

$$\mathcal{R}(f + \mathcal{P}g) = f - \mathcal{P}g, \qquad f, g \in \mathfrak{L}.$$

One can verify that $\mathcal{R}$ is a fundamental symmetry in $(\mathfrak{H}, (\cdot, \cdot))$ and that it anticommutes with $\mathcal{P}$. The decomposition (8.19) turns out to be fundamental in the *new* Kreĭn space $(\mathfrak{H}, [\cdot, \cdot]_{\mathcal{R}})$ with the indefinite metric $[f, g]_{\mathcal{R}} \equiv (\mathcal{R}f, g)$ [see (8.12)]. The original fundamental decomposition (8.13) is transformed to the orthogonal sum of hypermaximal neutral subspaces (8.19) in the Kreĭn space $(\mathfrak{H}, [\cdot, \cdot]_{\mathcal{R}})$.

Example 8.2. *Kreĭn space* $(L_2(\mathbb{R}), (\cdot, \cdot)_{\mathcal{PT}})$. The space-reflection operator $\mathcal{P}f(x) = f(-x)$ is an example of fundamental symmetry in the Hilbert space $\mathfrak{H} = L_2(\mathbb{R})$ with inner product (8.2). The $\mathcal{PT}$ inner product[4] (8.1) coincides with the indefinite metric (8.12) and the space $(L_2(\mathbb{R}), (\cdot, \cdot)_{\mathcal{PT}})$ is a Kreĭn space. The fundamental decomposition of $(L_2(\mathbb{R}, (\cdot, \cdot)_{\mathcal{PT}})$ has the form

$$L_2(\mathbb{R}) = L_2^{even}(\mathbb{R}) \oplus L_2^{odd}(\mathbb{R}), \tag{8.20}$$

where $L_2^{even}(\mathbb{R})$ and $L_2^{odd}(\mathbb{R})$ are the subspaces of even and odd functions in $L_2(\mathbb{R})$.

Denote $\mathcal{R}f(x) = (\operatorname{sgn} x)f(x)$, $f \in L_2(\mathbb{R})$. The operator $\mathcal{R}$ is a fundamental symmetry in $L_2(\mathbb{R})$ that anticommutes with $\mathcal{P}$. The collection of operators $T_\varepsilon = \varepsilon \mathcal{R}$, $|\varepsilon| < 1$, is an example of transition operators. Due to (8.16), the subspaces

$$\mathfrak{L}_+^\varepsilon = (I + \varepsilon \mathcal{R})L_2^{even}(\mathbb{R}) \quad \text{and} \quad \mathfrak{L}_-^\varepsilon = (I + \varepsilon \mathcal{R})L_2^{od}(\mathbb{R})$$

are maximal uniformly positive and maximal uniformly negative. The limits of subspaces $\mathfrak{L}_\pm^\varepsilon$ (as $\varepsilon \to 1$) coincide with the same hypermaximal neutral subspace $\mathfrak{L} = (I + \mathcal{R})L_2(\mathbb{R})$. Similarly, if $\varepsilon \to -1$, then $\mathfrak{L}_\pm^\varepsilon$ 'tend' to the dual hypermaximal neutral subspace $\mathcal{P}\mathfrak{L} = (I - \mathcal{R})L_2(\mathbb{R})$.

Example 8.3. *Kreĭn space* $(L_2(\Gamma), (\cdot, \cdot)_{\mathcal{PT}})$. Let $\eta(x)$ be a real-valued differentiable function on $\mathbb{R}$ such that $|\eta'(x)| < C$ for all $x \in \mathbb{R}$. The function $\eta(\cdot)$ defines a contour in the complex plane

$$\Gamma = \{\xi(x) = x + i\eta(x) : x \in \mathbb{R}\}. \tag{8.21}$$

A line integral along Γ is defined as

$$\int_\Gamma d\xi\, f(\xi) \equiv \int_\mathbb{R} dx\, f(\xi(x))|\xi'(x)| = \int_\mathbb{R} dx\, f(\xi(x))\sqrt{1 + \eta'^2(x)}.$$

---

[4]We preserve the traditional notation $(\cdot, \cdot)_{\mathcal{PT}}$ for this indefinite inner product.

Let $C_0(\Gamma)$ be the set of complex-valued functions that are continuous along Γ and have compact support. The Hilbert space $L_2(\Gamma)$ is the completion of $C_0(\Gamma)$ with respect to the norm

$$\|f\|_1^2 = \int_\Gamma d\xi |f(\xi)|^2 = \int_{\mathbb{R}} dx |f(\xi(x))|^2 \sqrt{1 + \eta'^2(x)}.$$

This norm is equivalent to

$$\|f\|^2 = \int_{\mathbb{R}} dx |f(\xi(x))|^2$$

because $1 \leq \sqrt{1 + \eta'^2(x)} \leq \sqrt{1 + C^2}$ for all $x \in \mathbb{R}$. Thus, it does not matter which norm will be considered as the original norm of $L_2(\Gamma)$. For simplicity, we work with $\| \cdot \|$. Then, the original inner product of $L_2(\Gamma)$ is

$$(f, g) = \int_{\mathbb{R}} dx f^*(\xi(x)) g(\xi(x)).$$

If we take the function $\eta(x)$ in (8.21) to be even, $\eta(x) = \eta(-x)$, then the contour Γ has mirror symmetry with respect to the imaginary axis. Denote $\mathcal{P}f(z) = f(-z^*)$, $z \in \mathbb{C}$. The restriction of $\mathcal{P}$ onto $L_2(\Gamma)$ determines the operator of fundamental symmetry in $L_2(\Gamma)$. Indeed, the condition $\mathcal{P}^2 = I$ is trivial. Along Γ we have $z = \xi(x) = x + i\eta(x)$ and $-z^* = -x + i\eta(x) = -x + i\eta(-x) = \xi(-x)$ because $\eta(x)$ is even. Therefore,

$$(\mathcal{P}f, \mathcal{P}g) = \int_{\mathbb{R}} dx f^*(\xi(-x)) g(\xi(-x)) = \int_{\mathbb{R}} dx' f^*(\xi(x')) g(\xi(x')) = (f, g).$$

Because of (8.11), $\mathcal{P}$ is a fundamental symmetry in the Hilbert space $L_2(\Gamma)$.

The fundamental symmetry $\mathcal{P}$ generates the indefinite inner product (8.12) in $L_2(\Gamma)$ known as the $\mathcal{PT}$ inner product [see (8.1)]:

$$(f, g)_{\mathcal{PT}} = (\mathcal{P}f, g) = \int_{\mathbb{R}} dx [\mathcal{PT}f(\xi(x))] g(\xi(x)), \qquad f, g \in L_2(\Gamma).$$

The Hilbert space $L_2(\Gamma)$ endowed with a $\mathcal{PT}$ inner product is the Kreĭn space $(L_2(\Gamma), (\cdot, \cdot)_{\mathcal{PT}})$.

8.3.2 Definition of $\mathcal{C}$ operators

Given a maximal positive subspace $\mathfrak{L}_+$, its $\mathcal{P}$-orthogonal complement $\mathfrak{L}_- = \mathfrak{L}_+^{[\perp]}$ is a maximal negative subspace and we can consider the dense set $\mathcal{D}$ defined by (8.15). Each vector $f \in \mathcal{D}$ admits the decomposition $f = f_{\mathfrak{L}_+} + f_{\mathfrak{L}_-}$, where $f_{\mathfrak{L}_\pm} \in \mathfrak{L}_\pm$ and the operator $\mathcal{C}$ associated with the $\mathcal{P}$-orthogonal direct sum (8.15) is defined as follows:

$$\mathcal{C}f = \mathcal{C}(f_{\mathfrak{L}_+} + f_{\mathfrak{L}_-}) = f_{\mathfrak{L}_+} - f_{\mathfrak{L}_-}. \qquad (8.22)$$

Remark 8.2. The operator C cannot be properly defined with the use of maximal nonnegative subspaces $\mathfrak{L}_+$. It is crucial that the nonnegative subspaces contain isotropic vectors. Therefore, there exists the common part $\mathfrak{L}_+ \cap \mathfrak{L}_- \neq \{0\}$ and the algebraic sum $\mathfrak{L}_+ + \mathfrak{L}_-$ is nondense in $\mathfrak{H}$.

The set of operators C in the Kreĭn space $(\mathfrak{H}, [\cdot, \cdot])$ is in one-to-one correspondence with the set of all possible $\mathcal{P}$-orthogonal direct sums (8.15). If C is given, then the corresponding maximal definite subspaces $\mathfrak{L}_\pm$ in (8.15) are recovered by the formula

$$\mathfrak{L}_\pm = \tfrac{1}{2}(I \pm C)\mathcal{D}. \tag{8.23}$$

The operator C is a closed operator in $\mathfrak{H}$ (see, for example, Lemma 6.2.2 in [Albeverio and Kuzhel (2015)]).

Proposition 8.1. *The operator C is bounded if and only if the corresponding subspaces $\mathfrak{L}_\pm$ in (8.15) are maximal uniformly definite.*

Proof. If $\mathfrak{L}_\pm$ are maximal uniformly definite subspaces, then $\mathfrak{L}_+[\dotplus]\mathfrak{L}_- = \mathfrak{H}$. Therefore, the closed operator C is defined on the whole space $\mathfrak{H}$. This implies that C is bounded. Conversely, if C is bounded, then its domain $\mathcal{D}$ coincides with $\mathfrak{H}$ (since C is closed). Therefore, the direct sum (8.15) is a decomposition of $\mathfrak{H}$. This is possible only in the case where $\mathfrak{L}_\pm$ are maximal uniformly definite.

Proposition 8.2. *Let C be associated with the $\mathcal{P}$-orthogonal direct sum (8.15). Then its adjoint $C^\dagger$ is determined by the dual direct sum (8.17).*

Proof. The proof is based on the expression of C in terms of the corresponding transition operator T,

$$C = P_{\mathfrak{L}_+} - P_{\mathfrak{L}_-} = (I - T)^{-1}(I + T)\mathcal{P}, \tag{8.24}$$

which follows directly from (8.14), (8.18), and (8.22). As a consequence of (8.24), the adjoint of C has the form

$$C^\dagger = \mathcal{P}(I + T)(I - T)^{-1} = (I - T)(I + T)^{-1}\mathcal{P} = (I + T)^{-1}(I - T)\mathcal{P}.$$

This means that $C^\dagger$ is determined by (8.24) with the transition operator $-T$ instead of T. The transition operator $-T$ corresponds to the dual direct sum (8.17). Hence $C^\dagger$ is determined by (8.17). This proves proposition 8.2.

Defining C by using (8.22) is not always convenient because it requires that we know $\mathfrak{L}_\pm$. An additional analysis of (8.24) allows one to establish the following theorem.

Theorem 8.1. *The following statements are equivalent:*

(i) *an operator $\mathcal{C}$ is determined by the direct sum (8.15) with the use of (8.22);*

(ii) *an operator $\mathcal{C}$ has the form $\mathcal{C} = e^{\mathcal{Q}}\mathcal{P}$, where $\mathcal{Q}$ is a Hermitian operator in $(\mathfrak{H}, (\cdot, \cdot))$ such that $\mathcal{P}\mathcal{Q} = -\mathcal{Q}\mathcal{P}$;*

(iii) *$\mathcal{C}$ is an operator in $\mathfrak{H}$ such that $\mathcal{C}^2 f = f$ for all $f \in \mathcal{D}(\mathcal{C})$ and $\mathcal{C}\mathcal{P}$ is a positive Hermitian operator.*

The particular cases (that is, the proof of equivalence of various relations between items $(i) - (iii)$ of this theorem) have been considered in [Kuzhel (2009); Bender and Kuzhel (2012); Albeverio and Kuzhel (2015)].

In what follows we use $(i) - (iii)$ as *equivalent definitions* of the operator $\mathcal{C}$. This means that each operator $\mathcal{C}$ has the form $\mathcal{C} = e^{\mathcal{Q}}\mathcal{P}$ and, simultaneously, the operator $\mathcal{C}$ can be determined by (8.22) for a special choice of the direct sum (8.15). Let us find the relationship between the operator of transition T from the fundamental decomposition (8.13) to (8.15) and the Hermitian operator $\mathcal{Q}$. From (8.24) we get

$$e^{\mathcal{Q}} = (I - T)^{-1}(I + T)$$

and, after simple manipulation, $T = (e^{\mathcal{Q}} - I)(e^{\mathcal{Q}} + I)^{-1}$. Hence,

$$T = (e^{\mathcal{Q}/2} - e^{-\mathcal{Q}/2})(e^{\mathcal{Q}/2} + e^{-\mathcal{Q}/2})^{-1} = \frac{\sinh \mathcal{Q}/2}{\cosh \mathcal{Q}/2} = \tanh \mathcal{Q}/2. \quad (8.25)$$

Example 8.4. *General form of $\mathcal{C}$ in the Kreĭn space $(\mathbb{C}^2, [\cdot, \cdot])$.* The matrix operator $\mathcal{P} = \begin{bmatrix} 0 & 1 \\ 1 & 0 \end{bmatrix}$ is a fundamental symmetry in the Hilbert space $\mathbb{C}^2$ with the ordinary inner product $(X, Y) = x_1^* y_1 + x_2^* y_2$, where $X = (x_1, x_2)^t$ and $Y = (y_1, y_2)^t$. The corresponding indefinite metric is

$$[X, Y] = (\mathcal{P}X, Y) = x_2^* y_1 + x_1^* y_2 \qquad (x_i, y_i \in \mathbb{C})$$

and the fundamental decomposition (8.13) of the Kreĭn space $(\mathbb{C}^2, [\cdot, \cdot])$ has the form $\mathbb{C}^2 = \mathfrak{H}_+ \oplus \mathfrak{H}_-$, where

$$\mathfrak{H}_+ = \{X_+ = (x, x)^t : x \in \mathbb{C}\}, \qquad \mathfrak{H}_- = \{X_- = (x, -x)^t : x \in \mathbb{C}\}.$$

According to statement (ii) of Theorem 8.1, every operator $\mathcal{C}$ acting in $(\mathbb{C}^2, [\cdot, \cdot])$ has the form $\mathcal{C} = e^{\mathcal{Q}}\mathcal{P}$, where $\mathcal{Q}$ is a Hermitian operator in $(\mathbb{C}^2, (\cdot, \cdot))$ that anticommutes with $\mathcal{P}$. The operator $\mathcal{Q}$ is a 2×2-matrix that admits the decomposition

$$\mathcal{Q} = q_0 \sigma_0 + q_1 \sigma_1 + q_2 \sigma_2 + q_3 \sigma_3, \qquad q_j \in \mathbb{C}, \quad (8.26)$$

where $\sigma_0 = \begin{bmatrix} 1 & 0 \\ 0 & 1 \end{bmatrix}$, $\sigma_1 = \begin{bmatrix} 0 & 1 \\ 1 & 0 \end{bmatrix}$, $\sigma_2 = \begin{bmatrix} 0 & -i \\ i & 0 \end{bmatrix}$, and $\sigma_3 = \begin{bmatrix} 1 & 0 \\ 0 & -1 \end{bmatrix}$ are the Pauli matrices. The Hermiticity of $\mathcal{Q}$ means that all coefficients q_j in (8.26) are real. Anticommutating $\mathcal{Q}$ with $\mathcal{P} = \sigma_1$ leads to the conclusion that $q_0 = q_1 = 0$ (because $\sigma_j\sigma_k = -\sigma_k\sigma_j$, $j \neq k$ and $\sigma_j^2 = \sigma_0$). Hence,

$$\mathcal{Q} = q_2\sigma_2 + q_3\sigma_3 = \rho(\cos\xi\sigma_2 + \sin\xi\sigma_3) = \rho Z,$$

where $\rho = \sqrt{q_2^2 + q_3^2}$, $\cos\xi = \frac{q_2}{\sqrt{q_2^2+q_3^2}}$, $\sin\xi = \frac{q_3}{\sqrt{q_2^2+q_3^2}}$, and

$$Z = \cos\xi\sigma_2 + \sin\xi\sigma_3$$

is a fundamental symmetry in the Hilbert space $(\mathbb{C}^2, (\cdot, \cdot))$ (because $\sigma_2\sigma_3 = -\sigma_3\sigma_2$). Therefore, every operator $\mathcal{C}$ acting in the Kreĭn space $(\mathbb{C}^2, [\cdot, \cdot])$ is determined by the expression

$$\begin{aligned} \mathcal{C} = e^{\mathcal{Q}}\mathcal{P} = e^{\rho Z}\mathcal{P} &= (\cosh\rho\sigma_0 + \sinh\rho Z)\mathcal{P} \\ &= \cosh\rho\sigma_1 + \sinh\rho(\cos\xi\sigma_2\sigma_1 + \sin\xi\sigma_3\sigma_1) \\ &= \cosh\rho\sigma_1 + i\sinh\rho\sin\xi\sigma_2 - i\sinh\rho\cos\xi\sigma_3 \\ &= \begin{bmatrix} -i\sinh\rho\cos\xi & \cosh\rho + \sinh\rho\sin\xi \\ \cosh\rho - \sinh\rho\sin\xi & i\sinh\rho\cos\xi \end{bmatrix}. \end{aligned} \quad (8.27)$$

For a similar description of $\mathcal{C}$ for the case in which $\mathcal{P}$ is an arbitrary fundamental symmetry in $(\mathbb{C}^2, (\cdot, \cdot))$ see [Albeverio and Kuzhel (2015)].

8.3.3 *Bounded and unbounded $\mathcal{C}$ operators*

There is an essential difference between the properties of bounded and unbounded operators $\mathcal{C}$ [Albeverio and Kuzhel (2015); Bender and Kuzhel (2012)]. In particular, if $\mathcal{C}$ is bounded, its spectrum consists of the eigenvalues $\{-1, 1\}$, but if $\mathcal{C}$ is unbounded, its spectrum has the additional continuous part $\mathbb{C} \setminus \{-1, 1\}$. The appearance of a continuous spectrum does not relate to an inherent structure of $\mathcal{C}$, but rather indicates a poor choice of the underlying Hilbert space $(\mathfrak{H}, (\cdot, \cdot))$.

If $\mathcal{C}$ is *bounded*, then $\mathcal{C} = e^{\mathcal{Q}}\mathcal{P}$, where $\mathcal{Q}$ is a bounded Hermitian operator in $\mathfrak{H}$ that anticommutes with $\mathcal{P}$; the subspaces $\mathfrak{L}_\pm$ in (8.15) are *maximal uniformly* definite in the Kreĭn space $(\mathfrak{H}, [\cdot, \cdot])$; the direct sum (8.15) gives the decomposition of the whole space $\mathfrak{H}$:

$$\mathfrak{H} = \mathfrak{L}_+[\dot{+}]\mathfrak{L}_-. \quad (8.28)$$

The given indefinite inner product $[\cdot, \cdot]$ generates infinitely many *equivalent* inner products $(\cdot, \cdot)_\mathcal{C}$ in the Hilbert space $(\mathfrak{H}, (\cdot, \cdot))$, which are determined by bounded operators $\mathcal{C}$:

$$(\cdot, \cdot)_\mathcal{C} = [\mathcal{C}\cdot, \cdot] = (\mathcal{PC}\cdot, \cdot) = (\mathcal{P}e^{\mathcal{Q}}\mathcal{P}\cdot, \cdot) = (e^{-\mathcal{Q}}\cdot, \cdot). \quad (8.29)$$

The original inner product $(\cdot,\cdot)$ coincides with $(\cdot,\cdot)_\mathcal{C}$ for $\mathcal{C} = \mathcal{P}$ and the subspaces $\mathfrak{L}_\pm$ in (8.28) are mutually orthogonal with respect to the new inner product $(\cdot,\cdot)_\mathcal{C}$. The operator $\mathcal{C}$ is a fundamental symmetry in the Hilbert space $(\mathfrak{H},(\cdot,\cdot)_\mathcal{C})$; $\mathcal{C}$ determines the original Kreĭn space $(\mathfrak{H},[\cdot,\cdot])$ because, by analogy with (8.12),

$$(\mathcal{C}f,g)_\mathcal{C} = [\mathcal{C}^2 f, g] = [f,g], \qquad f,g \in \mathfrak{H}.$$

If $\mathcal{C}$ is *unbounded*, then $\mathcal{C} = e^\mathcal{Q}\mathcal{P}$, where $\mathcal{Q}$ is an unbounded Hermitian operator in $\mathfrak{H}$ that anticommutes with $\mathcal{P}$. The subspaces $\mathfrak{L}_\pm$ in (8.15) are *maximal* definite in $(\mathfrak{H},[\cdot,\cdot])$ and the direct sum (8.15) determines a dense subset $\mathcal{D}$ of $\mathfrak{H}$ that depends on the choice of $\mathfrak{L}_\pm$. The formula (8.29) determines infinitely many *inequivalent* inner products $(\cdot,\cdot)_\mathcal{C}$ defined on various sets $\mathcal{D} = \mathcal{D}(\mathcal{C})$; the linear spaces $\mathcal{D}$ endowed with inner products $(\cdot,\cdot)_\mathcal{C}$ are pre-Hilbert spaces because $(\cdot,\cdot)_\mathcal{C}$ and $(\cdot,\cdot)$ are not equivalent. The completion $\mathfrak{H}_\mathcal{C}$ of $\mathcal{D}$ with respect to $(\cdot,\cdot)_\mathcal{C}$ does not coincide[5] with $\mathfrak{H}$. The Hilbert space $(\mathfrak{H}_\mathcal{C},(\cdot,\cdot)_\mathcal{C})$ has the decomposition

$$\mathfrak{H}_\mathcal{C} = \hat{\mathfrak{L}}_+ \oplus_\mathcal{C} \hat{\mathfrak{L}}_-, \tag{8.30}$$

where the mutually orthogonal[6] subspaces $\hat{\mathfrak{L}}_\pm$ are the completion of $\mathfrak{L}_\pm$ with respect to $(\cdot,\cdot)_\mathcal{C}$. The decomposition (8.30) gives rise to the new Kreĭn space $(\mathfrak{H}_\mathcal{C},[\cdot,\cdot])$ with the indefinite inner product

$$[f,g] = (f_{\hat{\mathfrak{L}}_+}, g_{\hat{\mathfrak{L}}_+})_\mathcal{C} - (f_{\hat{\mathfrak{L}}_-}, g_{\hat{\mathfrak{L}}_-})_\mathcal{C}, \quad f = f_{\hat{\mathfrak{L}}_+} + f_{\hat{\mathfrak{L}}_-}, \quad g = g_{\hat{\mathfrak{L}}_+} + g_{\hat{\mathfrak{L}}_-},$$

which coincides with the original indefinite inner product $[\cdot,\cdot]$ on $\mathcal{D}$.

8.3.4 *Linear operators having C-symmetry*

Definition 8.1. A densely defined operator H in $\mathfrak{H}$ has the property of $\mathcal{C}$-*symmetry* if there exists an operator $\mathcal{C}$ with properties described in Theorem 8.1 and such that

$$H\mathcal{C}f = \mathcal{C}Hf, \qquad \forall f \in \mathcal{D}(H). \tag{8.31}$$

The commutation identity (8.31) requires an additional explanation because $\mathcal{C}$ may be unbounded. To be precise, if (8.31) is satisfied, then

$$\mathcal{D}(\mathcal{C}) \supset \mathcal{D}(H), \quad \mathcal{C} : \mathcal{D}(H) \to \mathcal{D}(H), \quad H : \mathcal{D}(H) \to \mathcal{D}(\mathcal{C}).$$

If $\mathcal{C}$ is a bounded operator, then the first and third relations are trivial because $\mathcal{D}(\mathcal{C}) = \mathfrak{H}$. In summary, the property of $\mathcal{C}$-symmetry for H means

---

[5] An additional analysis shows that $\mathfrak{H} \cap \mathfrak{H}_\mathcal{C}$ coincides with the domain $\mathcal{D}(e^{-\mathcal{Q}/2})$.
[6] With respect to the inner product $(\cdot,\cdot)_\mathcal{C}$.

that there exists a $\mathcal{P}$-orthogonal direct sum (8.15) that decomposes the operator H into the sum

$$H = H_+ \dot{+} H_-, \quad \mathcal{D}(H) = \mathcal{D}(H_+) \dot{+} \mathcal{D}(H_-), \quad \mathcal{D}(H_\pm) \subset \mathfrak{L}_\pm$$

of operators $H_\pm : \mathcal{D}(H_\pm) \to \mathfrak{L}_\pm$ acting in $\mathfrak{L}_\pm$. We say that H has *bounded (unbounded)* $\mathcal{C}$-symmetry if the corresponding operator $\mathcal{C}$ is bounded (unbounded).

Proposition 8.3. *Let H be a closed operator in a Hilbert space $(\mathfrak{H}, (\cdot, \cdot))$ with nonempty resolvent set $\rho(H) \neq \emptyset$. Then H may only have a property of bounded $\mathcal{C}$-symmetry.*

Proof. Let a closed operator H have the property of $\mathcal{C}$-symmetry. Then the operator $H - \lambda I$, where $\lambda \in \rho(H)$, commutes with $\mathcal{C}$ and $\mathcal{D}(\mathcal{C}) \supseteq \mathcal{R}(H - \lambda I) = \mathfrak{H}$. Therefore, $\mathcal{C}$ is a bounded operator. This completes the proof.

If $\mathcal{C}$ is unbounded, then the new inner product $(\cdot, \cdot)_\mathcal{C}$ is not equivalent to the original inner product $(\cdot, \cdot)$. Therefore, the original closedness of the operator H (with $\rho(H) \neq \emptyset$) in the Hilbert space $(\mathfrak{H}, (\cdot, \cdot))$ cannot be preserved if we consider H in the new Hilbert space $(\mathfrak{H}_\mathcal{C}, (\cdot, \cdot)_\mathcal{C})$ with inequivalent inner product $(\cdot, \cdot)_\mathcal{C}$. For this reason the property of unbounded $\mathcal{C}$-symmetry is more natural for the case of closable (but not closed) operators.

Remark 8.3. The condition $\rho(H) \neq \emptyset$ in Proposition 8.3 is crucial. Indeed, every unbounded operator $H = \mathcal{C}$ is closed and its spectrum coincides with $\mathbb{C}$. Thus, H has the property of unbounded $\mathcal{C}$-symmetry.

8.3.5 $\mathcal{P}$-symmetric and $\mathcal{P}$-Hermitian operators

Let H be a densely defined operator acting in a Kreĭn space $(\mathfrak{H}, [\cdot, \cdot])$. The adjoint of H calculated in terms of the indefinite inner product $[\cdot, \cdot]$ is called $\mathcal{P}$-*adjoint* and it is denoted by H^+. A vector $g \in \mathfrak{H}$ belongs to the domain $\mathcal{D}(H^+)$ if and only if there exists a vector $\gamma \in \mathfrak{H}$ such that the identity

$$[Hf, g] = [f, \gamma] \tag{8.32}$$

holds for all $f \in \mathcal{D}(H)$. If $g \in \mathcal{D}(H^+)$, then $H^+g = \gamma$, where γ is uniquely determined by (8.32). Therefore,

$$[Hf, g] = [f, H^+g], \quad f \in \mathcal{D}(H), \ g \in \mathcal{D}(H^+).$$

The simple relationship between the adjoint operator $H^\dagger$ and the $\mathcal{P}$-adjoint operator H^+ immediately follows from (8.8), (8.12), and (8.32):

$$H^+ = \mathcal{P} H^\dagger \mathcal{P}. \tag{8.33}$$

An operator H is said to be $\mathcal{P}$-*symmetric* if $H \subset H^+$ or equivalently if

$$[Hf, g] = [f, Hg], \qquad f, g \in \mathcal{D}(H). \tag{8.34}$$

The identity (8.34) shows that any $\mathcal{P}$-symmetric operator H is symmetric with respect to the indefinite inner product $[\cdot, \cdot]$ (that is, H is symmetric in the Kreĭn space $(\mathfrak{H}, [\cdot, \cdot])$).

An operator H is said to be $\mathcal{P}$-*Hermitian* if $H = H^+$. This means that H is Hermitian in the Kreĭn space $(\mathfrak{H}, [\cdot, \cdot])$. By virtue of (8.33), H is $\mathcal{P}$-Hermitian if and only if

$$\mathcal{P}H = H^\dagger \mathcal{P}, \tag{8.35}$$

or equivalently if $\mathcal{P}H$ is Hermitian in the Hilbert space $(\mathfrak{H}, (\cdot, \cdot))$.

An operator $\mathcal{C}$ determined by the $\mathcal{P}$-orthogonal direct sum (8.15) is $\mathcal{P}$-Hermitian. Indeed, from Theorem 8.1, the operator $\mathcal{P}\mathcal{C} = \mathcal{P}e^{\mathcal{Q}}\mathcal{P} = e^{-\mathcal{Q}}\mathcal{P}^2 = e^{-\mathcal{Q}}$ is Hermitian. Therefore, $\mathcal{C}$ is $\mathcal{P}$-Hermitian.

If they are bounded, $\mathcal{P}$-Hermitian and $\mathcal{P}$-symmetric operators are equivalent. More generally, $\mathcal{P}$-Hermitian operators are $\mathcal{P}$-symmetric but the converse is not true. The spectrum of a $\mathcal{P}$-Hermitian operator H may be complex and spectrum is symmetric with respect to the real line; that is, $\lambda \in \sigma(H) \iff \lambda^* \in \sigma(H)$. By analogy to the case of Hilbert spaces, the statement that an operator H is *essentially Hermitian in the Kreĭn space* $(\mathfrak{H}, [\cdot, \cdot])$ is equivalent to saying that H is *essentially* $\mathcal{P}$-*Hermitian* if its closure[8] $\overline{H}$ is a $\mathcal{P}$-Hermitian operator.

Proposition 8.4. *A $\mathcal{P}$-symmetric operator H is essentially $\mathcal{P}$-Hermitian if and only if the sets $\mathcal{R}(\mathcal{P}H + iI)$ and $\mathcal{R}(\mathcal{P}H - iI)$ are dense in $(\mathfrak{H}, (\cdot, \cdot))$.*

Proof. An operator H is essentially Hermitian in the Kreĭn space $(\mathfrak{H}, [\cdot, \cdot])$ if and only if the operator $\mathcal{P}H$ is essentially Hermitian in the Hilbert space $(\mathfrak{H}, (\cdot, \cdot))$. By Lemma 8.1 the latter statement is equivalent to $\mathcal{R}(\mathcal{P}H \pm iI)$ being dense in $\mathfrak{H}$.

Proposition 8.5. *If a real eigenvalue λ of a $\mathcal{P}$-symmetric operator H is not semi-simple, then the kernel subspace $\ker(H - \lambda I)$ is degenerate.*

Proof. If λ is not semi-simple, then $m_\lambda^a > m_\lambda^g$. Therefore, there exist a root vector f and an eigenvector g of H such that $(H - \lambda I)f = g$. For any $\gamma \in \ker(H - \lambda I)$,

$$[g, \gamma] = [(H - \lambda I)f, \gamma] = [Hf, \gamma] - \lambda[f, \gamma] = [f, H\gamma] - \lambda[f, \gamma] = 0.$$

---

[8]With respect to the initial inner product $(\cdot, \cdot)$.

Thus, $g[\perp] \ker(H - \lambda I)$, and g is an isotropic vector for $\ker(H - \lambda I)$.

Proposition 8.6. *If λ is a nonreal eigenvalue of a $\mathcal{P}$-symmetric operator H, then the kernel subspace $\ker(H - \lambda I)$ is neutral (and hence, degenerate).*

Proof. Let $f \in \ker(H - \lambda I)$, where λ is nonreal. Then,

$$\lambda[f,f] = [f, \lambda f] = [f, Hf] = [Hf, f] = [\lambda f, f] = \lambda^*[f,f]$$

is possible only if $[f,f] = 0$.

Propositions 8.5 and 8.6 show that the existence of isotropic eigenvectors of the kernel subspace $\ker(H - \lambda I)$ of a $\mathcal{P}$-symmetric operator H means either the the eigenvalue λ is nonreal or that λ is real but not semi-simple (that is, the root vectors associated with λ form a Jordan chain).

Proposition 8.7 (cf. Theorem 2.5 in [Bognar (1974)]). *Let H be a $\mathcal{P}$-symmetric operator. If λ and μ are eigenvalues of H such that $\lambda \neq \mu^*$, then the root subspaces $\mathcal{L}_\lambda(H)$ and $\mathcal{L}_\mu(H)$ are $\mathcal{P}$-orthogonal.*

Proof. If $f \in \mathcal{L}_\lambda(H)$ and $g \in \mathcal{L}_\mu(H)$, then

$$(H - \lambda I)^r f = 0 \text{ and } (H - \mu I)^s g = 0 \text{ for some } r, s \in \mathbb{N}. \tag{8.36}$$

To complete the proof it suffices to show that (8.36) implies that $[f,g] = 0$. Let $r = s = 1$. Then the required $\mathcal{P}$-orthogonality of f and g follows from the identity $\lambda[g,f] = [g, \lambda f] = [g, Hf] = [Hg, f] = [\mu g, f] = \mu^*[g, f]$. We assume that (8.36) implies that $[f,g] = 0$ for $r + s < n$ $(n \geq 2)$ and prove that $[f, g] = 0$ for $r + s = n$. To this end we denote $f_1 = (H - \lambda I)f$ and $g_1 = (H - \mu I)g$. Then $(H - \lambda I)^{r-1} f_1 = (H - \mu I)^{s-1} g_1 = 0$. Hence, by assumption, $[f, g_1] = [f_1, g] = [f_1, g_1] = 0$. Therefore, we obtain

$$\lambda[g, f] = [g, Hf - f_1] = [g, Hf] = [Hg, f] = [\mu g + g_1, f] = \mu^*[g, f],$$

which is possible only if $[g, f] = 0$. This proves Proposition 8.7.

It follows from Proposition 8.7 that any root subspace corresponding to a nonreal eigenvalue of a $\mathcal{P}$-symmetric operator is neutral (cf. Proposition 8.6). This statement is not true for real eigenvalues.

Proposition 8.8. *Let a $\mathcal{P}$-symmetric operator H have a complete set of eigenvectors $\{f_n\}$ corresponding to real eigenvalues λ_n. Then all eigenvalues λ_n are semi-simple.*

Proof. Suppose that an eigenvalue $\lambda \in \{\lambda_n\}$ is not semi-simple. Then $\ker(H - \lambda I)$ contains at least one isotropic vector f (Proposition 8.5). By Proposition 8.7, the vector $g = \mathcal{P}f$ is orthogonal to all eigenvectors $\{f_n\}$.

Therefore, $g = 0$ (since $\{f_n\}$ is complete; see Example 8.1 for the definition), which contradicts the existence of an isotropic vector f.

Example 8.5. *$\mathcal{P}$-Hermitian operators in the Kreĭn space $(\mathbb{C}^2, [\cdot, \cdot])$.* Let $(\mathbb{C}^2, [\cdot, \cdot])$ be the Kreĭn space defined in Example 8.4. According to (8.35), a matrix operator $H = \begin{bmatrix} a & b \\ c & d \end{bmatrix}$ is $\mathcal{P}$-Hermitian if

$$\begin{bmatrix} 0 & 1 \\ 1 & 0 \end{bmatrix} \begin{bmatrix} a & b \\ c & d \end{bmatrix} = \begin{bmatrix} a^* & c^* \\ b^* & d^* \end{bmatrix} \begin{bmatrix} 0 & 1 \\ 1 & 0 \end{bmatrix}.$$

The obtained identity holds if $a = d^*$ and b, c are real. The eigenvalues of a $\mathcal{P}$-Hermitian $H = \begin{bmatrix} a & b \\ c & a^* \end{bmatrix}$ are the roots of the polynomial

$$\det(H - \lambda I) = \lambda^2 - \lambda \mathrm{Tr}(H) + \det(H) = 0,$$

where $\mathrm{Tr}(H) = a + a^*$ and $\det(H) = |a|^2 - bc$ are real numbers. The eigenvalues are

$$\lambda_\pm = \tfrac{1}{2}\mathrm{Tr}(H) \pm \sqrt{\Delta} = \mathrm{Re}(a) \pm \sqrt{\Delta},$$

where $\Delta = [\mathrm{Tr}(H)]^2/4 - \det(H) = bc - [\mathrm{Im}(a)]^2$.

If $\Delta > 0$, then H has two real eigenvalues $\lambda_\pm$. The corresponding eigenvectors,

$$X_+ = [b, \ \sqrt{\Delta} - i\,\mathrm{Im}(a)]^t \text{ and } X_- = [-b, \ \sqrt{\Delta} + i\,\mathrm{Im}(a)]^t,$$

are $\mathcal{P}$-orthogonal and $[X_\pm, X_\pm] = \pm b\sqrt{\Delta}$. Note that $b \neq 0$. Indeed, if $b = 0$, then $\Delta = -[\mathrm{Im}(a)]^2$, which is impossible because $\Delta > 0$. Hence, $X_\pm$ are positive/negative vectors depending on the sign of b.

If $\Delta < 0$, then H has two nonreal eigenvalues $\lambda_\pm = \mathrm{Re}(a) \pm i\sqrt{|\Delta|}$. The corresponding eigenvectors $X_\pm$ are neutral. These vectors cannot be $\mathcal{P}$-orthogonal. For instance, if $b \neq 0$, then

$$X_+ = [b, \ i\sqrt{|\Delta|} - i\,\mathrm{Im}(a)]^t, \qquad X_- = [-b, \ i\sqrt{|\Delta|} + i\,\mathrm{Im}(a)]^t,$$

and $[X_+, X_-] = 2bi\sqrt{|\Delta|} \neq 0$.

If $\Delta = 0$, then H has a unique real eigenvalue $\lambda = \mathrm{Re}(a)$. The root subspace $\mathcal{L}_{\mathrm{Re}(a)}(H)$ coincides with $\mathbb{C}^2$. The eigenvalue $\mathrm{Re}(a)$ is semi-simple (that is, $\ker(H - \mathrm{Re}(a)I) = \mathcal{L}_{\mathrm{Re}(a)}(H)$) if and only if $H = aI$, where $a \in \mathbb{R}$. For other cases, the eigenvalue $\mathrm{Re}(a)$ has a Jordan chain of root vectors and $\ker[H - \mathrm{Re}(a)I]$ is degenerate (see Proposition 8.5).

8.3.6 *C-symmetries for P-Hermitian operators*

Let a $\mathcal{P}$-symmetric operator H have the property of $\mathcal{C}$-symmetry. Then, from (8.29) and (8.34),

$$(Hf, g)_{\mathcal{C}} = [\mathcal{C}Hf, g] = [H\mathcal{C}f, g] = [\mathcal{C}f, Hg] = (f, Hg)_{\mathcal{C}}, \qquad f, g \in \mathcal{D}(H).$$

Therefore, H is a symmetric operator in the Hilbert space $(\mathfrak{H}, (\cdot, \cdot)_{\mathcal{C}})$.

If H is $\mathcal{P}$-Hermitian, it is closed and, from Proposition 8.3, *the corresponding operator of $\mathcal{C}$-symmetry should be bounded.* This means that the inner product $(\cdot, \cdot)_{\mathcal{C}}$ is equivalent to $(\cdot, \cdot)$ and the space $\mathfrak{H}$ endowed with $(\cdot, \cdot)_{\mathcal{C}}$ is a Hilbert space.

From the definition of $\mathcal{P}$-adjoint operators, H is $\mathcal{P}$-Hermitian if and only if for any pair of elements g, γ satisfying (8.32), the vector g belongs to $\mathcal{D}(H)$ and $Hg = \gamma$. Given that H has $\mathcal{C}$-symmetry, the operator $\mathcal{C}$ maps $\mathcal{D}(H)$ onto itself.[7] Therefore, we use $\mathcal{C}f$ instead of f in the principal relation (8.32), and it can be rewritten in the form

$$(Hf, g)_{\mathcal{C}} = (f, \gamma)_{\mathcal{C}}, \qquad f \in \mathcal{D}(H) \tag{8.37}$$

because $[H\mathcal{C}f, g] = [\mathcal{C}Hf, g] = (Hf, g)_{\mathcal{C}}$ and $[\mathcal{C}f, \gamma] = (f, \gamma)_{\mathcal{C}}$. The relation (8.37) is crucial for determining the adjoint operator with respect to the inner product $(\cdot, \cdot)_{\mathcal{C}}$ [see (8.8)]. The $\mathcal{P}$-Hermitian operator H is simultaneously a Hermitian operator in the Hilbert space $(\mathfrak{H}, (\cdot, \cdot)_{\mathcal{C}})$. To summarize,

> *The description of $\mathcal{C}$-symmetry for a $\mathcal{P}$-Hermitian ($\mathcal{P}$-symmetric) operator H leads to the explicit construction of the new inner product $(\cdot, \cdot)_{\mathcal{C}}$ which ensures the Hermiticity (that is, the symmetry) of H.*

An immediate consequence of this statement is that the reality of eigenvalues of $\mathcal{P}$-Hermitian ($\mathcal{P}$-symmetric) operators is a necessary condition for the existence of $\mathcal{C}$-symmetry.

The next result is a direct consequence of Theorem 1.9 in [Albeverio and Kuzhel (2015)] and Proposition 8.3.

Theorem 8.2. *Let H be a $\mathcal{P}$-Hermitian operator. The following are equivalent:*

> (i) *H has the property of $\mathcal{C}$-symmetry;*
> (ii) *H is a Hermitian operator with respect to a new (equivalent to $(\cdot, \cdot)$) inner product in the Hilbert space $\mathfrak{H}$;*
> (iii) *H is similar[8] to a Hermitian operator in $(\mathfrak{H}, (\cdot, \cdot))$.*

---

[7]The mapping is onto because $\mathcal{C}^2 = I$.

[8]An operator H is similar to a Hermitian operator A if there exists a bounded and boundedly invertible operator Z such that $ZA = HZ$.

Item (*iii*) can be verified by using the integral-resolvent criterion of similarity (see [Naboko (1984)]). To be precise, a $\mathcal{P}$-Hermitian operator H is similar to a Hermitian one if and only if $\sigma(H) \subset \mathbb{R}$ and there exists a constant M such that

$$\sup_{\varepsilon > 0} \varepsilon \int_{-\infty}^{\infty} \|(H - zI)^{-1} f\|^2 d\xi \leq M \|f\|^2, \quad \forall f \in \mathfrak{H}, \tag{8.38}$$

where the integral is taken along the line $z = \xi + i\varepsilon$ ($\varepsilon > 0$ is fixed).

A $\mathcal{P}$-Hermitian operator H with (8.38) is guaranteed to have bounded $\mathcal{C}$-symmetry as a consequence of Theorem 8.2. However, neither result explains how to construct the corresponding operator $\mathcal{C}$.

Example 8.6. Let $(\mathbb{C}^2, [\cdot, \cdot])$ be the Kreĭn space defined in Example 8.4 and let H be the $\mathcal{P}$-Hermitian operator considered in Example 8.5. The operator H has two real eigenvalues in the case where $\Delta > 0$. Assume without loss of generality that $b > 0$. Then X_+ is a positive eigenvector of H and X_- is negative. The Kreĭn space $(\mathbb{C}^2, [\cdot, \cdot])$ has the decomposition

$$\mathbb{C}^2 = \mathfrak{L}_+[\dotplus]\mathfrak{L}_-, \text{ where } \mathfrak{L}_\pm = \text{span}\{X_\pm\}. \tag{8.39}$$

The $\mathcal{C}$ operator associated with the $\mathcal{P}$-orthogonal direct sum (8.39) commutes with H because $X_\pm$ are eigenvectors of H. Therefore, H has the property of $\mathcal{C}$-symmetry with the operator $\mathcal{C}$ determined by (8.39). Specifically, the operator $\mathcal{C} = [c_{ij}]$ satisfies $\mathcal{C}X_+ = X_+$ and $\mathcal{C}X_- = -X_-$. The explicit form of eigenvectors $X_\pm$ in Example 8.5 allows one to determine the entries c_{ij}:

$$c_{11} = i\frac{\text{Im}(a)}{b}, \quad c_{22} = -i\frac{\text{Im}(a)}{b}, \quad c_{12} = \frac{b}{\sqrt{\Delta}}, \quad c_{21} = \frac{c}{\sqrt{\Delta}}.$$

Comparing c_{ij} with the matrix representation of $\mathcal{C}$ obtained in Example 8.4, we conclude that the operator $\mathcal{C}$ is defined by (8.27), where the parameters ρ, ξ are determined by

$$\cosh \rho = \frac{b + c}{2\sqrt{\Delta}}, \quad \frac{b - c}{2} \cos \xi = -\text{Im}(a) \sin \xi.$$

The $\mathcal{P}$-Hermitian operator H is Hermitian with respect to the new inner product $(\cdot, \cdot)_\mathcal{C}$.

8.3.7   *Bounded $\mathcal{C}$ operators and Riccati equation*

Let H_0 be a $\mathcal{P}$-Hermitian operator in $\mathfrak{H}$ that commutes with the fundamental symmetry $\mathcal{P}$, $H_0\mathcal{P} = \mathcal{P}H_0$, and let V be a bounded $\mathcal{P}$-Hermitian operator in $\mathfrak{H}$ that anticommutes with $\mathcal{P}$, $\mathcal{P}V = -V\mathcal{P}$. If we impose the condition (8.35), then H_0 is Hermitian and V is skew-Hermitian (that is, $V^\dagger = -V$). Furthermore, the operator

$$H = H_0 + V$$

is $\mathcal{P}$-Hermitian.

The commutation of H_0 and the anticommutation of V with $\mathcal{P}$ allows one to rewrite H as an operator block matrix

$$H = \begin{bmatrix} H_+ & V_0 \\ V_1 & H_- \end{bmatrix}, \qquad \mathcal{D}(H) = \mathcal{D}(H_+) \oplus \mathcal{D}(H_-) \qquad (8.40)$$

with respect to the fundamental decomposition (8.13). Here, $H_\pm = P_\pm H_0 P_\pm$ are Hermitian operators in $\mathfrak{H}_\pm$ and $V_0 = P_+ V P_- : \mathfrak{H}_- \to \mathfrak{H}_+$, $V_1 = P_- V P_+ : \mathfrak{H}_+ \to \mathfrak{H}_-$ are bounded operators.

To a block operator matrix H of the form (8.40), we associate the operator Riccati equation

$$KH_+ - H_- K + KV_0 K = V_1, \qquad (8.41)$$

where K is a linear operator mapping $\mathfrak{H}_+$ to $\mathfrak{H}_-$. A bounded operator $K : \mathfrak{H}_+ \to \mathfrak{H}_-$ is called *a strong solution* of the operator Riccati equation (8.41) if $K : \mathcal{D}(H_+) \to \mathcal{D}(H_-)$ and

$$KH_+ f - H_- K f + KV_0 K f = V_1 f, \quad \text{for all} \quad f \in \mathcal{D}(H_+).$$

Theorem 8.3. *A $\mathcal{P}$-Hermitian operator $H = H_0 + V$ has the property of $\mathcal{C}$-symmetry if and only if the Riccati equation (8.41) has a strong solution K such that $\|K\| < 1$.*

Remark 8.4. The proof that a strong solution K implies the similarity of H to a Hermitian operator (this is equivalent to the existence of bounded $\mathcal{C}$-symmetry of H; see Theorem 8.2) is given in [Albeverio *et al.* (2009), Theorem 5.2]. The converse is established in [Grod *et al.* (2011), Theorem 3.4].

The bounded operator $\mathcal{C}$ in Theorem 8.3 is determined via the direct sum (8.15), which is characterized by the transition operator $T = KP_+ + K^\dagger P_-$. Here, the operator $K : \mathfrak{H}_+ \to \mathfrak{H}_-$ is the strong solution to (8.41) and $K^\dagger : \mathfrak{H}_- \to \mathfrak{H}_+$ is the strong solution of the adjoint Riccati equation

$$K^\dagger H_- - H_+ K^\dagger - K^\dagger V_0^\dagger K^\dagger = -V_1^\dagger.$$

Since $V^\dagger = -V$, the relation $V_1 = -V_0^\dagger$ holds and the adjoint Riccati equation takes the form

$$K^\dagger H_- - H_+ K^\dagger + K^\dagger V_1 K^\dagger = V_0. \tag{8.42}$$

Combining (8.41) and (8.42), we conclude that the transition operator $T = KP_+ + K^\dagger P_-$ is the strong solution[9] of the extended Riccati equation

$$TH_0 - H_0 T + TVT = V. \tag{8.43}$$

Corollary 8.1. *A $\mathcal{P}$-Hermitian operator $H = H_0 + V$ has the property of bounded $\mathcal{C}$-symmetry $\mathcal{C} = e^{\mathcal{Q}}\mathcal{P}$ if and only if $T = \tanh \mathcal{Q}/2$ is the strong solution of (8.43).*

The proof of this corollary follows from the above and (8.25). The unperturbed operator H_0 has the $\mathcal{C}$-symmetry $\mathcal{C} = \mathcal{P}$ because H_0 commutes with $\mathcal{P}$. One would expect that the perturbed $\mathcal{P}$-Hermitian operator H will preserve the property of $\mathcal{C}$-symmetry for sufficiently small V.

The next statement is the reformulation of Theorem 5.8 in [Albeverio *et al.* (2009)].

Theorem 8.4. *Let the spectra of the operators $H_\pm$ in (8.40) satisfy the condition $d = \mathrm{dist}(\sigma(H_+), \sigma(H_-)) > 0$ and let $\|V\| < \frac{d}{\pi}$. Then, the $\mathcal{P}$-Hermitian operator $H = H_0 + V$ has the property of $\mathcal{C}$-symmetry and the corresponding operator $\mathcal{C}$ is unique.*

Example 8.7. [Albeverio *et al.* (2009)] The one-dimensional quantum harmonic oscillator

$$H_0 = -\frac{1}{2}\frac{d^2}{dx^2} + \frac{1}{2}x^2$$

with the domain

$$\mathcal{D}(H_0) = \{f \in W_2^2(\mathbb{R}) : \int_{\mathbb{R}} dx\ x^4 |f(x)|^2 < \infty\},$$

where $W_2^2(\mathbb{R})$ denotes the Sobolev space of those $L_2(\mathbb{R})$-functions that have their second derivatives in $L_2(\mathbb{R})$, commutes with the space reflection operator $\mathcal{P}f(x) = f(-x)$. The operator H_0 is Hermitian in the Hilbert space $L_2(\mathbb{R})$ and is simultaneously $\mathcal{P}$-Hermitian in the Kreĭn space $(L_2(\mathbb{R}), (\cdot, \cdot)_{\mathcal{PT}})$. (See Example 8.2.)

The decomposition $H_0 = H_+ + H_-$ with respect to the fundamental decomposition (8.20) consists of the Hermitian operators $H_+ = H_0 \restriction_{L_2^{even}(\mathbb{R})}$

---

[9] *Strong solution of the extended Riccati equation means that $T : \mathcal{D}(H_0) \to \mathcal{D}(H_0)$ and $TH_0 f - H_0 T f + TVT f = Vf$ for $f \in \mathcal{D}(H_0)$.*

and $H_- = H_0 \restriction_{L_2^{odd}(\mathbb{R})}$ acting in the spaces $L_2^{even}(\mathbb{R})$ and $L_2^{odd}(\mathbb{R})$, respectively. The spectrum of H_0 coincides with the set of simple eigenvalues $\{n + 1/2 : n = 0, 1 \ldots\}$. The corresponding eigenvectors are even (odd) functions if n is even (odd). Therefore, $\sigma(H_+) = \{n + 1/2 : n = 0, 2, 4 \ldots\}$, $\sigma(H_-) = \{n + 1/2 : n = 1, 3, 5 \ldots\}$ and $d = \text{dist}(\sigma(H_+), \sigma(H_-)) = 1$.

Let $v(\cdot) \in L_\infty(\mathbb{R})$ be an odd real-valued function. Then the operator of multiplication $Vf = iv(x)f(x)$ is a bounded operator in $L_2(\mathbb{R})$ with the norm $\|V\| = \text{ess sup}_{x \in \mathbb{R}} |v(x)|$. By construction, the operator V anticommutes with $\mathcal{P}$ and is $\mathcal{PT}$ symmetric; that is, $\mathcal{PT}V = V\mathcal{PT}$. This relation and (8.1) implies that V is symmetric with respect to the $\mathcal{PT}$ inner product $(\cdot, \cdot)_{\mathcal{PT}}$ and hence is $\mathcal{P}$-Hermitian (because V is bounded). According to Theorem 8.4, the $\mathcal{P}$-Hermitian operator

$$H = -\frac{1}{2}\frac{d^2}{dx^2} + \frac{1}{2}x^2 + iv(x)$$

has a unique $\mathcal{C}$-symmetry when $\|V\| < 1/\pi$. The corresponding operator $\mathcal{C} = e^{\mathcal{Q}}\mathcal{P}$ is determined via the solution of the Riccati equation (8.43).

8.4 $\mathcal{P}$-Symmetric Operators with Complete Set of Eigenvectors

8.4.1   *Preliminaries: Best-choice problem*

Let a $\mathcal{P}$-symmetric operator H have a complete set of eigenvectors corresponding to the eigenvalues $\{\lambda_n\}$. We seek conditions that are sufficient to guarantee the existence of $\mathcal{C}$-symmetry for H.

As mentioned in Subsec. 8.3.6, the reality of eigenvalues λ_n is necessary for the existence of $\mathcal{C}$-symmetry. Furthermore, the assumed completeness of the eigenvectors implies that all eigenvalues λ_n are semi-simple (Proposition 8.8). Another requirement is the mutual J-orthogonality of the kernel subspaces $\ker(H - \lambda_n I)$ and $\ker(H - \lambda_m I)$ corresponding to the different eigenvectors (Proposition 8.7). It is important that the subspaces $\ker(H - \lambda_n I)$ be nondegenerate. Indeed, if one of $\ker(H - \lambda_n I)$ is degenerate, then there exists a nonzero element $f \in \ker(H - \lambda_n I)$ such that $[f, g] = 0$ for all $g \in \ker(H - \lambda_n I)$. The latter means that the vector $\mathcal{P}f$ is orthogonal [with respect to the inner product $(\cdot, \cdot)$] to the $\mathcal{P}$-orthogonal sum

$$S = \sum_{n=1}^{\infty} [\dot{+}] \ker(H - \lambda_n I).$$

Hence, $\mathcal{P}f = 0$ (since S is a dense set in $\mathfrak{H}$). This contradicts the assumption that $f \neq 0$.

For simplicity, in the following we assume that *all eigenvalues* $\{\lambda_n\}$ of H *are simple*. Let $\{f_n\}$ be the corresponding set of eigenvectors. Then the linear span of f_n coincides with the one-dimensional subspace $\ker(H - \lambda_n I)$. This means that $[f_n, f_n] \neq 0$ [because $\ker(H - \lambda_n I)$ is nondegenerate]. Separating the sequence of eigenvectors $\{f_n\}$ by the signs of $[f_n, f_n]$,

$$f_n = \begin{cases} f_n^+ & \text{if } [f_n, f_n] > 0, \\ f_n^- & \text{if } [f_n, f_n] < 0, \end{cases}$$

we obtain two sequences of positive $\{f_n^+\}$ and negative $\{f_n^-\}$ elements of the Kreĭn space $(\mathfrak{H}, [\cdot, \cdot])$.

Let $\mathfrak{L}_+^0$ and $\mathfrak{L}_-^0$ be the closure with respect to the initial inner product $(\cdot, \cdot)$ of the linear spans of $\{f_n^+\}$ and $\{f_n^-\}$, respectively. By construction, $\mathfrak{L}_\pm^0$ are $\mathcal{P}$-orthogonal definite subspaces in $(\mathfrak{H}, [\cdot, \cdot])$ and the direct sum

$$\mathcal{D}_0 = \mathfrak{L}_+^0 [\dotplus] \mathfrak{L}_-^0 \tag{8.44}$$

is a dense set in $\mathfrak{H}$ because $\mathcal{D}_0$ contains $\mathcal{S}$.

Let $\mathcal{C}_0$ be an operator associated with (8.44):

$$\mathcal{C}_0 f = f_{\mathfrak{L}_+^0} - f_{\mathfrak{L}_-^0}, \qquad f \in \mathcal{D}(\mathcal{C}_0) = \mathcal{D}_0. \tag{8.45}$$

It follows from (8.45) that $\mathcal{C}_0$ is densely defined and that $\mathcal{C}_0^2 f = f$ for any $f \in \mathcal{D}(\mathcal{C}_0)$. However, $\mathcal{C}_0$ may not be an operator of $\mathcal{C}$-symmetry defined in Subsec. 8.3.2 because the subspaces $\mathfrak{L}_\pm^0$ in (8.44) may be *proper subspaces of maximal definite subspaces* $\mathfrak{L}_\pm$.

If at least one of the subspaces $\mathfrak{L}_\pm^0$ loses the property of being maximal in the classes of definite subspaces (positive or negative), then the sum (8.44) *cannot be the domain of definition* of an operator of $\mathcal{C}$-symmetry with the properties described in Theorem 8.1. In other words, the operator $\mathcal{C}_0$ cannot have the form $e^{\mathcal{Q}}\mathcal{P}$, where $e^{\mathcal{Q}}$ is a *Hermitian operator*. Of course, this phenomenon is possible only in the case of unbounded operators.

In general, the subspaces $\mathfrak{L}_\pm^0$ in (8.44) can be extended to *different pairs of $\mathcal{P}$-orthogonal maximal definite subspaces*[10] $\mathfrak{L}_\pm$, which implies the nonuniqueness of the operators of $\mathcal{C}$-symmetry $\mathcal{C} \supset \mathcal{C}_0$ associated with (8.44). This suggests the natural question, initially posed in [Bender and Klevansky (2009)], regarding the *best choice* of unbounded $\mathcal{C}$-operator. This question can be answered easily for the particular cases considered below.

---

[10]This fact was discovered in [Langer (1970a,b)].

8.4.2   *Riesz basis of eigenvectors*

A sequence $\{f_n\}$ is called a *Riesz basis* of a Hilbert space $\mathfrak{H}$ if there exists a bounded operator R with bounded inverse R^{-1} and an orthonormal basis $\{\psi_n\}$ in $\mathfrak{H}$ such that $f_n = R\psi_n$. Consider the following theorem [Mostafazadeh (2010); Bender and Kuzhel (2012)].

Theorem 8.5. *Assume that H is a $\mathcal{P}$-symmetric operator with a complete set of normalized eigenvectors $\{f_n : \|[f_n, f_n]\| = 1\}$ corresponding to real simple eigenvalues $\{\lambda_n\}$ and the domain of definition of H coincides with the linear span of $\{f_n\}$. Then, the following are equivalent:*

   (i) *the set of eigenvectors $\{f_n\}$ is a Riesz basis in the Hilbert space $(\mathfrak{H}, (\cdot, \cdot))$;*

   (ii) *the operator H has the property of bounded $\mathcal{C}$-symmetry, the corresponding bounded operator $\mathcal{C}$ is determined uniquely, and for any $f \in \mathfrak{H}$*

$$\mathcal{C}f = \sum_{n=1}^{\infty} [f_n, f]f_n. \tag{8.46}$$

Proof. $(i) \to (ii)$. If $\{f_n\}$ is a Riesz basis, then the direct sum (8.44) gives the decomposition of the whole space $\mathfrak{H}$. In this case the $\mathcal{P}$-orthogonal subspaces $\mathfrak{L}_{\pm}^0$ are maximal uniformly definite in the Kreĭn space $(\mathfrak{H}, [\cdot, \cdot])$. Therefore, the decomposition (8.44) coincides with (8.28) and it determines a bounded operator $\mathcal{C}$ that commutes with H. The subspaces $\mathfrak{L}_{\pm}^0$ are uniquely determined by the sequences of positive $\{f_n^+\}$ and negative $\{f_n^-\}$ eigenvectors. This guarantees the uniqueness of the operator $\mathcal{C}$, which acts as $\mathcal{C}f_n^+ = f_n^+$, $\mathcal{C}f_n^- = -f_n^-$ or, more compactly, as

$$\mathcal{C}f_n = [f_n, f_n]f_n. \tag{8.47}$$

The sequence of eigenvectors $\{f_n\}$ forms an orthonormal basis in $(\mathfrak{H}, (\cdot, \cdot)_{\mathcal{C}})$, where the new inner product $(\cdot, \cdot)_{\mathcal{C}}$ in $\mathfrak{H}$ is defined by (8.29). Therefore, any $f \in \mathfrak{H}$ has the decomposition $f = \sum_{n=1}^{\infty}(f_n, f)_{\mathcal{C}}f_n$ and

$$\mathcal{C}f = \sum_{n=1}^{\infty}(f_n, f)_{\mathcal{C}}\mathcal{C}f_n = \sum_{n=1}^{\infty}[\mathcal{C}f_n, f][f_n, f_n]f_n = \sum_{n=1}^{\infty}[f_n, f]f_n$$

because $[\mathcal{C}f_n, f][f_n, f_n] = [f_n, f]([f_n, f_n])^2 = [f_n, f]$.

   $(ii) \to (i)$. Let $\mathcal{C}$ be the operator of bounded $\mathcal{C}$-symmetry for H and let $\mathcal{C}$ be defined by (8.46). Then (8.47) holds. This means that the normalized eigenvectors $\{f_n\}$ form an orthonormal system in the Hilbert space

$(\mathfrak{H}, (\cdot, \cdot)_{\mathcal{C}})$. The completeness of $\{f_n\}$ with respect to $(\cdot, \cdot)$ yields the completeness of $\{f_n\}$ in $(\mathfrak{H}, (\cdot, \cdot)_{\mathcal{C}})$. Hence, $\{f_n\}$ is an orthonormal basis of $(\mathfrak{H}, (\cdot, \cdot)_{\mathcal{C}})$.

By virtue of Theorem 8.1, the bounded operator $\mathcal{C}$ can be written as $\mathcal{C} = e^{\mathcal{Q}}\mathcal{P}$, where $\mathcal{Q}$ is a bounded self-adjoint operator, which anticommutes with $\mathcal{P}$. Denote $\psi_n = e^{-\mathcal{Q}/2}f_n$. Then

$$(\psi_n, \psi_m) = (e^{-\mathcal{Q}/2}f_n, e^{-\mathcal{Q}/2}f_m) = (e^{-\mathcal{Q}}f_n, f_m) = [\mathcal{C}f_n, f_m] = (f_n, f_m)_{\mathcal{C}}.$$

Therefore, the sequence $\{\psi_n\}$ is an orthonormal basis of $(\mathfrak{H}, (\cdot, \cdot))$ and $f_n = R\psi_n$, where $R = e^{\mathcal{Q}/2}$. Hence, $\{f_n\}$ is the Riesz basis of the initial Hilbert space $(\mathfrak{H}, (\cdot, \cdot))$.

Corollary 8.2. *Assume that H satisfies the conditions of Theorem 8.5 and has bounded $\mathcal{C}$-symmetry. Then H is essentially Hermitian in the Hilbert space $(\mathfrak{H}, (\cdot, \cdot)_{\mathcal{C}})$ and its spectrum coincides with the closure of $\{\lambda_n\}$.*

Proof. If H has a bounded $\mathcal{C}$-symmetry, then the new inner product $(\cdot, \cdot)_{\mathcal{C}}$ is equivalent to the original one. The normalized eigenvectors $\{f_n\}$ of H are an orthonormal basis in $(\mathfrak{H}, (\cdot, \cdot)_{\mathcal{C}})$. It follows from Subsec. 8.3.6 that H is a symmetric operator in $(\mathfrak{H}, (\cdot, \cdot)_{\mathcal{C}})$. Moreover, taking into account that the domain $\mathcal{D}(H)$ of H is a linear span of $\{f_n\}$, we conclude that $\mathcal{R}(H \pm iI)$ are dense sets in $(\mathfrak{H}, (\cdot, \cdot)_{\mathcal{C}})$ because $\mathcal{R}(H \pm iI) = \mathcal{D}(H)$. This means that [see item (iii) in Lemma 8.1] H is essentially Hermitian with respect to $(\cdot, \cdot)_{\mathcal{C}}$ and $\sigma(H)$ coincides with the closure of $\{\lambda_n\}$.

8.4.3 *Schauder basis of eigenvectors*

A sequence $\{f_n\}$ is called *a Schauder basis* of a Hilbert space $\mathfrak{H}$ if for each $f \in \mathfrak{H}$, there exist uniquely determined scalar coefficients $\{c_n\}$ such that $f = \sum_{n=1}^{\infty} c_n f_n$. The coefficients $\{c_n\}$ can be specified easily by using a biorthogonal sequence[11] $\{g_n\}$. The sequence $\{g_n\}$ is defined uniquely (because $\{f_n\}$ is a complete set) and $c_n = (g_n, f)$. Therefore,

$$f = \sum_{n=1}^{\infty} (g_n, f)f_n, \qquad \forall f \in \mathfrak{H}. \tag{8.48}$$

Theorem 8.6. *Assume that H satisfies conditions of Theorem 8.5 and the set of its eigenvectors $\{f_n\}$ is a Schauder basis (not a Riesz basis) in the Hilbert space $(\mathfrak{H}, (\cdot, \cdot))$. Then, the operator H has the property of unbounded $\mathcal{C}$-symmetry, the corresponding unbounded operator $\mathcal{C}$ is determined*

---

[11]Biorthonormality of sequences $\{f_n\}$ and $\{g_n\}$ means that $(f_n, g_m) = \delta_{nm}$.

uniquely, and for any $f \in \mathcal{D}(\mathcal{C}) = \mathcal{D}_0$ *(8.46) holds. The operator H is essentially Hermitian in the new Hilbert space* $(\mathfrak{H}_{\mathcal{C}}, (\cdot, \cdot)_{\mathcal{C}})$ *and its spectrum coincides with the closure of* $\{\lambda_n\}$.

Proof. If $\{f_n\}$ is a Schauder basis (but not a Riesz basis), then the subspaces $\mathfrak{L}^0_\pm$ in (8.44) are maximal definite (but not uniformly definite) in the Kreĭn space $(\mathfrak{H}, [\cdot, \cdot])$ [Azizov and Iokhvidov (1998), Statement 10.12] and the decomposition (8.44) determines an unbounded operator $\mathcal{C}$ that commutes with H and satisfies (8.47). Uniqueness of $\mathcal{C}$ follows from the fact that the subspaces $\mathfrak{L}^0_\pm$ are uniquely determined by $\{f_n\}$.

Since the eigenvectors $\{f_n\}$ are mutually $\mathcal{P}$-orthogonal and normalized, the biorthogonal sequence $\{g_n\}$ associated with $\{f_n\}$ has the form $g_n = [f_n, f_n]\mathcal{P}f_n$. Therefore, (8.48) can be rewritten as

$$f = \sum_{n=1}^{\infty} [f_n, f_n][f_n, f]f_n, \qquad \forall f \in \mathfrak{H}. \tag{8.49}$$

Let $f \in \mathcal{D}(\mathcal{C}) = \mathcal{D}_0$. Substituting $\mathcal{C}f$ instead of f into (8.49) and taking (8.47) into account, we obtain

$$\mathcal{C}f = \sum_{n=1}^{\infty} [f_n, f_n][f_n, \mathcal{C}f]f_n = \sum_{n=1}^{\infty} [f_n, f_n][\mathcal{C}f_n, f]f_n = \sum_{n=1}^{\infty} [f_n, f]f_n.$$

Let $\mathfrak{H}_{\mathcal{C}}$ be the completion of $\mathcal{D}(\mathcal{C})$ with respect to $(\cdot, \cdot)_{\mathcal{C}}$ (Subsec. 8.3.3). It follows from (8.29) and (8.47) that the normalized eigenvectors $\{f_n\}$ form an orthonormal system in the Hilbert space $(\mathfrak{H}_{\mathcal{C}}, (\cdot, \cdot)_{\mathcal{C}})$.

The inner product $(\cdot, \cdot)_{\mathcal{C}}$ considered on $\mathfrak{L}^0_\pm$ coincides with $\pm[\cdot, \cdot]$ and satisfies the evaluation

$$(f_{\mathfrak{L}^0_\pm}, f_{\mathfrak{L}^0_\pm})_{\mathcal{C}} = |(\mathcal{P}f_{\mathfrak{L}^0_\pm}, f_{\mathfrak{L}^0_\pm})| \leq \|\mathcal{P}f_{\mathfrak{L}^0_\pm}\| \|f_{\mathfrak{L}^0_\pm}\| = \|f_{\mathfrak{L}^0_\pm}\|^2 = (f_{\mathfrak{L}^0_\pm}, f_{\mathfrak{L}^0_\pm})$$

for any $f_{\mathfrak{L}^0_\pm} \in \mathfrak{L}^0_\pm$. Therefore, each element $f_{\mathfrak{L}^0_\pm} \in \mathfrak{L}^0_\pm$ can be approximated by $\mathrm{span}\{f_n^\pm\}$ with respect to $(\cdot, \cdot)_{\mathcal{C}}$ since $\mathfrak{L}^0_\pm$ are the closure of $\mathrm{span}\{f_n^\pm\}$ with respect to $(\cdot, \cdot)$. Thus, the sets $\mathrm{span}\{f_n^\pm\}$ are dense in the subspaces $\hat{\mathfrak{L}}_\pm$ of the decomposition (8.30). Hence, the set $\{f_n\} = \{f_n^+\} \cup \{f_n^-\}$ of eigenvectors of H forms an orthonormal basis in $(\mathfrak{H}_{\mathcal{C}}, (\cdot, \cdot)_{\mathcal{C}})$. Arguing similarly to the proof of Corollary 8.2, we deduce that H is essentially Hermitian in $(\mathfrak{H}_{\mathcal{C}}, (\cdot, \cdot)_{\mathcal{C}})$ and that its spectrum coincides with the closure of $\{\lambda_n\}$.

8.4.4 *Complete set of eigenvectors and quasi basis*

In this subsection we assume that the complete set of eigenvectors $\{f_n\}$ does not form a basis of $(\mathfrak{H}, (\cdot, \cdot))$. Due to Theorem 8.5, this implies that operators of $\mathcal{C}$-symmetry associated with various (possible) extensions of (8.44) to $\mathcal{P}$-orthogonal direct sums of maximal definite subspaces $\mathfrak{L}_\pm$ are unbounded. To describe all such extensions we introduce the operator $G_0 = \mathcal{P}\mathcal{C}_0$, where $\mathcal{C}_0$ is defined by (8.45). It follows from Lemma 6.2.3 in [Albeverio and Kuzhel (2015)] that $\mathcal{C}_0$ is a closed operator in $\mathfrak{H}$. Hence, the operator G_0 is also closed. Furthermore, $\mathcal{P}G_0\mathcal{P}G_0 = I$ on $\mathcal{D}(\mathcal{C}_0)$ because $\mathcal{C}_0^2 f = f$ for all $f \in \mathcal{D}(\mathcal{C}_0)$. In view of (8.12) and (8.45),

$$(G_0 f, f) = [\mathcal{C}_0 f, f] = [f_{\mathfrak{L}_+^0}, f_{\mathfrak{L}_+^0}] - [f_{\mathfrak{L}_-^0}, f_{\mathfrak{L}_-^0}] > 0 \qquad (8.50)$$

for all nonzero $f \in \mathcal{D}(\mathcal{C}_0)$. Therefore, G_0 is a positive symmetric operator in $(\mathfrak{H}, (\cdot, \cdot))$.

Lemma 8.3. *The set of of all positive Hermitian extensions G of G_0 satisfying the additional 'boundary condition'*

$$\mathcal{P}G\mathcal{P}Gf = f, \qquad \forall f \in \mathcal{D}(G) \qquad (8.51)$$

is in one-to-one correspondence with the set of all possible extensions of the subspaces $\mathfrak{L}_\pm^0$ in (8.44) to $\mathcal{P}$-orthogonal sums (8.15), where $\mathfrak{L}_\pm$ are maximal definite subspaces in the Kreĭn space $(\mathfrak{H}, [\cdot, \cdot])$.

Proof. Let G be a positive Hermitian extension of G_0 and $\mathcal{P}G\mathcal{P}G = I$ on $\mathcal{D}(G)$. Then the operator $\mathcal{C} = \mathcal{P}G$ is an extension of $\mathcal{C}_0$ and it satisfies the conditions of item (iii) in Theorem 8.1. Therefore, $\mathcal{C}$ is an operator of $\mathcal{C}$-symmetry and it determines the maximal definite subspaces $\mathfrak{L}_\pm$ in (8.15) by the formula (8.23). The subspaces $\mathfrak{L}_\pm$ are extensions of $\mathfrak{L}_\pm^0$ since $\mathcal{C} \supset \mathcal{C}_0$.

Conversely, each extension of (8.44) to the $\mathcal{P}$-orthogonal direct sum (8.15) of maximal definite subspaces $\mathfrak{L}_\pm$ determines by (8.22) an operator of $\mathcal{C}$-symmetry $\mathcal{C}$ associated with (8.15). This operator is an extension of $\mathcal{C}_0$, and hence $G = \mathcal{P}\mathcal{C}$ is an extension of $G_0 = \mathcal{P}\mathcal{C}_0$. Using Theorem 8.1 again, we find that G is a positive Hermitian operator and that (8.51) holds, which completes the proof.

Lemma 8.3 reduces the description of all possible operators of $\mathcal{C}$-symmetry associated with the original $\mathcal{P}$-orthogonal sum (8.44) to the description of positive Hermitian extensions G of G_0 satisfying the additional condition (8.51). Each operator of $\mathcal{C}$-symmetry $\mathcal{C} = \mathcal{P}G$ constructed this way is an extension of $\mathcal{C}_0$ and its choice determines the new Hilbert space $(\mathfrak{H}_\mathcal{C}, (\cdot, \cdot)_\mathcal{C})$, where $\mathfrak{H}_\mathcal{C}$ is the completion of $\mathcal{D}(\mathcal{C})$ with respect to $(\cdot, \cdot)_\mathcal{C}$.

Sometimes, $\mathcal{C}_0$ can be extended to $\mathcal{C}$ by continuity in the Hilbert space $(\mathfrak{H}_\mathcal{C}, (\cdot, \cdot)_\mathcal{C})$. The following cases are possible [Kamuda *et al.* (2018)]:

(A) the extension by continuity of C_0 to an operator of C symmetry is unique. There are no other extensions;

(B) there are infinitely many extensions by continuity of C_0 to operators of C-symmetry in the corresponding Hilbert spaces $(\mathfrak{H}_C, (\cdot, \cdot)_C)$. Simultaneously, there are infinitely many extensions $C \supset C_0$ that generate Hilbert spaces $\mathfrak{H}_C$ containing the closure of $\mathcal{D}(C_0)$ as a proper subspace;

(C) there is no extension by continuity of C_0 to an operator of C-symmetry. The total number of possible extensions $C \supset C_0$ is not specified. (A unique extension is possible as well as infinitely many.)

The case (A) holds when the Friedrichs extension G_μ of G_0 satisfies (8.51).[12] This case is uniquely characterized by the condition [Kuzhel and Sudilovskaja (2017), Corollary 4.4]

$$\inf_{f \in D(G_0)} \frac{(G_0 f, f)}{|(f, g)|^2} = 0$$

for all nonzero vectors $g \in \ker(I + G_0^\dagger)$.

The case (B) deals with the existence of the so-called *extremal extensions* G of G_0 satisfying (8.51) [Kamuda *et al.* (2018), Theorem 4.2]. The Hilbert spaces $\mathfrak{H}_C$ in which C_0 can be extended to C by continuity are related by a unitary operator W such that $Wf = f$ for $f \in \mathcal{D}(G_0)$. For this reason, we can identify them.

Each constructed operator $C \supset C_0$ is an operator of C-symmetry for the $\mathcal{P}$-symmetric operator H because $\mathcal{D}(H) = \text{span}\{f_n\}$. Thus, $CHf = HCf$ for all $f \in \mathcal{D}(H)$.

A complete set of normalized eigenvectors $\{f_n\}$ of the $\mathcal{P}$-symmetric operator H is called *a quasi basis* if there exists an operator of C symmetry $C \supset C_0$ such that $\text{span}\{f_n\}$ remains dense in the new Hilbert space $(\mathfrak{H}_C, (\cdot, \cdot)_C)$.

Proposition 8.9. *If $\{f_n\}$ is a quasi basis, then there exists an operator of C-symmetry such that the $\mathcal{P}$-symmetric operator H is essentially Hermitian in the Hilbert space $(\mathfrak{H}_C, (\cdot, \cdot)_C)$.*

Proof. The operator H is symmetric in $(\mathfrak{H}_C, (\cdot, \cdot)_C)$ and it is essentially Hermitian if the sets $\mathcal{R}(H \pm iI)$ are dense in $\mathfrak{H}_C$. The last is true since

---

[12]See [Arlinskiĭ and Tsekanovskiĭ (2009)] for the properties of Friedrichs extensions.

$\mathcal{R}(H \pm iI) = \mathcal{D}(H) = \mathrm{span}\{f_n\}$ and $\{f_n\}$ is a quasi basis. This proves Proposition 8.9.

The sequence $\{f_n\}$ is a quasi basis if and only if there exists a self-adjoint operator $\mathcal{Q}$ in the Hilbert space $\mathfrak{H}$, which anticommutes with $\mathcal{P}$ and such that the sequence $\{g_n = e^{-\mathcal{Q}/2} f_n\}$ is an orthonormal basis of $\mathfrak{H}$ and each g_n belongs to one of the subspaces $\mathfrak{H}_\pm$ of the fundamental decomposition (8.13) [Kamuda *et al.* (2018), Theorem 6.3]. This relationship allows one to construct examples of quasi bases. Indeed, assume that g_n is an orthonormal basis of $(\mathfrak{H}, (\cdot, \cdot))$ such that each g_n belongs to one of the subspaces $\mathfrak{H}_\pm$ of (8.13). Let $\mathcal{Q}$ be a self-adjoint operator in $\mathfrak{H}$, which anticommutes with $\mathcal{P}$. If all g_n belong to $\mathcal{D}(e^{\mathcal{Q}/2})$, then $f_n = e^{\mathcal{Q}/2} g_n$ is a $\mathcal{P}$-orthonormal system of the Kreĭn space $(\mathfrak{H}, [\cdot, \cdot])$. If $\{f_n\}$ is complete in $\mathfrak{H}$, then $\{f_n\}$ is a quasi basis.

Example 8.8. *The shifted harmonic oscillator.* The Hermite functions

$$g_n(x) = \frac{1}{\sqrt{2^n n! \sqrt{\pi}}} H_n(x) e^{-x^2/2}, \quad H_n(x) = e^{x^2/2} (x - \frac{d}{dx})^n e^{-x^2/2}$$

form an orthonormal basis of $L_2(\mathbb{R})$. The functions g_n belong to one of the subspaces $L_2^{even}(\mathbb{R})$ or $L_2^{odd}(\mathbb{R})$ of the fundamental decomposition (8.20). Since g_n are entire functions, the complex shift can be defined:

$$f_n(x) = g_n(x + i\varepsilon), \quad \varepsilon \in \mathbb{R} \setminus \{0\}, \quad n = 0, 1, 2, \ldots$$

The sequence $\{f_n\}$ is complete in $L_2(\mathbb{R})$. Applying the Fourier transform $Ff = \frac{1}{\sqrt{2\pi}} \int_{-\infty}^{\infty} e^{-ix\xi} f(x) dx$ to f_n, we get $Ff_n = e^{-\varepsilon\xi} Fg_n$. Therefore, $f_n = F^{-1} e^{-\varepsilon\xi} Fg_n$. The last relation can be rewritten as $f_n = e^{\mathcal{Q}/2} g_n$, where $\mathcal{Q} = 2i\varepsilon \frac{d}{dx}$ anticommutes with $\mathcal{P}$. Therefore, $\{f_n\}$ is a quasi basis of $L_2(\mathbb{R})$. The functions $\{f_n\}$ are simple eigenfunctions of the $\mathcal{P}$-symmetric operator $H = -\frac{1}{2} \frac{d^2}{dx^2} + \frac{1}{2} x^2 + i\varepsilon x$.

Remark 8.5. In [Kretschmer and Szymanowski (2004)] it was proposed to define an appropriate (physical) Hilbert space $\hat{\mathfrak{H}}$ for H without an explicit construction of the operator of $\mathcal{C}$-symmetry (or the metric operator). The idea was to identify the inner product on $\mathcal{D}(H)$ with the one making the eigenvectors $\{f_n\}$ of H orthonormal. The natural candidate for this role is the sesquilinear form $[\mathcal{C}_0 \cdot, \cdot]$ in (8.50). In fact, it was believed that each complete $\mathcal{P}$-orthonormal sequence of eigenvectors $\{f_n\}$ is a quasi basis.

8.5   Property of $\mathcal{PT}$ Symmetry

8.5.1   *$\mathcal{PT}$-symmetric operators*

I. A bounded operator $\mathcal{T}$ acting in a Hilbert space $(\mathfrak{H}, (\cdot, \cdot))$ is called *conjugation* if

$$(i) \quad \mathcal{T}^2 = I, \qquad (ii) \quad (\mathcal{T}f, \mathcal{T}g) = (g, f), \quad f, g \in \mathfrak{H}.$$

Each conjugation operator is *antilinear* in the sense that

$$\mathcal{T}(\alpha f + \beta g) = \alpha^* \mathcal{T}f + \beta^* \mathcal{T}g, \qquad \alpha, \beta \in \mathbb{C}, \quad f, g \in \mathfrak{H}.$$

The time-reversal operator $\mathcal{T}f = f^*$ is an example of a conjugation operator in $L_2(\mathbb{R})$. Let $\mathcal{P}$ be a fundamental symmetry and let $\mathcal{T}$ be a conjugation operator in $(\mathfrak{H}, (\cdot, \cdot))$. In the following we assume that $\mathcal{P}$ and $\mathcal{T}$ commute: $\mathcal{PT} = \mathcal{TP}$.

Definition 8.2. A densely defined linear operator H in $\mathfrak{H}$ is called $\mathcal{PT}$ *symmetric* if

$$\mathcal{PT}Hf = H\mathcal{PT}f \quad \text{for all} \quad f \in \mathcal{D}(H). \tag{8.52}$$

It follows from (8.52) that the point σ_p, the residual σ_r, and the continuous σ_c parts of the spectrum of a $\mathcal{PT}$-symmetric operator H are symmetric with respect to the real axis:

$$\lambda \in \sigma_\alpha(H) \iff \lambda^* \in \sigma_\alpha(H), \qquad \alpha \in \{p, r, c\}.$$

This property illustrates the difference between $\mathcal{PT}$-symmetric and $\mathcal{P}$-Hermitian operators: In the latter case, the whole spectrum is symmetric with respect to the real line, but the symmetry between particular parts $\sigma_\alpha(H)$ may be violated.

The definition of $\mathcal{PT}$ symmetry is quite general:[13]

> *Each Hermitian operator in a separable Hilbert space $\mathfrak{H}$ is $\mathcal{PT}$ symmetric for a suitable choice of $\mathcal{P}$ and $\mathcal{T}$.*

Usually, $\mathcal{PT}$-symmetric operators are symmetric with respect to an appropriately chosen indefinite inner product.

Example 8.9. Let the eigenvalue problem

$$\tau(f)(x) = -f''(x) + v(x)f(x) = \lambda f, \qquad x \in \mathbb{R}, \quad \lambda \in \mathbb{C}$$

---

[13][Albeverio and Kuzhel (2015), Proposition 6.4.3].

have the solutions $f_n \in L_2(\mathbb{R})$, which are uniquely determined up to a constant multiple and which correspond to the eigenvalues $\lambda_n, n \in \mathbb{N}$. We also assume that the set of eigenfunctions $\{f_n\}$ is complete in $L_2(\mathbb{R})$.

Let $v(x) = v^*(-x)$. Then the operator

$$Hf = \tau(f)(x), \qquad f \in \mathcal{D}(H) = \mathrm{span}\{f_n\}$$

is $\mathcal{PT}$ symmetric and its spectrum contains simple eigenvalues $\{\lambda_n\}$. The $\mathcal{PT}$-symmetric operator H is symmetric in the Kreĭn space $(L_2(\mathbb{R}), (\cdot, \cdot)_{\mathcal{PT}})$. Indeed, due to (8.1),

$$(Hf, g)_{\mathcal{PT}} = \int_{\mathbb{R}} dx [\mathcal{PT} H f(x)] g(x) = \int_{\mathbb{R}} dx [H \mathcal{PT} f(x)] g(x)$$
$$= \int_{\mathbb{R}} dx [\mathcal{PT} f(x)] H g(x) = (f, Hg)_{\mathcal{PT}}.$$

Therefore, H is a $\mathcal{P}$-symmetric operator.

Assume that the $\mathcal{PT}$-symmetry of H is unbroken.[14] Then all eigenvalues $\{\lambda_n\}$ should be real [Bender (2005)], and the corresponding eigenfunctions $\{f_n\}$ determine the direct sum (8.44) and the operator C_0 (8.45). Let us assume that the set of eigenfunctions $\{f_n\}$ is a quasi basis. Then there exists an operator of C-symmetry $C \supset C_0$ such that $\{f_n\}$ is an orthonormal basis in the Hilbert space $(\mathfrak{H}_C, (\cdot, \cdot)_C)$. Here, $\mathfrak{H}_C$ is the completion of $\mathrm{span}\{f_n\}$ with respect to the CPT inner product $(\cdot, \cdot)_C$ (8.3).

According to Proposition 8.9, the operator H is essentially Hermitian in the Hilbert space $\mathfrak{H}_C$. The operator H is also essentially Hermitian in the Kreĭn space $(\mathfrak{H}_C, [\cdot, \cdot])$, where the indefinite metric $[\cdot, \cdot]$ is defined according to (8.12): $[f, g] = (Cf, g)_C$, $f, g \in \mathfrak{H}_C$.

Note that $\mathcal{PT}$-symmetric operators may have curious spectral properties. For example, cases where the spectrum of a $\mathcal{PT}$-symmetric operator H coincides with $\mathbb{C}$ or the residual spectrum $\sigma_r(H)$ is not empty are possible.

Example 8.10. [Azizov and Trunk (2012)] The differential expression

$$\tau(f)(x) = -f''(x) - x^{4n+4} f(x), \qquad x \in \mathbb{R}, \quad n \in \mathbb{N} \cup \{0\}$$

is in the limit-circle case[15] at $+\infty$ and $-\infty$. The operators

$$H_1 f = \tau(f), \qquad \mathcal{D}(H_1) = \{f \in W_2^2(\mathbb{R}) : \tau(f) \in L_2(\mathbb{R})\},$$

---

[14] We say that the $\mathcal{PT}$ symmetry of H is *unbroken* if all eigenfunctions $\{f_n\}$ of H are simultaneously eigenfunctions of $\mathcal{PT}$.

[15] The differential expression $\tau(f)$ is in the limit-circle case at $+\infty$ (at $-\infty$) if all solutions of the equation $\tau(f) - \lambda f = 0$, $\lambda \in \mathbb{C}$, are in $L_2(a, +\infty)$ ($L_2(-\infty, a)$) respectively for some $a \in \mathbb{R}$.

$$H_2 f = \tau(f), \qquad \mathcal{D}(H_2) = \{f \in W_2^2(\mathbb{R}) : \ f \text{ has compact support}\}$$

are both $\mathcal{PT}$ symmetric in $L_2(\mathbb{R})$ and $\sigma(H_1) = \sigma(H_2) = \mathbb{C}$. The operator H_1 is closed and its point spectrum $\sigma_p(H_1)$ coincides with $\mathbb{C}$. The operator H_2 is closable in $L_2(\mathbb{R})$, its closure $\overline{H_2}$ coincides with $H_1^\dagger$, and the nonreal part $\mathbb{C} \setminus \mathbb{R}$ of the complex plane belongs to the residual spectrum $\sigma_r(H_2)$ of H_2.

It is easy to check that

$$(H_j f, g)_{\mathcal{PT}} = (f, H_j g)_{\mathcal{PT}}, \qquad f, g \in \mathcal{D}(H_j),$$

where the $\mathcal{PT}$ inner product $(\cdot, \cdot)_{\mathcal{PT}}$ is defined by (8.1). Therefore, the operators H_j are $\mathcal{P}$-symmetric [that is, symmetric in the Kreĭn space $(L_2(\mathbb{R}), (\cdot, \cdot)_{\mathcal{PT}})$; see Example 8.2]. However, neither H_1 nor H_2 are $\mathcal{P}$-Hermitian or essentially $\mathcal{P}$-Hermitian operators.

To avoid such marginal cases, it seems natural to suppose that

> *Every physically meaningful $\mathcal{PT}$-symmetric operator H should be Hermitian (or essentially Hermitian) in a suitable chosen Kreĭn space $(\mathfrak{H}, [\cdot, \cdot])$.*

This condition guarantees the absence of the residual spectrum for $\mathcal{PT}$-symmetric operators [Albeverio and Kuzhel (2015), Proposition 1.20].

Remark 8.6. $\mathcal{P}$-Hermitian operators associated with the differential expression $\tau(\cdot)$ in Example 8.10 can be defined with the use of the extension theory of symmetric operators since H_2 is a symmetric operator in $L_2(\mathbb{R})$ and $H_2^\dagger = (H_1^\dagger)^\dagger = H_1$. The required $\mathcal{P}$-Hermitian operators are constructed as intermediate extensions between the 'smaller' H_2 and the 'bigger' H_1 $\mathcal{PT}$-symmetric operators. This approach gives infinitely many $\mathcal{P}$-Hermitian extensions of H_2 that are determined as the restrictions of H_1 onto the subsets of functions $f \in \mathcal{D}(H_1)$ satisfying special boundary conditions at $\pm\infty$. The obtained $\mathcal{P}$-Hermitian operators are defined on $\mathbb{R}$ and extra conditions are required for the determination of the unique $\mathcal{P}$-Hermitian operator associated with $\tau(\cdot)$.

Another approach deals with the analytic continuation of $\tau(\cdot)$ onto the complex plane:

$$\tau(f)(z) = -f''(z) - z^{4n+4} f(z), \qquad z \in \mathbb{C}, \quad n \in \mathbb{N} \cup \{0\}.$$

The idea is to consider the eigenvalue differential equation

$$\tau(f)(z) = \lambda f(z) \tag{8.53}$$

on a contour Γ in $\mathbb{C}$, which is placed symmetrically with respect to the imaginary axis (see Example 8.3), and to seek solutions $f(z)$ such that

$$|f(z)| \to 0 \quad \text{as } z \text{ moves off to infinity along } \Gamma. \qquad (8.54)$$

The choice of Γ is arbitrary except that it is required to stay in the Stokes wedges S_j for $|z| \to \infty$.

The Stokes wedges of (8.53) are

$$S_j = \left\{ z \in \mathbb{C} \; : \; \frac{(2j-1)\pi}{4n+6} - \frac{\pi}{2} < \arg z < \frac{(2j+1)\pi}{4n+6} - \frac{\pi}{2} \right\}, \quad j \in \mathbb{Z}.$$

There are $4n+6$ disjoint Stokes wedges S_j with the opening angle $\frac{\pi}{2n+3}$. The wedges S_j are bounded by the Stokes lines

$$\left\{ z \in \mathbb{C} \; : \; \arg z = \frac{(2j-1)\pi}{4n+6} - \frac{\pi}{2} \right\}, \quad j \in \mathbb{Z}.$$

of the differential equation (8.53). The positive $\mathbb{R}_+$ and the negative $\mathbb{R}_-$ semi-axes do not belong to the Stokes wedges and they are the Stokes lines for any choice of $n \in \mathbb{N} \cup \{0\}$.

It follows from the general theory[16] that (*i*) *for each S_j, there exist solutions $f_\pm(z, \lambda)$ of (8.53) such that $f_+(z, \lambda)$ exponentially decays to zero and $f_-(z, \lambda)$ blows up in S_j when $|z| \to \infty$. An arbitrary solution of (8.53) can be expressed as a linear combination of $f_+(\cdot, \lambda)$ and $f_-(\cdot, \lambda)$; (ii) every solution of (8.53) decays to zero algebraically as $z \to 0$ along a Stokes line.*

The property (*ii*) explains why the spectrum of the operator H_1 in Example 8.10 coincides with $\mathbb{C}$. Indeed, the real axis is the Stokes line for $\tau(\cdot)$ and if the contour Γ coincides with $\mathbb{R}$, then solutions $f(x, \lambda)$ of (8.53) belong to $L_2(\mathbb{R})$ for any λ. This means that the boundary condition (8.54) with $\Gamma = \mathbb{R}$ does not lead to the discrete spectrum.

Let us now consider the contour Γ defined by (8.21) with the even function $\eta(\cdot)$. The contour Γ is symmetric with respect to the imaginary axis. We assume that Γ belongs to the closed subsectors of the Stokes wedges $\tilde{S}_j$ and S_j as x approaches $-\infty$ and ∞, respectively.

Let λ be an eigenvalue of the eigenvalue problem (8.53)–(8.54) and let $f(z)$ ($z = \xi(x) \in \Gamma$) be the corresponding eigenfunction. It follows from (*i*) that $f(z)$ is defined up a multiplicative constant. Moreover, the function $f(z)$ admits different presentations $f(x) = c_1 f_+(z, \lambda)$ and $f(z) = c_2 \tilde{f}_+(z, \lambda)$, where $f_+, \tilde{f}_+$ are decaying solutions in the Stokes wedges S_j and $\tilde{S}_j$. An analysis of this fact leads to the conclusion that the

---

[16][Hille (1969); Sibuya (1975)].

eigenvalues λ should be the zeros of a certain entire function. The latter ensures the quantization of the eigenvalues.

Let $L_2(\Gamma)$ be the Hilbert space considered in Example 8.3. In the space $L_2(\Gamma)$, the differential expression $\tau(\cdot)$ determines the operator

$$Hf = \tau(f), \qquad f \in \mathcal{D}(H),$$

where the domain $\mathcal{D}(H)$ consists of smooth functions $f(z)$ that exponentially decay to zero as $|z| \to \infty$ along Γ. It follows from the above that H is $\mathcal{PT}$ symmetric in $L_2(\Gamma)$ and its spectrum has a countable set of simple eigenvalues. Detailed studies show that these eigenvalues are real and positive [Dorey *et al.* (2001a); Shin (2005b)].

If a $\mathcal{PT}$-symmetric operator H is Hermitian (or essentially Hermitian) in the Kreĭn space $(\mathfrak{H}, [\cdot, \cdot])$, then the corresponding indefinite inner product $[\cdot, \cdot]$ *is not necessarily determined by the fundamental symmetry* $\mathcal{P}$. In many cases, the indefinite metric $[\cdot, \cdot]$ is defined by another fundamental symmetry $\mathcal{P}'$ that commutes with $\mathcal{PT}$. The commutation between $\mathcal{P}'$ and $\mathcal{PT}$ preserves the property of $\mathcal{PT}$ to be a conjugation operator with respect to $[\cdot, \cdot]$:

$$[\mathcal{PT}f, \mathcal{PT}g] = (\mathcal{P}'\mathcal{PT}f, \mathcal{PT}g) = (\mathcal{PT}\mathcal{P}'f, \mathcal{PT}g) = (g, \mathcal{P}'f) = [g, f].$$

Example 8.11. *$\mathcal{PT}$-symmetric operators in the Kreĭn space* $(\mathbb{C}^2, [\cdot, \cdot])$. Let $(\mathbb{C}^2, [\cdot, \cdot])$ be the Kreĭn space defined in Example 8.4. According to (8.52), a matrix operator $H = \begin{bmatrix} a & b \\ c & d \end{bmatrix}$ is $\mathcal{PT}$ symmetric if

$$\begin{bmatrix} 0 & 1 \\ 1 & 0 \end{bmatrix} \begin{bmatrix} a^* & b^* \\ c^* & d^* \end{bmatrix} = \begin{bmatrix} a & b \\ c & d \end{bmatrix} \begin{bmatrix} 0 & 1 \\ 1 & 0 \end{bmatrix}.$$

The above relation holds if $a = d^*$ and $b = c^*$. Therefore, $H = \begin{bmatrix} a & b \\ b^* & a^* \end{bmatrix}$. Comparing these relations with the conditions of $\mathcal{P}$-Hermiticity, $a = d^*$ and $b, c \in \mathbb{R}$ (Example 8.5), we decide that H is $\mathcal{P}$-Hermitian if and only if b is real.

In the general case $b \in \mathbb{C}$, so we must modify the fundamental symmetry $\mathcal{P}$ by taking into account the possible nonreality of b. Denote

$$\mathcal{P}_\phi = (\cos \phi \sigma_0 + i \sin \phi \sigma_3)\mathcal{P} = e^{i\phi\sigma_3}\mathcal{P}, \qquad \phi \in [0, 2\pi].$$

The operator $\mathcal{P}_\phi$ is a fundamental symmetry in $(\mathbb{C}, (\cdot, \cdot))$ and $\mathcal{P}_\phi$ commutes with $\mathcal{PT}$ (since $\mathcal{P}\sigma_3 = -\sigma_3\mathcal{P}$).

It follows from Lemma 6.4.9 in [Albeverio and Kuzhel (2015)] that a $\mathcal{PT}$-symmetric operator H is $\mathcal{P}_\phi$-Hermitian where the parameter ϕ is determined by the relation $\mathrm{Re}(b) \sin \phi = \mathrm{Im}(b) \cos \phi$.

Example 8.12. [Günther and Kuzhel (2010)] A Hamiltonian with a general regularized zero-range potential at the point $x = 0$ is determined by

$$\tau(f) = -\frac{d^2 f}{dx^2} + a < \delta, f > \delta + b < \delta', f > \delta + c < \delta, f > \delta' + d < \delta', f > \delta',$$

where a, b, c, d are complex numbers, the operator $-\frac{d^2}{dx^2}$ acts on $f \in W_2^2(\mathbb{R} \setminus \{0\})$ in the distributional sense, and the regularized delta function δ and its derivative δ' are defined on piecewise continuous functions $f \in W_2^2(\mathbb{R} \setminus \{0\})$:

$$< \delta, f > = \tfrac{1}{2}[f(0+) + f(0-)], \qquad < \delta', f > = -\tfrac{1}{2}[f'(0+) + f'(0-)].$$

The formal expression $\tau(\cdot)$ gives rise to the operator H_T, $T = \begin{bmatrix} a & b \\ c & d \end{bmatrix}$ in the Hilbert space $L_2(\mathbb{R})$, which is defined as follows:

$$H_T f = \tau(f), \qquad \mathcal{D}(H_T) = \{f \in W_2^2(\mathbb{R} \setminus \{0\}) : \ \tau(f) \in L_2(\mathbb{R})\}.$$

For $\mathcal{P}$ and $\mathcal{T}$ we mean the standard space-reflection and conjugation operators. The operator H_T is $\mathcal{PT}$ symmetric if and only if the entries of T satisfy the conditions $a, d \in \mathbb{R}$ and $b, c \in i\mathbb{R}$. Only a part of $\mathcal{PT}$-symmetric operators H_T can be realized as $\mathcal{P}$-Hermitian ones. To be precise, H_T is $\mathcal{P}$-Hermitian if and only if the coefficients $b, c \in i\mathbb{R}$ satisfy the additional condition $b = c$.

In the general case, the $\mathcal{P}$-Hermiticity must be modified toward a $\mathcal{P}_\phi$-Hermiticity with the Clifford-rotated fundamental symmetries

$$\mathcal{P}_\phi = e^{i\phi \mathcal{R}} \mathcal{P} = e^{i\phi \mathcal{R}/2} \mathcal{P} e^{-i\phi \mathcal{R}/2}, \qquad \mathcal{R}f(x) = (\mathrm{sgn}\ x)f(x)$$

so that an appropriate Kreĭn space can be constructed for any parameter combination $b \neq c$. The fundamental symmetries $\mathcal{P}_\phi$ commute with $\mathcal{PT}$ because $\mathcal{P}$ anticommutes with $\mathcal{R}$. It is easy to check that a $\mathcal{PT}$-symmetric operator H_T is $\mathcal{P}_\phi$-Hermitian, where the angle ϕ is fixed by the constraint

$$i[\det T + 4] \sin \phi = 2(c - b) \cos \phi.$$

The fundamental symmetries $\mathcal{P}$ and $\mathcal{R}$ can be interpreted as basis (generating) elements of the complex Clifford algebra $\mathcal{C}l_2 \equiv \mathrm{span}\{I, \mathcal{P}, \mathcal{R}, \mathcal{PR}\}$. The set of nontrivial fundamental symmetries $\mathcal{P}'$ ($\mathcal{P}' \neq \pm I$) constructed in terms of the Clifford algebra $\mathcal{C}l_2$ consists of operators of the form

$$\mathcal{P}' = \alpha_1 \mathcal{P} + \alpha_2 \mathcal{R} + \alpha_3 i \mathcal{RP}, \tag{8.55}$$

where $(\alpha_1, \alpha_2, \alpha_3)$ is an arbitrary vector of the unit sphere $\mathbb{S}^2$ in $\mathbb{R}^3$. The additional requirement of $\mathcal{PT}$ symmetry of a fundamental symmetry $\mathcal{P}'$ in (8.55) extracts the subclass of $\mathcal{PT}$-symmetric fundamental symmetries $\mathcal{P}_\phi$ considered above [Albeverio and Kuzhel (2012), Lemma 2.5].

Interestingly, the (hypothetical) possibility of a $\mathcal{PT}$-symmetric operator H_T being a $\mathcal{P}'$-Hermitian operator with the non-$\mathcal{PT}$-symmetric fundamental symmetry $\mathcal{P}'$ specified by the general formula (8.55) leads to catastrophic spectral consequences. To be precise, *if a $\mathcal{PT}$-symmetric $\mathcal{P}_\phi$-Hermitian operator H_T can additionally be regarded as $\mathcal{P}'$-Hermitian with a non-$\mathcal{PT}$-symmetric fundamental symmetry $\mathcal{P}'$, then $\sigma(H_T) = \mathbb{C}$* [Kuzhel and Patsyuk (2012), Theorem 3.2].

8.5.2 *$\mathcal{PT}$-symmetric operators having $\mathcal{C}$-symmetry*

Proving that a $\mathcal{PT}$-symmetric operator H is Hermitian in some Kreĭn space is mathematically significant, but H does not have any obvious relevance to physics until it can be shown that H can serve as a basis for a theory of quantum mechanics. To do so one must demonstrate that the operator H is Hermitian on a Hilbert space (not a Kreĭn space!) that is endowed with an inner product whose associated norm is positive definite.

This problem can be overcome for a $\mathcal{PT}$-symmetric $\mathcal{P}$-Hermitian Hamiltonian H by finding a new $\mathcal{C}$-symmetry represented by an operator $\mathcal{C}$ (Theorem 8.2). Constructing a $\mathcal{C}$ operator is the key step in showing that the time evolution for the Hamiltonian H is unitary.

Usually, the definition of $\mathcal{C}$-symmetry for $\mathcal{PT}$-symmetric operators involves the additional requirement of $\mathcal{PT}$ symmetry of $\mathcal{C}$, that is, $\mathcal{CPT} = \mathcal{PTC}$. This condition arises as a direct consequence of the unbroken $\mathcal{PT}$ symmetry of a non-Hermitian Hamiltonian (see [Bender *et al.* (2002); Bender (2007)]). For this reason, we specify the general definition 8.1 for the case of $\mathcal{PT}$-symmetric operators.

Definition 8.3. A $\mathcal{PT}$-symmetric operator H in $\mathfrak{H}$ has *the property of $\mathcal{C}$-symmetry* if there exists an operator $\mathcal{C}$ with the properties described in Theorem 8.1 and such that

$$H\mathcal{C}f = \mathcal{C}Hf, \quad \mathcal{PTC}f = \mathcal{CPT}f, \qquad \forall f \in \mathcal{D}(H).$$

In many cases, the condition of $\mathcal{PT}$ symmetry imposed on $\mathcal{C}$ leads to the simplification of the corresponding expressions of $\mathcal{C}$.

Example 8.13. Let $(\mathbb{C}^2, [\cdot, \cdot])$ be the Kreĭn space defined in Example 8.4. The set of all possible operators $\mathcal{C}$ acting in $(\mathbb{C}^2, [\cdot, \cdot])$ is defined by (8.27), where ρ and ξ are independent real parameters. The additional condition of $\mathcal{PT}$ symmetry of $\mathcal{C}$ allows one to specify ξ. Indeed, the $\mathcal{PT}$ symmetry of $\mathcal{C}$ means that

$$\mathcal{PTC} = \begin{bmatrix} 0 & 1 \\ 1 & 0 \end{bmatrix} \cdot \begin{bmatrix} i\sinh\rho\cos\xi & \cosh\rho + \sinh\rho\sin\xi \\ \cosh\rho - \sinh\rho\sin\xi & -i\sinh\rho\cos\xi \end{bmatrix} \mathcal{T}$$

$$= \begin{bmatrix} -i\sinh\rho\cos\xi & \cosh\rho + \sinh\rho\sin\xi \\ \cosh\rho - \sinh\rho\sin\xi & i\sinh\rho\cos\xi \end{bmatrix} \cdot \begin{bmatrix} 0 & 1 \\ 1 & 0 \end{bmatrix} \mathcal{T}$$

is possible when $\sinh\rho\sin\xi = 0$. The case $\sinh\rho = 0$ gives the trivial $\mathcal{PT}$ operator $\mathcal{C} = \mathcal{P} = \sigma_1$. If $\sin\xi = 0$, then $|\cos\xi| = 1$ and

$$\mathcal{C} = \begin{bmatrix} -i\sinh\rho\cos\xi & \cosh\rho \\ \cosh\rho & i\sinh\rho\cos\xi \end{bmatrix} = \cosh\rho\,\sigma_1 - i\sinh\rho\cos\xi\,\sigma_3$$

$$= [\cosh\rho\,\sigma_0 + \sinh\rho\cos\xi\,\sigma_2]\sigma_1 = e^{\chi\sigma_2}\mathcal{P},$$

where $\chi = \mathrm{sgn}(\cos\xi)\rho$. Therefore, every $\mathcal{PT}$-symmetric operator $\mathcal{C}$ in the Kreĭn space $(\mathbb{C}^2, [\cdot,\cdot])$ has the form $\mathcal{C} = e^{\chi\sigma_2}\mathcal{P}$, where χ is an arbitrary real number.

In light of Subsec. 8.3.6, the property of $\mathcal{C}$-symmetry for a $\mathcal{PT}$-symmetric $\mathcal{P}$-Hermitian ($\mathcal{P}$-symmetric) operator guarantees its Hermiticity (symmetry) with respect to the new inner product $(\cdot,\cdot)_{\mathcal{C}}$ that preserves the conjugation property of $\mathcal{PT}$:

$$(\mathcal{PT}f, \mathcal{PT}g)_{\mathcal{C}} = [\mathcal{CPT}f, \mathcal{PT}g] = [\mathcal{PT}f\mathcal{C}, \mathcal{PT}g] = [g, \mathcal{C}f] = [\mathcal{C}g, f] = (g, f)_{\mathcal{C}}.$$

By analogy with Theorem 8.1, we can formulate (equivalent) definitions of the $\mathcal{PT}$-symmetric operators $\mathcal{C}$ acting in a Kreĭn space $(\mathfrak{H}, [\cdot,\cdot])$.

Theorem 8.7 ([Bender and Kuzhel (2012)]). *The following are equivalent:*

   (i) *an operator $\mathcal{C}$ is determined by the direct sum (8.15), where the subspaces $\mathfrak{L}_{\pm}$ are $\mathcal{PT}$ invariant (that is, $\mathcal{PT}\mathfrak{L}_{\pm} = \mathfrak{L}_{\pm}$);*

   (ii) *an operator $\mathcal{C}$ has the form $\mathcal{C} = e^{\mathcal{Q}}\mathcal{P}$, where $\mathcal{Q}$ is a Hermitian operator in $(\mathfrak{H}, (\cdot,\cdot))$ such that $\mathcal{PQ} = -\mathcal{QP}$ and $\mathcal{TQ} = -\mathcal{QT}$;*

   (iii) *$\mathcal{C}$ is a $\mathcal{PT}$-symmetric operator in $\mathfrak{H}$ such that $\mathcal{C}^2 f = f$ for all $f \in \mathcal{D}(\mathcal{C})$ and $\mathcal{CP}$ is a positive Hermitian operator.*

Example 8.14. *The imaginary cubic oscillator.* The differential expression

$$\tau(f)(x) = -f''(x) + ix^3 f(x), \qquad x \in \mathbb{R}$$

is $\mathcal{PT}$ symmetric and the eigenvalue problem $\tau(f)(x) = \lambda f$ has the solutions $f_n \in L_2(\mathbb{R})$, which are uniquely determined up a multiplicative constant and they correspond to real eigenvalues λ_n, $n \in \mathbb{N}$ [Dorey *et al.* (2001a)].

Moreover, the set of eigenfunctions $\{f_n\}$ is complete in $L_2(\mathbb{R})$ [Siegl and Krejčiřík (2012)].

The differential expression $\tau(\cdot)$ gives rise to the following $\mathcal{PT}$-symmetric operators in $L_2(\mathbb{R})$:

$$H_1 f = \tau(f), \qquad \mathcal{D}(H_1) = \{f \in L_2(\mathbb{R}) : \tau(f) \in L_2(\mathbb{R})\},$$

$$H_2 f = \tau(f), \qquad \mathcal{D}(H_2) = \text{span}\{f_n\}.$$

The operators H_j $(j = 1, 2)$ have simple eigenvalues $\{\lambda_n\}$ and their $\mathcal{PT}$ symmetry is unbroken.

As with Example 8.9, it is easy to check that both operators H_j are symmetric in the Kreĭn space $(L_2(\mathbb{R}), (\cdot, \cdot)_{\mathcal{PT}})$ (that is, $\mathcal{P}$-symmetric). Repeating the arguments of Example 8.9, we conclude that the eigenfunctions $\{f_n\}$ uniquely determine the direct sum (8.44) and the operator $\mathcal{C}_0$ (8.45). By construction, the operator $\mathcal{C}_0$ commutes with H_2 because $\mathcal{D}(\mathcal{C}_0) \supset \mathcal{D}(H_2)$, and $\mathcal{C}_0$ is $\mathcal{PT}$ symmetric because the subspaces $\mathfrak{L}_\pm^0$ are $\mathcal{PT}$ invariant (the latter, due to the unbroken $\mathcal{PT}$ symmetry). Therefore, by extending (if necessary) the operator $\mathcal{C}_0$ to an operator of $\mathcal{C}$-symmetry $\mathcal{C}$, we prove the existence of $\mathcal{C}$-symmetry for H_2.

The operator H_1 is totally different from H_2 because H_1 has no operators of $\mathcal{C}$-symmetry. Indeed, H_1 is $\mathcal{P}$-Hermitian in the Hilbert space $(L_2(\mathbb{R}), (\cdot, \cdot))$ and, by virtue of Proposition 8.3, H_1 may have only a bounded operator $\mathcal{C}$ of $\mathcal{C}$-symmetry in $(L_2(\mathbb{R}), (\cdot, \cdot))$. In this case H_1 is a Hermitian operator in the Hilbert space $(L_2(\mathbb{R}), (\cdot, \cdot)_{\mathcal{C}})$ with the $\mathcal{CPT}$ inner product (8.3), which is equivalent to the initial one $(\cdot, \cdot)$. Therefore, the norm estimate for the resolvent

$$\|(H_1 - zI)^{-1}\| \le \frac{C}{|\text{Im}(z)|}, \quad C > 0, \quad z \in \mathbb{C} \setminus \mathbb{R}$$

should hold, which is impossible for the operator H_1 [Siegl and Krejčiřík (2012)]. This contradiction shows that H_1 has no operators of $\mathcal{C}$-symmetry.

This result is natural and it is a good illustration of the nature of unbounded operators $\mathcal{C}$. Indeed, for unbounded $\mathcal{C}$, the corresponding $\mathcal{CPT}$ inner product $(\cdot, \cdot)_{\mathcal{C}}$ is inequivalent to the original inner product $(\cdot, \cdot)$, and the closure of the imaginary cubic oscillator initially determined on the linear span of eigenfunctions $\{f_n\}$ (that is, the closure of H_2) *should be realized with respect to* $(\cdot, \cdot)_{\mathcal{C}}$ *in the new Hilbert space* $\mathfrak{H}_{\mathcal{C}}$. The closure of H_2 with respect to the 'wrong inner product' $(\cdot, \cdot)$ gives rise to new functions from $\mathcal{D}(H_1) \setminus \mathcal{D}(H_2)$ that do not belong to the Hilbert space $\mathfrak{H}_{\mathcal{C}}$, where the operator of $\mathcal{C}$-symmetry is defined.

$\mathcal{PT}$-Symmetric Deformations of Nonlinear Integrable Systems

Contributed by Andreas Fring

"The excessive increase of anything causes
a reaction in the opposite direction."
—$\mathcal{Pla To}$

Integrable models occupy a special place in the set of classical as well as quantum systems. Classically they are characterized by the special feature that they possess enough conserved quantities that allow one to construct exact analytical solutions to their equations of motion. In quantum field-theoretic settings, integrability is usually identified with the factoring of multiparticle scattering amplitudes into a product of sequential two-particle amplitudes that can be determined to all orders in perturbation theory. Integrable systems can often be used as cornerstones or benchmarks around which new, often more realistic systems can be constructed in a perturbative or deformed fashion.

In this chapter we investigate whether the remarkable properties of integrable systems, or some parts of them, survive $\mathcal{PT}$-symmetric deformations. To appreciate what is special about classical integrable systems, we review some of their basic properties. Section 9.1 describes how to construct conserved quantities, how to decide whether a system is integrable, and how to construct solutions, often of solitonic type, to their nonlinear equations of motion. Section 9.2 explains possible ways to construct $\mathcal{PT}$-symmetric deformations of well-known classical integrable systems, such as the Burgers equation, the Korteweg–de Vries equation, compacton equations, and supersymmetric wave equations.

In Sec. 9.3 we study the properties of these new $\mathcal{PT}$-symmetric systems. We perform the Painlevé test for them and find examples that pass the test and are therefore integrable, examples that fail the test, and examples for which the test is inconclusive. For all of these models, integrable or not, we address the question of the role that PT symmetry plays in determining the reality of their conserved quantities, such as the energy. When the model is not integrable, we find that PT symmetry guarantees that the energies are real. When the model is integrable, we invoke that property to ensure that multisoliton solutions have real energies. We also investigate the role of PT symmetry in dictating the shape of trajectories in part of the phase space. We find that the distinctions between periodic orbits and those that possess asymptotic fixed points often coincide with the breaking of PT symmetry. In general, it is not the PT symmetry that determines these characteristic behaviors, but rather the nature of the fixed point. We discuss shock waves in PT-symmetric systems and the mechanism that leads to peaked solutions. The section concludes with a description of how the solutions of some PT-symmetric deformed nonlinear wave equations can be related to well-studied PT-symmetric quantum systems.

In Sec. 9.4 we discuss various versions of PT-deformed multiparticle systems of Calogero type. We elaborate on how the complex poles of some nonlinear wave equations are described by Calogero-type systems. When restricting the phase space to the region determined by the gradient of some of the conserved quantities, we naturally obtain complex soliton solutions. The different variants of PT-symmetry can be interpreted as deformed Weyl reflections. For a simple example of a PT-deformed Calogero model, we demonstrate how the corresponding quantum-mechanical model that is the Schrödinger equation can be solved, leading to new energy spectra.

9.1 Basics of Classical Integrable Systems

At the heart of the study of classical integrable Hamiltonian systems lies the notion of Liouville integrability [Liouville (1855)], which is summarized in the following theorem.

Liouville theorem: *Let $H(p,q)$ with $p = \{p_1, \ldots, p_n\}$, $q = \{q_1, \ldots, q_n\}$ be a Hamiltonian for a Hamiltonian system on the Euclidian phase space $\mathbb{R}^{2n} = \{p, q\}$ with a Poisson bracket structure. If the system possesses n independent integrals of motion I_i (that is, conserved quantities with $\dot{I}_i = 0$) for $i \in \{1, 2, \ldots, n\}$ in involution, then Hamilton's equations*

$$\frac{dq_i}{dt} = \frac{\partial H(p,q)}{\partial p_i} \quad and \quad \frac{dp_i}{dt} = -\frac{\partial H(p,q)}{\partial q_i} \tag{9.1}$$

can be solved by quadrature.

Let us explain some terminology and notation used in this theorem: The set of constants of motion I_i is said to be *in involution* when all their mutual Poisson brackets vanish: $\{I_i, I_j\} = \sum_{k=1}^{n}(\partial I_i/\partial q_k \partial I_j/\partial p_k - \partial I_i/\partial p_k \partial I_j/\partial q_k) = 0$ for all $i, j \in \{1, 2, \ldots n\}$ and *independent* if none of the quantities I_i can be expressed in terms of the other I_j for $i \neq j$. A differential equation is said to be *solvable by quadrature* if its solutions can be obtained by a finite number of algebraic operations, that is, by calculating integrals, inverting functions, and so on. For brevity we abbreviate total time derivatives by a dot: $\dot{I} \equiv dI/dt$.

There are more formal definitions of integrability, involving extensions to fields or dynamical systems that are not Hamiltonian, but for our purposes the above will suffice. We keep in mind that we must have enough conserved quantities available that allow for the system to be solved.

We illustrate the theorem for $n = 1$. Consider a Hamiltonian system that is potential with $H(p, q) = p^2/2 + V(q)$. Hamilton's equations (9.1) (also called the *equations of motion*) are simply $\dot{q} = p$ and $\dot{p} = -\partial V/\partial q$. Trajectories in phase space may then be found by solving these two equations for q and p. Noting that $I_1 = H$ is conserved, that is, $\dot{H} = 0$, this becomes a simple task. Denoting the constant of integration for a particular trajectory by the total energy E, the solution for this trajectory in phase space is then found by solving the equation containing the conserved quantity $p = \dot{q} = \pm\sqrt{2E - 2V(q)}$.

For larger values of n we use more conserved quantities and in principle solve the system as we did for $n = 1$. The central question arising is then how to construct more conserved quantities? One option is to use the *Poisson–Jacobi theorem*, which states that the Poisson bracket of two conserved quantities is also conserved. However, the method we shall outline in the following section is more powerful and systematic.

9.1.1 *Isospectral deformation method (Lax pairs)*

Suppose that we have a dynamical system, possibly a Hamiltonian system of the form (9.1), which can be written alternatively in operator form as

$$\dot{L} = [M, L].\tag{9.2}$$

Here L and M are noncommuting objects, such as matrices or differential operators, forming a *Lax pair* [Lax (1968)]. The Lax equation (9.2) holds if and only if the equations of motion hold. The key advantage of this reformulation is that one has converted partial differential equations, even nonlinear ones, into linear ordinary differential equations. We stress that not all equations of motion can be rewritten as Lax pairs, so this property is special. However, if this is possible, then the system is said to be *integrable* because this reformulation allows for the construction of infinitely many conserved quantities.

Time-independent eigenvalues of L: We must establish that the eigenvalues λ of L, defined through $L\phi = \lambda\phi$, are time independent. This property led to the name of the method as the spectra of L at different times, say t_1 and t_2, are isospectral; that is, they are the same for $L(t_1)$ and $L(t_2)$ with $t_1 \neq t_2$. In order to see this, we may solve equation (9.2) by

$$L(t) = u(t)L(0)u^{-1}(t) \qquad \text{and} \qquad M(t) = \dot{u}(t)u^{-1}(t), \qquad (9.3)$$

where $u(t)$ is as yet unknown. This is verified by direct calculation:

$$\dot{L}(t) = \dot{u}(t)L(0)u^{-1}(t) + u(t)\dot{L}(0)u^{-1}(t) + u(t)L(0)\dot{u}^{-1}(t),$$
$$= \dot{u}(t)u^{-1}(t)u(t)L(0)u^{-1}(t) - u(t)L(0)u^{-1}(t)\dot{u}(t)u^{-1}(t) = [M, L].$$

We have used here $\dot{u}^{-1}(t) = -u^{-1}(t)\dot{u}(t)u^{-1}(t)$, which follows from differentiating $u(t)u^{-1}(t) = 1$.

Since $L(t)$ and $L(0)$ in (9.3) are related by a similarity transformation, it follows that their eigenvalues, say λ, are the same,

$$L(0)\phi = \lambda\phi \Rightarrow u^{-1}(t)L(t)u(t)\phi = \lambda\phi \Rightarrow L(t)\left(u(t)\phi\right) = \lambda\left(u(t)\phi\right), \quad (9.4)$$

so the eigenvalues of $L(t)$ are indeed time independent.

An important consequence of $\dot{\lambda} = 0$ is that we can rewrite the Lax equation as a compatibility equation of two *linear* equations that serve as the starting point for several powerful methods in the field of classical integrable models, such as the AKNS method [Ablowitz *et al.* (1973)]. With λ a fixed time-independent constant, the two linear equations

$$L\phi = \lambda\phi \quad \text{and} \quad \dot{\phi} = M\phi \qquad (9.5)$$

are equivalent to the Lax equation (9.2). To verify this we differentiate the first equation in (9.5) with respect to time. Combining the resulting left side $\frac{d}{dt}(L\phi) = \dot{L}\phi + L\dot{\phi} = \dot{L}\phi + LM\phi$ and right side $\lambda\dot{\phi} = \lambda M\phi = M\lambda\phi = ML\phi$, we obtain $\dot{L}\phi = [M, L]\phi$.

Thus, we now have three different but equivalent ways to express our integrable dynamical system: (i) in the form of the equations of motion, such as (9.1) when the system is Hamiltonian, (ii) as the Lax pair (9.2), or (iii) in the form of the two linear equations (9.5).

We can use the operator L to construct conserved quantities. As we have seen from Liouville's theorem, when we have sufficiently many conserved quantities the system is integrable. If we can express the system in the form of a Lax pair (9.2), all quantities of the form

$$I_k = \tfrac{1}{k} tr(L^k) \qquad (9.6)$$

are conserved in time. To see this we calculate

$$\dot{I}_k = \tfrac{1}{k} tr(k L^{k-1} \dot{L}) = tr(L^{k-1}[M, L]) = tr(L^{k-1}ML) - tr(L^k M) = 0,$$

where we have used the cyclic property of the trace $tr(AB) = tr(BA)$. This is a powerful property because in principle we can generate from formula (9.6) as many conserved quantities as we need to match the number of degrees of freedom. When these integrals of motion I_k are all in involution and independent, the theory is integrable according to Liouville's theorem.

The existence of the Lax pair (9.2) serves as a criterion to decide whether or not a system is integrable. However, we must be cautious when drawing conclusions. Often it is technically difficult to construct the Lax pair explicitly, but that does not imply that the Lax pair does not exist and that the system is not integrable. It is not straightforward to show that a Lax pair does not exist for a particular system. However, the assertion that a Hamiltonian dynamical system is integrable is conclusive when one can rewrite the equation of motion in the form of a Lax pair. So how can we construct the operators L and M?

Construction of L and M: If one seeks the Lax pair in the form of differential operators, one can often make a suitable general *ansatz* in powers and derivatives of the fields, or if the rank of the matrices representing L and M is not too large, one might succeed in finding L and M in a brute-force fashion for such operators. Alternatively, one can rewrite L and M in terms of some new functions $u(t)$ and $Q(t)$ as

$$M = \dot{u}u^{-1} \qquad \text{and} \qquad L = \dot{Q} - MQ + QM, \qquad (9.7)$$

which can be calculated from a third function w whose second derivative with respect to time vanishes:

$$w = u^{-1}(t)Q(t)u(t) \quad \text{with} \quad \ddot{w} = 0. \qquad (9.8)$$

We can see that $\ddot{w}$ vanishes if and only if the Lax equation holds as follows: Taking the first derivative of (9.8) with respect to time, we obtain

$$\dot{w} = \dot{u}^{-1}(t)Q(t)u(t) + u^{-1}(t)\dot{Q}(t)u(t) + u^{-1}(t)Q(t)\dot{u}(t),$$
$$= u^{-1}(t)(\dot{Q} - MQ + QM)u(t),$$

where M comes from (9.7). Next, using the expression for L in terms of u and Q, the second derivative of w with respect to time is

$$\ddot{w} = \dot{u}^{-1}(t)L(t)u(t) + u^{-1}(t)\dot{L}(t)u(t) + u^{-1}(t)L(t)\dot{u}(t),$$
$$= u^{-1}(t)(\dot{L} - ML + LM)u(t).$$

Hence $\ddot{w} = 0$ if and only if the Lax equation (9.2) holds. So, solving (9.8) gives explicit expressions for the operators L and M.

Example 1 — free particle: The simplest example that illustrates the working of the above formulas is the *free particle* in two dimensions describing the motion of a particle in the (q_1, q_2)-plane with the Hamiltonian

$$H(p_1, p_2) = \tfrac{1}{2}p_1^2 + \tfrac{1}{2}p_2^2. \tag{9.9}$$

We start by identifying a quantity whose second derivative vanishes. The equations of motion resulting from Hamilton's equations (9.1) are $\dot{q}_i = \partial H/\partial p_i = p_i$, $\dot{p}_i = -\partial H/\partial q_i = 0$ for $i = 1, 2$. Therefore, we have obtained two quantities whose second derivatives vanish, since it follows that $\ddot{q}_i = 0$ for $i = 1, 2$. Next, we express q_i in terms of u and Q as defined in (9.7). For this purpose we convert the qs to matrix form:

$$(q_1, q_2) \to w = \begin{pmatrix} q_1 & -q_2 \\ -q_2 & -q_1 \end{pmatrix}. \tag{9.10}$$

This procedure appears to be somewhat *ad hoc* because some guesswork is involved. We could also have taken different types of matrices and different signs, but our choice has the desired property $\ddot{w} = 0$, so we proceed.

Converting to different variables allows us to factor the matrix in (9.10). Using the canonical variables (r, ϕ, p_r, p_ϕ) with $q_1 = r \cos \phi$, $q_2 = r \sin \phi$, $p_1 = p_r \cos \phi - p_\phi \sin(\phi)/r$ and $p_2 = p_r \sin \phi + p_\phi \cos(\phi)/r$, we convert the Hamiltonian (9.9) into the form

$$H(r, \phi, p_r, p_\phi) = \frac{1}{2}p_r^2 + \frac{p_\phi^2}{2r^2}, \tag{9.11}$$

so that the equations of motion become

$$\dot{\phi} = \frac{\partial H}{\partial p_\phi} = \frac{p_\phi}{r^2}, \ \dot{r} = \frac{\partial H}{\partial p_r} = p_r, \ \dot{p}_\phi = -\frac{\partial H}{\partial \phi} = 0, \ \dot{p}_r = -\frac{\partial H}{\partial r} = \frac{p_\phi^2}{r^3}. \tag{9.12}$$

To apply (9.8) we must bring w into the form $u^{-1}(t)Q(t)u(t)$ as in the first equation in (9.8). With w as a matrix we view this form as diagonalizing w with some similarity transformation. Computing

$$w = u^{-1}(t)Q(t)u(t), \ Q(t) = \begin{pmatrix} r & 0 \\ 0 & -r \end{pmatrix}, \ u(t) = \begin{pmatrix} \cos\frac{\phi}{2} & -\sin\frac{\phi}{2} \\ \sin\frac{\phi}{2} & \cos\frac{\phi}{2} \end{pmatrix},$$

where $r = \sqrt{q_1^2 + q_2^2}$, we now have precisely the form (9.8) from which we can determine the operators L and M. Noting that $[\dot{u}, u^{-1}] = 0$, we find from the above formulas that the M operator is

$$M = \dot{u}u^{-1} = \frac{p_\phi}{2r^2} \begin{pmatrix} 0 & -1 \\ 1 & 0 \end{pmatrix}, \tag{9.13}$$

and the L operator is

$$L = \dot{Q} - MQ + QM = \begin{pmatrix} p_r & -p_\phi/r \\ -p_\phi/r & -p_r \end{pmatrix}.$$

The Lax equation (9.2) is satisfied for this pair of operators.

Next, we calculate the conserved quantities from L using the formula (9.6). We find that

$$I_1 = tr(L) = tr \begin{pmatrix} p_r & -p_\phi/r \\ -p_\phi/r & -p_r \end{pmatrix} = 0,$$

$$I_2 = \frac{1}{2}tr(L^2) = \frac{1}{2}tr \begin{pmatrix} p_r^2 + p_\phi^2/r^2 & 0 \\ 0 & p_r^2 + p_\phi^2/r^2 \end{pmatrix} = 2H,$$

$$I_3 = \frac{1}{3}tr(L^3) = \frac{1}{3}tr \left[\begin{pmatrix} p_r & -p_\phi/r \\ -p_\phi/r & -p_r \end{pmatrix} \begin{pmatrix} 2H & 0 \\ 0 & 2H \end{pmatrix} \right] = 0.$$

Continuing in this way we deduce that $I_{2k} = 2H^k$ and $I_{2k-1} = 0$ for $k = 1, 2, \ldots$. This means that we do not obtain any new independent conserved quantities when calculating the trace of higher powers of L. However, in addition to the Hamiltonian H, p_ϕ is conserved. These two conserved quantities are sufficient to establish Liouville integrability as we have $n = 2$ for the degrees of freedom for the free-particle system.

Example 2 — Korteweg–de Vries equation: A well-studied prototype nonlinear integrable system is the Korteweg–de Vries (KdV) wave equation [Korteweg and de Vries (1895)],

$$u_t - 6uu_x + u_{xxx} = 0, \tag{9.14}$$

which arises as a small-amplitude approximate description of shallow water waves. This equation is equivalent to the Lax equation (9.2) with

$$L = \partial_x^2 + \tfrac{1}{6}u \quad \text{and} \quad M = 4\partial_x^3 + u\partial_x + \tfrac{1}{2}u_x, \tag{9.15}$$

which one may verify by direct calculation. (We denote partial derivatives by subscripts: $\phi_x \equiv \frac{\partial \phi}{\partial x}$, $\phi_{xx} \equiv \partial^2\phi/\partial x^2$.) Observe that the eigenvalue equation for the L operator (9.4) is identical to the time-independent Schrödinger equation in this example with u playing the role of the potential and λ being the time-independent energy.

Example 3 — **Calogero model:** The Calogero models are an important class of integrable systems [Calogero (1969a,b)]. (We discuss their $\mathcal{PT}$-deformed variants in Sec. 9.4.) In general, these models are multiparticle systems, with the exception of its simplest variant, a one-particle system described by the Hamiltonian

$$H_C(p,q) = \frac{1}{2}p^2 + \frac{g^2}{2q^2} \qquad (9.16)$$

with g denoting a coupling constant. The equations of motion are $\dot{q} = \partial H/\partial p = p$ and $\dot{p} = -\partial H/\partial q = g^2/q^3$. One can verify by substitution the corresponding Lax equation (9.2) for the operators

$$L = \begin{pmatrix} p & g/q \\ g/q & -p \end{pmatrix} \quad \text{and} \quad M = \frac{g}{2}\begin{pmatrix} 0 & q^{-2} \\ -q^{-2} & 0 \end{pmatrix}.$$

9.1.2 *Painlevé test*

When we do not know enough conserved quantities as specified by Liouville's theorem or an explicit form of the Lax pair, it is not possible to decide *a priori* if a nonlinear dynamical system is integrable. When this information is unavailable, one may employ the *Painlevé test*, which has been devised to discriminate between integrable and nonintegrable models. The Painlevé test is based on the feature that every dynamical nonlinear system that admits soliton solutions is Liouville integrable. (Solitons are discussed in more detail in Subsec. 9.1.3.) There is no proof of this assertion, but only heuristic arguments and support from many examples. For almost all known integrable dynamical systems it has been verified that the Painlevé test [Painlevé (1900); Ablowitz *et al.* (1980); Weiss *et al.* (1983); Grammaticos and Ramani (1997); Kruskal *et al.* (2004)] is passed, and this test was even used to identify new models previously not known to be integrable [Bountis *et al.* (1982); Dorizzi *et al.* (1983); Assis and Fring (2009b, 2010)]. We use the test here to find new integrable systems.

The principal steps in using this test, proposed originally in [Weiss *et al.* (1983)], are as follows: For a dynamical system in the form of a partial differential equation involving a field $u(x,t)$ and its derivatives, the starting point is the *Painlevé expansion*, which takes the form

$$u(x,t) = \sum_{k=0}^{\infty} \lambda_k(x,t)\phi(x,t)^{k+\alpha}. \qquad (9.17)$$

Here, α is a negative integer characterizing the leading-order singularity in the field equation when the newly introduced field approaches zero: $\phi(x,t) \to 0$. The coefficient functions $\lambda_k(x,t)$ are assumed to be analytic.

The procedure of the Painlevé test consists of substituting the expansion (9.17) into the partial differential equation under investigation. This leads to recurrence relations for the functions $\lambda_k(x, t)$ of the form

$$g(k, \phi_t, \phi_x, \phi_{xx}, \ldots)\lambda_k = h(\lambda_{k-1}, \lambda_{k-2}, \ldots, \lambda_1, \lambda_0, \phi_t, \phi_x, \phi_{xx}, \ldots), \quad (9.18)$$

with g and h being model-dependent functions. Solving the equations (9.18) recursively may lead at some value of k, say $\tilde{k}$, to $g = 0$. For that k we can then compute the right side of (9.18) and find that either $h \neq 0$ or that $h = 0$. In the former case *the Painlevé test fails* for the system under investigation so we conclude that the system is not integrable. In the latter case, also called a *resonance*, the $\lambda_{\tilde{k}}$ on the left side is found to be a free parameter. Subsequently, the vanishing of g may occur at various other values $k > \tilde{k}$. If the number of these resonances equals the order of the partial differential equation, *the Painlevé test is passed* by the system under investigation; in that scenario the expansion (9.17) has enough free parameters to accommodate all possible initial conditions. A stronger statement is made when the series is also shown to converge. In that case one says that the partial differential equation has the *Painlevé property*. If one does not have enough additional free parameters to match the order of the differential equation, one has a *defective Painlevé expansion*.

When the system under investigation passes the Painlevé test it is integrable, and when it fails the test it is not integrable. When the Painlevé expansion is defective, the test delivers an inclusive result.

9.1.3 *Transformation methods*

When trying to solve nonlinear equations it often useful to transform them into other types of equations, which could be a different kind, the same nonlinear equation, or possibly even a linear equation. Frequently this transformation procedure allows us to construct new solutions from simpler ones. As we see below, the transformation methods also shed some light on what type of equation in the nonlinear case replaces the standard linear superposition principle (which allows one to add two solutions to obtain a third). We present some examples to illustrate the general ideas.

Miura transformation: Two nonlinear wave equations, the KdV equation $u_t = 6uu_x - u_{xxx}$ in (9.14) and the modified KdV (mKdV) equation

$$v_t = 6v^2 v_x - v_{xxx} \quad (9.19)$$

are related by the *Miura transformation*

$$u = v^2 + v_x. \qu(9.20)$$

We verify this by substituting (9.20) into the KdV equation. Thus, if we solve the mKdV equation for some v, we obtain a solution to the KdV equation (albeit the reverse is not true). (Of course, we must solve this new nonlinear differential equation in order to solve the KdV equation.)

Hopf–Cole transformation: The *Hopf–Cole transformation*

$$v(x,t) = \exp\left[-\tfrac{1}{2\sigma}\int^x u(s,t)ds\right] \tag{9.21}$$

relates the nonlinear Burgers equation,

$$u_t + uu_x = \sigma u_{xx} \quad \text{with} \ \sigma \in \mathbb{R}, \tag{9.22}$$

to the linear diffusion equation

$$v_t = \sigma v_{xx}. \tag{9.23}$$

So, $v(x,t)$ in (9.21) solves (9.23) if $u(x,t)$ solves (9.22) and vice versa.

Bäcklund transformations: The two examples above are model specific but *Bäcklund transformations* are more universally applicable. They relate solutions of a given nonlinear equation to other solutions of the same or a different equation by means of a pair of first-order partial differential equations. Thus, this type of transformation reduces the order of the differential equation. An important application is that it can be used to generate new types of solutions in consecutive order, e.g. from a one-soliton solution one can systematically construct multisoliton solutions.

For example, solutions to the *Liouville equation* and the wave equation

$$u_{xy} = e^u \quad \text{and} \quad v_{xy} = 0 \tag{9.24}$$

are related by the Bäcklund transformation

$$u_x + v_x = \sqrt{2}\exp\left[\tfrac{1}{2}(u-v)\right], \quad u_y - v_y = \sqrt{2}\exp\left[\tfrac{1}{2}(u+v)\right]. \tag{9.25}$$

We verify this by deriving (9.24) by cross-differentiation of (9.25). We can construct solutions to the Liouville equation from solutions to the simpler wave equation. From the solution $v = 0$ to the wave equation, the remaining two first-order equations in (9.25) are solved by separating variables,

$$u(x,y) = -2\ln\left[f(y) - x/\sqrt{2}\right], \quad u(x,y) = -2\ln\left[\tilde{f}(x) - y/\sqrt{2}\right],$$

where $f(y)$ and $\tilde{f}(x)$ are integration constants. Comparing these two solutions we find that a solution to the Liouville equation is

$$u(x,y) = \ln\left[2/(x+y)^2\right].$$

We can construct more complicated solutions to the *nonlinear* Liouville equation from other solutions of the *linear* wave equation.

The Bäcklund transformation for the KdV equation is obtained as follows: Substituting $u \to w_x$ in (9.14) converts it to $w_{xt} - 3(w_x^2)_x + w_{xxxx} = 0$, which when integrated with respect to x becomes

$$w_t - 3w_x^2 + w_{xxx} = 0. \tag{9.26}$$

The Bäcklund transformations relate two different solutions u, w and u', w':

$$u + u' = \tfrac{1}{2}(w - w')^2 - \kappa, \tag{9.27}$$

$$w_t + w_t' = -(w - w')(w_{xx} - w'_{xx}) + 2[u^2 + uu' + (u')^2], \tag{9.28}$$

where κ is a constant. The transformation is verified by using (9.26) to replace w_t and w_t' in (9.28) and converting the resulting equation into (9.27) by integrating twice with respect to x.

Remarkably, when combining four Bäcklund transformations involving four solutions w_0, w_1, w_2, and w_{12}, and two constants κ_1 and κ_2, one can derive a functional relation amongst them in which all derivatives vanish. The solutions in (9.27) and (9.28) must be paired, as shown in Fig. 9.1.

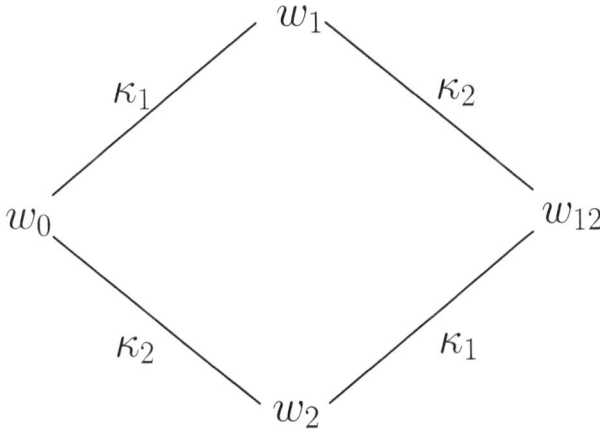

Fig. 9.1   Lamb diagram relating four solutions w_0, w_1, w_2 and w_{12} of the KdV equation (9.26). Each side of the diagram corresponds to a Bäcklund transformation (9.27) and (9.28) for the two solutions at the vertices and the constant at the link.

Diagrams of this type that relate different solutions by means of Bäcklund transformations are called *Lamb diagrams* [Lamb Jr (1971)] or *Bianchi diagrams* [Bianchi (1927)]. There is no *a priori* reason that ensures that the diagram actually closes when the solutions are related as indicated. However, closure is guaranteed by *Bianchi's theorem of permutability*, which states that for two known solutions w_1 and w_2 the Bäcklund transformation will always generate the same solution when the roles of the constants involved are reversed.[1]

The four Bäcklund transformations (9.27) from Fig. 9.1 are

$$(w_0)_x + (w_1)_x = \tfrac{1}{2}(w_0 - w_1)^2 - \kappa_1, \qquad (9.29)$$

$$(w_0)_x + (w_2)_x = \tfrac{1}{2}(w_0 - w_2)^2 - \kappa_2, \qquad (9.30)$$

$$(w_1)_x + (w_{12})_x = \tfrac{1}{2}(w_1 - w_{12})^2 - \kappa_2, \qquad (9.31)$$

$$(w_2)_x + (w_{12})_x = \tfrac{1}{2}(w_2 - w_{12})^2 - \kappa_1. \qquad (9.32)$$

Computing the differences (9.29)–(9.30) and (9.31)–(9.32), we obtain

$$(w_1)_x - (w_2)_x = \kappa_2 - \kappa_1 + \tfrac{1}{2}(w_0 - w_1)^2 - \tfrac{1}{2}(w_0 - w_2)^2, \qquad (9.33)$$

$$(w_1)_x - (w_2)_x = \kappa_1 - \kappa_2 + \tfrac{1}{2}(w_{12} - w_1)^2 - \tfrac{1}{2}(w_{12} - w_2)^2, \qquad (9.34)$$

and therefore equating (9.33) and (9.34) yields

$$w_{12} = w_0 + 2(\kappa_1 - \kappa_2)/(w_1 - w_2). \qquad (9.35)$$

This relation can be viewed as the *analog of the superposition principle* for linear equations, but instead of taking two solutions to construct a third, we require three solutions combined as in (9.35) to construct a new fourth solution. Also, unlike the linear superposition principle, the nonlinear superposition principle is model dependent; that is, the functional relation (9.35) is different in each nonlinear system.

Let us use (9.35) to construct a *multisoliton solution*. An N-soliton solution consists of N well-localized waves that travel at constant speeds c_i with $i = 1, \dots, N$ while keeping their shape. Whenever two of these waves scatter, either in a head-on collision or by a faster wave overtaking a slower one, they regain their original shape at sufficiently large time with a small displacement or equivalently with a shift in time. Identifying, for example, w_1 and w_2 with two different types of one-soliton solutions,

$$u_1(x, t) = -\tfrac{1}{2}c_1 \mathrm{sech}^2 \left[\tfrac{1}{2}\sqrt{c_1}(x - c_1 t + \delta_1)\right], \qquad (9.36)$$

$$u_2(x, t) = \tfrac{1}{2}c_2 \mathrm{csch}^2 \left[\tfrac{1}{2}\sqrt{c_2}(x - c_2 t + \delta_2)\right] \qquad (9.37)$$

with δ_1, δ_2 constant, and setting $w_0 = 0$ we find from (9.27) that the constants in (9.35) must be chosen as $\kappa_1 = \tfrac{1}{2}c_1$ and $\kappa_2 = \tfrac{1}{2}c_2$. [We derive the solutions (9.36) and (9.37) in Subsec. 9.3.4.] A new solution w_{12} is

---

[1] A proof of this theorem is beyond the scope of this book. We justify our assumptions in hindsight by explicitly constructing a solution.

obtained from the nonlinear superposition (9.35), which after differentiation with respect to x yields the two-soliton solution

$$u_{12}(x,t) = \frac{(c_1 - c_2)\left[c_2\mathrm{csch}^2\left(\frac{1}{2}\sqrt{c_2}\left(x - c_2 t\right)\right) + c_1\mathrm{sech}^2\left(\frac{1}{2}\sqrt{c_1}\left(x - c_1 t\right)\right)\right]}{2\left[\sqrt{c_2}\coth\left(\frac{1}{2}\sqrt{c_2}\left(x - c_2 t\right)\right) - \sqrt{c_1}\tanh\left(\frac{1}{2}\sqrt{c_1}\left(x - c_1 t\right)\right)\right]^2}$$

with $\gamma_1 = \gamma_2 = 0$. In principle, we can continue iterating in this way to construct a three-soliton solution from a two- and a one-soliton solution.

We conclude this section by observing that the Miura transformation is a Bäcklund transformation. This is seen by introducing the quantity $v := \frac{1}{2}(w - w')$, and rewriting (9.27) as

$$u = w_x = v^2 + v_x - \frac{1}{2}\kappa, \tag{9.38}$$

which is the Miura transformation (9.20) with u shifted by $\kappa/2$. This relation shows that the Bäcklund transformation is more general than the Miura transformation.

Finally, we mention that (9.38) is a starting point for the *inverse scattering method*.[2] Defining a new function ψ via the relation $v = \psi_x/\psi$, we can convert (9.38) into

$$u = \left(\frac{\psi_x}{\psi}\right)^2 + \frac{\psi\psi_{xx} - \psi_x^2}{\psi^2} - \frac{\kappa}{2},$$

and therefore

$$-\psi_{xx} + u\psi = E\psi, \tag{9.39}$$

where $E = -\frac{1}{2}\kappa$. This is a time-independent Schrödinger equation with u being the potential, E the energy, and ψ the wave function. In quantum mechanics the usual approach is to solve the scattering problem; that is, for a given potential u one finds the energy levels E and corresponding wave functions ψ. For the inverse problem one starts from ψ and E and uses this information to construct a solution u to the nonlinear KdV wave equation by means of (9.38).

9.2 $\mathcal{PT}$ Deformation of Nonlinear Wave Equations

There are various possibilities for complexifying or deforming real nonlinear wave equations in a $\mathcal{PT}$-symmetric manner. One may start with either the wave equation or the Hamiltonian density provided that the model under consideration is Hamiltonian. Our long range objective is to see which of

---

[2]For an explanation of the inverse-scattering method see [Ablowitz and Segur (1981); Ablowitz and Clarkson (1991)].

the properties discussed in the previous section survive when we $\mathcal{PT}$-deform an integrable system.

We consider here generic nonlinear wave equations of the form

$$u_t = f(\alpha, \beta, \gamma, \ldots, u, u_x, u_{xx}, u_{xxx}, \ldots), \qquad \alpha, \beta, \gamma, \ldots, u \in \mathbb{R}. \quad (9.40)$$

where f is some function that depends on a real-valued field $u(x,t)$, its derivatives $u_x, u_{xx}, u_{xxx}, \ldots$, and some real constants $\alpha, \beta, \gamma, \ldots$. We assume that the system is Hamiltonian; that is, that a Hamiltonian $H(u) = \int \mathcal{H} dx$ with density $\mathcal{H}$ exists and the equation of motion of the form (9.40) is computed from the variational principle as

$$u_t = \left[\frac{\delta H(u)}{\delta u} \right]_x = \left[\frac{\delta \int \mathcal{H} dx}{\delta u} \right]_x = \left[\sum_{n=0}^{\infty} (-1)^n \frac{d^n}{dx^n} \frac{\partial \mathcal{H}}{\partial u_{nx}} \right]_x. \quad (9.41)$$

At the Hamiltonian level we see immediately the usefulness of $\mathcal{PT}$ symmetry: As with the quantum-mechanical models discussed in previous chapters, $\mathcal{PT}$ symmetry can ensure the reality of the energy despite the complexity of the Hamiltonian density. We argue simply that if the Hamiltonian density is $\mathcal{PT}$ symmetric, that is, $\mathcal{H}^\dagger[u(x)] = \mathcal{H}[u(-x)]$, then the energy on the interval $[-a, a]$ is guaranteed to be real:

$$E = \int_{-a}^{a} \mathcal{H}[u(x)] dx = -\int_{a}^{-a} \mathcal{H}[u(-x)] dx = \int_{-a}^{a} \mathcal{H}^\dagger[u(x)] dx = E^\dagger. \quad (9.42)$$

We now discuss five distinct ways to deform the nonlinear wave equation (9.40) or the Hamiltonian that lead to new $\mathcal{PT}$-symmetric models [Fring (2007); Bagchi and Fring (2008); Fring (2009, 2013)]. (Combinations of the following options are also possible.)

(i) The simplest possibility is to *complexify the field* u by replacing it with $u = p + iq$, where p and q are real-valued fields. In addition, one can *complexify the constants* by replacing the real constants $\alpha, \beta, \gamma, \ldots \in \mathbb{R}$, if available, with complex ones $\tilde{\alpha}, \tilde{\beta}, \tilde{\gamma}, \ldots \in \mathbb{C}$ in such a way that the new system remains $\mathcal{PT}$ symmetric if the original system is $\mathcal{PT}$ symmetric.

(ii) One may *extend* the model by adding a $\mathcal{PT}$-symmetric term to the original wave equation $u_t = f(\alpha, \beta, \gamma, \ldots, u, u_x, u_{xx}, u_{xxx}, \ldots) + \tilde{f}(\ldots)$, or to the Hamiltonian $\mathcal{H} + \tilde{\mathcal{H}}$, where $\tilde{f} \in \mathbb{C}$ and $\tilde{\mathcal{H}} \in \mathbb{C}$ respect the same $\mathcal{PT}$ symmetry as f and $\mathcal{H}$.

(iii) One may introduce additional free parameters into the models by *deforming $\mathcal{PT}$-symmetric quantities*. For example, if the system is invariant under a transformation $\mathcal{PT} : \phi(x,t) \mapsto -\phi(x,t)$ for some field, one can define a deformation map $\delta_\varepsilon : \phi(x,t) \mapsto -i[i\phi(x,t)]^\varepsilon$ with $\varepsilon \in \mathbb{R}$ [Fring (2007)]. The undeformed case is recovered in the limit $\varepsilon = 1$. The new deformed quantity remains anti-$\mathcal{PT}$-symmetric with the crucial difference that the overall minus sign is generated from the antilinear nature of the $\mathcal{PT}$ operator; that is, from $i \mapsto -i$ rather than from $\phi(x, t) \mapsto -\phi(x, t)$. This option allows for various deformations of the wave equation (9.40) or the Hamiltonian by defining the deformation maps

$$\delta_\varepsilon^+ : u_x \mapsto u_{x,\varepsilon} := -i(iu_x)^\varepsilon, \quad \delta_\varepsilon^- : u \mapsto u_\varepsilon := -i(iu)^\varepsilon, \tag{9.43}$$

when $\phi \equiv u_x$, $\phi \equiv u$, respectively.[3] We define higher-order derivatives by successive actions of the standard derivative ∂_x:

$$\delta_\varepsilon^+ : u_{xx,\varepsilon} := (u_{x,\varepsilon})_x = \varepsilon u_{xx} (iu_x)^{\varepsilon - 1},$$
$$u_{xxx,\varepsilon} := (u_{x,\varepsilon})_{xx} = i\varepsilon (iu_x)^{\varepsilon - 2} \left[u_x u_{xxx} + (\varepsilon - 1)u_{xx}^2 \right],$$
$$u_{nx,\varepsilon} := \partial_x^{n-1} u_{x,\varepsilon},$$
$$\delta_\varepsilon^- : u_{x,\varepsilon} := (u_\varepsilon)_x = \varepsilon u_x (iu)^{\varepsilon - 1},$$
$$u_{xx,\varepsilon} := (u_\varepsilon)_{xx} = i\varepsilon(iu)^{\varepsilon - 2} \left[(\varepsilon - 1)u_x^2 + uu_{xx} \right],$$
$$u_{nx,\varepsilon} := \partial_x^n u_\varepsilon.$$

Similarly, we can deform the time derivative $\delta_\varepsilon^t : u_t \mapsto u_{t,\varepsilon} := -i(iu_t)^\varepsilon$ to get a new $\mathcal{PT}$-symmetric model.

(iv) Another possibility is to apply the deformation (iii) to *all* terms in the model; this corresponds to a *variable transformation*.

(v) One can identify some quantities in the original model that become complex in the *reduction to a constrained system*. In this kind of deformation we use a famous theorem by Airault, McKean, and Moser [Airault *et al.* (1977)]: *Given a Hamiltonian $H(p, q)$ with a flow described by (9.1) and conserved quantities I_i in involution with H, the locus of $\nabla I = 0$ is invariant with regard to time evolution.* Thus, we may restrict the flow in phase space to that region provided that it is not empty.

9.2.1 *PT-deformed supersymmetric equations*

We can extend the above ideas to wave equations involving superfields

$$\Phi(x, \theta) = \xi(x) + \theta u(x)$$

defined in terms of the fermionic (anticommuting) field $\xi(x)$, the bosonic (commuting) field $u(x)$, and the anticommuting superspace variable θ. The

---

[3] One may, of course, also deform the higher derivatives, either by successively applying the map δ_ε^+ or by replacing $u_{xx} \to u_{x,\varepsilon} \circ u_{x,\varepsilon}$ and similarly for higher derivatives. We do not explore these possibilities here because δ_ε^+ is already a nonlinear map, so compositions of this map will be highly nonlinear.

superderivative acting on these fields is defined as $D := \theta \partial_x + \partial_\theta$, such that $D\Phi(x,\theta) = \theta \partial_x \xi(x) + u(x)$. Let us see how the different types of symmetries are realized on these fields. If η is an anticommuting constant, the supersymmetry transformation in superspace is

$$\mathcal{SUSY} : \Phi(x,\theta) \to \Phi(x - \eta\theta, \theta + \eta) = \Phi(x,\theta) - \eta\theta\partial_x\Phi(x,\theta) + \eta\partial_\theta\Phi(x,\theta),$$

$$u \to u + \eta\xi_x, \quad \xi \to \xi + \eta u, \tag{9.44}$$

so the bosonic and fermionic fields are related.

To implement $\mathcal{PT}$ symmetry, we can assume an even bosonic field and demand that $\mathcal{PT} : u \to u, \ t \to -t, \ x \to -x, \ i \to -i$. This means that we must have $\mathcal{PT} : \eta\xi_x \to \eta\xi_x$ for the supersymmetric transformation to remain invariant. [The possibility $\xi \to \pm\xi$, $\eta \to \mp\eta$ is not compatible with the last relation in (9.44).] Thus, we require that

$$\mathcal{PT} : u \to u, \ \xi \to i\xi, \ \eta \to i\eta, \ \theta \to i\theta, \ \Phi \to i\Phi, \ D \to -iD. \tag{9.45}$$

This $\mathcal{PT}$ transformation is an automorphism satisfying $(\mathcal{PT})^2 = 1$.

It is clear how to generalize the five deformation options above except for option (iii). We therefore discuss the deformation of superderivatives. Thinking of the map δ_ε^+ as a deformed derivative $\partial_{x,\varepsilon}$ that acts on some arbitrary function $f(x)$ as $\partial_{x,\varepsilon} f(x) = -i(if_x)^\varepsilon$, and higher derivatives as $\partial_{x,\varepsilon}^n := \partial_x^{n-1}\partial_{x,\varepsilon}$, we define a deformed version of the superderivative as

$$D_\varepsilon := \theta \partial_{x,\varepsilon} + \partial_\theta.$$

Having different types of fields, there are more options for deforming higher derivatives. Some choices lead to the deformation of all derivatives, but there are also possibilities that lead only to the deformation of the derivatives acting on one type of field. The options leading to deformations of the fermionic fields are less interesting because $\partial_{x,\varepsilon}^n \xi = 0$ for $\varepsilon \geq 2$ when $\varepsilon \in \mathbb{N}$.

Defining, for instance, higher deformed superderivatives of bosonic-fermionic type as

$$D_\varepsilon^n := D^{n-2} D_\varepsilon^2 \quad (n \geq 2),$$

the action on the superfield $\Phi(x,\theta)$ yields $D_\varepsilon^{2n-1}\Phi = \theta\partial_{x,\varepsilon}^n\xi + \partial_{x,\varepsilon}^{n-1}u$, $D_\varepsilon^{2n}\Phi = \theta\partial_{x,\varepsilon}^n u + \partial_{x,\varepsilon}^n\xi$ for $n \in \mathbb{N}$. Similarly, defining deformed superderivatives of fermionic type as

$$\hat{D}_\varepsilon^n := D^{n-1} D_\varepsilon \quad (n \geq 1),$$

the action on the superfield $\Phi(x,\theta)$ gives $\hat{D}_\varepsilon^{2n-1}\Phi = \theta\partial_{x,\varepsilon}^n\xi + \partial_x^{n-1}u$, $\hat{D}_\varepsilon^{2n}\Phi = \theta\partial_x^n u + \partial_{x,\varepsilon}^n\xi$. More interesting are deformed superderivatives of bosonic type for which we define higher derivatives as

$$\tilde{D}_\varepsilon^n := D^{n-2} D_\varepsilon^2 \quad (n \geq 2).$$

Now we obtain nontrivial expressions with deformed bosonic fields from the action on the superfield $\Phi(x,\theta)$ as $\tilde{D}_\varepsilon^{2n-1}\Phi = \theta\partial_x^n\xi + \partial_{x,\varepsilon}^{n-1}u$, $\tilde{D}_\varepsilon^{2n}\Phi = \theta\partial_{x,\varepsilon}^n u + \partial_x^n\xi$.

Another possibility is to keep lower-order derivatives undeformed and let the deformation set in at higher order, say at order 3, by defining

$$\check{D}_\varepsilon^n := \begin{cases} D^n & (n = 1, 2), \\ -i(iD^3\Phi)^\varepsilon = \partial_{x,\varepsilon}u + i\theta\varepsilon\partial_{x,\varepsilon-1}u\xi_{xx} & (n = 3), \\ D^{n-3}\check{D}_\varepsilon^3 & (n > 3). \end{cases}$$

We now introduce some concrete examples of PT deformations of integrable systems. [We investigate these examples further in Sec. 9.3.]

9.2.2 *PT-deformed Burgers equation*

A simple nonlinear system is the extensively studied Burgers equation (9.22) in fluid dynamics. This equation is a special one-dimensional case of the Navier-Stokes equation applied to gas flow. The Burgers equation is invariant under the transformation $u \to u$, $x \to -x$, $t \to -t$ and $\sigma \to -\sigma$. Thus, taking $\sigma = i\kappa$ in (9.22) with $\kappa \in \mathbb{R}$ according to the deformation option (i), we interpret this symmetry as a PT symmetry. The field u is even under this PT symmetry, so we may also deform the derivatives by applying the deformation map δ_ε^+, as defined in (9.43) for option (iii). We obtain the PT-symmetrically-deformed Burgers equation [Assis and Fring (2009b)]

$$u_t + uu_{x,\varepsilon} = i\kappa u_{xx,\mu} \quad (\kappa \in \mathbb{R}, \ \varepsilon, \mu \in \mathbb{N}), \tag{9.46}$$

where we have two deformation parameters ε and μ. In Sec. (9.3) we show that this equation remains integrable under certain conditions.

Having implemented options (i) and (iii), we could also add a term according to option (ii) and obtain [Cavaglia *et al.* (2011)]

$$u_t + uu_x = i\kappa u_{xx} + \frac{i\kappa u}{2u_x^2}\left(u_x u_{xxx} - u_{xx}^2\right). \tag{9.47}$$

The additional term and therefore the entire equation is PT symmetric. (The form of the added term may appear to be somewhat arbitrary and unmotivated, but we see in Subsec. 9.2.4 that it has a natural origin.)

When $\kappa = 0$, the deformed Burgers equation (9.47) reduces to the inviscid Burgers equation, $u_t + uu_x = 0$, also called the *Riemann–Hopf* or *Euler–Monge* equation. A deformed equation studied in [Bender and Feinberg (2008)] is obtained by setting $\kappa = 0$ in (9.46). At first, it seems that this is a deformation of type (iii), but it was noted in [Curtright and Fairlie (2008)] that this deformation is just a variable transformation and is of type (iv). This is seen by making successive transformations:

$$w_t = -ww_x \;\rightarrow\; z_t = i(iz_x)^\varepsilon \;\rightarrow\; v_t = iv(iv_x)^\varepsilon \;\rightarrow\; u_t = if(u)(iu_x)^\varepsilon, \quad (9.48)$$

using

$$w = \varepsilon(iz_x)^{\varepsilon-1}, \quad z = \frac{\varepsilon-1}{\varepsilon}v^{\frac{\varepsilon}{\varepsilon-1}}, \quad u = g(v), \quad \text{with } v = \frac{f\,[g(v)]}{[g'(v)]^{1-\varepsilon}},$$

for any function g.[4] We can also take a slightly different starting point in (9.48) and still arrive at the $\mathcal{PT}$-deformed inviscid Burgers equation:

$$w_t + f(w)w_x = 0 \;\;\rightarrow\;\; u_t - if(u)(iu_x)^\varepsilon = 0. \quad (9.49)$$

For instance, for $f(w) = w^n$ this is achieved by

$$w = \sqrt[n]{\varepsilon u(iu_x)^{\varepsilon-1}}. \quad (9.50)$$

Because we have explicit transformations, we can directly translate properties from the undeformed Burgers equation (9.22) to the deformed equation.

9.2.3 $\mathcal{PT}$-deformed KdV equation

If we rescale the KdV equation (9.14) by $t \to \gamma t$, $x \to x$, and $u \to -\beta/(6\gamma)u$, we obtain a two-parameter version of the KdV equation:

$$u_t + \beta uu_x + \gamma u_{xxx} = 0 \quad (\beta, \gamma \in \mathbb{R}). \quad (9.51)$$

This system is in fact Hamiltonian. The variational principle (9.41) for the Hamiltonian density

$$\mathcal{H}_{\mathrm{KdV}}[u] = -\tfrac{1}{6}\beta u^3 + \tfrac{1}{2}\gamma u_x^2 \quad (9.52)$$

leads to the equation of motion (9.51). Both the wave equation (9.51) and the Hamiltonian density (9.52) are $\mathcal{PT}$ symmetric under

$$\mathcal{PT}_+ : x \mapsto -x,\; t \mapsto -t,\; i \mapsto -i,\; u \mapsto u \quad (\beta, \gamma \in \mathbb{R}).$$

If we use the deformation option (i) and replace $\beta \to i\beta$, we observe a second type of $\mathcal{PT}$ symmetry:

$$\mathcal{PT}_- : x \mapsto -x,\; t \mapsto -t,\; i \mapsto -i,\; u \mapsto -u \quad (\beta \in i\mathbb{R}),$$

while keeping $\gamma \in \mathbb{R}$. According to deformation option (iii), we may use either the map δ_ε^+ or δ_ε^- in (9.43). The deformed models, suitably normalized, are then defined by the following Hamiltonian densities and corresponding equations of motion:

$$\mathcal{H}_\varepsilon^+ = -\frac{\beta}{6}u^3 - \frac{\gamma}{1+\varepsilon}(iu_x)^{\varepsilon+1} \;\;\Rightarrow\;\; u_t + \beta uu_x + \gamma u_{xxx,\varepsilon} = 0, \quad (9.53)$$

$$\mathcal{H}_\varepsilon^- = \frac{\beta}{(1+\varepsilon)(2+\varepsilon)}(iu)^{\varepsilon+2} + \tfrac{\gamma}{2}u_x^2 \Rightarrow u_t + i\beta u_\varepsilon u_x + \gamma u_{xxx} = 0, \quad (9.54)$$

with $\beta, \gamma \in \mathbb{R}$.[5]

---

[4] A generalization from the v equation to the u equation was done in [Cavaglia and Fring (2012)] to study shock-wave formations. See Subsec. 9.3.4.

[5] The Hamiltonian density $\mathcal{H}_\varepsilon^+$ was proposed in [Fring (2007)], whereas $\mathcal{H}_\varepsilon^-$ corresponds to a complex version of the generalized KdV equation [Rosenau and Hyman (1993)]

9.2.4 $\mathcal{PT}$-deformed compacton equations

The three-deformation-parameter $\mathcal{PT}$-symmetric Hamiltonian density

$$\mathcal{H}_{l,m,p} = -\frac{u^l}{l(l-1)} - \frac{g}{m-1}u^p(iu_x)^m$$

[Cavaglia $et\ al.$ (2011)] leads to the $\mathcal{PT}$-symmetric equation of motion

$$u_t + u^{l-2}u_x + gi^m u^{p-2}u_x^{m-3}\left[p(p-1)u_x^4\right.$$
$$\left. +2pmuu_x^2u_{xx} + m(m-2)u^2u_{xx}^2 + mu^2u_xu_{xxx}\right] = 0$$

proposed in [Bender $et\ al.$ (2009)]. This equation is a deformation of the generalized KdV equations [Rosenau and Hyman (1993)] and is known to possess $compacton$ solutions, which are solitary-wave solutions with compact support. The model contains various prominent systems: A modification of the Hamiltonian description of the generalized KdV equations is obtained for $\mathcal{H}_{l,2,p}$ [Cooper $et\ al.$ (1993); Khare and Cooper (1993)]. We obtain a special case of the $\mathcal{PT}^+$-symmetric deformation of the KdV equation from $\mathcal{H}_{3,\varepsilon+1,0}(g) = \mathcal{H}_\varepsilon^+ [\beta = 1, \gamma = g(1+\varepsilon)/\varepsilon]$. The $\mathcal{PT}$-symmetric deformation of Burgers equation (9.47) is obtained for $\mathcal{H}_{3,1,1}(g = \kappa/2)$, which makes the added $\mathcal{PT}$-symmetric term in (9.47) appear more natural.

9.2.5 $\mathcal{PT}$-deformed supersymmetric equations (revisited)

The Hamiltonian

$$H_\varepsilon = \int dx\, d\theta \left[\Phi(D\Phi)^2 + \tfrac{1}{1+\varepsilon}D^2\Phi\check{D}_\varepsilon^3\Phi\right] \tag{9.55}$$

describes a supersymmetric $\mathcal{PT}$-deformed system (see Subsec. 9.2.1). Integrating over θ and using the properties of the Berezin integral $\int d\theta = 0$ and $\int d\theta\,\theta = 1$ converts (9.55) into the component version

$$H_\varepsilon = \int dx \left[u^3 - 2\xi\xi_x u - \tfrac{1}{1+\varepsilon}(iu_x)^{\varepsilon+1} - \tfrac{\varepsilon}{1+\varepsilon}(iu_x)^{\varepsilon-1}\xi_x\xi_{xx}\right]. \tag{9.56}$$

In the limit $\varepsilon \to 1$ this Hamiltonian reduces to the supersymmetric KdV Hamiltonian in [Mathieu (1988)] and in the limit $\xi \to 0$ it reduces to the $\mathcal{PT}$-deformed Hamiltonian H_ε^+ in (9.53). By construction, H_ε respects the

analyzed in [Cavaglia $et\ al.$ (2011)]. Option (iii) allows for the deformation of other terms in the equation of motion, such as the dispersion term uu_x in (9.53) or the derivatives u_x and u_{xxx} in (9.54). However, deformations such as (4.30) are not Hamiltonian systems and it is less evident how to utilize the $\mathcal{PT}$ symmetry. The $\mathcal{PT}$-symmetric extensions of the KdV equation proposed in [Bender $et\ al.$ (2007a)] cannot be obtained from (9.53) or (9.54) because they correspond to non-Hamiltonian systems.

PT symmetry as specified in (9.45), and it is also supersymmetric with respect to the transformation (9.44), which is most easily verified for the component version (9.56):

$$SUSY : H_\varepsilon \to H_\varepsilon + \eta \int dx \partial_x \left(\xi u^2 + i^{\varepsilon-1} \frac{1}{1+\varepsilon} u_x^\varepsilon \xi_x \right) = H_\varepsilon.$$

The variational principle applied to (9.55) yields the equation of motion

$$\Phi_t = 4D\Phi D^2\Phi + 2\Phi D^3\Phi - \tfrac{1}{1+\varepsilon} \left[\breve{D}_\varepsilon^6 \Phi + i\varepsilon D^4 (D^2\Phi \breve{D}_{\varepsilon-1}^3 \Phi) \right],$$

which in component form reads

$$u_t = 6uu_x - \partial_{x,\varepsilon}^3 u - 2\xi\xi_{xx}$$
$$+\varepsilon \tfrac{1-\varepsilon}{1+\varepsilon} \left[\partial_{x,\varepsilon-2}^3 u\xi_x\xi_{xx} + \partial_{x,\varepsilon-2}^2 u\xi_x\xi_{xxx} + \partial_x(\partial_{x,\varepsilon-2} u\xi_x\xi_{xxx}) \right],$$
$$\xi_t = 4u\xi_x + 2\xi u_x - \tfrac{i\varepsilon}{1+\varepsilon} \left(3\partial_{x,\varepsilon-1}^2 u\xi_{xx} + 2\partial_{x,\varepsilon-1} u\xi_{xxx} + \partial_{x,\varepsilon-1}^3 u\xi_x \right).$$

Thus, supersymmetry can be preserved under a PT-symmetric deformation although one must be careful about the choice of deformed derivatives; some of the choices discussed in Sec. 9.2.1 break the supersymmetry.

9.3   Properties of PT-Deformed Nonlinear Wave Equations

Let us now investigate some of the properties of PT-deformed models. Most importantly, we would like to find out whether integrable systems remain integrable when PT-symmetrically-deformed. For this purpose we carry out the Painlevé test as described in Sec. 9.1.2.

9.3.1   *Painlevé tests of PT-deformed Burgers equations*

PT-deformed Burgers equation (9.47): We start with the deformed Burgers equation of the type (9.47) corresponding to the $H_{3,1,1}$ model. The crucial first step in the analysis consists of determining the constant α in the Painlevé expansion (9.17) that characterizes the leading-order singularity. It suffices to consider the contributions from the first term in the expansion. If we substitute $u(x,t) \to \lambda_0(x,t)\phi(x,t)$ into (9.47), we can identify the leading-order singularities from each of the terms as $u_t \sim \phi^{\alpha-1}$, $uu_x \sim \phi^{2\alpha-1}$, and all remaining terms are proportional to $\phi^{\alpha-2}$.

For a solution to exist, at least two of these terms must balance while the remaining term must be of the same or higher order. Let us explore all possibilities. Assuming that the first two terms are of equal power in leading order gives the constraints $\alpha - 1 = 2\alpha - 1 \le \alpha - 2$, which has no solution. Equating the first and the last terms leads to $\alpha - 1 = \alpha - 2 \le 2\alpha - 1$, which again has no solution. A balance between the second and the remaining terms would be possible if and only if $2\alpha - 1 = \alpha - 2 \le \alpha - 1$. The unique solution to these latter constraints is $\alpha = -1$. We stress that if one does not find a solution to this leading-order problem, the test terminates here and the system has failed the test. One would conclude that the system under investigation is not integrable.

However, in our case we proceed and read off the coefficients of powers of ϕ with $\alpha = -1$. The ϕ^{-1} and ϕ^0 terms give the first two coefficients:

$$\lambda_0 = -3i\kappa\phi_x \quad \text{and} \quad \lambda_1 = \frac{3}{2}\frac{i\kappa\phi_{xx} - \phi_t}{\phi_x}.$$

If we use these values, the term proportional to ϕ^1 vanishes, suggesting that λ_2 is a free parameter; that is, a resonance. At order ϕ^2 all terms involving λ_3 are vanishing, but the entire coefficient is zero only when

$$\phi_x^2\phi_{tt} + \phi_x^2\phi_{tt} = 2\phi_{xt}\phi_x\phi_t \tag{9.57}$$

holds. Restricting $\phi(x,t)$ in such a way that (9.57) is satisfied allows one to fix all the remaining λ_i with $i > 3$; we get λ_4 from setting the coefficient of ϕ^3 to zero, λ_5 from setting the coefficient of ϕ^4 to zero, and so on. Because we have two free parameters, λ_2 and λ_3, the $\mathcal{PT}$-deformed system described by (9.47) passes the Painlevé test provided that the series expansion converges. Remarkably, despite having started with a perturbative expansion (9.17), we obtain an *exact* solution of (9.47): $u(x,t) = \frac{3i\kappa}{\omega t - x} + \frac{3}{2}\omega$.

The solution to (9.47) is obtained by solving the constraining equation (9.57) with a traveling wave $\phi(x,t) = x - \omega t$ and setting $\lambda_i = 0$ for $i \ge 3$. Another solution to (9.57) is $\phi(x,t) = \sinh(x - \omega t)$; in this case there exists no choice for the free parameters that forces the series to terminate. Nonetheless, even in this latter case, all coefficients apart from the free one may be calculated recursively. One may view equation (9.57) as analogous to the transformations as described in Subsec. 9.1.3 in the sense that each of its solutions generate a solution for the original equation (9.47), although often in the form of an infinite series.

$\mathcal{PT}$-**deformed Burgers equation (9.46):** Next, we perform the Painlevé test on the $\mathcal{PT}$-deformed Burgers equation (9.46) following [Assis and Fring (2009b)]. Once again, we first determine the constant α in the Painlevé expansion (9.17) by substituting $u(x,t) \to \lambda_0(x,t)\phi(x,t)$ into (9.46). The leading-order singularities from each of the terms are then identified as $u_t \sim \phi^{\alpha-1}$, $uu_{x,\varepsilon} \sim \phi^{\alpha\varepsilon+\alpha-\varepsilon}$ and $u_{xx,\mu} \sim \phi^{\alpha\mu-\mu-1}$. We can match the powers of two of the terms. Balancing the first term with the second and third yields $0 < \varepsilon < 1$ and $0 < \mu < 1$, whereas equating the powers in the last two terms yields $\alpha = (\varepsilon - \mu - 1)/(\varepsilon - \mu + 1)$. We discard the first two cases because they lead to undesirable branch cuts when $\varepsilon, \mu \notin \mathbb{N}$ and deduce from the latter equation that $\alpha = -1$ and $\mu = \varepsilon$. With these values for α and μ the coefficients of powers in ϕ are

$$
\begin{aligned}
\phi^{-2\varepsilon-1} &: & \lambda_0 + 2i\varepsilon\kappa\phi_x &= 0, \\
\phi^{-2\varepsilon} &: & \phi_t\delta_{\varepsilon,1} + \lambda_1\phi_x - i\kappa\varepsilon\phi_{xx} &= 0, \\
\phi^{-2\varepsilon+1} &: & \partial_x(\phi_t\delta_{\varepsilon,1} + \lambda_1\phi_x - i\kappa\varepsilon\phi_{xx}) &= 0, \\
&\vdots & &\vdots \\
\phi^{-2\varepsilon-1+r} &: & g\lambda_r &= h.
\end{aligned}
\tag{9.58}
$$

Note that the undeformed case $\varepsilon = 1$ is special; only in that case do the terms involving ϕ_t enter the recurrence relations. Furthermore, the third equation in (9.58) does not lead to a constraint for λ_2 as it is automatically satisfied when the second equation in (9.58) holds. Thus, we obtain

$$
\lambda_0 = -i2\varepsilon\kappa\phi_x, \quad \lambda_1 = (i\varepsilon\kappa\phi_{xx} - \phi_t\delta_{\varepsilon,1})/\phi_x, \quad \lambda_2 \text{ arbitrary}, \tag{9.59}
$$

so we have already found one of the desired free parameters (resonances).

Alternatively, one can determine all possible resonances from a generic argument. For this we need to know when the function g in the recurrence relations (9.18) vanishes. We have already seen that the term $g\lambda_r$ is determined by the coefficient of $\phi^{-2\varepsilon-1+r}$. The contributions to g come from just two terms in the Painlevé expansion:

$$
\tilde{u}(x,t) = \lambda_0\phi^{-1} + \lambda_r\phi^{r-1}. \tag{9.60}
$$

Using the expression for λ_0 in (9.59), we substitute $\tilde{u}(x,t)$ into (9.46) and identify the terms of highest order, that is $\phi^{-2\varepsilon-1+r}$, as

$$
i2^{\varepsilon-1}\varepsilon^\varepsilon(r+1)(r-2)\kappa^\varepsilon\phi_x^{2\varepsilon}\lambda_r = 0. \tag{9.61}
$$

Thus, the necessary condition for a resonance to exist is either $r = 2$ or $r = -1$. Since the right side of (9.61) vanishes for these values [$h = 0$ in (9.18)], this system cannot fail the Painlevé test. The condition for $g = 0$ is necessary and sufficient. The first resonance is the free parameter we already found. The second, at $r = -1$, is called a *universal resonance* because it is always present. Thus, in higher order we do not obtain more free parameters for any value of the deformation parameter ε.

At this point we have established that we can calculate the Painlevé expansion for (9.46) and that we have enough free parameters to match the order of the differential equation. However, we do not yet know yet if the expansion makes sense, that is, whether it converges. It is difficult to establish this in complete generality, so we make specific choices that simplify the system and present the arguments for that situation. We choose the deformation parameter to be $\varepsilon = 2$, the coefficient functions $\lambda_k(x,t) = \lambda_k(t)$ to depend only on t, and $\phi(x,t) = x - \xi(t)$, where $\xi(t)$ is an arbitrary function of t. In this case the recurrence relations for the coefficient functions are

$$8\kappa^2 \left[8\kappa\delta_{0,j} + i(j-2)(j+1)\lambda_j\right] = \sum_{n,m=1}^{j} i(1-m)(n-1)\lambda_m\lambda_{j-m-n}\lambda_n$$

$$+ \sum_{n=1}^{j-1} \left[2\kappa(n-1)\left(n^2 - n - j(n-2) + 2\right)\lambda_{j-n}\lambda_n\right] + (j-4)\lambda_{j-3}\dot{\xi} - \ddot{\lambda}_{j-4}.$$

Solving this equation recursively, we find that the first terms in the Painlevé expansion are

$$u(x,t) = -\frac{4i\kappa}{\phi} + \lambda_2\phi + \frac{\dot{\xi}}{8\kappa}\phi^2 - \frac{i\lambda_2^2}{20\kappa}\phi^3 - \frac{i\lambda_2\dot{\xi}}{96\kappa^2}\phi^4 + \mathcal{O}(\phi^5),$$

which agrees with the previously computed expressions in (9.59) if we adopt the above choices for ε, λ_k, and ϕ. In principle, we can carry out the recursive iteration up to any finite order. For the undeformed case, corresponding to the ordinary Burgers equation with $\varepsilon = 1$, there is an obvious choice for the free parameter λ_2 for which the expansion terminates. However, in the deformed case such a choice most likely does not exist. We therefore investigate the convergence of the expansion for another special value, $\lambda_2 = 0$. For this value the expansion becomes

$$u(x,t) = -\frac{4i\kappa}{\phi} + \frac{\dot{\xi}\phi^2}{2^3\kappa} - \frac{i\dot{\xi}^2\phi^5}{7 \times 2^8\kappa^3} + \frac{i\ddot{\xi}\phi^6}{5 \times 2^9\kappa^3} - \frac{\dot{\xi}^3\phi^8}{35 \times 2^{13}\kappa^5} - \frac{23\dot{\xi}\ddot{\xi}\phi^9}{385 \times 2^{13}\kappa^5}$$

$$- \frac{\dddot{\xi}\phi^{10}}{135 \times 2^{14}\kappa^5} + \frac{19i\dot{\xi}^4\phi^{11}}{3185 \times 2^{18}\kappa^7} - \frac{51i\dot{\xi}^2\ddot{\xi}\phi^{12}}{385 \times 2^{19}\kappa^7} + \mathcal{O}(\phi^{13}). \quad (9.62)$$

We have we have obtained a suitable form for the expansion whose convergence we can analyze. We note that (9.62) has the general form

$$u(x,t) = -\frac{4i\kappa}{\phi} + \phi\sum_{n=1}^{\infty} \alpha_n\phi^n. \quad (9.63)$$

To establish the convergence of this series we employ Cauchy's root test, which states that $\sum_{n=1}^{\infty} \gamma_n$ converges if and only if $\lim_{n\to\infty} |\gamma_n|^{1/n} \leq 1$. We can find an upper bound for the real part of α_n in (9.63),

$$|\mathrm{Re}(\alpha_{3n-\nu})| \leq \frac{\left|\mathrm{Re}\left[p_{3n-\nu}(\dot{\xi}, \ddot{\xi}, \dddot{\xi}, \ldots)\right]\right|}{2^{3n+4-\nu}\Gamma\left(\frac{3n-\nu}{2}\right)|\kappa|^{2n-1}} \quad \text{for } \nu = 0,1,2, \quad (9.64)$$

where the $p_n(\dot{\xi}, \ddot{\xi}, \dddot{\xi}, \ldots)$ are polynomials of finite order in t; that is, $\sum_{n=0}^{\ell} \omega^n t^n$ with $\ell < \infty$ and $\omega \in \mathbb{C}$. The same expression holds when we replace the real part by the imaginary part on both sides of this inequality. We have verified the estimate (9.64) up to order thirty. (We do not present arguments to establish this more rigorously.) Approximating the gamma function in (9.64) by Stirling's formula

$$\Gamma(n/2) \sim \sqrt{2\pi}e^{-n/2}(n/2)^{(n-1)/2} \quad (n \to \infty),$$

we deduce that

$$\lim_{n\to\infty} |\operatorname{Re}\alpha_{3n-\nu}|^{\frac{1}{2}} \sim \frac{|\operatorname{Re}p_{3n-\nu}|^{1/n}}{2^{3+\frac{4-\nu}{n}}(2\pi)^{\frac{1}{2n}}e^{-\frac{1}{2}}\left(\frac{3n-\nu}{2}\right)^{\frac{1}{2}-\frac{1}{2n}}|\kappa|^{2-\frac{1}{n}}} = 0.$$

The same argument holds for the imaginary part, so by Cauchy's root test the series (9.63) converges for any value of κ and choice for $\xi(t)$ that gives finite polynomials $p_n(\dot{\xi}, \ddot{\xi}, \dddot{\xi}, \ldots)$. This establishes that the PT-deformed Burgers equation (9.46) has the Painlevé property; it passes the Painlevé test and the series expansion converges for $\varepsilon = 2$ and $\lambda_2 = 0$. To establish that the expansion converges for other values of the deformation parameter ε, one must repeat this lengthy argument.

9.3.2 Painlevé procedure for deformed KdV equation

We now perform the Painlevé test for the previously PT-deformed KdV equation (9.53) with the parameters $\beta = -6$ and $\gamma = 1$:[6]

$$u_t - 6uu_{x,\varepsilon} + u_{xxx,\mu} = 0 \quad (\varepsilon, \mu \in \mathbb{R}). \tag{9.65}$$

As in Subsec. 9.3.1, we substitute $u(x,t) \to \lambda_0(x,t)\phi(x,t)^\alpha$ into (9.65) in order to determine the leading-order term. From $u_t \sim \phi^{\alpha-1}$, $uu_{x,\varepsilon} \sim \phi^{\alpha+\alpha\varepsilon-\varepsilon}$ and $u_{xxx,\mu} \sim \phi^{\alpha\mu-\mu-2}$ we deduce that $\alpha = (\varepsilon-\mu-2)/(\varepsilon-\mu+1) \in \mathbb{Z}_-$, so the only solution is $\alpha = -2$ with $\varepsilon = \mu$ provided that ε and μ are integers. This means that neither $\mu = 1$ and ε generic nor $\varepsilon = 1$ and μ generic can pass the Painlevé test, but the doubly-deformed system with $\varepsilon = \mu$ has a chance of being integrable.

Substituting the Painlevé expansion (9.17) for $u(x,t)$ with $\alpha = -2$ into (9.65) with $\varepsilon = \mu$ and identifying powers in $\phi(x,t)$ gives the recursion relations for the λ_k. We focus on integer values of ε and find that

$$\phi^{-3\varepsilon-2} : \lambda_0 = \tfrac{1}{2}\varepsilon(3\varepsilon+1)\phi_x^2,$$
$$\phi^{-3\varepsilon-1} : \lambda_1 = -\tfrac{1}{2}\varepsilon(3\varepsilon+1)\phi_{xx},$$
$$\phi^{-3\varepsilon} : \lambda_2 = \frac{\varepsilon(3\varepsilon+1)}{24}\left(\frac{4\phi_x\phi_{xxx}-3\phi_{xx}^2}{\phi_x^2}\right) + \delta_{\varepsilon,1}\frac{\phi_t}{6\phi_x},$$
$$\phi^{-3\varepsilon+1} : \lambda_3 = \frac{\varepsilon(3\varepsilon+1)}{24}\left(\frac{4\phi_x\phi_{xx}\phi_{xxx}-3\phi_{xx}^3-\phi_x^2\phi_{4x}}{\phi_x^4}\right) + \delta_{\varepsilon,1}\frac{\phi_t\phi_{xx}-\phi_x\phi_{xt}}{6\phi_x^3}.$$

---

[6]For (9.65) the models corresponding to $\mu = 1$ with ε generic and $\varepsilon = 1$ with μ generic were considered in [Bender et al. (2007a)] and [Fring (2007)]. The Painlevé test for the general doubly-deformed model was done in [Assis and Fring (2009b)].

The next order at $\phi^{-3\varepsilon+2}$ is special. For $\varepsilon \neq 1$ we find that

$$\lambda_4 = \frac{\varepsilon(3\varepsilon+1)}{24}\left(\frac{6\phi_x\phi_{xx}^2\phi_{xxx} - \frac{15}{4}\phi_{xx}^4 - \frac{3}{2}\phi_x^2\phi_{xx}\phi_{4x}}{\phi_x^6} + \frac{\phi_x\phi_{5x} - 5\phi_{xxx}^2}{5\phi_x^4}\right),$$

but for $\varepsilon = 1$ this order contains no λ_4 and it vanishes identically when we use the expressions already found for λ_k with $k < 4$. Thus, λ_4 is a resonance. The expressions in higher order become lengthy, so we repeat the shorter argument used in the discussion of the PT-deformed Burgers equation (9.46). We substitute the general *ansatz* (9.60) into (9.65) and compute all values of r for which λ_r becomes a resonance. From the highest-order term $\phi^{-3\varepsilon-2+r}$, we find the necessary condition to have a resonance:

$$\varepsilon^\varepsilon(-i)^{\varepsilon-1}(3\varepsilon+1)^{\varepsilon-1}(r+1)(r-r_-)(r-r_+)\lambda_r\phi_x^{3\varepsilon} = 0,$$

with $r_\pm = -(2+3\varepsilon) \pm \sqrt{9\varepsilon^2 - 6\varepsilon - 2}$. Once again, we find the universal resonance at $r = -1$, and for $r_\pm \in \mathbb{Z}$ we require that $9\varepsilon^2 - 6\varepsilon - 2 = n^2$ with $n \in \mathbb{N}$. For the solution $\varepsilon_\pm = (1 \pm \sqrt{n^2+3})/3$ of this equation to be an integer we need to solve the diophantine equation $3 + n^2 = m^2$ $(n, m \in \mathbb{N})$. The only solution is $n = 1$, $m = 2$. Thus $r_\pm$ are only integers in the undeformed case with $\varepsilon = 1$ when they take the values $r_+ = 6$ and the previously identified value $r_- = 4$. Hence, only the *undeformed* KdV equation, which we know is integrable, can fully pass the Painlevé test.

Nonetheless, we may still be able to obtain a *defective series* if all remaining coefficients λ_j can be computed recursively. This is indeed the case, as we demonstrate for one choice of the deformation parameter; namely, $\varepsilon = \mu = 2$. For this choice the deformed KdV equation (9.65) becomes

$$u_t - 6iuu_x^2 + 2iu_{xx}^2 + 2iu_xu_{xxx} = 0.$$

The expressions for the expansion become lengthy at higher order so we present here only the solution of the recursive equation when $\lambda_k(x,t) = \lambda_k(t)$ and $\phi(x,t) = x - \xi(t)$, with $\xi(t)$ being an arbitrary function:

$$u(x,t) = \frac{7}{\phi^2} + \frac{i\xi'\phi^3}{156} + \frac{(\xi')^2\phi^8}{192192} - \frac{\xi''\phi^9}{681408} + \frac{i(\xi')^3\phi^{13}}{73081008} - \frac{725i\xi'\xi''\phi^{14}}{216449705472}$$

$$+ \frac{i\xi'''\phi^{15}}{20262348288} - \frac{340915(\xi')^4\phi^{18}}{23989859332927488} + \frac{1867(\xi')^2\xi''\phi^{19}}{758331543121152} + \mathcal{O}(\phi^{20}).$$

Thus, we have obtained a solution of Painlevé type for the deformed KdV equation, albeit without enough free parameters to accommodate all possible initial values. The Painlevé test is inconclusive in this situation. As in the case of the deformed Burgers equation, we may simplify this further by specifying $\xi(t)$ and then analyzing whether the expansion is convergent.

9.3.3   *Conserved quantities*

In Subsec. 9.3.2 we showed that the Painlevé test for the $\mathcal{PT}$-deformed KdV equation (9.65) is inconclusive, so this model might not be integrable. Nevertheless, we may still construct some nontrivial conserved quantities corresponding to physical observables, such as the energy [Fring (2007)]. The time evolution of any quantity $\mathcal{I}$ is governed by its Poisson bracket with the Hamiltonian H: $\frac{d\mathcal{I}}{dt} = \{\mathcal{I}, H\}$. Thus, if we can find an expression $\mathcal{I}$ such that the Poisson bracket vanishes, we have found a conserved quantity; that is, one that does not change with time.

We assume that the conserved quantity is a functional of u and that it can be expressed as the integral $\mathcal{I}[u] = \int \mathbf{T}[u]dx$. We then compute

$$\frac{d\mathcal{I}}{dt} = \int \frac{\delta \mathbf{T}}{\delta u} u_t dx = \int \frac{\delta \mathbf{T}}{\delta u} \left(\frac{\delta H}{\delta u} \right)_x dx = \{\mathcal{I}, H\}$$

and set the result to zero. For example, for the Hamiltonian H_ε^+ in (9.53) we can confirm that the quantities

$$\mathcal{I}^{(1)} = \int u dx, \qquad \mathcal{I}^{(2)} = \int u^2 dx, \qquad \text{and} \qquad \mathcal{I}^{(3)} = H_\varepsilon^+(u) \qquad (9.66)$$

are conserved in time when we neglect surface terms. This is justified by imposing standard boundary conditions (demanding that the field u and its derivatives either vanish at infinity for the noncompact case or be periodic for the compact case). To verify that $\mathcal{I}^{(n)}$ in (9.66) are conserved we use the equation of motion (9.53) to replace the time derivatives of u:

$$\frac{d\mathcal{I}^{(1)}}{dt} = -\int \left[\frac{\beta}{2} u^2 + \gamma \varepsilon (iu_x)^{\varepsilon-1} u_{xx} \right]_x dx = 0,$$

$$\frac{d\mathcal{I}^{(2)}}{dt} = -2 \int \left[\frac{1}{3}\beta u^3 + \frac{\varepsilon}{1+\varepsilon}\gamma(iu_x)^{\varepsilon+1} + \varepsilon\gamma u(iu_x)^{\varepsilon-1}u_{xx} \right]_x dx = 0,$$

$$\frac{d\mathcal{I}^{(3)}}{dt} = \left\{ \mathcal{I}^{(3)}, H \right\} = -\left\{ H, \mathcal{I}^{(3)} \right\} = 0.$$

The last conservation law reflects the antisymmetry of Poisson brackets.

We can also construct the generic *local* conservation laws

$$\mathbf{T}_t^{(n)} + \mathcal{X}_x^{(n)} = 0, \qquad (9.67)$$

where $-\mathcal{X}^{(n)}$ is the *flux*. [The negative sign in the flux as used here differs from the conventional sign of the current in (1.2).] Equation (9.67) guarantees that the Poisson brackets vanish and that $\mathcal{I}^{(n)}$ is a conserved quantity if we neglect surface terms. The case $n = 1$ in (9.67) is just the equation of motion because we can rewrite (9.53) as $u_t + \left[\beta u^2/2 + \gamma\varepsilon(iu_x)^{\varepsilon-1}u_{xx} \right]_x = 0$.

For the case $n = 2$ we compute $(u^2)_t$, use the equation of motion to replace all expressions involving u_t, and then integrate the result with respect to x. We then obtain the local conservation law

$$\left(u^2\right)_t + \left(\tfrac{2}{3}\beta u^3 + \tfrac{2\varepsilon}{1+\varepsilon}\gamma(iu_x)^{\varepsilon+1} + 2\varepsilon\gamma u(iu_x)^{\varepsilon-1}u_{xx}\right)_x = 0.$$

To construct the flux for $n = 3$ we compute $\mathbf{T}_t^{(3)} = (\mathcal{H}_\varepsilon^+)_t$ and get

$$\mathcal{X}^{(3)} = \gamma^2\left(\tfrac{1}{2}\varepsilon^2 - \varepsilon\right)(iu_x)^{2\varepsilon-2}u_{xx}^2 + \beta\gamma\left(u_x^2 - \tfrac{1}{2}\varepsilon u u_{xx}\right)u(iu_x)^{\varepsilon-1}$$
$$-i\varepsilon\gamma^2(iu_x)^{2\varepsilon-1}u_{xxx} - \tfrac{1}{8}\beta^2 u^4.$$

Higher conserved quantities are unlikely to exist because we have established that the deformed KdV equation (9.53) is not integrable.

We can also derive conserved quantities for the deformed version of the inviscid Burgers equation (9.49). We multiply the undeformed equation by $\kappa f(w)^\kappa$ and rearrange it into the local conservation law $[f(w)^\kappa]_t + \tfrac{\kappa}{\kappa+1}\left[f(w)^{\kappa+1}\right]_x = 0$. Thus

$$I_\kappa(w) = \int_{-\infty}^{\infty} f[w(x,t)]^\kappa dx \tag{9.68}$$

is conserved in time for any asymptotically vanishing function $f(w)$ and constant $\kappa \in \mathbb{R}\backslash\{-1\}$. Using the transformation (9.50), we obtain a conserved quantity for the transformed system:

$$I_\kappa(u) = \int_{-\infty}^{\infty} f[\varepsilon f(u)(iu_x)^{\varepsilon-1}]^\kappa dx.$$

9.3.4   *Solutions of PT-deformed nonlinear equations*

To see whether the conserved quantities obtained in Subsec. 9.3.3 are real despite being computed from complex fields, we need some concrete solutions of the equations of motion. Here we explain how to construct solutions to the deformed equations of motion using arguments from [Fring (2007); Cavaglia *et al.* (2011)]. The simplest solutions are *traveling-wave* solutions for which the field $u(x,t)$ depends on the combination $x - ct =: \zeta$, with the constant c denoting the wave speed

$$u(x,t) = u(\zeta) = u(x - ct). \tag{9.69}$$

A traveling-wave solution to a partial differential equation satisfies an ordinary differential equation for $u(\zeta)$.

PT-deformed KdV system $\mathcal{H}_\varepsilon^+$: Using $u(x,t)$ in (9.69), the deformed equation of motion (9.53) for the $\mathcal{H}_\varepsilon^+$ model becomes

$$-cu_\zeta + \tfrac{1}{2}\beta(u^2)_\zeta + \gamma\left(u_{\zeta,\varepsilon}\right)_{\zeta\zeta} = 0. \tag{9.70}$$

We integrate each term in (9.70) with respect to ζ and obtain

$$-cu + \tfrac{1}{2}\beta u^2 + \gamma \left(u_{\zeta,\varepsilon}\right)_\zeta = \kappa_1, \tag{9.71}$$

where κ_1 is the integration constant. Multiplying (9.71) by u_ζ, we can rewrite the resulting equation once more as a derivative:

$$-\tfrac{1}{2}c\left(u^2\right)_\zeta + \tfrac{1}{6}\beta\left(u^3\right)_\zeta + \tfrac{\varepsilon\gamma}{1+\varepsilon}\left(-iu_\zeta\right)_\zeta^{1+\varepsilon} = \kappa_1 u_\zeta. \tag{9.72}$$

Integrating (9.72) with respect to ζ then yields

$$\left(u_\zeta\right)^{1+\varepsilon} = \tfrac{1+\varepsilon}{\gamma\varepsilon} \exp\left[\tfrac{1}{2}i\pi(1-\varepsilon)\right]\left(\kappa_1 u + \kappa_2 + \tfrac{1}{2}cu^2 - \tfrac{1}{6}\beta u^3\right) \tag{9.73}$$

with an additional integration constant κ_2. Finally, taking the $(1+\varepsilon)$th root of (9.73) and separating variables gives the traveling-wave solution

$$\zeta - \zeta_0 = \exp\left[\tfrac{i\pi(\varepsilon-1)}{2(\varepsilon+1)}\right]\int du\,[\lambda_\varepsilon \mu P(u)]^{-1/(1+\varepsilon)}, \tag{9.74}$$

where we have introduced the constant $\lambda_\varepsilon := (1+\varepsilon)/(\gamma\varepsilon)$ and the polynomial $P(u)$ multiplied by a constant μ:

$$\kappa_1 u + \kappa_2 + \tfrac{1}{2}cu^2 - \tfrac{1}{6}\beta u^3 := \mu P(u). \tag{9.75}$$

When $(1+\varepsilon)$ is a rational number, the zeros of $P(u)$ are branch points. To obtain $u(\zeta)$ rather than $\zeta(u)$ we still must calculate the integral in (9.75) and then solve for u. Assuming special forms for the polynomial $P(u)$ leads to different types of solutions. When all three roots coincide, we obtain rational solutions for $u(\zeta)$; when two are identical, the solutions are expressed in terms of trigonometric functions; when all three roots are different, the integral in (9.74) produces elliptic functions. These different solutions have varied characteristic behaviors and boundary conditions.

Let us consider a few specific cases and discuss the role PT symmetry plays in guaranteeing that the conserved quantities are real. Starting with the KdV equation with deformation option (i), where the field is complex, we take $\varepsilon = 1$, $\beta = -6$, $\gamma = 1$, and $\kappa_1 = \kappa_2 = 0$ corresponding to vanishing asymptotic boundary conditions $\lim_{\zeta\to\pm\infty} u = \lim_{\zeta\to\pm\infty} u_\zeta = 0$. Therefore, we have $\mu = 2$ and $P(u) = u^2(u + c/2)$, which yields by (9.74) the one-soliton solutions (9.36) and (9.37) with $c_1 = c_2 = c$. Using these solutions to calculate the conserved quantities in (9.66), we obtain

$$\mathcal{I}^{(1)} = -2c^{1/2}, \qquad \mathcal{I}^{(2)} = \tfrac{2}{3}c^{3/2}, \qquad \mathcal{I}^{(3)} = H = -\tfrac{1}{5}c^{5/2}$$

in both cases. These expressions hold whether the constant δ is real or complex. Both solutions (9.36) and (9.37) are PT_+ symmetric only when $\mathrm{Re}\,\delta = 0$. However, as argued in [Cen and Fring (2016); Cen et al. (2017)], in the expressions for the conserved quantities we can shift in x and absorb the real part of δ in the integral limits, or in t, as the expressions hold for any time. For the two-soliton solution we can compensate the real parts from two different shift parameters by a simultaneous shift in x and t. For the N-soliton solutions with $N > 2$ we cannot compensate for this type of PT_+-symmetry breaking by shifts as we have only two parameters, x and t, to shift. We can, however, use the fact that the N-soliton solutions separate asymptotically as single solitons and compute the conserved quantities for each of these one-soliton contributions.

Next we study the case when all three zeros coincide. We use the integral

$$\int du (u - A)^{-3/(1+\varepsilon)} = \begin{cases} \frac{\varepsilon+1}{\varepsilon-2} (u - A)^{\frac{\varepsilon-2}{\varepsilon+1}} & (\varepsilon \neq 2), \\ \ln(u - A) & (\varepsilon = 2). \end{cases}$$

Considering (9.75) we find that this factorization of $P(u)$ is indeed possible when we select the model-dependent constants as

$$\mu = -\frac{\beta}{6}, \quad \kappa_1 = -\frac{c^2}{2\beta}, \quad \kappa_2 = \frac{c^3}{6\beta^2} \quad \text{and} \quad A = \frac{c}{\beta}. \tag{9.76}$$

The boundary conditions, characterized by the constants κ_1 and κ_2, cannot be chosen freely and are fixed by the nonvanishing values in (9.76). Solving the resulting equation (9.74) for u at $\varepsilon = 2$, we obtain

$$u(\zeta) = \frac{c}{\beta} + \exp\left[-i \left(\frac{\beta}{4\gamma}\right)^{1/3} (\zeta - \zeta_0)\right]. \tag{9.77}$$

For the remaining cases with $\varepsilon \neq 2$ we compute

$$u(\zeta) = \frac{c}{\beta} + e^{\frac{i\pi(1-\varepsilon)}{2(\varepsilon-2)}} \left(\frac{\varepsilon-2}{\varepsilon+1}\right)^{\frac{\varepsilon+1}{\varepsilon-2}} \left(-\frac{\beta(\varepsilon+1)}{\gamma\varepsilon}\right)^{\frac{1}{\varepsilon-2}} (\zeta - \zeta_0)^{\frac{\varepsilon+1}{\varepsilon-2}}. \tag{9.78}$$

For the solutions (9.77) and (9.78) we observe that the PT_+ symmetry maps different branches into one another. Moreover, because these solutions are not asymptotically vanishing, we cannot use (9.66) to compute the conserved quantities.

PT-deformed Burgers equation: In addition to solutions in terms of rational, trigonometric, and elliptic functions, we can construct complex solutions to the deformed Burgers equation (9.46) for $\varepsilon = 2$ in terms of Airy functions. Using the traveling-wave *ansatz* (9.69), we get

$$-cu_\zeta + iuu_\zeta^2 + 2\kappa u_\zeta u_{\zeta\zeta} = 0.$$

When $u_\zeta \neq 0$ we can rewrite this equation as

$$\frac{d}{d\zeta}\left(\delta - c\zeta + \frac{i}{2}u^2 + 2\kappa u_\zeta\right) = 0.$$

When integrated, this equation becomes

$$u(\zeta) = e^{i\pi 5/3}(2c\kappa)^{1/3}\frac{\tilde{\delta}\,\mathrm{Ai}'(\chi) + \mathrm{Bi}'(\chi)}{\tilde{\delta}\,\mathrm{Ai}(\chi) + \mathrm{Bi}(\chi)},$$

where δ, $\tilde{\delta}$ are constants, $\chi = e^{i\pi/6}(c\zeta - \delta)(2v\kappa)^{-2/3}$, and $\mathrm{Ai}(\chi)$, $\mathrm{Bi}(\chi)$ are Airy functions.

$\mathcal{H}_\varepsilon^-$ **models:** Next we discuss deformations that respect $\mathcal{PT}_-$ symmetry. Our approach is similar to that of the previous two cases. To construct solutions we integrate the equation of motion in (9.54) twice and obtain

$$u_\zeta^2 = \frac{2}{\gamma}\left(\kappa_2 + \kappa_1 u + \frac{c}{2}u^2 - \beta\frac{i^\varepsilon}{(1+\varepsilon)(2+\varepsilon)}u^{2+\varepsilon}\right) =: \lambda Q(u), \qquad (9.79)$$

where $\lambda = -2\beta i^\varepsilon/[\gamma(1+\varepsilon)(2+\varepsilon)]$.

A crucial difference between the $\mathcal{H}_\varepsilon^+$ and $\mathcal{H}_\varepsilon^-$ models is that the order of $Q(u)$ depends on ε, unlike the polynomial $P(u)$ in (9.75), which was of fixed order 3. This means that we have more possibilities to factor for larger ε. For instance, for a given integer value $n \in \mathbb{N}_0$, the factored form of $Q(u)$

$$Q(u) = (u - A_1)^{\varepsilon+2-n}\prod_{i=1}^{n}(u - A_{i+1}), \qquad (9.80)$$

admits solutions provided that $n - 2 \leq \varepsilon \leq n + 1$ and $\varepsilon \in \mathbb{N}$. This allows for yet another infinity of possibilities.

For vanishing boundary conditions, $\kappa_1 = \kappa_2 = 0$, we find a closed-form solution valid for all ε by integrating (9.79) and solving for u:

$$u(\zeta) = \left(\frac{c(\varepsilon+1)(\varepsilon+2)}{i^\varepsilon\beta\left[\cosh\left(\sqrt{c}\,\varepsilon(\zeta - \zeta_0)/\sqrt{\gamma}\right) + 1\right]}\right)^{1/\varepsilon}. \qquad (9.81)$$

Nonvanishing boundary conditions do not give such a compact answer.

The case $\mathcal{H}_2^-$ is especially interesting because it corresponds to a complex version of the mKdV equation already encountered in (9.19). If we specify (9.80) as $Q(u) = (u - A)^3(u - B)$, we can factor the polynomial in (9.79) with the choice

$$\kappa_1 = -\frac{2c^{3/2}}{3\sqrt{-\beta}}, \quad \kappa_2 = -\frac{c^2}{4\beta}, \quad A = -\frac{B}{3}, \quad \text{and} \quad B = -\frac{3\sqrt{c}}{\sqrt{-\beta}}.$$

This fixes all the boundary conditions for a given model. Solving (9.79) then yields a rational solution for the equation of motion in (9.54):

$$u(\zeta) = \sqrt{-\frac{c}{\beta}}\frac{2c\zeta^2 - 9\gamma}{3\gamma + 2c\zeta^2}. \qquad (9.82)$$

As discussed in Subsec. 9.1.3, for the real case one may construct solutions for the KdV equation from those of the modified KdV equation by means of a Miura transformation. We expect this also to hold for their complex versions. Indeed, using the transformation of the form

$$u_{\mathrm{KdV}}(\zeta) = \sqrt{6\gamma/\beta}\, u_\zeta - u^2 \tag{9.83}$$

yields the rational solution of the KdV equation in (9.82) when we identify $\zeta_0 = i\sqrt{3\gamma/2c}$.

Alternatively, if we assume $Q(u) = (u - A)^2(u - B)(u - C)$ in (9.80), we can factor the polynomial in (9.79) with the constraints

$$A = -(B+C)/2, \quad \kappa_1 = \left[\beta C^2(\vartheta - 5C\beta) + 9c(\vartheta - 3C\beta)\right]/(81\beta),$$
$$B = (2\vartheta - C\beta)/(3\beta), \quad \kappa_2 = \left[C(2\vartheta - C\beta)(C\beta + \vartheta)^2\right]/(324\beta^2),$$

where $\vartheta = \sqrt{-2\beta(\beta C^2 + 9c)}$. In this case one constant remains free. The integration of (9.79) yields a trigonometric solution, which the same Miura transformation (9.83) converts into a solution of the KdV equations.

For the case $\mathcal{H}_4^-$ the polynomial on the right side of (9.79) is of sixth order. We present here just one symmetric solution by assuming the factorization $Q(u) = u^2(u^2 - B^2)(u^2 - C^2)$, which is possible with the choice

$$\kappa_1 = \kappa_2 = 0, \qquad B = iC, \qquad \text{and} \qquad C^4 = 15c/\beta.$$

Thus, we have made contact with the solution (9.81). We can understand the reality of the conserved quantities in the $\mathcal{H}^-$ models by using the same arguments as for the $\mathcal{H}^+$ models.

We can also investigate the role that $\mathcal{PT}_-$ symmetry plays in governing the shapes of the trajectories. In Figs. 9.2 and 9.3 we plot u as functions of ζ for several of the trajectories for different values of the model parameters. Panels (a) show $\mathcal{PT}_-$-symmetric solutions as we can clearly observe with $\operatorname{Re} u \leftrightarrow -\operatorname{Re} u$, $\operatorname{Im} u \leftrightarrow \operatorname{Im} u$, and reversed time, whose direction is indicated by arrows for some curves. In panels (b) the $\mathcal{PT}$ symmetry is broken by choosing $\operatorname{Im}\beta \neq 0$. Observe that the overall shapes of the trajectories are slightly distorted. In Fig. 9.3 we have closed periodic orbits about fixed-point centers, whereas in Fig. 9.2 the trajectories leave or enter the fixed point, defined by $u_\zeta = 0$, at the origin in straight lines. Such fixed points are referred to as *star nodes*.

Hence it appears that the $\mathcal{PT}_-$ symmetry plays only a minor role in determining the overall shape; the distinction between trajectories with periodic orbits and those with asymptotic fixed points is controlled by a different principle. So how do we understand this dramatic change from the scenario in Fig. 9.2 to Fig. 9.3 and the similarity in behavior in the different $\mathcal{PT}$ regimes? Can we possibly predict the nature of the fixed points?

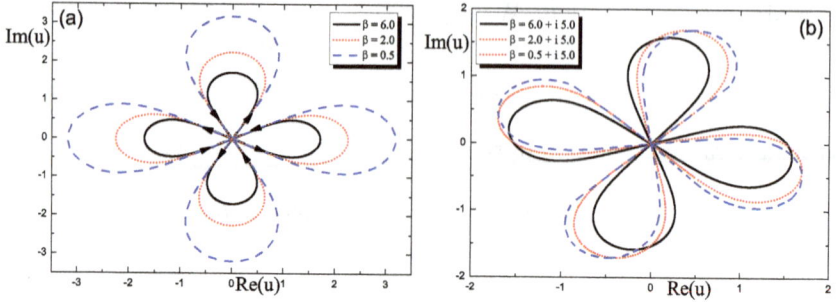

Fig. 9.2   Trajectories in the $\mathcal{H}_4^-$ model with a star-node fixed point at the origin for $c = 1$, $\gamma = 3$, $\zeta_0 = 0.5i$. For real values $\beta = 2$ the $\mathcal{PT}_-$-symmetry is unbroken (panel a) and broken (panel b) when $\mathrm{Im}\,\beta \neq 0$.

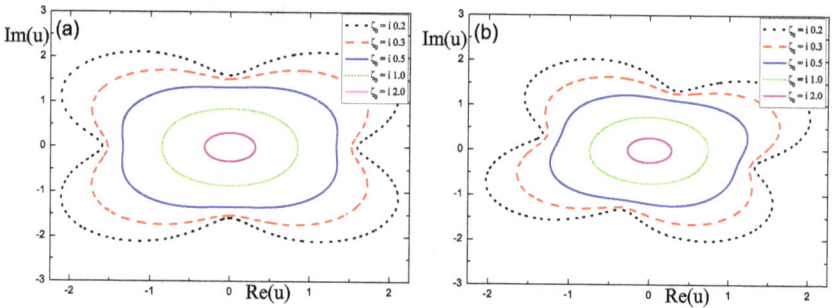

Fig. 9.3   Trajectories in the $\mathcal{H}_4^-$ model with closed periodic orbits (centers) around the origin for $c = 1$, $\gamma = -3$, and varying values of ζ_0. For real values of β the $\mathcal{PT}_-$-symmetry is unbroken (panel a) and broken (panel b) for $\beta = 2 + 3i$.

To answer these questions it is useful to look at the equations of motion in a slightly different way and to separate the complex function u into its real and imaginary part u^r and u^i. We then view (9.79) as a system of two coupled first-order differential equations in these variables rewritten as

$$u_\zeta^r = \pm \mathrm{Re}\left[\sqrt{\lambda}\sqrt{Q(u^r + iu^i)}\right], \quad u_\zeta^i = \pm \mathrm{Im}\left[\sqrt{\lambda}\sqrt{Q(u^r + iu^i)}\right].$$

For such types of systems many useful techniques have been developed that allow one to deduce the qualitative behavior even when the analytical solutions of the dynamical system under investigation are not known explicitly.[7] Here we use the *linearization theorem*, which allows one to extract information about a nonlinear system in the neighborhood of a fixed point u_f

---

[7]See, for instance, [Arrowsmith and Place (1990)] for the general theory.

by converting the nonlinear system to a linear system as

$$\begin{pmatrix} u_\zeta^r \\ u_\zeta^i \end{pmatrix} = J(u^r, u^i)\big|_{u=u_f} \begin{pmatrix} u_\zeta^r \\ u_\zeta^i \end{pmatrix},$$

with the Jacobian matrix

$$J(u^r, u^i)\big|_{u=u_f} = \begin{pmatrix} \pm\frac{\partial}{\partial u^r}\operatorname{Re}\sqrt{\lambda Q(u)} & \pm\frac{\partial}{\partial u^i}\operatorname{Re}\sqrt{\lambda Q(u)} \\ \pm\frac{\partial}{\partial u^r}\operatorname{Im}\sqrt{\lambda Q(u)} & \pm\frac{\partial}{\partial u^i}\operatorname{Im}\sqrt{\lambda Q(u)} \end{pmatrix}\Bigg|_{u=u_f}.$$

Recall that a *fixed point* is defined by the property $u_\zeta^r(u_f) = u_\zeta^i(u_f) = 0$. The linearization theorem states that if $\det J(u^r, u^i)\big|_{u=u_f} \neq 0$ and the linearized system is not a center, the phase portraits (that is, a collection of many trajectories) in some neighborhood of the fixed point of the nonlinear and linearized system are qualitatively equivalent.

Qualitative behaviors near fixed points of linear systems are classified according to the eigenvalues of the Jacobian. For 2×2 matrices there are ten similarity classes. Denoting the eigenvalues of $J(u = u_f)$ by j_1, j_2 there are the following types of behavior at the fixed point: stable (unstable) nodes when $j_1 < j_2 < 0$ ($j_1 > j_2 > 0$); saddle points when $j_1 < 0 < j_2$ (only unstable); stable (unstable) star nodes when $j_1 = j_2 < 0$ ($j_1 = j_2 > 0$) with diagonal J; stable (unstable) improper nodes when $j_1 = j_2 < 0$ ($j_1 = j_2 > 0$) with nondiagonal J; stable (unstable) foci when $\operatorname{Re} j_{1,2} < 0$ ($\operatorname{Re} j_{1,2} > 0$) and $\operatorname{Im} j_{1,2} \neq 0$; centers when $\operatorname{Re} j_{1,2} = 0$ and $\operatorname{Im} j_{1,2} \neq 0$.

Using the notation $z = r_z e^{i\theta_z}$ for the polar form of a complex number z, we compute the eigenvalues of the Jacobian when linearized about $u = 0$:

$$\begin{aligned}
j_1 &= \pm i\sqrt{r_\lambda}r_B^2 \exp\left[\tfrac{1}{2}i(4\theta_B + \theta_\lambda)\right] = \mp\sqrt{r_\lambda}r_B^2 \exp\left(-\tfrac{1}{2}i\theta_\gamma\right), \\
j_2 &= \mp i\sqrt{r_\lambda}r_B^2 \exp\left[-\tfrac{1}{2}i(4\theta_B + \theta_\lambda)\right] = \mp\sqrt{r_\lambda}r_B^2 \exp\left(\tfrac{1}{2}i\theta_\gamma\right), \quad (9.84)
\end{aligned}$$

where we have used $4\theta_B + \theta_\lambda = \pi - \theta_\gamma$. We deduce from these expressions that the nature of the fixed point is not governed by the $\mathcal{PT}$ symmetry, that is, by the parameters we tune to break it. The value of β is entirely irrelevant whereas changing the values of γ drastically alters the qualitative behavior. The eigenvalues in (9.84) suggest that for positive real γ with $\theta_\gamma = 0$ there are star nodes at the origin, whereas for complex γ with $\theta_\gamma = \pi$ there are centers (periodic orbits). This corresponds to the behavior we observe in Figs. 9.2 and 9.3.

According to the classification of fixed points above, we should therefore be able to produce a focus (trajectories that spiral into the origin) by choosing γ with a nonvanishing imaginary part. This is indeed possible, as illustrated in Fig. 9.4. The example displayed in the figure confirms that

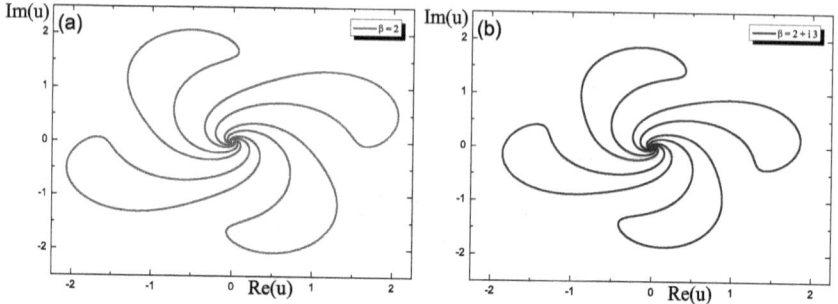

Fig. 9.4   Trajectories in the $\mathcal{H}_4^-$ model with a focus at the origin for $c = 1$, $\gamma = -0.5+2i$, $\zeta_0 = 2.5i$ with broken $\mathcal{PT}_-$ symmetry. There is no major change in behavior between real values of β (panel a) and Im $\beta \neq 0$ (panel b).

there is no qualitative difference in behavior when changing β from being real to having a nonvanishing complex part.

We may treat γ or β as *bifurcation* parameters. Varying them leads to qualitatively different fixed-point behaviors. This change in behavior is usually referred to as *Hopf bifurcation*. We have seen that some of the transitions from one characteristic fixed-point behavior to another coincide with a $\mathcal{PT}_-$ symmetry breaking (from Fig. 9.3 to Fig. 9.4), whereas others leave $\mathcal{PT}_-$ symmetry unchanged (from Fig. 9.2 to 9.3). Evidently the techniques explained here for the $\mathcal{H}_4^-$ model can be applied to a wider range of models, especially to all of the $\mathcal{H}_\varepsilon^-$ series.

From shock waves to peaked solutions: Let us now see how a different type of wave, a *shock wave*, is manifested in the deformed equations. For this purpose we first show how the undeformed equation in (9.49) is solved by using the *method of characteristics*. We note that

$$\frac{d}{dt}w = w_x \frac{dx}{dt} + w_t = 0 \quad \text{iff} \quad \frac{dx}{dt} = f(w). \tag{9.85}$$

This means that when the last equality holds, the field $w(x,t)$ is conserved in time. As $w(x,t)$ takes the same values at all times, we get

$$w(x,t) = w(x_0,t) = w_0(x_0) = w_0[x - f(w_0)t], \tag{9.86}$$

where in the last equality we note that the condition in (9.85) is solved by the *characteristic* $x = f(w_0)t + x_0$. Given an initial profile $w_0(x)$, the general solution to the first equation in (9.49) is (9.86).

When two characteristics cross, a shock wave (also called a *gradient catastrophe*) is formed because $w_x \to \infty$. This situation occurs when the crest of a wave overtakes the troughs (Fig. 9.5a). This happens because the wave speed at the crest is faster than the wave speed at the trough. This poses a problem because the wave profile is no longer a single-valued function, but instead becomes multivalued.

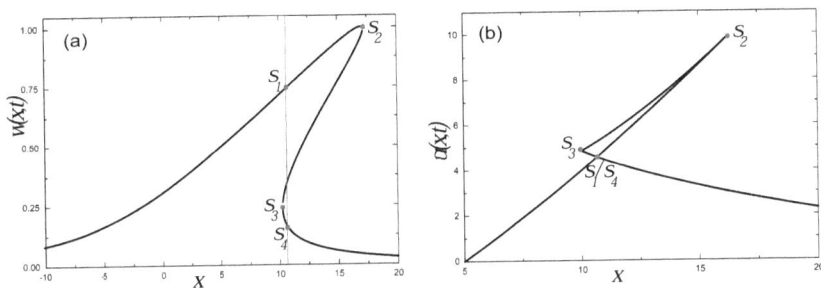

Fig. 9.5 (a) A multivalued self-avoiding wave $w(x, t_1)$. (b) Transformed multivalued self-crossing wave $\tilde{u}(x, t_1)$ for $\varepsilon = 3$. The vertical line in (a) indicates the shock front for the conservation law with $\kappa = (\varepsilon - 1)^{-1}$.

The smallest time for which this gradient catastrophe occurs is called the *breaking time*, or *shock time*. This is smallest time for which

$$w_x(x, t) = w_0'[x_0(x, t)]\frac{dx_0}{dx} = w_0'(x_0) \Big/ \left[1 + t\frac{d}{dx_0}f(w_0)\right]$$

becomes infinite. Thus, the shock time t_{shock} is

$$t_{shock} = \min_t\left\{-1 \Big/ \left(\frac{d}{dx_0}f[w_0(x_0)]\right)\right\}. \tag{9.87}$$

How does this expression translate to the deformed equation? For $f(w) = w^n$, the shock time (9.87) is [Cavaglia and Fring (2012)]

$$t_{shock}^w = -1 \Big/ \left\{\sqrt[n]{\varepsilon}\frac{d}{dx_0}\left[\sqrt[n]{u_0}(iu_{x_0})^{\frac{\varepsilon-1}{n}}\right]\right\}, \tag{9.88}$$

which for $n = 1$ agrees with the expression in [Bender and Feinberg (2008)]. This time is real when we replace $u_0 \to i^\alpha \hat{u}_0$ with $\hat{u}_0 \in \mathbb{R}$ and $\alpha = (4m \pm 1)n/\varepsilon$, $m \in \mathbb{Z}$. Consequently, for some values of n and ε, some shock times correspond to scenarios in which the undeformed wave is real and the deformed wave is complex, but we can engineer both waves to be real. So how is a shock wave in the undeformed case transformed to the deformed case? We compute w_x for $n = 1$ from (9.50):

$$w_x = i\varepsilon(iu_x)^{\varepsilon-2}\left[u_x^2 + (\varepsilon - 1)uu_{xx}\right].$$

We observe that a shock $w_x \to \infty$ for finite u might result from either $u_x \to \infty$ or $u_{xx} \to \infty$. Thus, any shock in the deformed system will lead to a shock in the undeformed system, but the latter might also result from a gradient catastrophe with $u_{xx} \to \infty$. In both cases the shock time is given by (9.88). We can see this more explicitly by expressing u in terms of w:

$$u(x,t) = (-i)^{1-\frac{1}{\varepsilon}}(\varepsilon - 1)^{\frac{1}{\varepsilon}-1}\varepsilon^{\frac{\varepsilon-2}{\varepsilon}}\left[\int^x w(q,t)^{\frac{1}{\varepsilon-1}}\,dq\right]^{\frac{\varepsilon-1}{\varepsilon}}.$$

Differentiating u twice yields

$$u_x(x,t) = (-i)^{1-\frac{1}{\varepsilon}}(\varepsilon - 1)^{\frac{1}{\varepsilon}}\varepsilon^{-\frac{2}{\varepsilon}}w(x,t)^{\frac{1}{\varepsilon-1}}\left[\int^x w(q,t)^{\frac{1}{\varepsilon-1}}\,dq\right]^{-\frac{1}{\varepsilon}}, \quad (9.89)$$

$$u_{xx}(x,t) = (-i)^{1-\frac{1}{\varepsilon}}(\varepsilon - 1)^{\frac{1}{\varepsilon-1}}\varepsilon^{\frac{\varepsilon-2}{\varepsilon}}\left(\int^x w(q,t)^{\frac{1}{\varepsilon-1}}\,dq\right)^{-\frac{\varepsilon+1}{\varepsilon}}w(x,t)^{\frac{2}{\varepsilon-1}}$$

$$\times \varepsilon\left(1 - \varepsilon + \int^x w(q,t)^{\frac{1}{\varepsilon-1}}\,dq\right)w_x(x,t)w(x,t)^{\frac{\varepsilon}{1-\varepsilon}}. \quad (9.90)$$

From (9.90) we observe that the direct dependence of u_{xx} on w_x implies that a shock in the undeformed system leads to the gradient catastrophe $u_{xx} \to \infty$. However, u_x depends only on w, so a shock in the deformed case cannot be traced back directly to a shock in the undeformed case.

We now demonstrate that a shock is converted into a peak by means of the PT deformation. To do this we consider the function

$$\tilde{u}(s,t) := iu(x,t)^{\frac{\varepsilon}{\varepsilon-1}} = (\varepsilon-1)^{-1}\varepsilon^{\frac{\varepsilon-2}{\varepsilon-1}}\int^s w(q,t)^{\frac{1}{\varepsilon-1}}\frac{dq}{d\tilde{s}}\,d\tilde{s},$$

instead of u and parametrize it by arc length s with arc length element $d\tilde{s} = \sqrt{dw^2 + dx^2}$. This converts the multivalued function $w(x,t)$ (like that in Fig. 9.5a) into a single-valued function. As an example, we calculate $\tilde{u}(s,t)$ for $\varepsilon = 3$ and the positive and negative square root for $s < s_3$ and $s > s_3$. If we transform back from s to x, this function becomes multivalued, crossing itself, as depicted in Fig. 9.5b for $t_1 > t_s$.

With this choice of branches we obtain a peaked function. (In general, possible choices for different branches might be restricted by specific boundary conditions to match the initial profile.) Remarkably we can eliminate the loop above the peak $s_1 \to s_2 \to s_3 \to s_4$ without violating the conservation laws. The conserved quantity I_κ with $\kappa = (\varepsilon - 1)^{-1}$ in (9.68) remains unchanged when we chop off the loop, since for that choice we have $\int_{-\infty}^{\infty} = \int_{-\infty}^{s_1} + \int_{s_4}^{\infty}$. However, this only holds for the specific value of κ. Changing from $\tilde{u}(s,t)$ to $u(s,t)$ will not change the main argument apart from introducing yet more possible branches. We conclude that shocks in the undeformed cases are converted into peaks in the deformed equations.[8]

---

[8]We might also have an entirely different scenario. We observe from (9.89) that we can have shocks in the deformed system, i.e. $u_x \to \infty$, whenever $\int^x w(q,t)^{1/(\varepsilon-1)}\,dq = 0$ for $\varepsilon > 0$. Such types of solutions are studied numerically in [Cavaglia and Fring (2012)].

9.3.5   *From wave equations to quantum mechanics*

We showed in (9.39) how solutions to nonlinear wave equations can be related to quantum-mechanical systems. We conclude this section by demonstrating a further possibility that arises for $\mathcal{H}_\varepsilon^-$ models. If we identify $u \to x$ and $\zeta \to t$ for the traveling-wave equation (9.79), then taking a derivative with respect to t yields the equation

$$\ddot{x} = \frac{1}{\gamma}\left(\kappa_1 + cx - \beta \frac{i^\varepsilon}{1+\varepsilon} x^{\varepsilon+1}\right).$$

This equation also results when combining Hamiltonian's equations into Newton's equation for the Hamiltonian

$$H_\varepsilon = \frac{p^2}{2} - \frac{\kappa_1}{\gamma}x - \frac{c}{2\gamma}x^2 + \frac{\beta}{\gamma}\frac{i^\varepsilon}{(1+\varepsilon)(2+\varepsilon)}x^{\varepsilon+2}.$$

For $\varepsilon = 1$, the special choices $\kappa_1 = 0$, $\beta = -6cg$, $\gamma = -c$, give the complex cubic oscillator $H_{\text{cubic}} = \frac{1}{2}p^2 + \frac{1}{2}x^2 + igx^3$ [Bender and Boettcher (1998b)], and for $\varepsilon = 1$ with $\kappa_1 = -\gamma\tau$, $\beta = -3\gamma g$, $c = -\gamma\omega^2$ the Hamiltonian density $\mathcal{H}_\varepsilon$ becomes the quartic oscillator $H_{\text{quartic}} = E_x = \frac{1}{2}p^2 + \tau x + \frac{1}{2}\omega^2 x^2 + \frac{1}{4}gx^4$ [Anderson *et al.* (2011b)]. This observation shows that one may translate some properties of nonlinear systems, such as trajectories, to quantum-mechanical models. Choosing $\kappa_2 = \gamma E_x$ gives asymptotically vanishing waves $\lim_{\zeta\to\infty} u(\zeta) = 0$ obeying the Neumann boundary condition $\lim_{\zeta\to\infty} u_x(\zeta) = \sqrt{2E_x}$. On the quantum-mechanical side this implies that we can identify $H_\varepsilon = E_x$ as the total energy of the system. In short, the energy in the classical analog of a complex classical particle corresponds to an integration constant in the nonlinear wave-equation context multiplied by one of the coupling constants.[9]

9.4 $\mathcal{PT}$-Deformed Calogero–Moser–Sutherland Models

In addition to integrable systems of fields obeying nonlinear wave equations, integrable multiparticle systems have also been well studied as Calogero-type models. This section discusses their $\mathcal{PT}$-symmetric deformations. The undeformed models all trace back to the seminal paper [Calogero (1969b)], which considers the behavior of three indistinguishable spinless particles moving along the real line. The particles interact pairwise via a combined short-range centrifugal-like potential $V_c(x_i, x_j) = g/(x_i - x_j)^2$ and a long-range quadratic attraction $V_q(x_i, x_j) = \frac{1}{8}\omega^2(x_i - x_j)^2$ with $1 \leq i, j \leq 3$

---

[9]This not the same as the energy in (9.42), which leads to different conclusions regarding the reality of these quantities resulting from the various $\mathcal{PT}$-symmetric scenarios.

and $g, \omega \in \mathbb{R}$. Remarkably, the model is separable and solvable, and it has become a central part of the canon of integrable systems.

A key feature of the quantum version of this model is the quantum impenetrability of its singular interaction $V_c(x_i, x_j)$. As a consequence of impenetrability, the domain of definition of the model must be specified in Weyl chambers characterized by a fixed and unchangeable ordering, such as $x_1 < x_2 < x_3$, of the particles on the real line. Of course, for a genuine physical model we would like the particles to be able to pass one another; for example, we would like to allow $x_1 < x_2$ and $x_2 < x_1$ at different times. The restriction $x_1 < x_2 < x_3$ is unappealing and in Subsec. 9.4.3 we discuss how $\mathcal{PT}$ symmetry can be utilized to resolve this issue.

The Calogero model has been generalized in many ways, such as by including more particles by changing 3 to generic $N \in \mathbb{N}$ in the formulas [Calogero (1969a)], allowing for nonpairwise interactions [Olshanetsky and Perelomov (1977, 1981)], using different types of potentials [Sutherland (1971a,b); Moser (1975)], including interacting spins [Krichever *et al.* (1995)], and allowing the particles to leave the real axis, as discussed here.

The Hamiltonian for this large class of Calogero–Moser–Sutherland (CMS) models is

$$H_{CMS} = \tfrac{1}{2}p^2 + \tfrac{1}{4}\omega^2 \sum_{\alpha \in \Delta}(\alpha \cdot q)^2 + \tfrac{1}{2}\sum_{\alpha \in \Delta} g_\alpha^2 V(\alpha \cdot q). \qquad (9.91)$$

Here, $\omega, g_\alpha, \tilde{g}_\alpha \in \mathbb{R}$ are coupling constants, $q, p \in \mathbb{R}^{\ell+1}$ are canonical variables, and the α's are $(\ell + 1)$-dimensional vectors in some vector space Δ. In general, the α's are taken to be roots belonging to a root system Δ that is invariant under the Coxeter group [Humphreys (1972)].

We focus here on the A_2 Coxeter group for which there are only six roots, which may be represented as three-dimensional vectors $\alpha_1 = \{1, -1, 0\}$, $\alpha_2 = \{0, 1, -1\}$, and $\alpha_3 = \alpha_1 + \alpha_2 = \{1, 0, -1\}$ together with their negatives [Fring (2006)]. The standard CMS models H_C, H_{CM}, H_{CM}, H_{CMS} are obtained by considering the singular potential $V(x)$ to be rational $V(x) = 1/x^2$, trigonometric $V(x) = 1/\sin^2(x)$, hyperbolic $V(x) = 1/\sinh^2(x)$, or elliptic $V(x) = 1/\operatorname{sn}^2(x)$. Spins may be introduced by including additional indices to the coupling constants [Krichever *et al.* (1995)].

9.4.1   *Extended Calogero–Moser–Sutherland models*

We can obtain $\mathcal{PT}$-deformed versions of the CMS models H_{CMS} by following option (ii) in Sec. 9.2 and adding a $\mathcal{PT}$-symmetric term $H_{\mathcal{PT}}$ to the original model. This was done for a specific A_n Calogero model in [Basu-Mallick and Kundu (2000)] and generalized in [Fring (2006)] to all Coxeter groups, representation independent of the roots and different potentials as

$$H_{\mathcal{PTCMS}}(p, q, g_\alpha, \tilde{g}_\alpha) := H_{CMS}(p, q, g_\alpha) + H_{\mathcal{PT}}(p, q, \tilde{g}_\alpha), \qquad (9.92)$$

where

$$H_{\mathcal{PT}} = \tfrac{1}{2} i \tilde{g}_\alpha f(\alpha \cdot q) \alpha \cdot p. \qquad (9.93)$$

with $f(x) := \sqrt{V(x)}$. When $f(x)$ is an odd function, both terms in $H_{\mathcal{PTCMS}}(p, q, g_\alpha, \tilde{g}_\alpha)$ are invariant under the antilinear symmetry $\mathcal{PT}$: $p_j \mapsto p_j$, $q_j \mapsto -q_j$, $i \mapsto -i$. For the rational case one may rewrite the Hamiltonian in (9.92) so that it becomes the standard Calogero Hamiltonian with momenta shifted by $i\mu := i/2 \sum_{\alpha \in \Delta} \tilde{g}_\alpha f(\alpha \cdot q) \alpha$ with the new constant $\hat{g}$ [Fring (2006)]:

$$H_{\mathcal{PTC}}(p, q, g_\alpha, \tilde{g}_\alpha) = H_C(p + i\mu, q, \hat{g}). \qquad (9.94)$$

The coupling constants have been redefined to $\hat{g}_\alpha^2 := g_{s,l}^2 + \alpha_{s,l}^2 \tilde{g}_{s,l}^2$ for $\alpha \in \Delta_{s,l}$, where Δ_l and Δ_s refer to the root system of the long and short roots. For the rational case only, the identity (9.94) is based on the identity $\mu^2 = \alpha_s^2 \tilde{g}_s^2 \sum_{\alpha \in \Delta_s} V(\alpha \cdot q) + \alpha_l^2 \tilde{g}_l^2 \sum_{\alpha \in \Delta_l} V(\alpha \cdot q)$.

Since the identities are nontrivial in general, we present here the details for the A_2 model. For this model we have

$$\mu = \tilde{g} \left(\frac{\alpha_1}{q_1 - q_2} + \frac{\alpha_2}{q_2 - q_3} + \frac{\alpha_3}{q_1 - q_3} \right),$$

from which we calculate

$$\mu^2 = \tilde{g}^2 \left[\frac{\alpha_1^2}{(q_1 - q_2)^2} + \frac{\alpha_2^2}{(q_2 - q_3)^2} + \frac{\alpha_3^2}{(q_1 - q_3)^2} \right] \qquad (9.95)$$
$$+ \frac{2\alpha_1 \cdot \alpha_2}{(q_1 - q_2)(q_2 - q_3)} + \frac{2\alpha_2 \cdot \alpha_3}{(q_2 - q_3)(q_1 - q_3)} + \frac{2\alpha_1 \cdot \alpha_3}{(q_1 - q_2)(q_1 - q_3)}.$$

The second line in (9.95) vanishes, so μ^2 can be expressed in terms of the Calogero potential, which establishes (9.94). The identities for other Coxeter groups are more involved, but the reasoning is similar.

Based on (9.94) it is straightforward to show that the complex extended Calogero model $H_{\mathcal{PTC}}$ is integrable because we can construct explicitly the Lax pair (9.2). The shift in p is generated from a similarity transformation $\mathcal{H}_{\mathcal{PTC}} = \eta^{-1} \mathcal{H}_C \eta$ with $\eta = e^{-q \cdot \mu}$. The Lax pair L_C and M_C for the Calogero models are well-known generalizations of (9.16), so the new $\mathcal{PT}$-deformed Lax operators are simply $L_{\mathcal{PTC}}(p) = \eta^{-1} L_C(p) \eta = L_C(p + i\mu)$ and $M_{\mathcal{PTC}}(p) = \eta^{-1} M_C(p) \eta = M_C(p + i\mu)$.[10]

---

[10]For the nonrational models the term $-\mu^2/2$ cannot be absorbed into the potential with a simultaneous redefinition of the coupling constant. However, if we add that term in (9.92), we also obtain new integrable $\mathcal{PT}$-deformed models for this type of potentials.

9.4.2   *From fields to particles*

So far we have not employed the deformation option (v) from Sec. 9.2. We now provide an example for this option and reveal some remarkable connections between the fields governed by nonlinear wave equations and the particle systems described by complex Calogero models. The idea of this approach is to seek solutions to the wave equations in the form of a terminated Painlevé expansion (9.17) with $\phi(x,t) = x - \zeta(t)$ and $\zeta(t) \in \mathbb{C}$. We have seen in Sec. 9.3.1 that for real $\zeta(t)$ this may happen at some resonance level, possibly with an additional constraint.

The simplest case for such a scenario involves the Benjamin–Ono equation, which describes internal waves in deep water [Chen *et al.* (1979)]:

$$u_t + uu_x + gHu_{xx} = 0, \qquad g \in \mathbb{R}. \tag{9.96}$$

Here $Hu(x) = \frac{P}{\pi}\int_{-\infty}^{\infty}\frac{u(x)}{z-x}dz$ is the Hilbert transform and P is Cauchy's principle value. For real $u(x,t)$ the truncated Painlevé expansion

$$u(x,t) = \frac{g}{2}\sum_{k=1}^{\ell}\left(\frac{i}{x - z_k(t)} - \frac{i}{x - z_k^*(t)}\right) \tag{9.97}$$

solves the Benjamin–Ono equation (9.96) subject to the ℓ constraints

$$\ddot{z}_k = \tfrac{1}{2}g^2\sum_{j\neq k}(z_j - z_k)^{-3} \qquad (k = 1,\dots\ell, z_k \in \mathbb{C}) \tag{9.98}$$

for the complex poles z_k in (9.97). Remarkably, (9.98) are the equations of motion for the complex Calogero Hamiltonian $H_{\mathcal{PT}}$ (9.93) for the A_ℓ-Coxeter group for a specific representation of the roots.

For the Benjamin–Ono equation we did not have to invoke the additional freedom in the theorem in option (v) about the locus related to conserved quantities. However, we do use this freedom for the Boussinesq equation

$$v_{tt} = (v^2)_{xx} + \tfrac{1}{12}v_{xxxx} + v_{xx}, \tag{9.99}$$

which models long water waves. The truncated Painlevé expansion

$$v(x,t) = -\tfrac{1}{2}\sum_{k=1}^{\ell}[x - z_k(t)]^{-2} \tag{9.100}$$

then solves (9.99) provided that the two sets of equations

$$\ddot{z}_k = 2\sum_{j\neq k}(z_j - z_k)^{-3} \quad \text{and} \quad \dot{z}_k^2 = 1 - \sum_{j\neq k}(z_j - z_k)^{-2} \tag{9.101}$$

hold. The first constraint in (9.101) is the equations of motion for the A_ℓ-Calogero model with $g = 2$. To understand the second constraint we need to invoke the part of the theorem that allows us to restrict the motion to the manifold described by the vanishing gradient of some conserved quantities.

Employing the A_ℓ-Calogero model equations of motion we can verify that $I_1 = \sum_{j=1}^{\ell} \dot{z}_j$ and $I_3 = \sum_{j=1}^{\ell} [\dot{z}_j^3/3 + \sum_{k \neq j} \dot{z}_j (z_j - z_k)^2]$ are two conserved quantities of the A_ℓ-Calogero model. Using these expressions we calculate $\mathrm{grad}(I_3 - I_1) = 0$, which coincides with the second constraint in (9.101).

In principle, the set of equations (9.101) describes a consistent dynamical system related to the A_ℓ-Calogero model as long as the set of solutions to these equations is not empty. The simplest two-particle solution for the real setting is [Airault *et al.* (1977)]

$$z_1 = \mu + \sqrt{(t+\lambda)^2 + 1/4}, \qquad z_2 = \mu - \sqrt{(t+\lambda)^2 + 1/4}, \qquad (9.102)$$

where $\mu, \lambda \in \mathbb{R}$, and the solution to the Boussinesq equation (9.100) is

$$v(x,t) = -\frac{(x-\mu)^2 + (t+\lambda)^2 + 1/4}{[(x-\mu)^2 - (t+\lambda)^2 - 1/4]^2}. \qquad (9.103)$$

Taking the constants μ and λ in (9.102) to be purely imaginary, the Boussinesq equation (9.99) and its solution (9.103) remain invariant under the symmetry $\mathcal{PT} : x \to -x, \, t \to -t, \, v \to v, \, i \to -i$. The main difference between the real and complex solutions is that the poles that occur in the real case have been regularized in the complex equation, leading to a well-defined two-soliton solution, as displayed in Fig. 9.6. We also observe that in the unbroken-$\mathcal{PT}$-symmetric case the two solitons exactly exchange their positions under time reversal, unlike the case for the broken regime when μ, λ are complex but not purely imaginary.

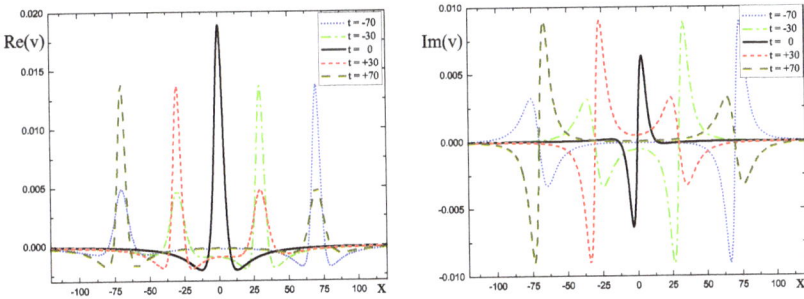

Fig. 9.6   Complex $\mathcal{PT}$-symmetric two-soliton solutions for the Boussinesq equation with $\mu = 2i$ and $\lambda = 8i$ at different times.

A three-particle solution exhibiting similar qualitative features to the two-soliton solution was found [Assis and Fring (2009a)].[11] We conjecture

---

[11]However, there is a crucial difference: In the two-soliton case the complexification is a convenient choice that regularizes the poles, but in the one-soliton case no real solution was found and one is forced to consider a complex particle system.

that solutions exist for systems with more particles, different types of algebras, and other types of nonlinear differential equations, but no such solutions have been constructed so far.

9.4.3   Deformed Calogero–Moser–Sutherland models

In this subsection we explain how to eliminate the aforementioned singularities in the potential and to overcome the restrictions on the defining regime of the CMS models. We focus first on the A_2 models and deform the three coordinates [Znojil and Tater (2001); Fring and Znojil (2008)]:

$$\begin{aligned}
q_1 &\to \tilde{q}_1 = q_1 \cosh \varepsilon + i 3^{-1/2} q_{23} \sinh \varepsilon, \\
q_2 &\to \tilde{q}_2 = q_2 \cosh \varepsilon + i 3^{-1/2} q_{31} \sinh \varepsilon, \\
q_3 &\to \tilde{q}_3 = q_3 \cosh \varepsilon + i 3^{-1/2} q_{12} \sinh \varepsilon,
\end{aligned} \qquad (9.104)$$

where $q_{ij} = q_i - q_j$ and $\varepsilon \in \mathbb{R}$ is a deformation parameter. In the new coordinates $\tilde{q}$ we calculate the arguments in the potential functions in (9.91):

$$\begin{aligned}
\alpha_1 \cdot \tilde{q} &= q_{12} \cosh \varepsilon - i 3^{-1/2}(q_{13} + q_{23}) \sinh \varepsilon, \\
\alpha_2 \cdot \tilde{q} &= q_{23} \cosh \varepsilon - i 3^{-1/2}(q_{21} + q_{31}) \sinh \varepsilon, \\
(\alpha_1 + \alpha_2) \cdot \tilde{q} &= q_{13} \cosh \varepsilon + i 3^{-1/2}(q_{12} + q_{32}) \sinh \varepsilon.
\end{aligned} \qquad (9.105)$$

Clearly, we have eliminated the singularities in the potential and no longer have to impose any ordering on the real coordinates q_i. At this point the deformation appears to be *ad hoc* and arbitrary because we can easily imagine other possibilities in (9.104). So let us try to understand what is special about the above deformation. First, we notice that the three inner products (9.105) are mapped into each other by the antilinear involutory symmetries

$$\mathcal{PT}_1 : q_1 \leftrightarrow q_2, \; q_3 \to q_3, \; i \to -i; \; \mathcal{PT}_2 : q_2 \leftrightarrow q_3, \; q_1 \to q_1, \; i \to -i. \quad (9.106)$$

Evidently, $\mathcal{PT}_{1,2}^2 = 1$.[12] Thus, we have produced a new complex $\mathcal{PT}$-symmetric Hamiltonian which, in the limit $\varepsilon \to 0$, reduces to the A_2-Calogero model. Here this convention is more convenient than identifying the undeformed model with $\varepsilon \to 1$ as in previous sections. While this deformation leads to nice potentials, we still do not have a good explanation for why the deformation (9.104) is special, and we do yet not know how to deform systems involving more particles and other algebras. To answer

---

[12]We may possibly get an overall minus sign when acting with these transformations, but this is irrelevant because negative terms would appear in the sum of the potential terms that also include negative roots.

these questions we consider deformations of the dual space of the inner product; that is, we deform the roots $\alpha \to \tilde{\alpha}$ instead of the coordinates $q \to \tilde{q}$ and demand that the two versions of this deformation produce the identical inner products $\alpha \cdot \tilde{q} = \tilde{\alpha} \cdot q$. This is achieved by deforming the two simple A_2-roots [Fring and Znojil (2008)]:

$$\tilde{\alpha}_1 = \alpha_1 \cosh \varepsilon + i3^{-1/2} \sinh \varepsilon (\alpha_1 + 2\alpha_2),$$
$$\tilde{\alpha}_2 = \alpha_2 \cosh \varepsilon - i3^{-1/2} \sinh \varepsilon (2\alpha_1 + \alpha_2).$$

Now we have an explanation: The two symmetries in (9.106) can be interpreted as natural mathematical objects, *deformed Weyl reflections*:

$$\mathcal{PT}_1 \equiv \sigma_1^\varepsilon : \tilde{\alpha}_1 \leftrightarrow -\tilde{\alpha}_1, \tilde{\alpha}_2 \leftrightarrow \tilde{\alpha}_1 + \tilde{\alpha}_2,$$
$$\mathcal{PT}_2 \equiv \sigma_2^\varepsilon : \tilde{\alpha}_2 \leftrightarrow -\tilde{\alpha}_2, \tilde{\alpha}_1 \leftrightarrow \tilde{\alpha}_1 + \tilde{\alpha}_2.$$

There is little flexibility to deform these reflections, so we have unraveled the deeper origin of the deformation (9.104). As they are reflections, they have the crucial property $(\sigma_1^\varepsilon)^2 = (\sigma_2^\varepsilon)^2 = 1$.[13]

Let us next see how the Calogero systems can be solved by comparing the deformed and undeformed cases. Usually, one introduces a new set of variables, which may appear to be arbitrary or unmotivated. However, they have a natural origin if we think in terms of generic roots. Note that a root system for a particular Coxeter group is fixed entirely in terms of the Cartan matrix $K_{ij} = 2\alpha_i \cdot \alpha_j / \alpha_j^2$ [Humphreys (1972)]. The A_2-Cartan matrix has entries $K_{11} = K_{22} = 2$, $K_{12} = K_{21} = -1$, which are obtained when $\alpha_1^2 = \alpha_2^2 = 2$ and $\alpha_1 \cdot \alpha_2 = -1$ are satisfied for the three-dimensional representation above. We may also satisfy these relations with a two-dimensional representation for the roots β with $\beta_1 = \{\sqrt{2}, 0\}$ and $\beta_2 = \{-1, \sqrt{3}\}/\sqrt{2}$. Clearly, $K_{ij} = 2\beta_i \cdot \beta_j / \beta_j^2$ is the same Cartan matrix. Demanding equality between the inner products in the different set of variables $\alpha_i \cdot q = \beta_i \cdot Q$ for $i = 1, 2$ with $q = \{q_1, q_2, q_3\}$ and $Q = \{X, Y\}$, we can express the new variables in terms of the old variables:

$$X = (q_1 - q_2)/\sqrt{2} \quad \text{and} \quad Y = (q_1 + q_2 - 2q_3)/\sqrt{6}. \tag{9.107}$$

As a third independent variable we select the center-of-mass coordinate $R = (q_1 + q_2 + q_3)/3$. We then express the new variables in (9.107) in polar coordinates $X = r \sin \phi$, $Y = r \cos \phi$. Assuming that all coupling constants

---

[13]To understand the entire construction would require more background on Weyl and Coxeter groups, which is beyond the scope of this book. For a general scheme regarding all Coxeter groups see [Fring and Smith (2010, 2011, 2012)].

$g_\alpha = g$ are the same, we make the variable transformations $(q_1, q_2, q_3) \rightarrow (X, Y, R) \rightarrow (r, \phi, R)$ for the A_2-CMS Hamiltonian (9.91):

$$H(q_1, q_2, q_3) = \frac{1}{2}\Delta_{q_1 q_2 q_3} + \frac{\omega^2}{2}\sum_{i=1}^{3}(\alpha_i \cdot q)^2 + g^2 \sum_{i=1}^{3}\frac{1}{(\alpha_i \cdot q)^2},$$

$$H(X, Y, R) = -\frac{1}{2}\left[\frac{1}{3}\partial_R^2 + \partial_X^2 + \partial_Y^2\right] + \frac{3}{2}\omega^2(X^2 + Y^2) + \frac{9}{2}\frac{g^2(X^2 + Y^2)^2}{(X^3 - 3XY^2)^2},$$

$$H(r, \phi, R) = -\frac{1}{2}\left[\frac{1}{3}\partial_R^2 + \partial_r^2 + \frac{1}{r^2}\partial_\phi^2 + \frac{1}{r}\partial_r\right] + \frac{3\omega^2}{2}r^2 + \frac{9}{2}\frac{g^2}{r^2 \sin(3\phi)}.$$

Let us now see how these translations manifest in the $\mathcal{PT}$-deformed system. In that case we demand the equality of the arguments of the potentials for the deformed quantities $\alpha_i \cdot \tilde{q} = \beta_i \cdot \tilde{Q}$ for $i = 1, 2$ with $q = \{\tilde{q}_1, \tilde{q}_2, \tilde{q}_3\}$ and some as yet unknown $\tilde{Q} = \{\tilde{X}, \tilde{Y}\}$. From the two equations we identify the new deformed variables

$$\tilde{X} = X\cosh\varepsilon + iY\sinh\varepsilon = r\sin(\phi + i\varepsilon),$$
$$\tilde{Y} = Y\cosh\varepsilon - iX\sinh\varepsilon = r\cos(\phi + i\varepsilon).$$

We can then express the deformed Hamiltonian as follows:

$$H(\tilde{q}_1, \tilde{q}_2, \tilde{q}_3) = H(\tilde{X}, \tilde{Y}, R) = H(r, \phi + i\varepsilon, R). \tag{9.108}$$

Schrödinger equation for deformed Hamiltonian (9.108): This chapter deals mainly with classical integrable systems, but here we briefly discuss the solution of the Schrödinger equation for the deformed Hamiltonian (9.108). For the undeformed case we assume that the wave functions for $H(r, \phi, R)$ factor as $\Psi(r, \phi, R) = \chi(r)f(\phi)\Phi(R)$. Separating off the center-of-mass motion by setting $\Phi''(R) = 0$ and rescaling $\omega \rightarrow \hat{\omega}/\sqrt{3}$, we obtain the two eigenvalue equations

$$-\chi''(r) - r^{-1}\chi'(r) + \left(\omega^2 r^2 + \lambda^2 r^{-2}\right)\chi(r) = E\chi(r),$$
$$-f''(\phi) + 9g^2\csc(3\phi)f(\phi) = \lambda^2 f(\phi).$$

These equations involve the inverse-square and the Pöschl–Teller potentials, so they can be solved for any real values of the parameters r, ϕ, g, ω as functions of the eigenvalues E and λ^2 [Flügge (1971)]:

$$\chi(r) = r^\lambda \exp\left(-\omega r^2/2\right) {}_1F_1\left[\tfrac{1}{2}(1 + \lambda) - \tfrac{E}{4\omega}; 1 + \lambda; \omega r^2\right], \tag{9.109}$$

$$f(\phi) = \sin^{2\kappa}(3\phi)\cos^{2\kappa}(3\phi) {}_2F_1\left[\kappa - \tfrac{\lambda}{6}, \kappa + \tfrac{\lambda}{6}; 2\kappa + \tfrac{1}{2}; \sin^2(3\phi)\right], \tag{9.110}$$

where $\kappa = \kappa^\pm = (1 \pm \sqrt{1 + 4g^2})/4$, ${}_1F_1$ is *Kummer's confluent hypergeometric function*, and ${}_2F_1$ is the *Gauss hypergeometric function*. The deformed system differs from the undeformed system by a simple shift of variables. These two systems are related by $H(r, \phi + i\varepsilon, R) = \eta_\phi H(r, \phi, R)$ and $\Psi(R, r, \phi + i\varepsilon) = \eta_\phi \Psi(R, r, \phi)$, with $\eta_\phi = \exp(p_\phi \varepsilon)$. The corresponding energy eigenvalues are the same,

$$E_{n\ell}^\pm = 2|\omega| \left[2n + 6(\kappa^\pm + \ell) + 1\right] \quad \text{for} \quad 6(\kappa^\pm + \ell) > 0; \ n, \ell \in \mathbb{N}_0, \quad (9.111)$$

but with different parametric constraints. In the undeformed case only κ^+ appears in (9.111), but in the deformed case both signs are allowed.

Let us see how these spectra result from the solution, paying special attention to the difference between the two cases. First, we impose the physical requirement that the wave function vanish as $r \to \infty$. For the solution $\chi(r)$ in (9.109) we must investigate the asymptotic expansion for Kummer's confluent hypergeometric function [MacDonald (1948)],

$$_1F_1 [\alpha; \gamma; z] \sim \frac{\Gamma(\gamma)}{\Gamma(\alpha)} e^z z^{\alpha-\gamma} G(1 - \alpha; \gamma - \alpha, z) \quad (\text{Re } z > 0), \quad (9.112)$$

$$_1F_1 [\alpha; \gamma; z] \sim \frac{\Gamma(\gamma)}{\Gamma(\gamma-\alpha)} (-z)^{-\alpha} G(\alpha; \alpha - \gamma - 1, -z) \quad (\text{Re } z < 0), \quad (9.113)$$

with $G(\alpha; \gamma, z) = 1 + \alpha/z + \alpha(\alpha+1)\gamma(\gamma+1)/(2!z^2) + \dots$. Observe that for $z = \omega r^2$ the function grows exponentially unless this growth is compensated by a diverging gamma function, either from the corresponding $\Gamma(\alpha)$ in (9.112) or from $\Gamma(\gamma - \alpha)$ in (9.113). As this is the case when the first argument in $_1F_1$ becomes a negative integer, that is, when the hypergeometric series terminates, the wave function $\chi(r)$ vanishes at infinity when we impose

$$E = 2 |\omega| (2n + \lambda + 1) \quad \text{for } n \in \mathbb{N}_0.$$

For these values the Kummer confluent hypergeometric function reduces to a generalized Laguerre polynomial $L_n^\alpha(z)$:

$$_1F_1 [-n; \alpha + 1; z] = \frac{\Gamma(n+1)\Gamma(\alpha+1)}{\Gamma(n+\alpha+1)} L_n^\alpha(z) \quad (\alpha \in \mathbb{R}, \ n \in \mathbb{N}_0),$$

and we get $\chi(r)$ up to the normalization already found in [Calogero (1971)].

A physical wave function must be finite on its domain. Thus, (9.109) suggests that we must demand that $\lambda > 0$. The difference between the undeformed and deformed cases now results from making the same demand on $f(\phi)$. In the undeformed case the prefactors in (9.110) diverge for $\phi = 0, \pi/3, \dots$ and $\phi = \pi/6, \pi/2, \dots$, so when $\kappa = \kappa^-$ we must exclude this possibility. However, when shifting $\phi \to \phi + i\varepsilon$, this restriction is no longer needed. Finally, we fix the value of λ. To do this we note that there are also divergences of $f(\phi)$ resulting from the Gauss hypergeometric function. For generic arguments the function $_2F_1(\alpha, \beta; \gamma; 1)$ is absolutely convergent when $\text{Re } \gamma > \text{Re}(\alpha + \beta)$, which for the values in (9.110) translates into

$\kappa < \frac{1}{4}$. Having already excluded κ^- in the undeformed case, this inequality can never be satisfied. However, when α becomes a negative integer, the hypergeometric series terminates and reduces to a Jacobi polynomial $P_\ell^{\alpha,\beta}(z)$:

$$_2F_1(-\ell, \alpha + \beta + \ell + 1; \alpha + 1; z) = \frac{\Gamma(\ell+1)\Gamma(\alpha+1)}{\Gamma(\ell+\alpha+1)} P_\ell^{\alpha,\beta}(1 - 2z)$$

for $\ell \in \mathbb{N}_0$ and $\alpha, \beta \in \mathbb{R}$. Since $P_\ell^{\alpha,\beta}(-1) = (\beta + 1)_\ell/\ell!$ with $(x)_n := x(x+1)(x+2)\ldots(x+n)$, the divergence is removed by imposing $\lambda = 6(\kappa + \ell)$.[14] Thus, in the deformed case the termination of the series appears even more natural than in the undeformed case and we see how the deformation weakens some of the physical constraints and leads to new spectra.[15]

We have seen in this chapter several examples of integrable models that remain integrable when suitably PT deformed. Supersymmetry was shown to be another property that can be maintained when PT-deforming a model. Even if the PT deformation breaks the integrability of the model, the conserved quantities in the new models that survive the deformation were shown to be real due to the PT symmetry. When the models remain integrable after the deformation, the PT symmetry was not sufficient to ensure the reality of the energy and also the integrability. This is synonymous with the asymptotic factorization of the N-soliton solution into N single solitons. We saw that PT symmetry alone cannot distinguish between different types of qualitative behaviors of trajectories in phase space.

Acknowledgment

Andreas Fring thanks Paulo Assis, Bijan Bagchi, Andrea Cavaglia, Julia Cen, Francisco Correa, Sanjib Dey, Carla Figueira de Morisson Faria, Thomas Frith, Laure Gouba, Monique Smith, and M. Znojil for collaborations related to this work.

---

[14] Alternatively, we could equate the second argument in (9.110) to an integer and deduce that $\lambda = -6(\kappa + \ell)$, which is excluded by the previous requirement that $\lambda > 0$. When $\text{Im}\,\phi \neq 0$, we even leave the unit circle $|z| \leq 1$, and convergence can be achieved unless we restrict the real part of ϕ depending on its imaginary part, which seems artificial.

[15] The key ideas in this section on using root systems to define deformed models in a systematic way and to employ them to introduce new sets of variables may be applied for other types of groups [Fring and Smith (2010, 2011, 2012)].

Chapter 10

$\mathcal{PT}$ Symmetry in Optics

Contribution from H. F. Jones

"The right question is usually more important than the right answer."
—*PlaTo*

The ideas of $\mathcal{PT}$ symmetry were originally introduced in the context of quantum mechanics, but in recent years they have led to rapid developments in the apparently unconnected field of classical optics, starting with the pioneering papers [El-Ganainy *et al.* (2007); Musslimani *et al.* (2008a); Makris *et al.* (2008); Klaiman *et al.* (2008); Guo *et al.* (2009); Makris *et al.* (2010); Rüter *et al.* (2010); Ramezani *et al.* (2010)]. The reason for this surprising connection is that when one makes a particular approximation, the *paraxial approximation*, for the equation of propagation of an electromagnetic wave, the resulting equation is formally identical to the Schrödinger equation, but with completely different interpretations for the symbols appearing therein. In particular, the role of the quantum-mechanical potential is taken by the refractive index. The complex potentials appearing in $\mathcal{PT}$ quantum mechanics thus correspond to a complex refractive index, whose imaginary part corresponds to loss or gain.

The distinguishing feature of $\mathcal{PT}$ quantum mechanics is that energy eigenvalues can be real even if the potential $V(x)$ is complex provided that $V(x)$ is $\mathcal{PT}$ symmetric. Translated into the optical realm this means that one can have real propagation constants even in the presence of loss and gain provided that these are balanced in a $\mathcal{PT}$-symmetric way. Thus loss, which in practice is always present in optical systems, can be used in a positive way when appropriately combined with gain, and it can be exploited to great effect in designing novel optical devices.

10.1   Paraxial Approximation

To explain the paraxial approximation let us consider a wave of frequency ω propagating in an optical medium whose refractive index n depends on the (transverse) direction x, while the propagation is predominantly in the (longitudinal) direction z. A given component of the electric field E satisfies the scalar Helmholtz equation, which reads

$$\frac{\partial^2}{\partial z^2}E(x,z) + \frac{\partial^2}{\partial x^2}E(x,z) + k^2 E(x,z) = 0,$$

where $k = n(x)\omega/c$. This set-up is shown in Fig. 10.1, in which an electromagnetic wave is incident on a block of material with refractive index $n(x)$. The incident wave is predominantly in the z direction, and any deviation from that direction is small — hence "paraxial".

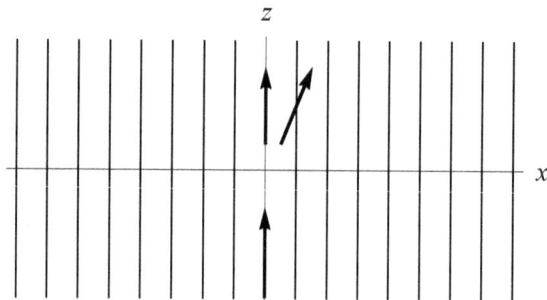

Fig. 10.1   Set-up for paraxial approximation. Propagation of an electromagnetic wave in a medium with refractive index $n(x)$. In this and subsequent diagrams the (usually periodic) variation in n is indicated as vertical stripes. The incident wave is predominantly in the z direction, and any deviation from that direction is small.

In normal circumstances the deviation of n from the background refractive index n_0 will be small, so we write n in the form $n = n_0[1 + v(x)]$, where $v \ll 1$. The idea is that the wave propagates broadly like $e^{ik_0 z}$, where $k_0 = n_0\omega/c$, but with small deviations. We therefore explicitly factor out this main z-dependence of E and write $E = e^{ik_0 z}\psi(x,z)$. The quantity ψ

is known as the *envelope function*. It satisfies the equation

$$\frac{\partial^2}{\partial z^2}\psi(x,z) + 2ik_0\frac{\partial}{\partial z}\psi(x,z) + \frac{\partial^2}{\partial x^2}\psi(x,z) + (k^2 - k_0^2)\psi = 0,$$

where $k^2 = k_0^2(n^2/n_0^2)$.

The paraxial approximation amounts to neglecting the first term in this equation under the assumption that $\partial^2\psi/\partial z^2 \ll \partial^2\psi/\partial x^2$, which is the case if the main dependence of $E(x,z)$ on z is given by the first factor $e^{ik_0 z}$. Given that v is small, we also make the approximation $k^2 - k_0^2 \approx 2k_0^2 v$.

Putting all this together, we get the paraxial equation for the envelope function ψ:

$$i\frac{\partial\psi}{\partial z} + \frac{1}{2k}\frac{\partial^2\psi}{\partial x^2} + k\,v(x)\psi = 0. \tag{10.1}$$

This has the same form as the time-dependent Schrödinger equation, although the quantities appearing in (10.1) are quite different from their quantum-mechanical counterparts. First, ψ is not the wave function, but is rather the envelope function of the electromagnetic wave. Second, the role of t is played by the longitudinal coordinate z, and finally the role of the quantum-mechanical potential $V(x)$ is played by $v(x)$, which we recall is given by the excess refractive index: $v(x) = n(x)/n_0 - 1$.

The crucial importance of this correspondence is that results obtained for PT quantum-mechanical systems can be taken over to PT-symmetric systems in classical optics with the appropriate interpretation. In optics it is extremely common for the refractive index to have a positive imaginary part, corresponding to loss or damping, but a negative imaginary part can also be implemented by optical pumping, leading to gain. So we are naturally led to consider systems with a non-Hermitian potential V in the analog Hamiltonian.

Imposing PT symmetry on $v(x)$ in (10.1) requires that loss and gain be balanced in a particular way, namely[1] that $n^*(-x) = n(x)$, or equivalently that $\text{Re}\,n(x)$ be an even function and $\text{Im}\,n(x)$ an odd function of x. When the PT symmetry is unbroken, the usual quantum-mechanical result of real eigenvalues corresponds in optics to the surprising result that one can have real propagation constants, that is, no exponential growth or decay, even in the presence of gain and loss. On the other hand, in some optical applications it could be advantageous to go to the broken-symmetry regime

---

[1] It has been emphasized [Zyablovsky *et al.* (2014)] that, because the refractive index is frequency dependent and must satisfy a dispersion relation, PT symmetry cannot hold exactly for a continuous range of frequencies, so in any practical application one has to restrict the range so that this frequency dependence is not important.

by varying one parameter or another. This opens the way to devising a variety of possible switching devices that exploit this broken-symmetry feature of $\mathcal{PT}$ symmetry.

It is important to note that once the connection between quantum mechanics and optics in the paraxial approximation was realized [El-Ganainy *et al.* (2007)], it was found that the ideas of $\mathcal{PT}$ symmetry could also be applied to various other optical systems where the paraxial approximation was either not true or not necessary, such as the set-up dealt with in Sec. 10.4. Since the original pioneering papers, the application of the ideas of $\mathcal{PT}$ symmetry to optics has undergone an exponential expansion, with a large number of papers on many different theoretical and experimental aspects of the subject published in high-profile journals. In this chapter we are only able to outline the main results and methods, with apologies to those whose work we have not been able to mention.

10.2   First Applications

$\mathcal{PT}$-symmetric optical systems governed by the paraxial equation have many interesting properties [Makris *et al.* (2008)]. For example, they can exhibit birefringence, as shown in Fig. 10.2 for $v(x) = v_0(\cos 2x + i\lambda \sin 2x)$. Here, $\lambda = 0.9$, which is below the critical value $\lambda = 1$, above which the symmetry is broken. Note that the system is not left-right symmetric; the beam deflects to the right. Instead the system is $\mathcal{PT}$ symmetric, with P representing $x \to -x$ and T implemented by complex conjugation, or effectively by $z \to -z$.

The power of the beam can oscillate in z, as shown more prominently in Fig. 10.3 for slightly different input parameters, because the medium is an active structure, so the power is not conserved. (The conserved quantity would be given by the $\mathcal{PT}$ norm, but that is not physically relevant here.)

Precisely at the symmetry-breaking point $\lambda = 1$ the total power grows linearly with z, as shown in Fig. 10.4. This is a typical phenomenon of $\mathcal{PT}$-symmetric systems and is due to the formation of Jordan blocks [Longhi (2010b)], where eigenvalues, together with their eigenfunctions, coalesce. In this case the amplitude of the beam becomes saturated after an initial rise, and the linear power growth is due to the lateral spread of the beam. The details of how this happens are given in [Graefe and Jones (2011)].

How are the intensity patterns shown above worked out? There are two main methods. The first is essentially the *method of stationary states*, which is employed in quantum mechanics to calculate $\psi(x, t)$ given an initial wave

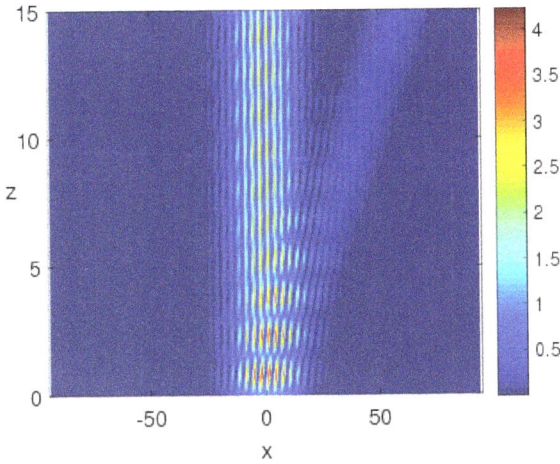

Fig. 10.2 Intensity pattern $|\psi(x,z)|^2$ for the optical potential $v(x) = v_0(\cos 2x + i\lambda \sin 2x)$ with $\lambda = 0.9$, showing bifurcation of the initial beam, which is a wave packet incident normally with a Gaussian spread in x.

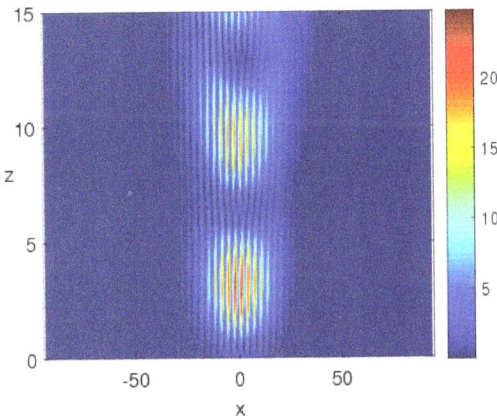

Fig. 10.3 Power oscillations: for slightly different input parameters, still with $\lambda = 0.9$, the beam does not bifurcate, but instead shows pronounced intensity oscillations.

function $\psi(x,0)$. The time dependence[2] is governed by the time-dependent Schrödinger equation $i\hbar\frac{\partial}{\partial t}\psi(x,t) = H\psi(x,t)$, with the formal solution

$$\psi(x,t) = e^{-iHt/\hbar}\psi(x,0).$$

---

[2]Here we are using the language of quantum mechanics, but with the understanding that in the analog Schrödinger equation t should be replaced by z.

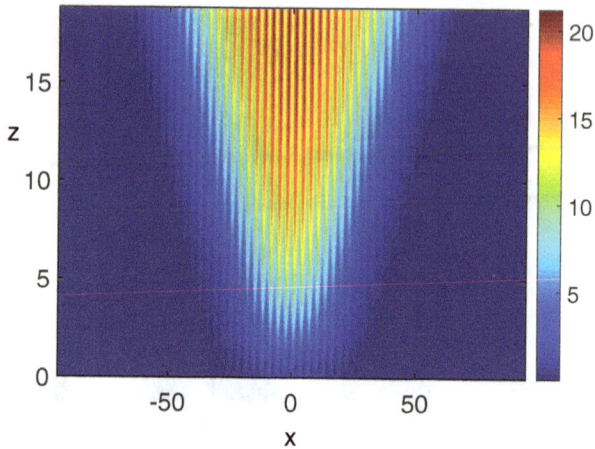

Fig. 10.4 Input at a specific angle at the symmetry-breaking point ($\lambda = 1$), exhibiting the spreading of the beam and saturation of the intensity $|\psi|^2$, which grows initially as a function of z but then levels off. As a result of these two effects, spreading and saturation, the total power grows linearly in z.

This formal series (H is an operator) can be evaluated if we know all the eigenvectors $\psi_n(x, t)$ satisfying the time-independent equation $H\psi_n = E_n\psi_n$, which we assume to be complete. Then we expand $\psi(x, 0)$ as

$$\psi(x, 0) = \sum_{n=0}^{\infty} c_n \psi_n(x, 0).$$

Therefore,

$$\psi(x, t) = e^{-iHt/\hbar} \sum_{n=0}^{\infty} c_n \psi_n(x, 0) = \sum_{n=0}^{\infty} c_n e^{-iE_n t/\hbar} \psi_n(x, 0), \qquad (10.2)$$

and the envelope function $\psi(x, t)$ can then be found by evaluating the final sum in (10.2), which in practice will have to be approximated by truncation at some large N.

For a periodic potential with period a in x, the eigenfunctions ψ_n must also be periodic up to a phase. These are the *Bloch functions* $\psi_k(x)$, which satisfy the condition

$$\psi_k(x + a) = e^{ika} \psi_k(x),$$

where k can be taken within the region $|k| \le \pi/a$, which is the first Brillouin zone. For a given E there may or may not be solutions to the eigenvalue equation. Where there are solutions, the plot of E versus k gives the band structure of solid-state physics. In Fig. 10.5 we show the band structure of the potential $V = 4(\cos(\pi x) + i\lambda \sin(\pi x))$ for values of λ below, at, and above the symmetry-breaking point $\lambda = 1$. Below the symmetry-breaking value we have a standard band structure of the kind seen in (Hermitian) solid-state physics. But the peculiar properties of PT-symmetric potentials are manifest at $\lambda = 1$ when the bands meet at $k = 0$ and at the Brillouin-zone boundaries $k = \pm 1$. These are the *exceptional points* where Jordan blocks occur, a phenomenon that cannot happen in the Hermitian case. Above $\lambda = 1$ the bands have an unusual structure, with the emergence of pairs of complex-conjugate values of E (not shown in the figure).

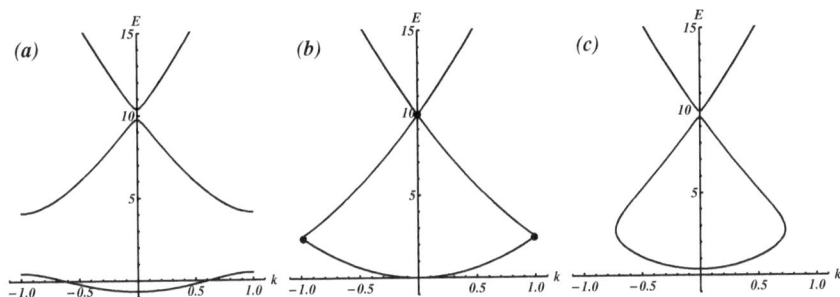

Fig. 10.5 Band structure of the potential $V = 4(\cos(\pi x) + i\lambda \sin(\pi x))$ for λ (a) below ($\lambda = 0.4$), (b) at ($\lambda = 1$), and (c) above ($\lambda = 1.2$), the symmetry-breaking point. Note the coalescence of eigenvalues in Fig. 10.5(b) at the marked points $k = 0, \pm 1$.

An alternative numerical method for working out the time dependence is the *split-operator* method, which is ideally suited for the standard situation when H is of the form $H = p^2/(2m) + V(x)$. That is, for a small interval δt the relevant operator is

$$e^{-iH\delta t/\hbar} = \exp\left\{-i(\delta t/\hbar)[p^2/(2m) + V(x)]\right\},$$

which can be approximated as

$$e^{-iH\delta t/\hbar} = e^{-\frac{1}{2}i(\delta t/\hbar)V(x)} e^{-\frac{1}{2}i(\delta t/\hbar)(p^2/(2m))} e^{-\frac{1}{2}i(\delta t/\hbar)V(x)} \tag{10.3}$$

with an error of order δt^3. This operator acts on $\psi(x,0)$. Now, the three exponential factors on the right side of (10.3) are functions of the single variables x, p, x. So working from the right, we first multiply $\psi(x,0)$ by the rightmost factor $e^{-\frac{1}{2}i(\delta t/\hbar)V(x)}$ in (10.3). Then we take the Fourier transform of the result and multiply it by the next factor. Finally, we take the inverse Fourier transform and multiply it by the last factor. This method is so effective because Fourier transforms can be evaluated very rapidly on a computer by bitwise operations (Fast Fourier Transform).

The stationary-states method breaks down and must be modified for $\mathcal{PT}$-symmetric systems at the symmetry-breaking point, where two or more eigenfunctions coalesce. In that situation the number of eigenfunctions is reduced, so they no longer form a complete set in which to expand $\psi(x,0)$. Instead, they must be supplemented by the associated Jordan functions $\varphi_n(x,0)$, which satisfy the equation $(H - E_n)\varphi_n = \psi_n$, rather than the eigenvalue equation $(H - E_n)\varphi_n = \varphi_n$. As a consequence, $(H - E_n)^2 \varphi_n = 0$, so the time development of the functions φ_n is given by

$$e^{-iHt/\hbar}\varphi_n = e^{-iE_n t/\hbar}e^{-i(H-E_n)t/\hbar}\varphi_n$$
$$= e^{-iE_n t/\hbar}\left[1 - i(H - E_n)t/\hbar\right]\varphi_n$$
$$= e^{-iE_n t/\hbar}\left(\varphi_n - it\psi_n/\hbar\right).$$

Thus, while the time dependence of the eigenstates ψ_n is oscillatory and proportional to $e^{-iE_n t/\hbar}$, the associated Jordan functions φ_n have an additional linear dependence on t, which leads to secular growth of amplitudes. This phenomenon (with some additional subtleties [Graefe and Jones (2011)]) is shown in Fig. 10.4.

10.3   A Simpler System: Coupled Waveguides

Let us consider a simpler situation, which consists of two single-mode parallel waveguides, one with gain γ and the other with equal loss [Rüter *et al.* (2010)], as shown in Fig. 10.6. This configuration can be analyzed completely and has led to one of the first verified predictions of $\mathcal{PT}$ symmetry.

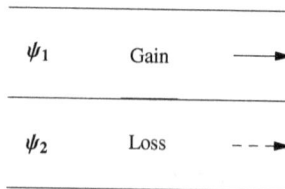

Fig. 10.6   Propagation down a pair of coupled waveguides, one with gain (ψ_1, solid arrow), the other with an equal amount of loss (ψ_2, dashed arrow).

The relevant equations for propagation down the waveguides are

$$i\frac{d\psi_1}{dz} - i\gamma\psi_1 + \kappa\psi_2 = 0, \qquad i\frac{d\psi_2}{dz} + i\gamma\psi_2 + \kappa\psi_1 = 0, \qquad (10.4)$$

where $\psi_i(z)$ represents the electric field in the respective guides. This is a stripped-down version of the set-up in the previous section, with the x coordinate discretized and limited to two elements. The solutions in the three different regimes of the parameter γ are follows

(1) Below threshold ($\gamma < \kappa$) (as shown in Fig. 10.7):

$$\psi_1(z) = \cos\mu z + (\gamma/\mu)\sin\mu z, \qquad \psi_2(z) = i(\kappa/\mu)\sin\mu z,$$

$$\text{or} \quad \psi_2(z) = \cos\mu z - (\gamma/\mu)\sin\mu z, \qquad \psi_1(z) = i(\kappa/\mu)\sin\mu z, \quad (10.5)$$

where $\mu = \sqrt{\kappa^2 - \gamma^2}$.

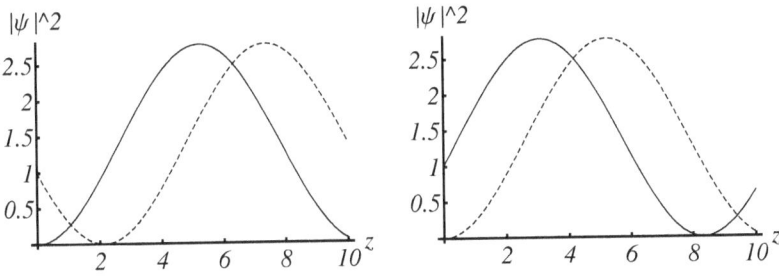

Fig. 10.7   Intensities $|\psi_1|^2$ (solid line) and $|\psi_2|^2$ (dotted line) below threshold ($\gamma < \kappa$) with $\kappa = \frac{1}{2}$. Left panel: input into loss channel, with $\psi_1(0) = 0$, $\psi_2(0) = 1$; right panel: input into gain channel, with $\psi_1(0) = 1$, $\psi_2(0) = 0$. Below threshold the solutions are oscillatory.

(2) At threshold ($\gamma = \kappa$) (as shown in Fig. 10.8):

$$\psi_1(z) = 1 + \kappa z, \qquad \psi_2(z) = i\kappa z,$$

$$\text{or} \quad \psi_2(z) = 1 - \kappa z, \qquad \psi_1(z) = i\kappa z.$$

Note the linear dependence of ψ on z, which tallies with our previous discussion of the dynamics at the symmetry-breaking point.

(3) Above threshold ($\gamma > \kappa$) (as shown in Fig. 10.9):

$$\psi_1(z) = \cosh\lambda z + (\gamma/\lambda)\sinh\lambda z, \qquad \psi_2(z) = i(\kappa/\lambda)\sinh\lambda z,$$

$$\text{or} \quad \psi_2(z) = \cosh\lambda z - (\gamma/\lambda)\sinh\lambda z, \qquad \psi_1(z) = i(\kappa/\lambda)\sinh\lambda z, (10.6)$$

where $\lambda = \sqrt{\gamma^2 - \kappa^2}$.

Note the characteristic square-root behavior of λ in (10.6), or equally of μ in (10.5). This is typical of a $\mathcal{PT}$ system near the symmetry-breaking

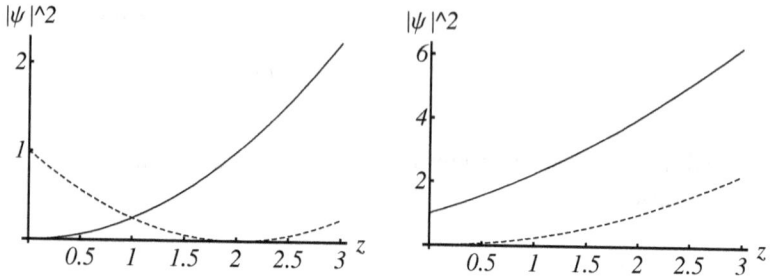

Fig. 10.8   Solutions at threshold ($\gamma = \kappa$), with $\kappa = 1/2$ and the same initial conditions as in Fig. 10.7. For this value of κ the amplitudes ψ_1 and ψ_2 depend linearly on z.

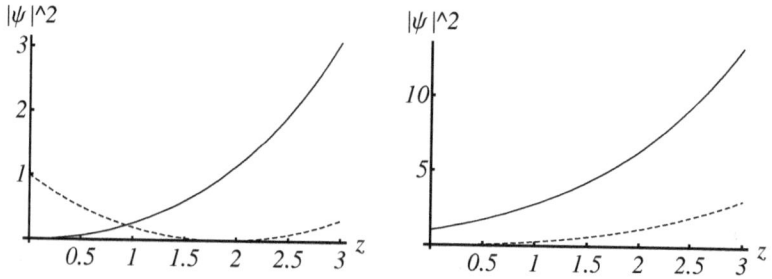

Fig. 10.9   Solutions above threshold ($\gamma > \kappa$), with $\kappa = \frac{1}{2}$ and the same initial conditions as in Fig. 10.7. The solutions are now exponential in nature.

point. Thus, if γ is fixed and the coupling κ is varied in the neighborhood of γ, the value of λ is very sensitive to small changes in κ. That is, if $\kappa = \gamma + \varepsilon$, where ε is small, then $\lambda \propto \sqrt{\varepsilon}$ rather than ε. This effect can be exploited to devise ultrasensitive sensors, for example, by using coupled ring lasers [Ren *et al.* (2017b)] or $\mathcal{PT}$ metasurfaces [Sakhdari *et al.* (2017)].

Returning to our coupled-waveguide example, the $\mathcal{PT}$ symmetry was modified in the experiment of [Guo *et al.* (2009)] by adding an overall loss to make a *passive* system, one with no gain, only loss. This is equivalent to adding a term $-i\gamma$ to H. In this set-up channel 1 is neutral and channel 2 has twice the amount of loss. The result is that the solutions (10.5)-(10.6) are multiplied by $e^{-\gamma z}$. So, for input into the neutral channel we get

$$\psi_1(z) = e^{-\gamma z}[\cos \mu z + (\gamma/\mu) \sin \mu z], \qquad \psi_2(z) = ie^{-\gamma z}(\kappa/\mu) \sin \mu z$$

for $\gamma < \kappa$ (and similarly for $\gamma > \kappa$). The plot of the total transmission $T \equiv |\psi_1(L)|^2 + |\psi_2(L)|^2$ as a function of γ in Fig. 10.10 shows an initial decrease with increasing γ, as one would expect. The surprise is that as γ increases further, the graph turns round and T increases again, even though γ continues to increase. The minimum of T occurs near, but not exactly at, the critical value $\gamma = \kappa$. This $\mathcal{PT}$ transition was predicted and verified experimentally in [Guo *et al.* (2009)].

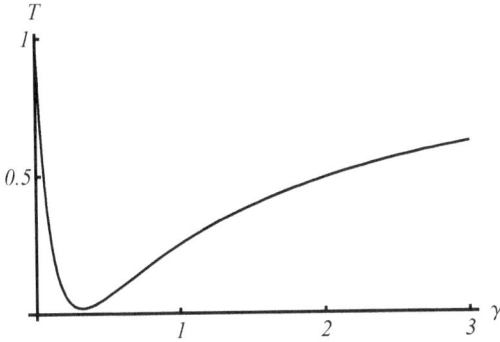

Fig. 10.10   Total transmission T versus loss γ for $\kappa = 0.5$, $L = 6$ and input into the passive channel. The initial decrease of T as γ increases is to be expected, but the subsequent increase is not immediately intuitive. The minimum occurs near, but not exactly at, the symmetry-breaking point $\gamma = \kappa$.

An important generalization of this system includes nonlinearity, which in practice is present to some degree in all optical systems. Nonlinearity arises because the response functions such as the refractive index are not strictly constant, but depend on the intensity of the electromagnetic fields. So for strong electromagnetic fields the equations (10.4) are modified to

$$i\frac{d\psi_1}{dz} - i\gamma\psi_1 + \kappa\psi_2 + \chi|\psi_1|^2\psi_1 = 0,$$
$$i\frac{d\psi_2}{dz} + i\gamma\psi_2 + \kappa\psi_1 + \chi|\psi_2|^2\psi_2 = 0. \tag{10.7}$$

In [Ramezani *et al.* (2010)] it is shown how such a system can act as an optical diode. (This is discussed further in Subsec. 10.7.2.)

10.4   Unidirectional Invisibility

Unidirectional invisibility, which would perhaps be better named unidirectional transparency, is the highly unusual property of a medium that permits light incident from one side to be completely transmitted, while none is reflected. Light from the other side is still completely transmitted, but also strongly reflected. These properties can be realized by a $\mathcal{PT}$ medium, as explained below. The one-sided nature of the medium is explained by the fact that a $\mathcal{PT}$-symmetric medium does distinguish between left and right; it is $\mathcal{PT}$ symmetric, but not $\mathcal{P}$ symmetric. Further, the surprising result that one can have perfect transmission together with nonzero reflection can be understood because the $\mathcal{PT}$ medium is an active medium incorporating gain in conjunction with loss.

Although the connection with $\mathcal{PT}$ symmetry was first made explicit in [El-Ganainy *et al.* (2007)], the exotic properties of materials with combined gain and loss were previously explored in [Poladian (1996); Berry (1997)]. In these papers the variation in the refractive index was in the direction of propagation z rather than in the transverse x direction, so the equation of propagation for a given component of the electric field is just the scalar Helmholtz equation

$$\left[d^2/dz^2 + k^2 (n(z)/n_0)^2 \right] E(z) = 0, \tag{10.8}$$

which can be compared[3] to the time-independent Schrödinger equation[4]

$$\left[d^2/dz^2 + (E - V(z)) \right] \psi(z) = 0. \tag{10.9}$$

We are particularly concerned with the situation where the refractive index $n(z)$ is periodic, in which case the optical medium constitutes a solid grating. The special case when the $\mathcal{PT}$-symmetric variation in $n(z)$ is a pure complex exponential proportional to $e^{2i\beta z}$ was first discussed in [Kulishov *et al.* (2005)] and later in [Lin *et al.* (2011)]. To a very good approximation this grating exhibits the phenomenon of unidirectional invisibility, with perfect transmission for incidence from the left or right and zero reflection for incidence from the left. However, for right incidence, the concomitant property is a greatly enhanced reflectivity, sharply peaked at $k = \beta$.

In this situation we are interested in the amplitudes for transmission and reflection, given respectively by

$$t_{\mathrm{L,R}} = 1/A_{\mathrm{L,R}} \quad \text{and} \quad r_{\mathrm{L,R}} = B_{\mathrm{L,R}}/A_{\mathrm{L,R}},$$

whose squared moduli $T_{\mathrm{L,R}} \equiv |t_{\mathrm{L,R}}|^2$ and $R_{\mathrm{L,R}} \equiv |r_{\mathrm{L,R}}|^2$ are the transmission and reflection coefficients. The subscripts L and R refer to incidence

---

[3]There is an unfortunate clash of notation here. In the quantum-mechanical context E stands for energy, whereas in optics it represents a component of the electric field.

[4]Here, we are considering a fixed frequency $\omega = kc$. If the frequency dependence of $n(z)$ is taken into account, a time-*dependent* Schrödinger equation emerges [Muga *et al.* (2004)] by Fourier transform in which the time t is the variable complementary to ω.

from the left and right respectively. In the next subsection we concentrate on the particular form $v(z) = \frac{1}{2}\alpha^2 e^{2i\beta z}$. This was the most interesting case discussed in [Kulishov *et al.* (2005); Lin *et al.* (2011)].

10.4.1 *Coupled-mode approximation*

In [Lin *et al.* (2011)] a particular approximation, the *coupled-mode approximation*, was used in order to obtain the transmission and reflection coefficients. The details of the approximation need not concern us here, but the results were as follows:

(1) Left incidence (configuration illustrated in Fig. 10.11)

$$T_{\mathrm{L}} = 1, \qquad R_{\mathrm{L}} = 0; \qquad (10.10)$$

that is, perfect transmission and no reflection: *invisibility* for left incidence.

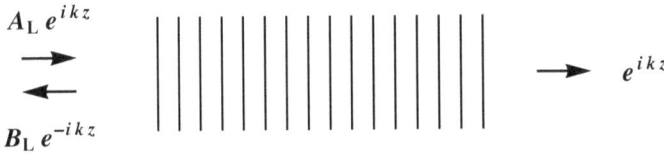

Fig. 10.11 Left incidence on a grating modulated in the z direction; that is, where $n = n(z)$. For convenience we have normalized the transmitted wave to 1. The transmission coefficient is then $T_{\mathrm{L}} = 1/|A_{\mathrm{L}}|^2$, while the reflection coefficient is $R_{\mathrm{L}} = |B_{\mathrm{L}}|^2/|A_{\mathrm{L}}|^2$.

(2) Right incidence (configuration illustrated in Fig. 10.12):

$$T_{\mathrm{R}} = 1, \qquad R_{\mathrm{R}} = \left(\tfrac{1}{2}k\alpha^2\ell\right)^2 [(\sin \ell\delta)/\ell\delta]^2. \qquad (10.11)$$

Here, δ is the *detuning parameter*, which is defined by $\delta = k - \beta$, and ℓ is the length of the grating. The right reflection coefficient R_{R} is proportional to the square of a sinc function and become very large ($\propto \ell^2$) near $\delta = 0$.

10.4.2 *Analytic solution for the scattering coefficients*

By a suitable change of variable the equation of propagation (10.8) can be transformed into Bessel's equation [Longhi (2011); Jones (2012)], and thanks to the recursion relations of the associated Bessel functions, the

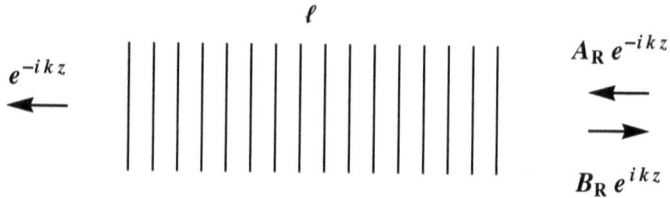

Fig. 10.12   Right incidence on a grating of length ℓ modulated in the z direction. The transmission coefficient is now $T_R = 1/|A_R|^2$ and the reflection coefficient is $R_R = |B_R|^2/|A_R|^2$.

coefficients $A_{L,R}$, $B_{L,R}$ can be cast in the relatively simple forms

$$A_L = \tfrac{1}{2}\alpha^2 \nu e^{ik\ell}[K_{\nu+1}(y_+)I_{\nu-1}(y_-) - I_{\nu+1}(y_+)K_{\nu-1}(y_-)],$$
$$B_L = \tfrac{1}{2}\alpha^2 \nu[-K_{\nu+1}(y_+)I_{\nu+1}(y_-) + I_{\nu+1}(y_+)K_{\nu+1}(y_-)],$$
$$A_R = \tfrac{1}{2}\alpha^2 \nu e^{ik\ell}[-K_{\nu-1}(y_-)I_{\nu+1}(y_+) + I_{\nu-1}(y_-)K_{\nu+1}(y_+)],$$
$$B_R = \tfrac{1}{2}\alpha^2 \nu[K_{\nu-1}(y_-)I_{\nu-1}(y_+) - I_{\nu-1}(y_-)K_{\nu-1}(y_+)], \qquad (10.12)$$

where $\nu = k/\beta$ and $y_\pm = (k\alpha/\beta)e^{\pm i\beta\ell/2}$. The transmission and reflection amplitudes are then given by $t = 1/A_{L,R}$ and $r_{L,R} = B_{L,R}/A_{L,R}$. From (10.12) it can be seen that $A_L = A_R$. This is a general result, not dependent on the particular form of $n(z)$, as is shown in Subsec. 10.4.3.

The numerical results from these formulas show that the coupled-mode approximation is actually very good, but that the transmission amplitude t is not exactly equal to one and r_L is not exactly equal to zero. In Figs. 10.13-10.15 the parameters are the same as those used in [Lin *et al.* (2011)]. We see from Fig. 10.13 that the transmission coefficient is very close to one (note the very small vertical range), and it is precisely one at $\delta = 0$.

Figure 10.15 shows that the shape of the right reflection coefficient is very close to that given by the simple formula (10.11). Interestingly, in the case of a wave-packet input the very large value of the reflectivity near $\delta = 0$ arises from an increased *length* (rather than the height) of the pulse [Jones (2012)]. Equivalent analytical results are obtained in [Longhi (2011)], where it is shown that if the length ℓ of the grating is made sufficiently large, the property of unidirectional invisibility (or unidirectional transparency) ceases to hold. The analytic expressions in (10.12) have recently been generalized [Jones and Kulishov (2016)] to cover the situations of oblique incidence and different background refractive indices on either side of the

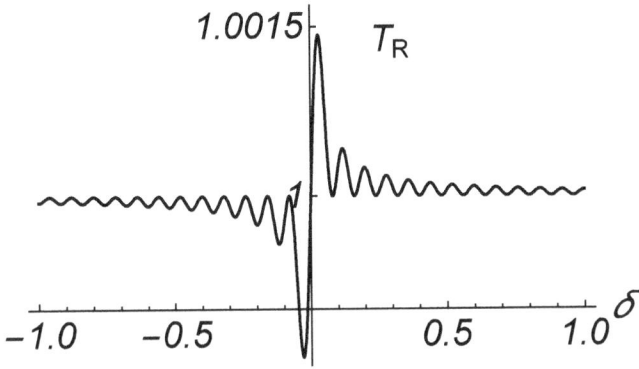

Fig. 10.13 Transmission coefficient (L or R) as a function of the detuning parameter δ. Note the very small vertical range, showing that T is very close to one.

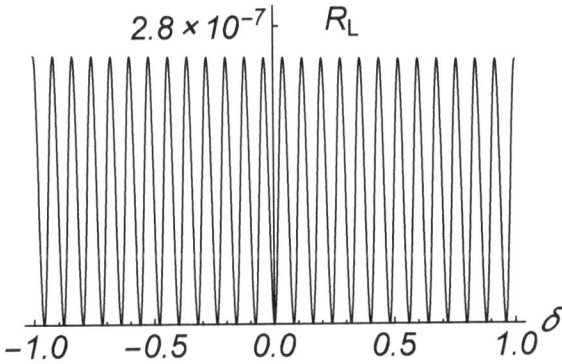

Fig. 10.14 Reflection coefficient for left incidence, R_L, as a function of the detuning parameter δ; here, R_L is not exactly zero, but is of the order of 10^{-7}.

grating. With oblique incidence the structures of Figs. 10.13–10.15 can be seen for fixed frequency as a function of the angle of incidence. With different refractive indices the clear features of unidirectional invisibility are degraded by multiple reflections at the boundaries.

10.4.3 *Wronskians and pseudo-unitarity*

In the Hermitian setting we are familiar with unitarity, meaning that the reflection and transmission coefficients add up to one, which corresponds in

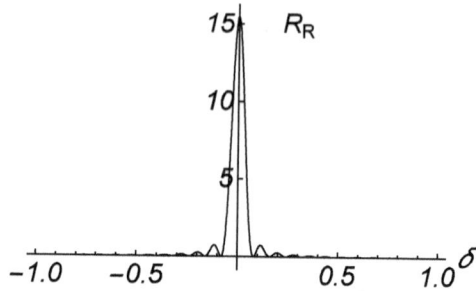

Fig. 10.15 Reflection coefficient for right incidence R_R as a function of the detuning parameter δ. The exact shape of the right reflection coefficient is very close to the square of a sinc function, as given by the simple formula (10.11) of [Lin et al. (2011)].

quantum-mechanical language to conservation of probability. This property is most easily derived from the constancy of the Wronskians of pairs of solutions to the Schrödinger equation coupled with the reality of the potential. Moreover, in the Hermitian case the left and right reflection coefficients are equal, $R_L = R_R$. For a PT-symmetric potential we have $V^*(z) = V(-z)$ instead of $V^*(z) = V(z)$, and this leads to a modified unitarity relation, which we can call *pseudo-unitarity*, involving R_L and R_R, which are no longer equal.

As is implied by Figs. 10.11 and 10.12,

$$\psi_L = \begin{cases} A_L e^{ikz} + B_L e^{-ikz} & (z < -L/2) \\ e^{ikz} & (z > L/2) \end{cases} \tag{10.13}$$

and

$$\psi_R = \begin{cases} e^{-ikz} & (z < -L/2) \\ A_R e^{-ikz} + B_R e^{ikz} & (z > L/2). \end{cases} \tag{10.14}$$

These are solutions to the differential equation (10.9) with $E = k^2$. In equations of this type the *Wronskian* $W[\psi_L, \psi_R] := \psi_L \psi_R' - \psi_R \psi_L'$ is a constant independent of z. Calculating W first on the left, we obtain

$$\begin{aligned} W_L[\psi_L, \psi_R] &= (A_L e^{ikz} + B_L e^{-ikz})(-ik)e^{-ikz} \\ &\quad -(A_L e^{ikz} - B_L e^{-ikz})(ik)e^{-ikz} \\ &= -2ik A_L, \end{aligned}$$

while on the right w is given by

$$\begin{aligned} W_R[\psi_L, \psi_R] &= -(A_R e^{-ikz} + B_R e^{ikz})(ik)e^{ikz} \\ &\quad -(A_R e^{-ikz} - B_R e^{ikz})(ik)e^{ikz} \\ &= -2ik A_R. \end{aligned}$$

Equating the two, we obtain

$$A_L = A_R, \tag{10.15}$$

and consequently the left and right transmission coefficients are equal.

This is a general result for an equation of the type (10.9). It does not depend on any symmetry, be it Hermiticity or PT-symmetry, of $V(z)$. We obtain more specific results by imposing one or the other of these symmetries. Thus, in the Hermitian case we know that $\psi^*(z)$ is also a solution of the equation. We therefore consider $W[\psi_L, \psi_R^*]$, which evaluates to

$$W_L[\psi_L, \psi_R^*] = 2ikB_L, \qquad W_R[\psi_L, \psi_R^*] = -2ikB_R^*,$$

showing that $B_R = -B_L^*$. Calculating $W[\psi_L, \psi_L^*]$ gives

$$W_L[\psi_L, \psi_L^*] = -2ik(|A_L|^2 - |B_L|^2), \qquad W_R[\psi_L, \psi_L^*] = -2ik.$$

Hence we derive the relation

$$|A_L|^2 - |B_L|^2 = 1, \tag{10.16}$$

and similarly $|A_R|^2 - |B_R|^2 = 1$. In terms of the transmission and reflection coefficients T and R, where we no longer need to distinguish between left and right, (10.16) corresponds to the usual unitarity relation

$$T + R = 1.$$

In the PT-symmetric case the other solution of (10.9) is $\psi^*(-z)$ rather than $\psi^*(z)$, so we now consider the Wronskian $W[\psi_L(z), \psi_L^*(-z)]$. This evaluates to

$$W_L[\psi_L(z), \psi_L^*(-z)] = 2ikB_L, \qquad W_R[\psi_L(z), \psi_L^*(-z)] = -2ikB_L^*,$$

showing that $B_L = -B_L^*$. Thus B_L is pure imaginary. Similarly, we can show that B_R is pure imaginary. A calculation of $W[\psi_L(z), \psi_R^*(-z)]$ on the left and on the right gives

$$W_L[\psi_L(z), \psi_R^*(-z)] = -2ik(A_L A_R^* - B_L B_R^*),$$
$$W_R[\psi_L(z), \psi_R^*(-z)] = -2ik.$$

Thus, instead of the standard unitarity relation (10.16) we have the pseudo-unitarity relation [Ge *et al.* (2012b)]

$$|A_L|^2 - B_L B_R^* = 1, \quad \text{or} \quad |A_L|^2 \pm |B_L||B_R| = 1. \tag{10.17}$$

In terms of T, R_L, and R_R (10.17) becomes

$$T - 1 = \pm\sqrt{R_L R_R}. \tag{10.18}$$

This pseudo-unitarity relation can provide an extremely useful check on numerical calculations. It also means that in the PT-symmetric case, as opposed to the Hermitian case, T can be greater than 1, which is understandable because the medium is an active medium that includes gain. Moreover, whereas in the Hermitian case $R = R_L = R_R$ must be less than 1, in the PT-symmetric case the transmission and reflection coefficients can actually go to infinity as long as the relation $T - 1 = \sqrt{R_L R_R}$ is maintained. In an optical system this corresponds to the onset of lasing. That is, the output of the device can become very large even as the input stays finite. However, at the onset of lasing the system is no longer classical, and a full description requires a quantum-mechanical treatment. The pseudo-unitarity relation (10.18) and the associated symmetry properties of the reflection and transmission amplitudes have recently been generalized to a multichannel situation[Ge *et al.* (2015)], which could be realized, for example, as a PT-symmetric multimode waveguide.

10.4.4   *Transfer matrix*

Until now we have described scattering in terms of the individual wave functions ψ_L and ψ_R. An alternative and more powerful way of treating scattering is in terms of the *transfer matrix M*. In this formulation we consider both incoming and outgoing waves on the left and right of the grating, as illustrated in Fig. 10.16.

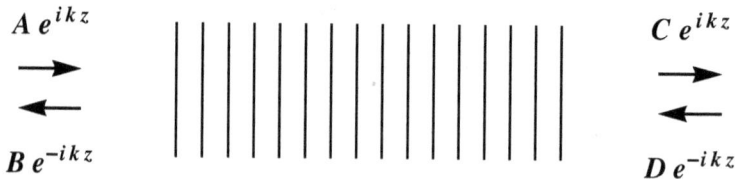

Fig. 10.16   Set-up for the transfer matrix M. Now we consider waves incoming from both left and right and M relates C and D to A and B.

Explicitly

$$\psi_L = Ae^{ikz} + Be^{-ikz}, \qquad \psi_R = Ce^{ikz} + De^{-ikz}. \tag{10.19}$$

The transfer matrix then relates the waves on the right to those on the left according to

$$\begin{pmatrix} C \\ D \end{pmatrix} = M \begin{pmatrix} A \\ B \end{pmatrix}. \tag{10.20}$$

That is,

$$C = M_{11}A + M_{12}B, \qquad D = M_{21}A + M_{22}B. \tag{10.21}$$

From the Wronskian of the two solutions u corresponding to $A = 0$, $B = 1$, and v corresponding to $A = 1$, $B = 0$, we find that M is *unimodular*; that is, $\det M = 1$. Hence, the inverse relation relating left to right is

$$A = M_{22}C - M_{12}D, \qquad B = -M_{21}C + M_{11}D,$$

from which we read off

$$M^{-1} = \begin{pmatrix} M_{22} & -M_{12} \\ -M_{21} & M_{11} \end{pmatrix}. \tag{10.22}$$

We can relate the elements of M to our previous coefficients A_{L}, A_{R}, B_{L}, and B_{R} in (10.13) and (10.14). We first consider the case when $D = 0$ and $C = 1$. This gives

$$A = A_{\mathrm{L}} = M_{22}, \qquad B = B_{\mathrm{L}} = -M_{21}. \tag{10.23}$$

Similarly, by taking $A = 0$ and $B = 1$ we obtain

$$D = A_{\mathrm{R}} = M_{22}, \qquad C = B_{\mathrm{R}} = M_{12}. \tag{10.24}$$

These results agree with the previously derived result $A_{\mathrm{L}} = A_{\mathrm{R}}$ in (10.15). The pseudo-unitarity relation (10.17) can be derived by combining the unimodularity of M with the symmetry relation [Longhi (2010a)] $M^* = M^{-1}$ implied[5] by $\mathcal{PT}$. From (10.23) and (10.24) we can write M in terms of the transmission and reflection amplitudes as

$$M = \begin{pmatrix} t - r_{\mathrm{L}}r_{\mathrm{R}}/t & r_{\mathrm{R}}/t \\ -r_{\mathrm{L}}/t & 1/t \end{pmatrix},$$

where we have determined the element M_{11} by using the unimodularity property of M.

In terms of the matrix elements of M, the signal for the onset of lasing is $M_{22} = 0$ for real k. The vanishing of M_{22} implies that $T \to \infty$, which can only happen for a non-Hermitian system with gain. But, as was pointed out in [Longhi (2010a)], for a $\mathcal{PT}$-symmetric system the symmetry relation

---

[5] Strictly speaking, the relation also involves taking the complex conjugate of the angular frequency ω, but here we consider ω to be real.

$M_{11} = M_{22}^*$, or equivalently the pseudo-unitarity relation (10.17), implies that whenever $M_{22} = 0$, we simultaneously have $M_{11} = 0$. The vanishing of M_{11} corresponds to the time-reversed situation, governed by M^{-1} in (10.22), where coherent incoming radiation can be completely absorbed with no reflection or transmission. This effect is termed *coherent perfect absorption*[Chong et al. (2010)]. In general, $M_{22} = 0$ occurs for different values of the parameters than $M_{11} = 0$, but the special feature of $\mathcal{PT}$ symmetry, where gain and loss are exactly balanced, is that the two occur for the same values of the parameters.

The phenomenon of lasing coexisting with coherent perfect absorption can also be discussed in the framework of the S matrix, as was done in [Chong et al. (2010)]. Whereas the transfer matrix relates the waves to the right and left, the scattering matrix S relates outgoing to incoming waves; that is, $\begin{pmatrix} B \\ C \end{pmatrix} = S \begin{pmatrix} A \\ D \end{pmatrix}$. We find the elements of S from (10.21) by using the unimodularity of M:

$$S = \frac{1}{M_{22}} \begin{pmatrix} -M_{12} & 1 \\ 1 & M_{21} \end{pmatrix} = \begin{pmatrix} r_{\mathrm{L}} & t \\ t & r_{\mathrm{R}} \end{pmatrix}.$$

The eigenvalues of S are

$$s_{\pm} = \frac{1}{2M_{22}} \left[M_{21} - M_{12} \pm \sqrt{(M_{21} + M_{12})^2 + 4} \right], \qquad (10.25)$$

whose product is $s_+ s_- = -M_{11}/M_{22}$. So if $M_{22} = M_{11}^*$ goes to zero, then one of the eigenvalues goes to infinity. Simultaneously, the other eigenvalue must go to zero[6] in order to satisfy $|s_+ s_-| = 1$. Thus, the phenomenon of coherent perfect absorption (CPA) can be characterized by the coincidence of a pole and a zero of the S matrix for real k.

The practical virtue of the transfer-matrix formalism is that if one has a succession of gratings, or gratings interspersed by a medium of constant refractive index, the M matrix for the entire set-up can be derived simply by multiplying in succession the M matrices of the separate components of the system. This property has been exploited in a series of papers summarized in [Mostafazadeh (2014a)]. The key point of these papers is that M can be identified with the evolution operator $\mathcal{U}$ associated with a 2×2 Hamiltonian $\mathcal{H}$.

The starting point of this approach is to realize that in quantum-mechanical language we can cast the second-order time-independent

---

[6]Using the unimodularity property of M, the quantity under the square root in (10.25) can be written as $(M_{21} + M_{12})^2 + M_{11}M_{22}$, so when $M_{22} = 0$ the square root is equal to $\pm(M_{21} - M_{12})$.

Schrödinger equation $-\psi'' + V(z)\psi = k^2\psi$ for wave functions whose asymptotic behavior is $e^{\pm ikz}$ as a first-order two-component equation in terms of the particular combinations

$$\Psi_1 \equiv \tfrac{1}{2}e^{-ikz}\left(\psi - i\psi'/k\right), \qquad \Psi_2 \equiv \tfrac{1}{2}e^{ikz}\left(\psi + i\psi'/k\right),$$

which are to be regarded as the upper and lower components of a two-component object Ψ. The particular feature of this choice is that outside of the support of the potential V, when ψ becomes the combinations ψ_L or ψ_R given in (10.19), Ψ_1 and Ψ_2 act as projection operators onto e^{ikz} and e^{-ikz}, respectively. Thus,

$$\Psi_1 \to \begin{cases} 1 & \text{for } \psi = e^{ikz}, \\ 0 & \text{for } \psi = e^{-ikz} \end{cases} \quad \text{and} \quad \Psi_2 \to \begin{cases} 0 & \text{for } \psi = e^{ikz}, \\ 1 & \text{for } \psi = e^{-ikz}. \end{cases}$$

Therefore, in view of (10.19), to the left of the scattering potential Ψ becomes $\Psi_L = \begin{pmatrix} A \\ B \end{pmatrix}$, while to the right Ψ has the form $\Psi_R = \begin{pmatrix} C \\ D \end{pmatrix}$. Finally, from (10.20) we see that $\Psi_R = M\Psi_L$.

The first-order equations obeyed by Ψ_1 and Ψ_2 are readily shown to be

$$\frac{i}{k}\frac{d\Psi_1}{dz} = \frac{V}{2k^2}\left(\Psi_1 + e^{-2ikz}\Psi_2\right), \qquad \frac{i}{k}\frac{d\Psi_2}{dz} = -\frac{V}{2k^2}\left(e^{2ikz}\Psi_1 + \Psi_2\right).$$

Hence, thinking of $\tau \equiv kz$ as a time variable, ψ_R evolves from ψ_L according to the Schrödinger-like equation

$$i\frac{d\Psi}{d\tau} = \mathcal{H}(\tau)\Psi,$$

where $\mathcal{H}$ is the matrix Hamiltonian

$$\mathcal{H} \equiv \frac{V}{2k^2}\begin{pmatrix} 1 & e^{-2i\tau} \\ -e^{2i\tau} & -1 \end{pmatrix}.$$

This "time" evolution is governed by the evolution operator $U(\tau, \tau_0)$, which is the time-ordered exponential of $-i\int\mathcal{H}(\tau)$, namely

$$U(\tau, \tau_0) = T\left(\exp(-i\int_{\tau_0}^{\tau}\mathcal{H}(\tau')d\tau')\right), \qquad (10.26)$$

where U satisfies the differential equation

$$i\frac{d}{d\tau}U(\tau, \tau_0) = \mathcal{H}(\tau)U(\tau, \tau_0). \qquad (10.27)$$

If we evolve Ψ from τ_1 to the left of the potential (possibly from $-\infty$) to τ_2 to the right of the potential (possibly ∞), we can make the identification $M = U(\tau_2, \tau_1)$. This identification has been used in [Mostafazadeh (2014a)] to derive a differential equation for M, and hence for the reflection and transition amplitudes, which has proved to be of great utility in constructing finite-range potentials with a variety of desired properties. Perturbative expressions for the transition and reflection coefficients can be obtained by applying standard perturbation theory to (10.26).

10.5 $\mathcal{PT}$ Lasers

As already mentioned in Sec. 10.4, lasing can occur in an optical cavity when the element M_{22} of the transfer matrix has a zero for real k, or equivalently when the reflection and transmission coefficients go to infinity. Standard lasers have a cavity in which the real part of the refractive index has some particular modulation whereby an effective imaginary part is applied to the whole cavity by optical pumping.

A common problem in conventional lasers is that the cavity typically has competing modes, so it is difficult to avoid mode-hopping, where the lasing action switches between the desired mode and a nearby one. Using the additional freedom to modify the imaginary part of n in a $\mathcal{PT}$-symmetric way, various proposals have been made to achieve single-mode and/or directional lasing.

One ingenious approach to achieve single-mode lasing is to exploit $\mathcal{PT}$-symmetry breaking [Feng *et al.* (2014)] within the context of ring lasers. Using azimuthal symmetry the ring can be configured so that lasing occurs for only a single pair of coupled modes. The wave vector of one mode then acquires a negative imaginary part, leading to lasing, while that of the other mode acquires a positive imaginary part, corresponding to absorption.

The set-up [Feng *et al.* (2014)] is a microring, of dimensions $9\mu m$, of InGaAsP on a substrate of InP, as shown in Fig. 10.17. The ring is doped periodically by alternate wedges of Cr and Ge. In the absence of doping the modes of the ring are degenerate-energy pairs of *whispering-gallery modes*, which propagate clockwise and anticlockwise in the ring. However, when the ring is doped in a periodic fashion, the modes become coupled, but the important thing is that this coupling only occurs for the two modes that have the same azimuthal periodicity as the perturbation. Moreover, as the perturbation involves only the imaginary part of the refractive index, the original $\mathcal{PT}$ symmetry is broken immediately; this is called *threshold-less symmetry breaking*. As was realized in [Ge and Stone (2014)], this phenomenon commonly occurs in higher-dimensional systems, and in particular in the two-dimensional circular geometry considered here, because of degeneracy.[7] The net result is that one and only one mode is excited, and single-mode lasing is achieved when the ring is pumped appropriately. However, the output from this device is not unidirectional.

---

[7]In fact the same phenomenon can occur for the purely imaginary one-dimensional optical potential $v(x) = ia \sin 2x$.

Fig. 10.17 Schematic of a microring cavity, with periodic doping on the upper surface. In the absence of doping the cavity would have many competing modes, but the doping eliminates all modes with a different azimuthal periodicity.

An alternative way of exciting only a single mode, again exploiting $\mathcal{PT}$-symmetry breaking, is to consider two coupled ring resonators [Hodaei et al. (2014)], as shown in Fig. 10.18. The time development of the mode amplitudes a_n and b_n in the two rings is described by the coupled equations

$$\dot{a}_n = -i\omega_n a_n + i\kappa_n b_n + \gamma_n a_n, \qquad \dot{b}_n = -i\omega_n b_n + i\kappa_n a_n - \gamma_n b_n.$$

Here, κ_n represents the coupling between the two modes and γ_n is the gain in one ring and an equal loss in the other. Note the great similarity between these equations and those governing the case of two coupled waveguides in (10.4). The results have the same general features, and the eigenvalues, in this case ω, are given by

$$\omega_n^\pm = \omega_n \pm \sqrt{\kappa_n^2 - \gamma_n^2}.$$

If it can be arranged that γ_n is greater than κ_n for one particular mode only, then the device will lase in only that one mode. This single-mode operation has been demonstrated experimentally in [Hodaei et al. (2014)].

Yet another alternative route to achieve single-mode lasing is to exploit the special properties of the $\mathcal{PT}$-symmetric "invisibility" optical potential $v(z) = v_0 e^{2i\beta z}$ of Sec. 10.5. There we saw in (10.10)-(10.11) that, although the reflectivity for incidence from the left side onto a slab of material with that particular refractive index is almost negligible, the reflectivity for incidence from the right is large and narrowly peaked, proportional to the square of a sinc function, as shown in Fig. 10.15. This feature has been used to eliminate other modes of an otherwise multimode cavity by placing a slab of material, that is, a solid grating, with the optical potential $v(z) = v_0 e^{2i\beta z}$ in the middle of the cavity [Kulishov et al. (2014)]. However,

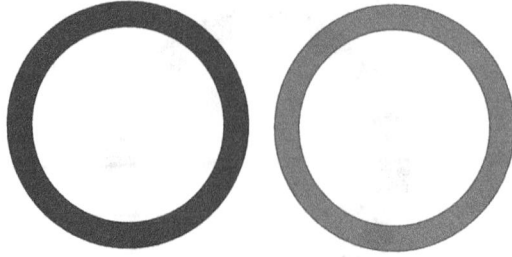

Fig. 10.18   Two coupled microrings, one with loss (gray, left), the other with equal gain (black, right). When the rings are sufficiently close together, they are coupled via the evanescent (exponentially decaying) fields outside each ring.

a simpler set-up exploiting the same idea [Jones *et al.* (2016)] consists of merely placing a mirror at the right end of the PT-symmetric grating, as shown in Fig. 10.19. In general, the refractive index n_1 to the left of the

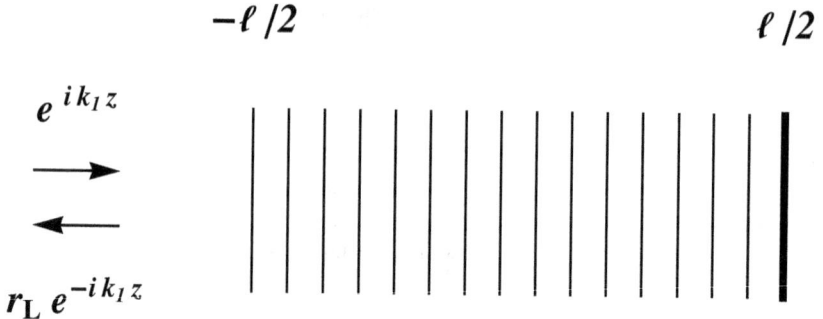

Fig. 10.19   Left incidence on a grating of length ℓ terminated by a mirror. The condition for the onset of lasing is $r_L \to \infty$.

grating will be different from the base refractive index n_2 of the material on which the grating is constructed. There is therefore a transfer matrix for the left boundary at $z = -\ell/2$, which is

$$M^{(12)} = \frac{1}{2} \begin{pmatrix} (1+\gamma)e^{-i(k_1-k_2)\ell/2} & (1-\gamma)e^{i(k_1+k_2)\ell/2} \\ (1-\gamma)e^{-i(k_1+k_2)\ell/2} & (1+\gamma)e^{i(k_1-k_2)\ell/2} \end{pmatrix},$$

where k_2 is the wave vector inside the grating and $\gamma = k_1/k_2 = n_1/n_2$. This initial transfer matrix $M^{(12)}$ is to be multiplied by the transfer matrix of the grating, which is

$$M^{\mathcal{PT}} = \begin{pmatrix} 1 & r_{\mathrm{R}}^{\mathcal{PT}} \\ 0 & 1 \end{pmatrix}, \quad \text{where} \quad r_{\mathrm{R}}^{\mathcal{PT}} = \tfrac{1}{2} i \alpha^2 k_2 \ell \left(\frac{\sin \ell \delta}{\ell \delta} \right). \tag{10.28}$$

Here we have used the sinc function of (10.11), except that $k \to k_2$, so that $\delta = k_2 - \beta$. We are also free to multiply $r_{\mathrm{R}}^{\mathcal{PT}}$ by a phase factor $e^{i\varphi}$, which corresponds to a shift in the origin of z in the expression $\tfrac{1}{2}\alpha^2 \exp(2i\beta z)$ for the $\mathcal{PT}$-symmetric perturbation $v(z)$.

We write the electric field within the grating as $E = E_f(z)e^{ik_2 z} + E_b(z)e^{-ik_2 z}$. Then multiplication of the two transfer matrices gives

$$\begin{pmatrix} E_f(\ell/2) \\ E_b(\ell/2) \end{pmatrix} = M^{\mathcal{PT}} M^{(12)} \begin{pmatrix} 1 \\ r_{\mathrm{L}} \end{pmatrix}. \tag{10.29}$$

At the mirror, which for simplicity we assume to be perfect, the boundary condition is that E should vanish. Imposing this condition on E then gives the following expression for r_{L}:

$$r_{\mathrm{L}} e^{-ik_1 \ell} = - \frac{(1+\gamma)e^{ik_2\ell} + (1-\gamma)(e^{-ik_2\ell} + r_{\mathrm{R}}^{\mathcal{PT}})}{(1-\gamma)e^{ik_2\ell} + (1+\gamma)(e^{-ik_2\ell} + r_{\mathrm{R}}^{\mathcal{PT}})}. \tag{10.30}$$

The condition for the onset of lasing is that $r_{\mathrm{L}} \to \infty$; that is, the denominator in this expression should vanish for a real value of k_2. Thus we require that

$$(1-\gamma)e^{ik_2\ell} + (1+\gamma)(e^{-ik_2\ell} + r_{\mathrm{R}}^{\mathcal{PT}}) = 0. \tag{10.31}$$

The real and imaginary parts of this equation take on different forms depending on whether or not we include the extra factor $e^{i\varphi}$ in the definition of $r_{\mathrm{R}}^{\mathcal{PT}}$. If we omit any such phase shift, $r_{\mathrm{R}}^{\mathcal{PT}}$ is as given in (10.28) and is pure imaginary. So the real and imaginary parts of (10.31) give

$$\cos k_2 \ell = 0, \tag{10.32a}$$

$$\frac{2\gamma}{1+\gamma} \sin k_2 \ell = \tfrac{1}{2}\alpha^2 k_2 \ell \left(\frac{\sin \ell \delta}{\ell \delta} \right). \tag{10.32b}$$

The first equation (10.32a) implies that $k_2 \ell = (n + 1/2)\pi$, so $\sin k_2 \ell = (-1)^n$. If we take the length of the grating to be an integral number of periods, that is, $\ell = m\Lambda = m\pi/\beta$, then the argument $\ell \delta$ of the sinc function in the second equation is $\ell \delta = (k_2 - \beta)\ell = (n - m + 1/2)$. Thus, (10.32b) has many possible solutions for α^2, occurring at half-integral values of $\ell \delta$, as shown in Fig. 10.20(a).

In order to achieve single-mode lasing, it is clear that we have to shift the allowed values of $k_2 \ell$ so that they correspond to *integral* values of $\ell \delta$, as shown in Fig. 10.20(b). Then the sinc function is equal to 1 when $\ell\delta = 0$, but it vanishes for all the other possible values. This shift can be achieved either by taking the grating length to be a half-integral number of periods or by choosing $\phi = \pi$. Choosing $\phi = \pi$ makes $r_R^{\mathcal{PT}}$ real rather than imaginary, in which case the equations for the imaginary and real parts of (10.30) become respectively

$$\sin k_2\ell = 0, \tag{10.33a}$$

$$\frac{2}{1+\gamma}\cos k_2\ell = \tfrac{1}{2}\alpha^2 k_2\ell\left(\frac{\sin\ell\delta}{\ell\delta}\right). \tag{10.33b}$$

The first equation (10.33a) requires that $k_2\ell = n\pi$, and then the second equation (10.33b) only has a solution for $n = m$, corresponding to $\delta = 0$. [Recall that $\delta = k_2 - \beta$ and $\beta = m\pi/\ell$, so $\ell\delta = (m-n)\pi$.] We have thus exploited both the central peak of the sinc function and its outlying zeros to obtain single-mode lasing.

The laser output of this device is highly directional. In the above we have considered only normal incidence, but we can generalize the analysis in both the coupled-wave method and the exact Bessel-function method [Jones *et al.* (2016)] to incidence at an angle in order to examine the angular distribution. If the parameters are chosen to satisfy (10.33a) and (10.33b), the angular distribution shows a very narrow peak around $\theta = 0$.

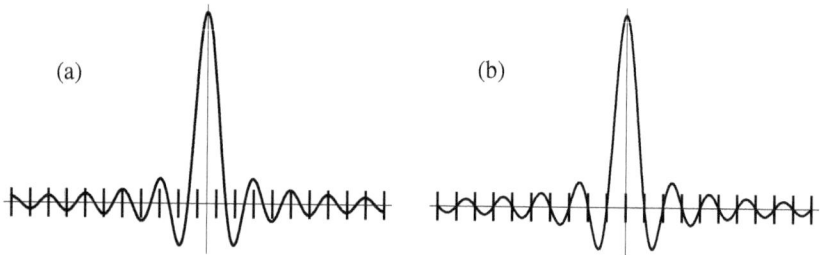

Fig. 10.20 Possible lasing modes for (a) $\phi = 0$, (b) $\phi = \pi/2$. The vertical bars indicate the possible values of $k_2\ell$, as given by the first lasing condition, (10.32a) or (10.33a), while the curve is the sinc function of (10.28). To satisfy the second lasing condition, (10.32b) or (10.33b), the sinc function must be nonzero at those allowed values of $k_2\ell$. In case (a) this condition is satisfied for all the allowed values of $k_2\ell$, while in case (b) it is satisfied at one and only one value of $k_2\ell$.

10.6 Supersymmetry in Quantum Mechanics and Optics

In particle physics and quantum field theory supersymmetry is a symmetry where each particle has a partner of the same mass but a different spin. For example, a spin-$\frac{1}{2}$ particle and a spin-1 particle are grouped together in the same multiplet and have the same mass in the absence of symmetry breaking. In quantum mechanics supersymmetry is different; it relates two different Hamiltonians that have the same spectrum except for one state. In standard Hermitian quantum mechanics that unpartnered state is always the ground state of one of the Hamiltonians, but in non-Hermitian, and in particular PT-symmetric, quantum mechanics it can instead be an excited state. Thus, in the context of non-Hermitian quantum mechanics (or classical optics in the paraxial approximation), supersymmetry can be used as a tool to modify Hamiltonians in order to remove one or more unwanted states from the spectrum while leaving the others intact.

To see how this works we start with a one-dimensional Hamiltonian $H_1 = -d^2/dx^2 + V_1(x)$. Then if $\chi(x)$ is a solution of the eigenvalue equation $(H_1 - E_1)\chi = 0$, that is, $\chi'' = (V_1 - E_1)\chi$, we see that V_1 can be written as

$$V_1 = W^2 + W' + E_1,$$

where W, defined as $W = \chi'/\chi$, is known as the *superpotential*. In standard quantum mechanics, χ is taken as the ground state, which has no nodes, in order that W be nonsingular.

Now if we define the generalized lowering and raising operators as

$$A = -\frac{d}{dx} + W, \qquad B = \frac{d}{dx} + W,$$

H_1 can be written as $H_1 = E_1 + BA$. Note that $A\chi = (-\chi' + W\chi) = 0$. Now the commutation relation between A and B is $[A, B] = -2W'$, so if we define $H_2 \equiv -d^2/dx^2 + V_2 = E_1 + AB$, then

$$V_2 = W^2 - W' + E_1 = V_1 - 2W'. \tag{10.34}$$

We now have a pair of isospectral Hamiltonians H_1 and H_2, apart from the state $|\chi\rangle$. Thus, if $|\psi\rangle$ is an eigenstate of H_1 with eigenvalue E, then $(BA + E_1)|\psi\rangle \equiv H_1|\psi\rangle = E|\psi\rangle$, and multiplying this on the left by A gives

$$(AB + E_1)(A|\psi\rangle) \equiv H_2(A|\psi\rangle) = E(A|\psi\rangle).$$

Thus, the eigenstates of H_2 are obtained from those of H_1 by applying the operator A, and H_2 has the same eigenvalues as H_1 except for E_1 because $A|\chi\rangle = 0$.

This is all a generalization of the well-known solution of the harmonic-oscillator $H_1 = p^2 + \omega^2 x^2$ using the raising and lowering operators $a^\dagger = d/dx - \omega x$ and $a = -d/dx - \omega x$. These operators satisfy the commutation relation $[a, a^\dagger] = 2\omega$, and H_1 can be written as

$$H_1 = aa^\dagger - \omega = a^\dagger a + \omega.$$

Then it is easily seen that the spectrum of H_1 is $E = (2n + 1)\omega$, with ground-state energy $E_1 = \omega$. When acting on an eigenstate, $a^\dagger$ produces the next-higher eigenstate, increasing the eigenvalue by 2ω. Similarly, a produces the next-lower eigenstate except when it acts on the ground-state wave function χ, in which case $a\chi = 0$.

For the harmonic oscillator the ground-state wave function is $\chi(x) = \exp(-\omega x^2/2)$, so $W = -\omega x$ and W' is just a constant: $W' = -\omega$. Then, according to (10.34), V_2 is simply $V_1 + 2\omega$, and the spectrum of H_2 is $E = (2n + 1)\omega$ for $n \geq 1$. That is, the spectrum is shifted up by 2ω. The result is that each eigenstate of H_1 has a partner eigenstate of H_2 except for the ground state, which remains unpaired.

Now consider an initial Hamiltonian H_1 with unbroken $\mathcal{PT}$ symmetry, so that χ is also $\mathcal{PT}$ symmetric, with $\chi^*(x) = \chi(-x)$. Then $W \equiv \chi'/\chi$ is $\mathcal{PT}$ antisymmetric, so W' is $\mathcal{PT}$ symmetric. But V_2 differs from V_1 by $2W'$, so V_2 is also $\mathcal{PT}$ symmetric. Thus, in the unbroken-symmetry regime, supersymmetry leads to a $\mathcal{PT}$-symmetric partner Hamiltonian H_2.

If H_1 is $\mathcal{PT}$ symmetric rather than Hermitian, one has the freedom to remove any given state from the spectrum, as has been emphasized in [Miri *et al.* (2013)]. That is, one can take χ to be an excited state of H_1 rather than just the ground state because the nodes of higher-order wave functions do not lie on the real-x axis. Then H_2 will have the same spectrum as H_1 except for the eigenstate corresponding to χ. An example of just such an excision is given in [Miri *et al.* (2013)]. Moreover, the process can be repeated to remove any number of energy levels (or modes, in an optical context) while leaving the remaining levels unaltered. This excision process can be very useful in the $\mathcal{PT}$-broken regime if only a few eigenvalues are complex-conjugate pairs, as it gives one the possibility of finding an equivalent potential containing only the real eigenstates.

It is even possible to construct partner Hamiltonians that while having a real spectrum are not $\mathcal{PT}$ symmetric, at least in the conventional sense, when $\mathcal{P}$ is the parity operator. Thus, starting with V_2 and thinking of (10.34) as a Riccati equation to be solved for W, we of course have the previous solution $W = \chi'/\chi$. But there is a more general solution $\tilde{W}$, which we write as $\tilde{W} = W + 1/v$. Then v must satisfy the linear equation $v' = 1 - 2Wv$, with the solution

$$v(x) = \frac{1}{\chi^2(x)} \left(\int_{-\infty}^{x} \chi^2(x')dx' + c \right),$$ (10.35)

where c is an arbitrary constant, possibly complex, chosen so that v does not vanish on the real axis. The resulting superpotential is

$$W = \chi'/\chi + \frac{\chi^2(x)}{\int_{-\infty}^{x} \chi^2(x')dx' + c}.$$ (10.36)

In this section we have been discussing confining potentials and their possible energy levels, which in optics could correspond to allowed modes in a multi-mode waveguide. However, the situation is subtly different if we consider periodic potentials [Dunne and Feinberg (1998)]. In a periodic potential, with $V(x) = V(x+a)$, we are primarily concerned with the *Bloch waves*, which are scattering solutions that have the same periodicity as the potential up to a phase; that is, $\psi_k(x + a) = \exp(ika)\psi_k(x)$. In this case there is no missing state, and the two potentials are strictly isospectral.

An interesting though not necessarily PT-symmetric application of these ideas is given in [Longhi and Della Valle (2013)], where a prescription based on supersymmetry is given to construct a completely transparent interface between two isospectral periodic crystals. In this case the initial function $\chi(x)$ is not a Bloch wave, but rather an *evanescent* wave for an energy in a forbidden region; that is, a wave falling off exponentially away from the interface. The eventual potential interpolates smoothly from the original potential $V_1(x)$ on the far left of the interface to an isospectral potential $V_2(x)$ on the far right. The construction is very similar to (10.36).

In [Midya (2014)] a sequence of supersymmetric partners of the PT-symmetric potential of Sec. 10.4 is considered, taking as a starting point the Bloch wave $\chi = I_{-\nu}(y)$ for a particular energy E_1. These supersymmetric partner potentials have an identical band structure and the same transmission coefficient as the original one, but it is possible in only a few steps to greatly enhance the right reflectance and correspondingly to suppress the left reflectance.

We close this section with an intriguing result [Makris *et al.* (2015)], which looks very much like supersymmetry but is not; there is a crucial factor of i in the definition of V in terms of W. If one is considering wave propagation in the z direction with a potential $V(x)$, the equation of propagation for the envelope function $\psi(z)$ can be written in suitable units as

$$i\frac{\partial \psi}{\partial z} + \frac{\partial^2 \psi}{\partial x^2} + V(x)\psi = 0.$$ (10.37)

If in this equation we take $V(x)$ to be given by $V(x) = W^2(x) - iW'(x)$, where W is *real*, then it is easy to see that the solution to the equation is

$$\psi(x, z) = A \exp\left[i \int^x dx'\, W(x')\right],$$

which crucially represents a constant-intensity wave. (Recall that the electric field itself is given by $E = e^{ik_0 z}\psi$.) If W is symmetric, then V is PT-symmetric, but for general W there is no obvious PT symmetry. However, V still represents an overall balance of loss and gain because

$$\int_{-\infty}^{\infty} dx\, \text{Im}\, V = -iW(x)\big|_{-\infty}^{\infty},$$

which is zero if W is either confined or periodic. Even more remarkable is that the property of constant intensity holds even in the presence of nonlinearity; that is, when there is an additional term $g|\psi|^2\psi$ on the left side of (10.37). In that case ψ merely acquires an additional unimodular factor $\exp(igA^2 z)$. The concept of constant-intensity waves can also be readily generalized to two transverse directions [Makris *et al.* (2015)].

10.7   Wave Propagation in Discrete PT Systems

In Sec. 10.3 we considered just two single-mode waveguides and found that PT symmetry led to interesting and unusual propagation features. A natural generalization is to consider a sequence of waveguides, each coupled to its two nearest neighbours by their evanescent tails. That sequence could be a small finite number or a periodic array of waveguides. In either case we may expect some exotic properties.

10.7.1   *Propagation in infinite systems*

An infinite sequence of waveguides effectively constitutes a discretization in the x direction of the type of systems considered in Sec. 10.2, where the refractive index varied continuously in x. The relevant set of equations[8] is

$$i\frac{d\psi_n}{dz} = \kappa_n\psi_{n-1} + \kappa_{n+1}\psi_{n+1} + V_n\psi_n. \tag{10.38}$$

The simplest possible model occurs when both κ_n and V_n are site independent; that is, when $\kappa_n = \kappa$ and $V_n = V$ for all n. In this model we

---

[8]These are essentially of the same form as the equations for electronic states in a one-dimensional atomic lattice in the tight-binding approximation (see, for example, [Hook and Hall (1991)]), where the κ_n are known as hopping parameters and the V_n are the on-site energies.

have a regular lattice of period 1 in the x-direction, whose band structure, the relation between E and k for the Bloch waves $\psi_n^0(\beta, k, z) = e^{i\beta z}e^{ikn}$ satisfying (10.38), is $\beta = -V - 2\kappa \cos k$. The propagation in z of any initial configuration at $z = 0$ can be obtained by expanding in the Bloch wave functions $\psi_n^0(\beta, k, z = 0)$ and using the method of stationary states, as explained in (10.2), after discretizing k as $k_m = 2\pi m/N$ for some large N.

In a discrete system there is the additional freedom of introducing local defects into the system by allowing either κ_n or V_n to differ from their constant values at a few particular sites. Centered on those defects one may then expect to find localized states, constituting special optical modes that can be excited.

Examples of such defects in both κ_n and V_n are given in [Longhi (2015)]. In the first example V_n is modified in a PT-symmetric manner by giving the nonzero values $V_{-1} = V_1^* = \sigma \equiv \Delta + ig$, while $V_n = 0$ for all other n. We then look for bound states of the form $\psi_n(\beta, q, z) = c_n(q)e^{i\beta z}$, where the coefficients $c_n(q)$ fall off exponentially like $e^{-q|n|}$ away from the defect region $n = \pm 1$. If we start by assuming that $c_n(q) = e^{-q|n|}$ for $n \leq -1$, it is sufficient to impose the condition $c_2(q) = e^{-q}c_1(q)$. Equivalently, we can look for poles of the transmission coefficient $t(k)$ for scattering in the x direction for $k = iq$. The denominator of $t(iq)$ for $\kappa = 1$ is

$$f(q) = \Delta - \sinh q - |\sigma|^2 e^{-2q} \cosh q,$$

which can be rewritten in terms of $x \equiv e^{-q}$ as $f = -h(x)/(2x)$, where

$$h(x) = 1 - x^2 - 2x\Delta + |\sigma|^2 x^2(1 + x^2).$$

Below the threshold for PT-symmetry breaking we are looking for real positive roots of $f(q)$, while above the threshold complex roots are acceptable provided that $\text{Re}(q) > 0$ so that the $c_n(q)$ do fall off exponentially.

For $g^2 < \Delta(1 - \Delta)$ (recall that $\sigma = \Delta + ig$) there is a standard bound state with β below the continuum band $-2\kappa \leq \beta \leq 2\kappa$. It disappears when $g^2 = \Delta(1 - \Delta)$, at which point $\beta = -2\kappa$ and q becomes zero, so there is no longer any exponential fall off. Then, when g exceeds the threshold value for PT-symmetry breaking, which is given by $g_{th}^2 = 1 - \Delta\sqrt{2 + \Delta^2}$, a pair of complex-conjugate roots appear, the real part of whose β values lies within the continuum band. These unusual states are known as *bound states in the continuum*. This unusual spectrum is illustrated in Fig. 10.21 for $\Delta = 0.3$. The presence of this localized state has a significant effect on the z-development of any given input at $z = 0$. In particular, precisely at $g = g_{th}$, the power grows linearly with z, which is the typical behavior associated with a PT phase transition.

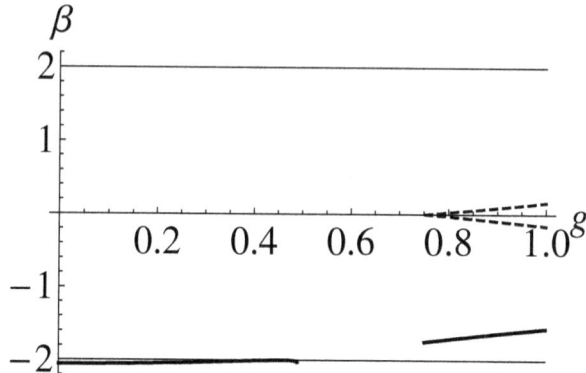

Fig. 10.21   Bound states of the lattice of (10.38) with $\kappa_n = 1$ and $V_n = 0$ apart from $V_{-1} = V_1^* = \Delta + ig$ as a function of g for $\Delta = 0.3$. The thick curves represent the real part, and the dashed curves the imaginary parts, of $\beta(q)$. Below $g^2 = \sqrt{\Delta(1 - \Delta)}$ there is a standard real solution for β outside the continuum region $|\beta| < 2$. However, above the symmetry-breaking point g_{th} given by $g_{th}^2 = 1 - \Delta\sqrt{2 + \Delta^2}$, a pair of complex-conjugate roots emerge, whose real part lies within the continuum region.

The second model considered in [Longhi (2015)] involved keeping $V_n = 0$ in (10.38), but modifying κ_n according to

$$\frac{\kappa_n}{\kappa} = \begin{cases} \sqrt{(n+1)/(n-1)} & n \text{ even}, \quad n \neq 0, \quad \kappa_0 = -ig, \\ \sqrt{(n-2)/n} & n \text{ odd}, \quad n \neq 1, \quad \kappa_1 = ig. \end{cases} \tag{10.39}$$

The lattice of (10.39) is PT symmetric, and it approaches the simple lattice $\kappa_n = \kappa$ for large n. The purely imaginary couplings κ_0 and κ_1 can be engineered experimentally by modulating the central waveguide at $n = 0$ in the z direction. The details are given in [Longhi and Della Valle (2014)].

This lattice supports a bound state in the continuum at $\beta = 0$, with an *algebraic*, rather than exponential, fall-off of the amplitudes ψ_n. Thus, the recursion relation $\kappa_n\psi_{n-1} + \kappa_n c_{n+1} = 0$ has the solution

$$\bar{\psi}_n = \begin{cases} 0 & n \text{ odd} \\ \kappa/g & n = 0 \\ \text{sign}(n)\dfrac{i^{n+1}}{\sqrt{n^2-1}} & n \text{ even}, \quad n \neq 0 \end{cases}$$

for all values of g, both below and above the PT-symmetry-breaking threshold, which turns out to be $g_{th} = \kappa$. Precisely at threshold, $\beta = 0$ is an *exceptional point in the continuum* [Andrianov and Sokolov (2011)] and $\bar{\psi}_n$ has an associated Jordan function $f_n = (-i/(2\kappa))\sin(n\pi/2)$ satisfying $\kappa_n f_{n-1} + \kappa_{n+1} f_{n+1} = \bar{\psi}_n$. Because of the presence of the Jordan function the amplitude grows linearly and the power quadratically with z.

10.7.2 *Finite systems: Dimers, trimers, quadrimers ...*

We have already encountered the dimer – two coupled waveguides or microcavities – in Sec. 10.3. Systems consisting of three units (trimers), four units (quadrimers) or more have also been studied intensively, and all have interesting and potentially useful properties, particularly when nonlinearity is present, that is, when the electromagnetic field is so strong that it locally modifies the refractive index itself.

Recall that the relevant coupled equations for the propagation of the fields in the two components of the dimer (10.7) are

$$i\frac{d\psi_1}{dz} - i\gamma\psi_1 + \kappa\psi_2 + \chi|\psi_1|^2\psi_1 = 0,$$
$$i\frac{d\psi_2}{dz} + i\gamma\psi_2 + \kappa\psi_1 + \chi|\psi_2|^2\psi_2 = 0. \tag{10.40}$$

The linear version of these equations, when $\chi = 0$, is worked out completely in Sec. 10.3, but in fact the greater interest, as with trimers or quadrimers, lies in the nonlinear regime. It is then useful to analyze the solution of (10.40) in terms of the Stokes variables

$$S_0 = |\psi_1|^2 + |\psi_2|^2, \ S_1 = \psi_1^*\psi_2 + \psi_1\psi_2^*, \ S_2 = i(\psi_1\psi_2^* - \psi_1^*\psi_2), \ S_3 = |\psi_1|^2 - |\psi_2|^2,$$

as was done in [Ramezani *et al.* (2010)]. It is easily seen that the vector $\mathbf{S} \equiv (S_1, S_2, S_3)$ satisfies $\mathbf{S}^2 = S_0^2$ so that the renormalized vector $\hat{\mathbf{S}} \equiv \mathbf{S}/S_0$ lives on the surface of the unit sphere (Bloch sphere), with polar angles θ, ϕ. The Bloch sphere is important insofar as it gives a pictorial representation of the properties of the solution as ψ_1 and ψ_2 develop as functions of z.

In [Ramezani *et al.* (2010)], $S_0(0)$ was fixed at 1 (so that $\theta = 0$ or π), and in that case the critical value χ_d of χ, above which the $\mathcal{PT}$ symmetry is broken and the output grows exponentially in the gain channel only, was found to be $\chi_d = 4 - 2\pi\gamma$. The same system was also treated using a different parametrization in [Sukhorukov *et al.* (2010)], where a wider variety of initial values of θ and ϕ was considered. But the general message was the same, namely that when χ in (10.40) becomes sufficiently large the system acts as an optical diode for z greater than a minimum length z_d if the input is in the lossy channel.

It is worth mentioning that an elegant alternative picture of the evolution of the (unnormalized) Stokes variables was subsequently given in [Barashenkov *et al.* (2013)]. There it was shown that these variables lie on the surfaces of a series of cylinders, and the two conserved quantities identified in [Ramezani *et al.* (2010)] were given a clear geometrical meaning.

The trimer, a set of three coupled waveguides or microcavities, can come in one of two $\mathcal{PT}$-symmetric configurations, as shown in Fig. 10.22. In the first, the couplings κ are between the neutral waveguide ψ_2 and those on either side with equal loss and gain, whereas in the second configuration all three components are coupled together with equal strength.

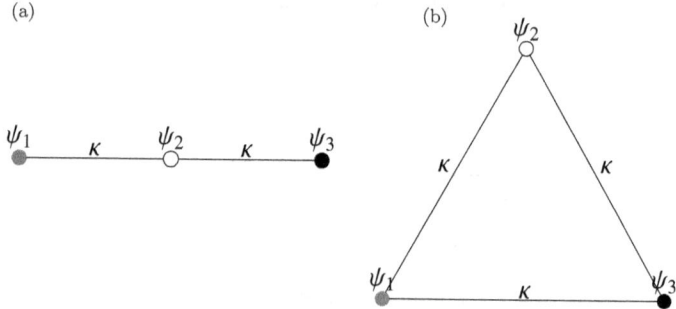

Fig. 10.22   Two possible $\mathcal{PT}$-symmetric configurations for a trimer: (a) open, (b) closed (periodic). The black and gray disks denote gain and loss respectively, while the open circles are neutral. The couplings between the units are all equal to κ. In both (a) and (b) the $\mathcal{P}$ operator is the interchange $1 \leftrightarrow 3$.

These configurations were studied intensively in [Li and Kevrekidis (2011); Li *et al.* (2013)]. The nonlinear evolution equations corresponding to Fig. 10.22(a) are

$$i\frac{d\psi_1}{dz} + i\gamma\psi_1 + \kappa\psi_2 + \chi|\psi_1|^2\psi_1 = 0,$$

$$i\frac{d\psi_2}{dz} + \kappa(\psi_1 + \psi_3) + \chi|\psi_1|^2\psi_1 = 0, \qquad (10.41)$$

$$i\frac{d\psi_3}{dz} - i\gamma\psi_3 + \kappa\psi_2 + \chi|\psi_3|^3\psi_2 = 0.$$

In the linear case ($\chi = 0$) any solution of (10.41) is given as a linear combination of the stationary states, that is, as solutions where all three ψ_r have a common z dependence $e^{i\mu z}$. The three possible values for μ are $\mu=0$ and $\pm\sqrt{2\kappa^2 - \gamma^2}$, so that the critical value of γ beyond which $\mathcal{PT}$ symmetry is broken and two of the eigenvalues become complex is $\gamma_{\mathcal{PT}} = \sqrt{2}\kappa$. In the nonlinear equations one can still look for stationary states, which will give some indication of the features of the system evolution, although the general solution is no longer a linear superposition of the stationary states. It turns out that there are three branches of stationary solutions, all of which have $|\psi_1| = |\psi_3|$, with different regions of stability and instability. One of the branches is unstable and can develop into runaway modes in which ψ_3 grows exponentially and ψ_1 decays away while, depending on the parameters, ψ_2 either decays away or grows together with ψ_3.

The quadrimer, a system of four coupled units, is discussed in [Li and Kevrekidis (2011); Li *et al.* (2012); Zezyulin and Konotop (2012)]. There are many different possible PT-symmetric couplings between the four sites; however, some of these can be transformed into each other. In Fig. 10.23 we show just two of the possible configurations — a rather natural "doubling-up" of the dimer and a plaquette (square) configuration. In [Zezyulin and Konotop (2012)] it was found that the linear configuration is equivalent to an alternative plaquette configuration in which the κ couplings alternate around the square. Not unexpectedly, quadrimers display a yet more complicated pattern of stationary modes and dynamics, which may be exploited in optical systems. Other discrete optical systems, such as radially symmetric multicore fibres and PT-symmetric necklaces of gain and loss waveguides, are discussed in the excellent review [Suchkov *et al.* (2016)].

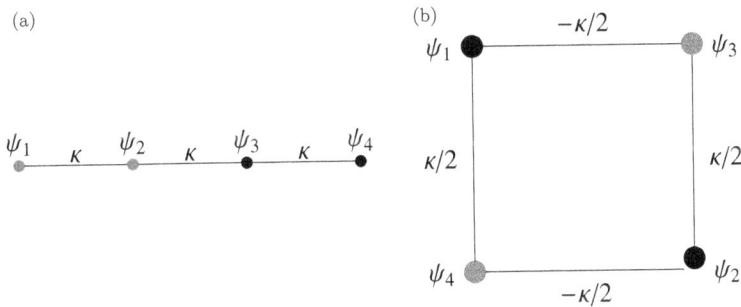

Fig. 10.23 Two possible PT-symmetric configurations for a quadrimer: (a) chain, (b) plaquette. The chain is equivalent to a plaquette with couplings $\kappa/2$, $-\kappa/2$, $-\kappa/2$, $\kappa/2$. In (a) the P operator is $(1, 2) \leftrightarrow (3, 4)$, while in (b) there are various possibilities, such as $(1, 3) \leftrightarrow (4, 2)$.

10.8 Optical Solitons

Having discussed nonlinearity for discrete PT-symmetric systems, we now do the same for continuous systems. The main interest here is in finding nonlinear potentials that can support *solitons*, which are solutions of the evolution equations where the light, instead of being diffracted, becomes self-trapped. In such solutions the beam is confined in the transverse direction and travels in the longitudinal direction without change in amplitude or shape.

In one transverse dimension the relevant equation governing wave propagation is a rescaled version of the paraxial equation (10.1) augmented by the nonlinear term $|\psi|^2\psi$:

$$i\frac{\partial\psi}{\partial z} + \frac{\partial^2\psi}{\partial x^2} + V(x)\psi + |\psi|^2\psi = 0. \tag{10.42}$$

An exact analytic solution of (10.42) can be found for the $\mathcal{PT}$-symmetric Scarff II potential (see Chap. 7)

$$V(x) = V_0\text{sech}^2(x) + iW_0\text{sech}(x)\tanh(x). \tag{10.43}$$

We are looking for a solution of the form $\psi(x, z) = \varphi(x)e^{i\lambda z}$ in which, provided that λ is real, the z dependence is just a phase. Inserting this expression for ψ into (10.42), we indeed find an exact solution for φ, namely [Musslimani *et al.* (2008a)]

$$\varphi(x) = \varphi_0\text{sech}(x)\exp\left\{i\mu\tan^{-1}[\sinh(x)]\right\}, \tag{10.44}$$

which satisfies (10.42) provided that $\lambda = 1$, $\mu = W_0/3$ and $\varphi_0 = \sqrt{2 - V_0 + \mu^2}$. The real and imaginary parts of the solution for $V_0 = 1$ and $W_0 = 1/2$ are shown in Fig. 10.24 below. It was proved in [Musslimani *et al.* (2008a)] that this solution is stable against small perturbations; that is, a small perturbation added to φ will die away as z increases, and the solution will revert to (10.44). Note that the solution (10.44) is in the regime of unbroken $\mathcal{PT}$ symmetry, where λ is real and $\varphi(x)$ is $\mathcal{PT}$-symmetric.

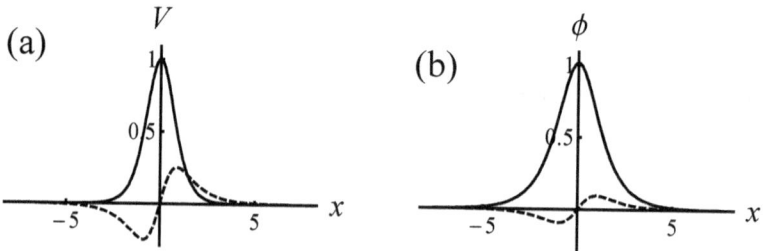

Fig. 10.24 Real (solid) and imaginary (dashed) parts of (a) the Scarff II potential and (b) the exact solution $\varphi(x)$ of (10.44) for $V_0 = 1$, $W_0 = 1/2$.

A wider class of potentials, based on Jacobi elliptic functions, was studied in [Musslimani *et al.* (2008b)]. In the limit where the elliptic modulus

$k \to 1$ the elliptic functions become standard hyperbolic trigonometric functions, yielding analytic soliton solutions for the PT-symmetric potentials

$$V_1(x) = V_0[\tanh^2(x) + W_0 \tanh^4(x)] + 4iW_0 \tanh(x)\text{sech}^2(x),$$
$$V_2(x) = V_0[\text{sech}^2(x) + W_0\text{sech}^4(x)] + 4iW_0 \tanh(x)\text{sech}^2(x).$$

The Gaussian potential $V(x) = V_0(1 + iW_0x)e^{-x^2}$ has been studied in [Hu *et al.* (2011)]. Here there is no analytical solution, but a whole range of solitons was found numerically and their stability investigated. Although this potential is different from the Scarf II potential of (10.43), the "fundamental" soliton solutions, that is, those with a single peak, look quite similar to those in Fig. 10.24(b). For larger potential strengths V_0, stable dipole and tripole soliton solutions were also found.

Soliton solutions have also been found numerically for periodic potentials, in particular in [Musslimani *et al.* (2008a)] for the potential $V = \cos^2(x) + iW_0 \sin(2x)$, which is essentially the potential discussed in Sec. 10.1 up to an additive constant. The soliton profiles of this potential are of the same general nature as those of Fig. 10.24. They are PT symmetric and stable below the linear symmetry-breaking threshold $W_0 = 1/2$. Moreover, a similar analysis can be carried out for periodic potentials in two transverse directions. The potential $V = \cos^2(x) + iW_0 \sin(2x)$ has been studied further to examine the motion of wave packets near the linear PT breaking point [Nixon *et al.* (2012b)], and the stability of the solitons in one and two transverse directions [Nixon *et al.* (2012a)].

10.9   Cloaking, Metamaterials, and Metasurfaces

Metamaterials are artificial materials, engineered on the scale of nanometers, that have exotic properties not found in nature, such as a negative refractive index. With the use of such materials, one can design lenses that overcome the resolution limits that apply to standard materials. Another fascinating application is *cloaking*, in which the material surrounding an object is designed to deflect any incoming light around the object in such a way that the pattern of light at large distances is the same as if the object were not there. The original studies of cloaking involved engineering the real parts of the electric and magnetic response functions, the permittivity ε, and the permeability μ. However, PT symmetry allows one to have complex response functions, that is, gain and loss, while still avoiding exponentially increasing or decaying fields, and thus PT symmetry widens considerably the range of metamaterials and their possible properties.

10.9.1 *One-way invisibility cloak*

The general methodology behind designs for cloaking is the transformational optics pioneered in [Pendry *et al.* (2006); Schurig *et al.* (2006)], whereby a given set of electromagnetic fields, sources, permittivity, and permeability in an initial virtual space is transformed into a different set in the final physical space with a different geometry. The transformation of coordinates induces a transformation of the fields, sources, permittivity and permeability such that Maxwell's equations remain satisfied in the new system. Typically the new geometry contains a forbidden region which the fields do not penetrate, thus providing cloaking for any object placed inside.

In the standard Hermitian context the cloaking is omnidirectional; however, in the $\mathcal{PT}$-symmetric context there is the possibility of designing cloaks that render the object invisible from one direction but not the other. This is related to, but different from the concept of unidirectional invisibility discussed earlier in Sec. 10.4. The transformation prescribes the distributions of the permittivity and permeability tensors in the physical space, which will generally require a metamaterial for their realization.

A nice example of a one-way invisiblity cloak exploiting $\mathcal{PT}$ symmetry is given in [Zhu *et al.* (2013)]. The initial configuration is a cylinder of cross section $r \leq b$ [see Fig 10.25(a)] in the virtual space (x, y, z), or (r, θ, z) in cylindrical coordinates, endowed with a permittivity $\varepsilon = 1 + \zeta \exp(i\beta r \cos\theta)$, which corresponds to the standard $\mathcal{PT}$-symmetric one-way potential. Then we transform to the physical space (x', y', z), or (r', θ, z), by means of the transformation

$$r = f(r') = b\frac{r' - a}{b - a}, \tag{10.45}$$

or equivalently

$$r' = a + r(1 - a/b). \tag{10.46}$$

This produces a space whose cross section is a circular hole of radius a inside the disk of radius b, as shown in Fig. 10.25(b).

For a transverse-electric (TE) wave propagating in the x direction with E in the z direction, the effective permittivity in the physical space given by (10.46) becomes

$$\varepsilon' = \frac{f(r')f'(r')\varepsilon[f(r'), \theta]}{r'} = \frac{1 - a/r'}{(1 - a/b)^2}\left[1 + \zeta \exp\left(i\beta x'\frac{1 - a/r'}{1 - a/b}\right)\right]. \tag{10.47}$$

This distorted permittivity in the annular region $a \leq r' \leq b$ (see Fig. 10.26) gives rise to one-way cloaking. That is, a perfectly conducting cylinder

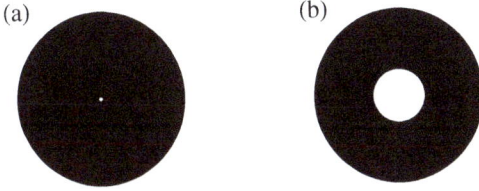

Fig. 10.25   Cross section of the initial virtual space (a) and the real physical space (b) after the transformation of (10.45). The physical space has an interior hole into which the electromagnetic fields do not penetrate.

placed inside the inner cavity of radius $r' = a'$ will be completely concealed when light is incident from one direction, but not from the other.

Fig. 10.26   Real part of permittivity ε_{zz} in the initial virtual space (a) and in the real physical space (b) after the transformation (10.47).

Note that the transformation (10.46) also produces a nontrivial (symmetric) permeability tensor. However, the only relevant components are the diagonal elements $\mu'_{r'}$ and $\mu'_{\theta'}$ because for a TE wave the magnetic field is in the plane of the disk. A suitable metamaterial would need to be designed to realize these components $\mu'_{r'}$ and $\mu'_{\theta'}$ as well as the required ε'_z. An earlier scheme for achieving one-way cloaking, which relies on the introduction of an external magnetic field rather than on PT symmetry, was previously given in [He *et al.* (2011)], but this scheme is considered less desirable for optical devices and applications [Zhu *et al.* (2013)].

10.9.2   *Cloaking by metasurface*

A promising alternative to the type of cloaking just described, which requires a bulk metamaterial, is to use a two-dimensional cloak, that is, a thin patterned metallic surface coating the object to be cloaked [Alú (2009)]. This approach has been promoted in [Sounas *et al.* (2015)], where

$\mathcal{PT}$-symmetric metasurfaces of both circular and rhomboidal cross sections were considered. The essential idea is that the coating on the front surface of the cloak should totally absorb the incoming power, which is then re-emitted by the coating on the rear surface. This is illustrated in Fig. 10.27, where the metallic cylinder has a radius a and the coating has a radius d, so that there is a small spacing $d - a$ between them. We are considering a transvers-electric wave $\boldsymbol{E}_{\text{inc}} = (0, 0, E_0)e^{ikx}$ incident on the cloaked cylinder from the left, with the $\boldsymbol{E}$-field in the z direction. The relevant property

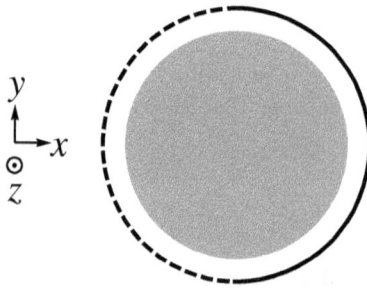

Fig. 10.27 $\mathcal{PT}$-symmetric surface coating a perfectly conducting cylinder. The dashed curve denotes loss while the solid curve denotes gain. The coating is designed so that light incident from the left is totally absorbed and then re-emitted on the right.

of the coating is its surface impedance Z_s, or equivalently its inverse, the surface admittance $Y_s \equiv 1/Z_s$, which relates the macroscopic (averaged over the unit cell) tangential electric field $\boldsymbol{E}_s$ on the surface to the induced averaged electric surface current density $\boldsymbol{J}_s$ according to $\boldsymbol{J}_s = Y_s \boldsymbol{E}_s$.

When the spacing between the cylinder and the metasurface is small compared with the wavelength, that is, when $d - a \ll \lambda$, the values of the real and imaginary parts of Y_s needed to give perfect absorption on the left are [Sounas *et al.* (2015)]

$$\text{Re}\, Y_s \approx -Y_0 \cos\varphi, \qquad \text{Im}\, Y_s \approx -\frac{Y_0}{k}\left(\frac{1}{d - a} - \frac{1}{2a}\right),$$

where $Y_0 = \sqrt{\varepsilon_0/\mu_0}$ is the admittance of free space and φ is the polar angle of the point $P = (x, y)$ on the left surface ($|\varphi| > \pi/2$). The same formulas apply on the right surface ($|\varphi| < \pi/2$), where the sign of $\text{Re}\, Y_s$ is reversed, which makes the system $\mathcal{PT}$ symmetric in the sense that $Y_s(\pi - \varphi) = -Y_s^*(\varphi)$. The right part of the cloak then reconstitutes the incoming wave, resulting in a total invisibility cloak, with zero reflection and full transmission. However, this absorption and subsequent re-emission works for light incident from the left, but not from the right, so this is again a one-way invisibility cloak, like the one discussed in Subsec. 10.9.1.

An even simpler cloak is a rhomboidal surface, as shown in Fig. 10.28, where the opening angle has been taken as 90^0. In this case the required values of $\mathrm{Re}\,Y_s$ and $\mathrm{Im}\,Y_s$ for perfect absorption on the left are given exactly [Sounas *et al.* (2015)] by

$$\mathrm{Re}\,Y_s = -Y_0/\sqrt{2}, \qquad \mathrm{Im}\,Y_s = -\frac{Y_0}{\sqrt{2}}\cot(kw),$$

where w is the spacing between the conductor and the metasurface. Again, for the full $\mathcal{PT}$-symmetric cloak, the sign of $\mathrm{Re}\,Y_s$ is reversed on the right. These approaches and other not necessarily $\mathcal{PT}$-symmetric approaches to

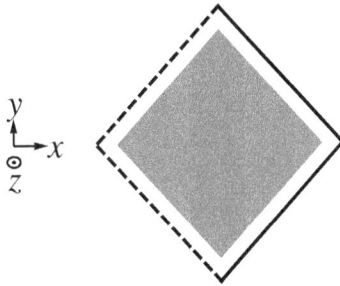

Fig. 10.28 $\mathcal{PT}$-symmetric surface coating a perfectly conducting rhomboid. The dashed lines denote loss and the solid lines denote gain. The coating is designed to produce the same one-way cloaking as that of Fig. 10.27.

cloaking are discussed in depth in the review [Fleury *et al.* (2015)].

Finally, a related topic is the use of $\mathcal{PT}$-symmetric metasurfaces [Fleury *et al.* (2014b)] to achieve negative refraction, something that is difficult to achieve with bulk metamaterials [Pendry (2000); Veselago (1968)]. Negative refraction allows one in principle to overcome the usual restrictions on the resolution of a lens and thereby to construct a perfect lens with unlimited resolution. Figure 10.29 illustrates the two different strategies. In Fig. 10.29(a) the arrows denote the flow of power, which is always forwards. However, because ε and μ are negative in the bulk material, the phase velocity in the material is in the opposite direction. In Fig. 10.29(b) the arrows again denote the flow of power, which in the space between the surfaces is backwards, going from the gain to the loss surface.

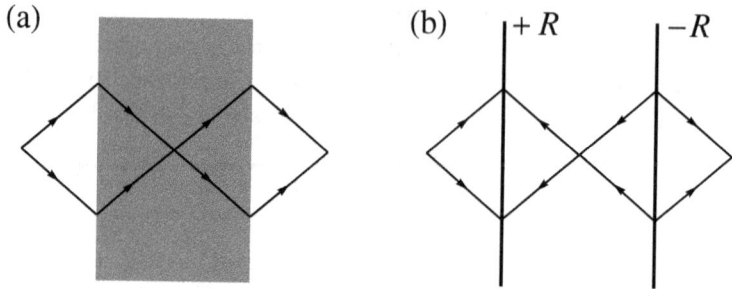

Fig. 10.29 Alternative strategies for achieving perfect lensing: (a) a bulk metamaterial with negative permittivity and permeability, (b) two metasurfaces with surface impedances $+R$ (loss) and $-R$ (gain), respectively. In (a) the arrows denote the flow of power, which is always forwards. In (b) the power flow is the same to the left and right of the two metasurfaces, but in between the metasurfaces the power flow is in the backward direction, going from the gain to the loss surface.

An interesting feature of this set-up is that it becomes reflectionless from the left, $R_{\rm L} = 0$, if the value of R is chosen appropriately as $R = \frac{1}{2} Y_0 \sec \theta$ for transverse-electric and $R = \frac{1}{2} Y_0 \cos \theta$ for transverse-magnetic waves. In that case the transmission coefficient $T = |t|^2$ is 1, but the phase of t is nonzero. In fact, t exhibits a phase advance, equal in magnitude to the phase lag it would have had in the absence of the metasurfaces.

10.10 Conclusion

This has been just a brief introduction to the large and rapidly developing subject of $\mathcal{PT}$ optics. The coverage has necessarily been rather selective, with apologies to those whose important contributions have not been mentioned because of space limitations. The hope is that this chapter has outlined and explained the basic ideas behind $\mathcal{PT}$-symmetric optics, thereby providing the background to enable the reader to tackle more complicated topics and future developments. Although it is strictly beyond the scope of this chapter, we should perhaps mention that many of the above ideas can also be applied to acoustics, including two-dimensional acoustic cloaks [Zhu *et al.* (2014)] and invisible acoustic sensors [Fleury *et al.* (2014a)].

Bibliography

Abdalla, E., Abdalla, M. C. B., and Rothe, K. D. (2001). *Non-Perturbative Methods in 2 Dimensional Quantum Field Theory* (World Scientific, Singapore), ISBN 978-981-02-4596-2.

Ablowitz, M., Ramani, A., and Segur, H. (1980). A connection between nonlinear evolution equations and ordinary differential equations of P-type. II, *Journal of Mathematical Physics* **21**, pp. 1006–1015, doi:10.1063/1.524548, https://doi.org/10.1063/1.524548.

Ablowitz, M. J. and Clarkson, P. A. (1991). *Solitons, Nonlinear Evolution Equations and Inverse Scattering*, London Mathematical Society Lecture Note Series, Vol. 149 (Cambridge University Press, Cambridge, UK), ISBN 978-0-51162-399-8, doi:10.1017/cbo9780511623998, https://doi.org/10.1017/cbo9780511623998.

Ablowitz, M. J., Kaup, D. J., Newell, A. C., and Segur, H. (1973). Nonlinear-evolution equations of physical significance, *Physical Review Letters* **31**, 2, pp. 125–127, doi:10.1103/PhysRevLett.31.125, https://doi.org/10.1103/PhysRevLett.31.125.

Ablowitz, M. J. and Segur, H. (1981). *Solitons and the Inverse Scattering Transform*, Studies in Applied and Numerical Mathematics (Society for Industrial and Applied Mathematics, Philadelphia), ISBN 978-0-89871-174-5, doi:10.1137/1.9781611970883, https://doi.org/10.1137/1.9781611970883.

Abramowitz, M. and Stegun, I. (1965). *Handbook of Mathematical Functions* (Dover Publications, New York), ISBN 0-486-61272-4.

Adamopoulou, P. and Dunning, C. (2014). Bethe ansatz equations for the classical $A_n^{(1)}$ affine Toda field theories, *Journal of Physics A: Mathematical and Theoretical* **47**, p. 205205, doi:10.1088/1751-8113/47/20/205205, https://doi.org/10.1088/1751-8113/47/20/205205.

Ahmed, Z. (2001a). Energy band structure due to a complex, periodic, $\mathcal{PT}$-invariant potential, *Physics Letters A* **286**, pp. 231–235, doi:10.1016/S0375-9601(01)00426-1, https://doi.org/10.1016/S0375-9601(01)00426-1.

Ahmed, Z. (2001b). Pseudo-Hermiticity of Hamiltonians under imaginary shift of the coordinate: Real spectrum of complex potentials, *Physics Letters*

    A **290**, pp. 19–22, doi:10.1016/S0375-9601(01)00622-3, `https://doi.org/` `10.1016/S0375-9601(01)00622-3`.

Ahmed, Z. (2001c). Real and complex discrete eigenvalues in an exactly solvable one-dimensional complex $\mathcal{PT}$-invariant potential, *Physics Letters A* **282**, pp. 343–348, doi:10.1016/S0375-9601(01)00218-3, `https://doi.org/` `10.1016/S0375-9601(01)00218-3`.

Ahmed, Z. (2002). Pseudo-Hermiticity of Hamiltonians under gauge-like transformation: real spectrum of non-Hermitian Hamiltonians, *Physics Letters A* **294**, pp. 287–291, doi:10.1016/S0375-9601(02)00124-X, `http://doi.org/` `10.1016/S0375-9601(02)00124-X`.

Ahmed, Z. (2003a). An ensemble of non-Hermitian Gaussian-random 2×2 matrices admitting the Wigner surmise, *Physics Letters A* **308**, pp. 140–142, doi:10.1016/S0375-9601(03)00053-7, `http://doi.org/10.1016/` `S0375-9601(03)00053-7`.

Ahmed, Z. (2003b). $\mathcal{C}$-, $\mathcal{PT}$- and $\mathcal{CPT}$-invariance of pseudo-Hermitian Hamiltonians, *Journal of Physics A: Mathematical and General* **36**, pp. 9711–9719, doi:10.1088/0305-4470/36/37/309, `http://doi.org/10.1088/0305-4470/` `36/37/309`.

Ahmed, Z. (2003c). $\mathcal{P}$-, $\mathcal{T}$-, $\mathcal{PT}$-, and $\mathcal{CPT}$-invariance of Hermitian Hamiltonians, *Physics Letters A* **310**, pp. 139–142, doi:10.1016/S0375-9601(03)00339-6, `http://doi.org/10.1016/S0375-9601(03)00339-6`.

Ahmed, Z. (2003d). Pseudo-reality and pseudo-adjointness of Hamiltonians, *Journal of Physics A: Mathematical and General* **36**, pp. 10325–10333, doi: 10.1088/0305-4470/36/41/005, `http://doi.org/10.1088/0305-4470/36/` `41/005`.

Ahmed, Z. (2004). Handedness of complex $\mathcal{PT}$-symmetric potential barriers, *Physics Letters A* **324**, pp. 152–158, doi:10.1016/j.physleta.2004.03.002, `https://doi.org/10.1016/j.physleta.2004.03.002`.

Ahmed, Z. (2009). Zero width resonance (spectral singularity) in a complex $\mathcal{PT}$-symmetric potential, *Journal of Physics A: Mathematical and Theoretical* **42**, p. 472005, doi:10.1088/1751-8113/42/47/472005, `https://doi.` `org/10.1088/1751-8113/42/47/472005`.

Ahmed, Z. (2013). Reciprocity and unitarity in scattering from a non-Hermitian complex $\mathcal{PT}$-symmetric potential, *Physics Letters A* **377**, pp. 957–959, doi:10.1016/j.physleta.2013.02.031, `https://doi.org/10.1016/` `j.physleta.2013.02.031`.

Ahmed, Z., Bender, C. M., and Berry, M. V. (2005). Reflectionless potentials and $\mathcal{PT}$ symmetry, *Journal of Physics A: Mathematical and General* **38**, p. L627, doi:10.1088/0305-4470/38/39/L01, `https://doi.org/10.` `1088/0305-4470/38/39/L01`.

Ahmed, Z., Ghosh, D., and Nathan, J. A. (2015a). A new solvable complex $\mathcal{PT}$-symmetric potential, *Physics Letters A* **379**, pp. 1639–1642, doi:10.1016/ j.physleta.2015.04.032, `https://doi.org/10.1016/j.physleta.2015.04.` `032`.

Ahmed, Z., Ghosh, D., Nathan, J. A., and Parkar, G. (2015b). Accidental crossings of eigenvalues in the one-dimensional complex $\mathcal{PT}$-symmetric Scarf-II potential, *Physics Letters A* **379**, pp. 2424–2429, doi:10.1016/j.physleta.2015.06.024, `https://doi.org/10.1016/j.physleta.2015.06.024`.

Ahmed, Z. and Jain, S. R. (2003a). Gaussian ensemble of 2×2 pseudo-Hermitian random matrices, *Journal of Physics A: Mathematical and General* **36**, pp. 3349–3362, doi:10.1088/0305-4470/36/12/327, `http://doi.org/10.1088/0305-4470/36/12/327`.

Ahmed, Z. and Jain, S. R. (2003b). Pseudounitary symmetry and the Gaussian pseudounitary ensemble of random matrices, *Physical Review E* **67**, p. 045106(R), doi:10.1103/PhysRevE.67.045106, `http://doi.org/10.1103/PhysRevE.67.045106`.

Ahmed, Z. and Jain, S. R. (2003c). Pseudounitary symmetry and the Gaussian pseudounitary ensemble of random matrices, *Physical Review A* **67**, p. 045106, doi:10.1103/physreve.67.045106, `https://doi.org/10.1103/physreve.67.045106`.

Airault, H., McKean, H. P., and Moser, J. (1977). Rational and elliptic solutions of the Korteweg-de Vries equation and a related many-body problem, *Communications on Pure and Applied Mathematics* **30**, 1, pp. 95–148, doi:10.1002/cpa.3160300106, `https://doi.org/10.1002/cpa.3160300106`.

Alaeian, H., Baum, B., Jankovic, V., Lawrence, M., and Dionne, J. A. (2016). Towards nanoscale multiplexing with parity-time-symmetric plasmonic coaxial waveguides, *Physical Review B* **93**, p. 205439, doi:10.1103/PhysRevB.93.205439, `https://doi.org/10.1103/PhysRevB.93.205439`.

Alaeian, H. and Dionne, J. A. (2014). Parity-time-symmetric plasmonic metamaterials, *Physical Review A* **89**, p. 033829, doi:10.1103/PhysRevA.89.033829, `https://doi.org/10.1103/PhysRevA.89.033829`.

Albeverio, S. and Kuzhel, S. (2012). On elements of the Lax-Phillips scattering scheme for $\mathcal{PT}$-symmetric operators, *Journal of Physics A: Mathematical and Theoretical* **45**, p. 444001, doi:10.1088/1751-8113/45/44/444001, `https://doi.org/10.1088/1751-8113/45/44/444001`.

Albeverio, S. and Kuzhel, S. (2015). $\mathcal{PT}$-symmetric operators in quantum mechanics: Kreĭn spaces methods, in F. Bagarello, J.-P. Gazeau, F. H. Szafraniec, and M. Znojil (eds.), *Non-Selfadjoint Operators in Quantum Physics: Mathematical Aspects* (John Wiley & Sons, Hoboken, NJ), ISBN 978-1-118-85528-7.

Albeverio, S., Motovilov, A. K., and Shkalikov, A. A. (2009). Bounds on variation of spectral subspaces under $\mathcal{J}$-self-adjoint perturbations, *Integral Equations and Operator Theory* **64**, pp. 455–486, doi:10.1007/s00020-009-1702-1, `https://doi.org/10.1007/s00020-009-1702-1`.

Alday, L. F., Gaiotto, D., and Maldacena, J. (2011). Thermodynamic bubble ansatz, *Journal of High Energy Physics* **9**, p. 032, doi:10.1007/JHEP09(2011)032, `https://doi.org/10.1007/JHEP09(2011)032`.

Alexandersson, P. (2012). On eigenvalues of the Schrödinger operator with an even complex-valued polynomial potential, *Computational Methods and Function Theory* **12**, 2, pp. 465–481, doi:10.1007/BF03321838, `https://doi.org/10.1007/BF03321838`.

Alexandersson, P. and Gabrielov, A. (2012). On eigenvalues of the Schrödinger operator with a complex-valued polynomial potential, *Computational Methods and Function Theory* **12**, 1, pp. 119–144, doi:10.1007/BF03321817, https://doi.org/10.1007/BF03321817.

Alhassid, Y., Gürsey, F., and Iachello, F. (1983). Group theory approach to scattering, *Annals of Physics* **148**, pp. 346–380, doi:10.1016/0003-4916(83)90244-0, https://doi.org/10.1016/0003-4916(83)90244-0.

Alú, A. (2009). Mantle cloak: Invisibility induced by a surface, *Physical Review B* **80**, p. 245115, doi:10.1103/PhysRevB.80.245115, https://doi.org/10.1103/PhysRevB.80.245115.

Anderson, A. G., Bender, C. M., and Morone, U. I. (2011a). Periodic orbits for classical particles having complex energy, *Physics Letters A* **375**, pp. 3399–3404, doi:10.1016/j.physleta.2011.07.051, https://doi.org/10.1016/j.physleta.2011.07.051.

Anderson, A. G., Bender, C. M., and Morone, U. I. (2011b). Periodic orbits for classical particles having complex energy, *Physics Letters A* **375**, 39, pp. 3399–3404, doi:10.1016/j.physleta.2011.07.051, https://doi.org/10.1016/j.physleta.2011.07.051.

Andrianov, A. A. (1982). The large N expansion as a local perturbation theory, *Annals of Physics* **140**, pp. 82–100, doi:10.1016/0003-4916(82)90336-0, https://doi.org/10.1016/0003-4916(82)90336-0.

Andrianov, A. A. and Sokolov, A. V. (2011). Resolutions of identity for some non-Hermitian Hamiltonians. I. Exceptional point in continuous spectrum, *SIGMA* **7**, p. 111, doi:10.3842/SIGMA.2011.111, https://doi.org/10.3842/SIGMA.2011.111.

Arlinskiĭ, Y. and Tsekanovskiĭ, E. (2009). M. Kreĭn's research on semi-bounded operators, its contemporary developments, and applications, in *The Mark Kreĭn Centenary Conference Volume 1: Operator Theory and Related Topics*, Operator Theory: Advances and Applications, Vol. 190 (Springer), pp. 65–112, doi:10.1007/978-3-7643-9919-{1_5}, https://doi.org/10.1007/978-3-7643-9919-{1_5}.

Arpornthip, T. and Bender, C. M. (2009). Conduction bands in classical periodic potentials, *Pramana* **73**, pp. 259–267, doi:10.1007/s12043-009-0117-5, https://doi.org/10.1007/s12043-009-0117-5.

Arrowsmith, D. K. and Place, C. M. (1990). *An Introduction to Dynamical Systems* (Cambridge University Press, Cambridge), ISBN 978-0521316507.

Ashida, Y., Furukawa, S., and Ueda, M. (2017). Parity-time-symmetric quantum critical phenomena, *Nature Communications* **8**, p. 15791, doi:10.1038/ncomms15791, https://doi.org/10.1038/ncomms15791.

Assawaworrarit, S., Yu, X., and Fan, S. (2017). Robust wireless power transfer using a nonlinear parity-time-symmetric circuit, *Nature* **546**, doi:10.1038/nature22404, https://doi.org/10.1038/nature22404.

Assis, P. E. G. and Fring, A. (2009a). From real fields to complex Calogero particles, *Journal of Physics A: Mathematical and Theoretical* **42**, p. 425206, doi:10.1088/1751-8113/42/42/425206, https://doi.org/10.1088/1751-8113/42/42/425206.

Assis, P. E. G. and Fring, A. (2009b). Integrable models from $\mathcal{PT}$-symmetric deformations, *Journal of Physics A: Mathematical and Theoretical* **42**, p. 105206, doi:10.1088/1751-8113/42/10/105206, `https://doi.org/10.1088/1751-8113/42/10/105206`.

Assis, P. E. G. and Fring, A. (2010). Compactons versus solitons, *Pramana-Journal of Physics* **74**, pp. 857–865, doi:10.1007/s12043-010-0078-8, `https://doi.org/10.1007/s12043-010-0078-8`.

Aurégan, Y. and Pagneux, V. (2017). $\mathcal{PT}$-symmetric scattering in flow duct acoustics, *Physical Review Letters* **118**, p. 174301, doi:10.1103/physrevlett.118.174301, `https://doi.org/10.1103/physrevlett.118.174301`.

Azizov, T. Y. and Iokhvidov, I. S. (1998). *Linear Operators in Spaces with Indefinite Metric* (Wiley, New York).

Azizov, T. Y. and Trunk, C. (2012). $\mathcal{PT}$-symmetric, Hermitian and $\mathcal{P}$-self-adjoint operators related to potentials in $\mathcal{PT}$ quantum mechanics, *Journal of Mathematical Physics* **53**, p. 012109, doi:10.1063/1.3677368, `https://doi.org/10.1063/1.3677368`.

Bagchi, B. and Fring, A. (2008). $\mathcal{PT}$-symmetric extensions of the supersymmetric Korteweg–de Vries equation, *Journal of Physics A: Mathematical and Theoretical* **41**, p. 392004, doi:10.1088/1751-8113/41/39/392004, `https://doi.org/10.1088/1751-8113/41/39/392004`.

Bagchi, B., Mallik, S., and Quesne, C. (2002). $\mathcal{PT}$-symmetric square well and the associated SUSY hierarchies, *Modern Physics Letters A* **17**, pp. 1651–1664, doi:10.1142/S0217732302008009, `https://doi.org/10.1142/S0217732302008009`.

Bagchi, B. and Quesne, C. (2000). sl(2, **C**) as a complex Lie algebra and the associated non-Hermitian Hamiltonians with real eigenvalues, *Physics Letters A* **273**, pp. 285–292, doi:10.1016/S0375-9601(00)00512-0, `https://doi.org/10.1016/S0375-9601(00)00512-0`.

Bagchi, B. and Quesne, C. (2002). Pseudo-Hermiticity, weak pseudo-Hermiticity and η-orthogonality condition, *Physics Letters A* **301**, pp. 173–176, doi:10.1016/S0375-9601(02)00929-5, `http://doi.org/10.1016/S0375-9601(02)00929-5`.

Bagchi, B. and Quesne, C. (2010). An update on the $\mathcal{PT}$-symmetric complexified Scarf II potential, spectral singularities and some remarks on the rationally extended supersymmetric partners, *Journal of Physics A: Mathematical and Theoretical* **43**, p. 305301, doi:10.1088/1751-8113/43/30/305301, `https://doi.org/10.1088/1751-8113/43/30/305301`.

Bagchi, B., Quesne, C., and Roychoudhury, R. (2005). Pseudo-Hermiticity and some consequences of a generalized quantum condition, *Journal of Physics A: Mathematical and General* **38**, pp. L647–L652, doi:10.1088/0305-4470/38/40/L01, `http://doi.org/10.1088/0305-4470/38/40/L01`.

Bagchi, B., Quesne, C., and Roychoudhury, R. (2009). Isospectrality of conventional and new extended potentials, second-order supersymmetry and role of $\mathcal{PT}$ symmetry, *Pramana-Journal of Physics* **73**, pp. 337–347, doi:10.1007/s12043-009-0126-4, `https://doi.org/10.1007/s12043-009-0126-4`.

Bagchi, B., Quesne, C., and Znojil, M. (2001). Generalized continuity equation and modified normalization in $\mathcal{PT}$-symmetric quantum mechanics, *Modern Physics Letters A* **16**, pp. 2047–2057, doi:10.1142/S0217732301005333, https://doi.org/10.1142/S0217732301005333.

Baranov, D. G., Krasnok, A., Shegai, T., Alù, A., and Chong, Y. (2017). Coherent perfect absorbers: linear control of light with light, *Nature Reviews: Materials* **2**, p. 17064, doi:10.1038/natrevmats.2017.64, https://doi.org/10.1038/natrevmats.2017.64.

Barashenkov, I. V., Jackson, G. S., and Flach, S. (2013). Blow-up regimes in the $\mathcal{PT}$-symmetric coupler and the actively coupled dimer, *Physical Review A* **88**, p. 053817, doi:10.1103/PhysRevA.88.053817, https://doi.org/10.1103/PhysRevA.88.053817.

Barton, G. (1963). *An Introduction to Advanced Field Theory* (Interscience (Wiley), New York).

Basu-Mallick, B. and Kundu, A. (2000). Exact solution to the Calogero model with competing long-range interactions, *Physical Review B* **62**, pp. 9927–9930, doi:10.1103/PhysRevB.62.9927, https://doi.org/10.1103/PhysRevB.62.9927.

Bateman, H. (1931). On dissipative systems and related variational principles, *Physical Review* **38**, pp. 815–819, doi:10.1103/PhysRev.38.815, https://doi.org/10.1103/PhysRev.38.815.

Batic, D., Williams, R., and Nowakowski, M. (2013). Potentials of the Heun class, *Journal of Physics A: Mathematical and Theoretical* **46**, p. 245204, doi:10.1088/1751-8113/46/24/245204, https://doi.org/10.1088/1751-8113/46/24/245204.

Baxter, R. (1971). Partition function of the eight-vertex lattice model, *Annals of Physics* **70**, pp. 193–228, doi:10.1016/0003-4916(72)90335-1, https://doi.org/10.1016/0003-4916(72)90335-1.

Bazhanov, V., Hibberd, A., and Khoroshkin, S. M. (2002). Integrable structure of $\mathcal{W}_3$ conformal field theory, quantum Boussinesq theory and boundary affine Toda theory, *Nuclear Physics B* **622**, pp. 475–547, doi:10.1016/S0550-3213(01)00595-8, https://doi.org/10.1016/S0550-3213(01)00595-8.

Bazhanov, V. V., Lukyanov, S. L., and Tsvelik, A. M. (2003a). Analytical results for the Coqblin-Schrieffer model with generalized magnetic fields, *Physical Review B* **68**, 9, p. 094427, doi:10.1103/PhysRevB.68.094427, https://doi.org/10.1103/PhysRevB.68.094427.

Bazhanov, V. V., Lukyanov, S. L., and Zamolodchikov, A. B. (1996). Integrable structure of conformal field theory, quantum KdV theory and thermodynamic Bethe ansatz, *Communications in Mathematical Physics* **177**, pp. 381–398, doi:10.1007/BF02101898, https://doi.org/10.1007/BF02101898.

Bazhanov, V. V., Lukyanov, S. L., and Zamolodchikov, A. B. (1997a). Integrable quantum field theories in finite volume: Excited state energies, *Nuclear Physics B* **489**, pp. 487–531, doi:10.1016/S0550-3213(97)00022-9, https://doi.org/10.1016/S0550-3213(97)00022-9.

Bazhanov, V. V., Lukyanov, S. L., and Zamolodchikov, A. B. (1997b). Integrable structure of conformal field theory II. Q-operator and DDV equation, *Communications in Mathematical Physics* **190**, pp. 247–278, doi: 10.1007/s002200050240, https://doi.org/10.1007/s002200050240.

Bazhanov, V. V., Lukyanov, S. L., and Zamolodchikov, A. B. (2001). Spectral determinants for Schrödinger equation and Q-operators of conformal field theory, *Journal of Statistical Physics* **102**, pp. 567–576, doi:10.1023/A:1004838616921, https://doi.org/10.1023/A:1004838616921.

Bazhanov, V. V., Lukyanov, S. L., and Zamolodchikov, A. B. (2003b). Higher-level eigenvalues of Q-operators and Schrödinger equation, *Advances in Theoretical and Mathematical Physics* **7**, 4, pp. 711–725, doi:10.4310/ATMP.2003.v7.n4.a4, https://doi.org/10.4310/ATMP.2003.v7.n4.a4.

Bencze, G. (1966). Analytical solution of the Schrödinger equation with optical model potential for S-wave neutrons, in *Commentationes Physico-Mathematicae 31* (University of Michigan), p. 1.

Bender, C. M. (2005). Introduction to $\mathcal{PT}$-symmetric quantum theory, *Contemporary Physics* **46**, pp. 277–292, doi:10.1080/00107500072632, https://doi.org/10.1080/00107500072632.

Bender, C. M. (2007). Making sense of non-Hermitian Hamiltonians, *Reports on Progress in Physics* **70**, pp. 947–1018, doi:10.1088/0034-4885/70/6/R03, https://doi.org/10.1088/0034-4885/70/6/R03.

Bender, C. M., Berntson, B. K., Parker, D., and Samuel, E. (2013a). Observation of $\mathcal{PT}$ phase transition in a simple mechanical system, *American Journal of Physics* **81**, pp. 173–179, doi:10.1119/1.4789549, https://doi.org/10.1119/1.4789549.

Bender, C. M. and Boettcher, S. (1998a). Quasi-exactly solvable quartic potential, *Journal of Physics A: Mathematical and General* **31**, pp. L273–L277, doi:10.1088/0305-4470/31/14/001, https://doi.org/10.1088/0305-4470/31/14/001.

Bender, C. M. and Boettcher, S. (1998b). Real spectra in non-Hermitian Hamiltonians having $\mathcal{PT}$ symmetry, *Physical Review Letters* **80**, pp. 5243–5246, doi:10.1103/PhysRevLett.80.5243, https://doi.org/10.1103/PhysRevLett.80.5243.

Bender, C. M., Boettcher, S., Jones, H. F., Meisinger, P. N., and Şimşek, M. (2001a). Bound states of non-Hermitian quantum field theories, *Physics Letters A* **291**, pp. 197–202, doi:10.1016/S0375-9601(01)00745-9, http://doi.org/10.1016/S0375-9601(01)00745-9.

Bender, C. M., Boettcher, S., Jones, H. F., and Savage, V. M. (1999a). Complex square well – a new exactly solvable quantum mechanical model, *Journal of Physics A: Mathematical and General* **32**, pp. 6771–6781, doi:10.1088/0305-4470/32/39/305, https://doi.org/10.1088/0305-4470/32/39/305.

Bender, C. M., Boettcher, S., and Lipatov, L. (1992). Almost zero-dimensional quantum field theories, *Physical Review D* **46**, pp. 5557–5573, doi:10.1103/PhysRevD.46.5557, https://doi.org/10.1103/PhysRevD.46.5557.

Bender, C. M., Boettcher, S., and Meisinger, P. N. (1999b). $\mathcal{PT}$-symmetric quantum mechanics, *Journal of Mathematical Physics* **40**, pp. 2201–2229, doi: 10.1063/1.532860, https://doi.org/10.1063/1.532860.

Bender, C. M., Boettcher, S., and Savage, V. M. (2000a). Conjecture on the interlacing of zeros in complex Sturm-Liouville problems, *Journal of Mathematical Physics* **41**, pp. 6381–6387, doi:10.1063/1.1288247, https://doi.org/10.1063/1.1288247.

Bender, C. M., Brandt, S. F., Chen, J. H., and Wang, Q. (2005a). The C operator in $\mathcal{PT}$-symmetric quantum field theory transforms as a Lorentz scalar, *Physical Review D* **71**, p. 065010, doi:10.1103/PhysRevD.71.065010, http://doi.org/10.1103/PhysRevD.71.065010.

Bender, C. M., Brandt, S. F., Chen, J. H., and Wang, Q. (2005b). Ghost busting: $\mathcal{PT}$-symmetric interpretation of the Lee model, *Physical Review D* **71**, p. 025014, doi:10.1103/PhysRevD.71.025014, http://doi.org/10.1103/PhysRevD.71.025014.

Bender, C. M., Brod, J., Refig, A., and Reuter, M. E. (2004a). The $\mathcal{C}$ operator in $\mathcal{PT}$-symmetric quantum theories, *Journal of Physics A: Mathematical and General* **37**, pp. 10139–10165, doi:10.1088/0305-4470/37/43/009, http://doi.org/10.1088/0305-4470/37/43/009.

Bender, C. M., Brody, D. C., Chen, J.-H., and Furlan, E. (2007a). $\mathcal{PT}$-symmetric extension of the Korteweg-de Vries equation, *Journal of Physics A: Mathematical and Theoretical* **40**, 5, pp. F153–F160, doi:10.1088/1751-8113/40/5/F02, https://doi.org/10.1088/1751-8113/40/5/F02.

Bender, C. M., Brody, D. C., Chen, J.-H., Jones, H. F., Milton, K. A., and Ogilvie, M. C. (2006a). Equivalence of a complex $\mathcal{PT}$-symmetric quartic Hamiltonian and a Hermitian quartic Hamiltonian with an anomaly, *Physical Review D* **74**, p. 025016, doi:10.1103/PhysRevD.74.025016, https://doi.org/10.1103/PhysRevD.74.025016.

Bender, C. M., Brody, D. C., and Hook, D. W. (2008a). Quantum effects in classical systems having complex energy, *Journal of Physics A: Mathematical and Theoretical* **41**, p. 352003, doi:10.1088/1751-8113/41/35/352003, https://doi.org/10.1088/1751-8113/41/35/352003.

Bender, C. M., Brody, D. C., Hughston, L. P., and Meister, B. K. (2016a). Geometric aspects of space-time reflection symmetry in quantum mechanics, in F. Bagarello, R. Passante, and C. Trapani (eds.), *Non-Hermitian Hamiltonians in Quantum Physics*, Vol. 184 (Springer, Cham), pp. 185–199, doi:10.1007/978-3-319-31356-6_12, https://doi.org/10.1007/978-3-319-31356-6_12.

Bender, C. M., Brody, D. C., and Jones, H. F. (2002). Complex extension of quantum mechanics, *Physical Review Letters* **89**, p. 270401, doi:10.1103/PhysRevLett.89.270401, http://link.aps.org/doi/10.1103/PhysRevLett.89.270401.

Bender, C. M., Brody, D. C., and Jones, H. F. (2003a). Must a Hamiltonian be Hermitian? *American Journal of Physics* **71**, pp. 1095–1102, doi:10.1119/1.1574043, http://doi.org/10.1119/1.1574043.

Bender, C. M., Brody, D. C., and Jones, H. F. (2004b). Complex extension of quantum mechanics (Erratum), *Physical Review Letters* **92**, p. 119902(E), doi:10.1103/PhysRevLett.92.119902, https://doi.org/10.1103/PhysRevLett.92.119902.

Bender, C. M., Brody, D. C., and Jones, H. F. (2004c). Extension of $\mathcal{PT}$-symmetric quantum mechanics to quantum field theory with cubic interaction, *Physical Review D* **70**, p. 025001, doi:10.1103/PhysRevD.70.025001, https://doi.org/10.1103/PhysRevD.70.025001.

Bender, C. M., Brody, D. C., and Jones, H. F. (2004d). Scalar quantum field theory with a complex cubic interaction, *Physical Review Letters* **93**, p. 251601, doi:10.1103/PhysRevLett.93.251601, http://doi.org/10.1103/PhysRevLett.93.251601.

Bender, C. M., Cavero-Palaez, I., Milton, K. A., and Shajesh, K. V. (2005c). $\mathcal{PT}$-symmetric quantum electrodynamics, *Physics Letters B* **613**, pp. 97–104, doi:10.1016/j.physletb.2005.03.032, https://doi.org/10.1016/j.physletb.2005.03.032.

Bender, C. M., Chen, J.-H., Darg, D. W., and Milton, K. A. (2006b). Classical trajectories for complex Hamiltonians, *Journal of Physics A: Mathematical and General* **39**, pp. 4219–4238, doi:10.1088/0305-4470/39/16/009, http://doi.org/10.1088/0305-4470/39/16/009.

Bender, C. M., Chen, J.-H., and Milton, K. A. (2006c). $\mathcal{PT}$-symmetric versus Hermitian formulations of quantum mechanics, *Journal of Physics A: Mathematical and General* **39**, pp. 1657–1668, doi:10.1088/0305-4470/39/7/010, http://doi.org/10.1088/0305-4470/39/7/010.

Bender, C. M., Cooper, F., Khare, A., Mihaila, B., and Saxena, A. (2009). Compactons in $\mathcal{PT}$-symmetric generalized Korteweg-de Vries equations, *Pramana-Journal of Physics* **73**, 2, pp. 375–385, doi:10.1007/s12043-009-0129-1, https://doi.org/10.1007/s12043-009-0129-1.

Bender, C. M., Cooper, F., Meisinger, P. N., and Savage, V. M. (1999c). Variational ansatz for $\mathcal{PT}$-symmetric quantum mechanics, *Physics Letters A* **259**, pp. 224–231, doi:10.1016/S0375-9601(99)00468-5, http://doi.org/10.1016/S0375-9601(99)00468-5.

Bender, C. M. and Darg, D. W. (2007). Spontaneous breaking of classical $\mathcal{PT}$ symmetry, *Journal of Mathematical Physics* **48**, p. 042703, doi:10.1063/1.2720279, http://doi.org/10.1063/1.2720279.

Bender, C. M. and Dunne, G. V. (1989a). Exact solutions to operator differential equations, *Physics Review D* **40**, pp. 2739–2742, doi:10.1103/PhysRevD.40.2739, http://doi.org/10.1103/PhysRevD.40.2739.

Bender, C. M. and Dunne, G. V. (1989b). Integration of operator differential equations, *Physics Review D* **40**, pp. 3504–3511, doi:10.1103/PhysRevD.40.3504, http://doi.org/10.1103/PhysRevD.40.3504.

Bender, C. M. and Dunne, G. V. (1996). Quasiexactly solvable systems and orthogonal polynomials, *Journal of Mathematical Physics* **37**, pp. 6–11, doi:10.1063/1.531373, https://doi.org/10.1063/1.531373.

Bender, C. M. and Dunne, G. V. (1999). Large-order perturbation theory for a non-Hermitian $\mathcal{PT}$-symmetric Hamiltonian, *Journal of Mathematical Physics* **40**, pp. 4616–4621, doi:10.1063/1.532991, https://doi.org/10.1063/1.532991.

Bender, C. M., Dunne, G. V., and Meisinger, P. N. (1999d). Complex periodic potentials with real band spectra, *Physics Letters A* **252**, pp.

272–276, doi:10.1016/S0375-9601(98)00960-8, https://doi.org/10.1016/S0375-9601(98)00960-8.

Bender, C. M., Dunne, G. V., Meisinger, P. N., and Şimşek, M. (2001b). Quantum complex Hénon-Heiles potentials, *Physics Letters A* **281**, pp. 311–316, doi:10.1016/S0375-9601(01)00146-3, http://doi.org/10.1016/S0375-9601(01)00146-3.

Bender, C. M. and Feinberg, J. (2008). Does the complex deformation of the Riemann equation exhibit shocks? *Journal of Physics A: Mathematical and Theoretical* **41**, 24, p. 244004, doi:10.1088/1751-8113/41/24/244004, https://doi.org/10.1088/1751-8113/41/24/244004.

Bender, C. M., Fring, A., and Komijani, J. (2014a). Nonlinear eigenvalue problems, *Journal of Physics A: Mathematical and Theoretical* **47**, p. 235204, doi:10.1088/1751-8113/47/23/235204, https://doi.org/10.1088/1751-8113/47/23/235204.

Bender, C. M. and Gianfreda, M. (2015). $\mathcal{PT}$-symmetric interpretation of the electromagnetic self-force, *Journal of Physics A: Mathematical and Theoretical* **48**, 34, p. 34FT01, doi:10.1088/1751-8113/48/34/34FT01, https://doi.org/10.1088/1751-8113/48/34/34FT01.

Bender, C. M., Gianfreda, M., Hassanpour, N., and Jones, H. F. (2016b). Comment on 'on the Lagrangian and Hamiltonian description of the damped linear harmonic oscillator' [J. Math. Phys. 48, 032701 (2007)], *Journal of Mathematical Physics* **57**, p. 084101, doi:10.1063/1.4960722, https://doi.org/10.1063/1.4960722.

Bender, C. M., Guralnik, G. S., Keener, R. W., and Olaussen, K. (1976). Numerical study of truncated Green's-function equations, *Physical Review D* **14**, pp. 2590–2595, doi:10.1103/PhysRevD.14.2590, https://doi.org/10.1103/PhysRevD.14.2590.

Bender, C. M., Holm, D. D., and Hook, D. W. (2007b). Complex trajectories of a simple pendulum, *Journal of Physics A: Mathematical and Theoretical* **40**, pp. F81–F89, doi:10.1088/1751-8113/40/3/F01, http://doi.org/10.1088/1751-8113/40/3/F01.

Bender, C. M., Holm, D. D., and Hook, D. W. (2007c). Complexified dynamical systems, *Journal of Physics A: Mathematical and Theoretical* **40**, 32, pp. F793–F804, doi:10.1088/1751-8113/40/32/F02, https://doi.org/10.1088/1751-8113/40/32/F02.

Bender, C. M. and Hook, D. W. (2008). Exact isospectral pairs of $\mathcal{PT}$ symmetric Hamiltonians, *Journal of Physics A: Mathematical and Theoretical* **41**, p. 24405, doi:10.1088/1751-8113/41/24/244005, https://doi.org/10.1088/1751-8113/41/24/244005.

Bender, C. M. and Hook, D. W. (2012). Universal spectral behavior of $x^2(ix)^\varepsilon$ potentials, *Physical Review A* **86**, p. 022113, doi:10.1103/PhysRevA.86.022113, https://doi.org/10.1103/PhysRevA.86.022113.

Bender, C. M., Hook, D. W., and Kooner, K. S. (2010a). Classical particle in a complex elliptic potential, *Journal of Physics A: Mathematical and Theoretical* **43**, p. 3165201, doi:10.1088/1751-8113/43/16/165201, https://doi.org/10.1088/1751-8113/43/16/165201.

Bender, C. M., Hook, D. W., Mavromatos, N. E., and Sarkar, S. (2014b). Infinite class of $\mathcal{PT}$-symmetric theories from one timelike Liouville Lagrangian, *Physical Review Letters* **113**, p. 231605, doi:10.1103/physrevlett. 113.231605, https://doi.org/10.1103/physrevlett.113.231605.

Bender, C. M., Ghatak, A., and Gianfreda, M. (2017a). $\mathcal{PT}$-symmetric model of immune response, *Journal of Physics A: Mathematical and Theoretical* **50**, doi:10.1088/1751-8121/50/3/035601, https://doi.org/10.1088/1751-8121/50/3/035601.

Bender, C. M., Hassanpour, N., Hook, D. W., Klevansky, S. P., Sünderhauf, C., and Wen, Z. (2017b). Behavior of eigenvalues in a region of broken-$\mathcal{PT}$ symmetry, *Physical Review A* **95**, p. 052113, doi:10.1103/PhysRevA. 95.052113, https://doi.org/10.1103/PhysRevA.95.052113.

Bender, C. M., Hook, D. W., Mavromatos, N. E., and Sarkar, S. (2016c). $\mathcal{PT}$-symmetric interpretation of unstable effective potentials, *Journal of Physics A: Mathematical and Theoretical* **49**, p. 45LT01, doi:10.1088/1751-8113/49/45/45LT01, https://doi.org/10.1088/1751-8113/49/45/45LT01.

Bender, C. M., Hook, D. W., and Mead, L. R. (2008b). Conjecture on the analyticity of $\mathcal{PT}$-symmetric potentials and the reality of their spectra, *Journal of Physics A: Mathematical and Theoretical* **41**, p. 392005, doi:10.1088/1751-8113/41/39/392005, https://doi.org/10.1088/1751-8113/41/39/392005.

Bender, C. M., Hook, D. W., Meisinger, P. N., and Wang, Q. (2010b). Complex correspondence principle, *Physical Review Letters* **104**, p. 061601, doi:10.1103/physrevlett.104.061601, https://doi.org/10.1103/physrevlett.104.061601.

Bender, C. M., Hook, D. W., Meisinger, P. N., and Wang, Q. (2010c). Probability density in the complex plane, *Annals of Physics* **325**, pp. 2332–2362, doi:10.1016/j.aop.2010.02.011, https://doi.org/10.1016/j.aop.2010.02.011.

Bender, C. M. and Jones, H. F. (2004). Semiclassical calculation of the $\mathcal{C}$ operator in $\mathcal{PT}$-symmetric quantum mechanics, *Physics Letters A* **328**, pp. 102–109, doi:10.1016/j.physleta.2004.05.063, http://doi.org/10.1016/j.physleta.2004.05.063.

Bender, C. M., Jones, H. F., and Rivers, R. J. (2005d). Dual $\mathcal{PT}$-symmetric quantum field theories, *Physics Letters B* **625**, pp. 333–340, doi:10.1016/j.physletb.2005.08.087, https://doi.org/10.1016/j.physletb.2005.08.087.

Bender, C. M. and Klevansky, S. (2009). Nonunique $\mathcal{C}$ operator in $\mathcal{PT}$ quantum mechanics, *Physics Letters A* **373**, pp. 2670–2674, doi:10.1016/j.physleta.2009.05.066, https://doi.org/10.1016/j.physleta.2009.05.066.

Bender, C. M. and Klevansky, S. P. (2010). Families of particles with different masses in $\mathcal{PT}$-symmetric quantum field theory, *Physical Review Letters* **105**, p. 031601, doi:10.1103/PhysRevLett.105.031601, https://doi.org/10.1103/PhysRevLett.105.031601.

Bender, C. M. and Komijani, J. (2015). Painlevé transcendents and $\mathcal{PT}$-symmetric hamiltonians, *Journal of Physics A: Mathematical and Theoretical* **48**, p. 475202, doi:10.1088/1751-8113/48/47/475202, https://doi.org/10.1088/1751-8113/48/47/475202.

Bender, C. M., Komijani, J., and Wang, Q. (2017c). Nonlinear eigenvalue problems, doi:10.1007/978-88-7642-613-1_2, https://doi.org/10.1007/978-88-7642-613-1_2.

Bender, C. M. and Kuzhel, S. (2012). Unbounded $\mathcal{C}$-symmetries and their nonuniqueness, *Journal of Physics: Mathematical and Theoretical* **45**, p. 444005, doi:10.1088/1751-8113/45/44/444005, https://doi.org/10.1088/1751-8113/45/44/444005.

Bender, C. M., Mandula, J. E., and McCoy, B. M. (1970). Does renormalized perturbation theory diverge? *Physical Review Letters* **24**, pp. 681–683, doi:10.1103/PhysRevLett.24.681, https://doi.org/10.1103/PhysRevLett.24.681.

Bender, C. M. and Mannheim, P. D. (2008). No-ghost theorem for the fourth-order derivative pais-uhlenbeck oscillator model, *Physical Review Letters* **100**, p. 110402, doi:10.1103/physrevlett.100.110402, https://doi.org/10.1103/physrevlett.100.110402.

Bender, C. M. and Mannheim, P. D. (2010). $\mathcal{PT}$ symmetry and necessary and sufficient conditions for the reality of energy eigenvalues, *Physics Letters A* **374**, pp. 1616–1620, doi:10.1016/j.physleta.2010.02.032, https://doi.org/10.1016/j.physleta.2010.02.032.

Bender, C. M. and Mannheim, P. D. (2011). $\mathcal{PT}$ symmetry in relativistic quantum mechanics, *Physical Review D* **84**, p. 105038, doi:10.1103/PhysRevD.84.105038, https://doi.org/10.1103/PhysRevD.84.105038.

Bender, C. M., Meisinger, P. N., and Wang, Q. (2003b). Calculation of the hidden symmetry operator in $\mathcal{PT}$-symmetric quantum mechanics, *Journal of Physics A: Mathematical and General* **36**, pp. 1973–1983, doi:10.1088/0305-4470/36/7/312, http://doi.org/10.1088/0305-4470/36/7/312.

Bender, C. M., Meisinger, P. N., and Yang, H. (2001c). Calculation of the one-point Green's function for a $-g\phi^4$ quantum field theory, *Physical Review D* **63**, p. 045001, doi:10.1103/PhysRevD.63.045001, https://doi.org/10.1103/PhysRevD.63.045001.

Bender, C. M. and Milton, K. A. (1999). A nonunitary version of massless quantum electrodynamics possessing a critical point, *Journal of Physics A: Mathematical and General* **32**, pp. L87–L92, doi:10.1088/0305-4470/32/7/001, https://doi.org/10.1088/0305-4470/32/7/001.

Bender, C. M., Milton, K. A., Pinsky, S. S., and Simmons Jr, L. M. (1989). A new perturbative approach to nonlinear problems, *Journal of Mathematical Physics* **30**, doi:10.1063/1.528326, https://doi.org/10.1063/1.528326.

Bender, C. M., Milton, K. A., and Savage, V. M. (2000b). Solution of Schwinger-Dyson equations for $\mathcal{PT}$-symmetric quantum field theory, *Physical Review D* **62**, p. 085001, doi:10.1103/PhysRevD.62.085001, https://doi.org/10.1103/PhysRevD.62.085001.

Bender, C. M., Moshe, M., and Sarkar, S. (2013b). $\mathcal{PT}$-symmetric interpretation of double-scaling, *Journal of Physics A: Mathematical and Theoretical* **46**, p. 102002, doi:10.1088/1751-8113/46/10/102002, https://doi.org/10.1088/1751-8113/46/10/102002.

Bender, C. M. and Orszag, S. A. (1999). *Advanced Mathematical Methods for Scientists and Engineers I* (Springer, New York).

Bender, C. M. and Sarkar, S. (2014). Double-scaling limit of the O(N)-symmetric anharmonic oscillator, *Journal of Physics A: Mathematical and Theoretical* **46**, p. 442001, doi:10.1088/1751-8113/46/44/442001, https://doi.org/10.1088/1751-8113/46/44/442001.

Bender, C. M. and Tan, B. (2006). Calculation of the hidden symmetry operator for a $\mathcal{PT}$-symmetric square well, *Journal of Physics A: Mathematical and General* **39**, pp. 1945–1953, doi:10.1088/0305-4470/39/8/011, https://doi.org/10.1088/0305-4470/39/8/011.

Bender, C. M. and Turbiner, A. (1993). Analytic continuation of eigenvalue problems, *Physics Letters A* **173**, pp. 442–446, doi:10.1016/0375-9601(93)90153-Q, https://doi.org/10.1016/0375-9601(93)90153-Q.

Bender, C. M. and Wang, Q. (2001). Comment on a recent paper by Mezincescu, *Journal of Physics A: Mathematical and General* **33**, pp. 3325–3328, doi:10.1088/0305-4470/34/15/401, https://doi.org/10.1088/0305-4470/34/15/401.

Bender, C. M. and Weniger, E. J. (2001). Numerical evidence that the perturbation expansion for a non-Hermitian $\mathcal{PT}$-symmetric Hamiltonian is Stieltjes, *Journal of Mathematical Physics* **42**, pp. 2167–2183, doi:10.1063/1.1362287, https://doi.org/10.1063/1.1362287.

Bender, C. M. and Wu, T. T. (1968). Analytic structure of energy levels in a field-theory model, *Physical Review Letters* **21**, pp. 406–409, doi:10.1103/PhysRevLett.21.406, https://doi.org/10.1103/PhysRevLett.21.406.

Bender, C. M. and Wu, T. T. (1969). Anharmonic oscillator, *Physical Review* **184**, pp. 1231–1260, doi:10.1103/PhysRev.184.1231, https://doi.org/10.1103/PhysRev.184.1231.

Bender, C. M. and Wu, T. T. (1973). Anharmonic oscillator. II. A study of perturbation theory in large order, *Physical Review D* **7**, pp. 1620–1636, doi:10.1103/PhysRevD.7.1620, https://doi.org/10.1103/PhysRevD.7.1620.

Berry, M. (1989). Uniform asymptotic smoothing of Stokes's discontinuities, *Proceedings of the Royal Society of London A* **422**, pp. 7–21, doi:10.1098/rspa.1989.0018, https://doi.org/10.1098/rspa.1989.0018.

Berry, M. V. (1997). Lop-sided diffraction by absorbing crystals, *Journal of Physics A: Mathematical and General* **31**, pp. 3493–3502, doi:10.1088/0305-4470/31/15/014, https://doi.org/10.1088/0305-4470/31/15/014.

Berry, M. V. and O'Dell, D. H. J. (1998). Diffraction by volume gratings with imaginary potentials, *Journal of Physics A: Mathematical and General* **31**, pp. 2093–2101, doi:10.1088/0305-4470/31/8/019, https://doi.org/10.1088/0305-4470/31/8/019.

Bessis, D. and Zinn-Justin, J. (1992). Private discussion, unpublished.

Bhattacharjie, A. and Sudarshan, E. (1962). A class of solvable potentials, *Nuovo Cimento* **25**, pp. 864–879, doi:10.1007/BF02733153, https://doi.org/10.1007/BF02733153.

Bianchi, L. (1927). *Vorlesungen über Differentialgeometrie* (Teubner, Leipzig).

Bittner, S., Dietz, B., Günther, U., Harney, H. L., Miski-Oglu, M., Richter, A., and Schäfer, F. (2012). $\mathcal{PT}$ symmetry and spontaneous symmetry breaking in a microwave billiard, *Physical Review Letters* **108**, p. 024101, doi:10.1103/PhysRevLett.108.024101, https://doi.org/10.1103/PhysRevLett.108.024101.

Blasi, A., Scolarici, G., and Solombrino, L. (2004). Pseudo-Hermitian Hamiltonians, indefinite inner product spaces and their symmetries, *Journal of Physics A: Mathematical and General* **37**, pp. 4335–4351, doi:10.1088/0305-4470/37/15/003, http://doi.org/10.1088/0305-4470/37/15/003.

Bleuler, K. (1950). Eine neue methode zur behandlung der longitudinalen und skalaren photonen, *Helvetica Physica Acta* **23**, pp. 567–586, doi:10.5169/seals-112124, http://doi.org/10.5169/seals-112124.

Bognar, J. (1974). *Indefinite Inner Product Spaces* (Springer, Berlin).

Bountis, T., Segur, H., and Vivaldi, F. (1982). Integrable Hamiltonian systems and the Painlevé property, *Physical Review A* **25**, pp. 1257–1264, doi:10.1103/PhysRevA.25.1257, https://doi.org/10.1103/PhysRevA.25.1257.

Bousso, R., Maloney, A., and Strominger, A. (2002). Conformal vacua and entropy in de Sitter space, *Physical Review D* **65**, p. 104039, doi:10.1103/PhysRevD.65.104039, https://doi.org/10.1103/PhysRevD.65.104039.

Brody, D. C. (2014). Biorthogonal quantum mechanics, *Journal of Physics A: Mathematical and Theoretical* **47**, p. 035305, doi:10.1088/1751-8113/47/3/035305, https://doi.org/10.1088/1751-8113/47/3/035305.

Brody, D. C. (2016). Consistency of $\mathcal{PT}$-symmetric quantum mechanics, *Journal of Physics A: Mathematical and Theoretical* **49**, p. 10LT03, doi:10.1088/1751-8113/49/10/10lt03, https://doi.org/10.1088/1751-8113/49/10/101t03.

Brody, D. C. and Graefe, E.-M. (2012). Mixed-state evolution in the presence of gain and loss, *Physical Review Letters* **109**, p. 230405, doi:10.1103/physrevlett.109.230405,
https://doi.org/10.1103/physrevlett.109.230405.

Brody, D. C. and Graefe, E.-M. (2013). Information geometry of complex hamiltonians and exceptional points, *Entropy* **15**, pp. 3361–3378, doi:10.3390/e15093361, https://doi.org/10.3390/e15093361.

Brower, R. C., Furman, M. A., and Moshe, M. (1978). Critical exponents for the Reggeon quantum spin model, *Physics Letters B* **76**, 2, pp. 213–219, doi:10.1016/0370-2693(78)90279-4, https://doi.org/10.1016/0370-2693(78)90279-4.

Buslaev, V. and Grecchi, V. (1993). Equivalence of unstable anharmonic oscillators and double wells, *Journal of Physics A: Mathematical and General* **26**, 20, pp. 5541–5549, doi:10.1088/0305-4470/26/20/035, https://doi.org/10.1088/0305-4470/26/20/035.

Caliceti, E., Graffi, S., and Maioli, M. (1980). Perturbation theory of odd anharmonic oscillators, *Communications in Mathematical Physics* **75**, 1, pp. 51–66, doi:10.1007/BF01962591, https://doi.org/10.1007/BF01962591.

Calogero, F. (1969a). Ground state of one-dimensional N-body system, *Journal of Mathematical Physics* **10**, pp. 2197–2200, doi:10.1063/1.1664821, https://doi.org/10.1063/1.1664821.

Calogero, F. (1969b). Solution of a three-body problem in one dimension, *Journal of Mathematical Physics* **10**, pp. 2191–2196, doi:10.1063/1.1664820, https://doi.org/10.1063/1.1664820.

Calogero, F. (1971). Solution of the one-dimensional N-body problems with quadratic and/or inversely quadratic pair potentials, *Journal of Mathematical Physics* **12**, pp. 419–436, doi:10.1063/1.1665604, https://doi.org/10.1063/1.1665604.

Camassa, R. and Holm, D. D. (1993). An integrable shallow-water equation with peaked solutions, *Physical Review Letters* **71**, pp. 1661–1664, doi:10.1103/PhysRevLett.71.1661, http://doi.org/10.1103/PhysRevLett.71.1661.

Cardy, J. L. (1985). Conformal invariance and the Yang-Lee edge singularity in two dimensions, *Physical Review Letters* **54**, pp. 1354–1356, doi:10.1103/PhysRevLett.54.1354, https://doi.org/;10.1103/PhysRevLett.54.1354.

Cardy, J. L. and Mussardo, G. (1989). S-matrix of the Yang-Lee edge singularity in two dimensions, *Physics Letters B* **225**, pp. 275–278, doi:10.1016/0370-2693(89)90818-6, https://doi.org/10.1016/0370-2693(89)90818-6.

Cardy, J. L. and Sugar, R. L. (1975). Reggeon field theory on a lattice, *Physical Review D* **12**, pp. 2514–2522, doi:10.1103/PhysRevD.12.2514, https://doi.org/10.1103/PhysRevD.12.2514.

Cavaglia, A. and Fring, A. (2012). $\mathcal{PT}$-symmetrically deformed shock waves, *Journal of Physics A: Mathematical and Theoretical* **45**, p. 444010, doi:10.1088/1751-8113/45/44/444010, https://doi.org/10.1088/1751-8113/45/44/444010.

Cavaglia, A., Fring, A., and Bagchi, B. (2011). $\mathcal{PT}$-symmetry breaking in complex nonlinear wave equations and their deformations, *Journal of Physics A: Mathematical and Theoretical* **44**, p. 325201, doi:10.1088/1751-8113/44/32/325201, https://doi.org/10.1088/1751-8113/44/32/325201.

Cen, J., Correa, F., and Fring, A. (2017). Time-delay and reality conditions for complex solitons, *Journal of Mathematical Physics* **58**, 3, p. 032901, doi:10.1063/1.4978864, https://doi.org/10.1063/1.4978864.

Cen, J. and Fring, A. (2016). Complex solitons with real energies, *Journal of Physics A: Mathematical and Theoretical* **49**, 36, p. 365202, doi:10.1088/1751-8113/49/36/365202, https://doi.org/10.1088/1751-8113/49/36/365202.

Cerjan, A., Raman, A., and Fan, S. (2016). Exceptional contours and band structure design in parity-time symmetric photonic crystals, *Physical Review Letters* **116**, p. 203902, doi:10.1103/physrevlett.116.203902, https://doi.org/10.1103/physrevlett.116.203902.

Cerveró, J. M. and Rodríguez, A. (2004). The band spectrum of periodic potentials with $\mathcal{PT}$-symmetry, *Journal of Physics A: Mathematical and General* **37**, pp. 10167–10177, doi:10.1088/0305-4470/37/43/010, https://doi.org/10.1088/0305-4470/37/43/010.

Chandrasekar, V. K., Senthilvelan, M., and Lakshmanan, M. (2007). On the Lagrangian and Hamiltonian description of the damped linear harmonic oscillator, *Journal of Mathematical Physics* **48**, p. 032701, doi:10.1063/1.2711375, https://doi.org/10.1063/1.2711375.

Chen, H. H., Lee, Y. C., and Pereira, N. R. (1979). Algebraic internal wave solutions and the integrable Calogero-Moser-Sutherland N-body problem, *Physics of Fluids* **22**, pp. 187–188, doi:10.1063/1.862457, https://doi.org/10.1063/1.862457.

Chen, W., Şahin Kaya Özdemir, Zhao, G., Wiersig, J., and Yang, L. (2017). Exceptional points enhance sensing in an optical microcavity, *Nature* **548**, pp. 192–196, doi:10.1038/nature23281, https://doi.org/10.1038/nature23281.

Chong, Y., Ge, L., Cao, H., and Stone, A. (2010). Coherent perfect absorbers: Time-reversed lasers, *Physical Review Letters* **105**, p. 053901, doi:10.1103/PhysRevLett.105.053901, https://doi.org/10.1103/PhysRevLett.105.053901.

Chong, Y. D., Ge, L., and Stone, A. D. (2011). $\mathcal{PT}$-symmetry breaking and laser-absorber modes in optical scattering systems, *Physical Review Letters* **106**, p. 093902, doi:10.1103/physrevlett.106.093902, https://doi.org/10.1103/physrevlett.106.093902.

Clarkson, P. A. (2003). The fourth Painlevé equation and associated special polynomials, *Journal of Mathematical Physics* **44**, 11, pp. 5350–5374, doi:https://doi.org/10.1063/1.1603958, 10.1063/1.1603958.

Cooper, F., Ginocchio, J. N., and Khare, A. (1987). Relationship between supersymmetry and solvable potentials, *Physical Review D* **36**, pp. 2458–2473, doi:10.1103/PhysRevD.36.2458, https://doi.org/10.1103/PhysRevD.36.2458.

Cooper, F., Khare, A., and Sukhatme, U. (1995). Supersymmetry and quantum mechanics, *Physics Reports* **251**, pp. 267–385, doi:10.1016/0370-1573(94)00080-M, https://doi.org/10.1016/0370-1573(94)00080-M.

Cooper, F., Shepard, H., and Sodano, P. (1993). Solitary waves in a class of generalized Korteweg-de Vries equations, *Physical Review E* **48**, pp. 4027–4032, doi:10.1103/PhysRevE.48.4027, https://doi.org/10.1103/PhysRevE.48.4027.

Cordero, P. and Salamo, S. (1991). Algebraic solution for the Natanzon confluent potentials, *Journal of Physics A: Mathematical and General* **24**, pp. 5299–5305, doi:10.1088/0305-4470/24/22/014, https://doi.org/10.1088/0305-4470/24/22/014.

Cummer, S. A., Christensen, J., and Alù, A. (2016). Controlling sound with acoustic metamaterials, *Nature Reviews: Materials* **1**, p. 16001, doi:10.1038/natrevmats.2016.1, https://doi.org/10.1038/natrevmats.2016.1.

Curtright, T. and Mezincescu, L. (2007). Biorthogonal quantum systems, *Journal of Mathematical Physics* **48**, p. 092106, doi:10.1063/1.2196243, https://doi.org/10.1063/1.2196243.

Curtright, T., Mezincescu, L., Ivanov, E., and Townsend, P. K. (2007). Planar super-Landau models revisited, *Journal of High Energy Physics*, p. 020 doi:

10.1088/1126-6708/2007/04/020, https://doi.org/10.1088/1126-6708/2007/04/020.

Curtright, T. and Veitia, A. (2007). Quasi-Hermitian quantum mechanics in phase space, *Journal of Mathematical Physics* **48**, p. 102112, doi:10.1063/1.2365716, https://doi.org/10.1063/1.2365716.

Curtright, T. L. and Fairlie, D. B. (2008). Euler incognito, *Journal of Physics A: Mathematical and Theoretical* **41**, 24, p. 244009, doi:10.1088/1751-8113/41/24/244009, https://doi.org/10.1088/1751-8113/41/24/244009.

Dabrowska, J. W., Khare, A., and Sukhatme, U. P. (1988). Explicit wavefunctions for shape-invariant potentials by operator techniques, *Journal of Physics A: Mathematical and General* **21**, pp. L195–L200, doi:10.1088/0305-4470/21/4/002, https://doi.org/10.1088/0305-4470/21/4/002.

Darboux, J. (1882). Sur une proposition relative aux equations lineaires, *Comptes rendus de l'Académie des Sciences, Serie I, Mathematics, Paris* **94**, pp. 1456–1459.

Davies, E. B. (2010). An indefinite convection-diffusion operator, *LMS Journal of Computation and Mathematics* **10**, pp. 288–306, doi:10.1112/S1461157000001418, https://doi.org/10.1112/S1461157000001418.

Delabaere, E. and Pham, F. (1998). Eigenvalues of complex Hamiltonians with $\mathcal{PT}$-symmetry. II, *Physics Letters A* **250**, pp. 29–32, doi:10.1016/S0375-9601(98)00792-0, https://doi.org/10.1016/S0375-9601(98)00792-0.

Delabaere, E. and Trinh, D. T. (2000). Spectral analysis of the complex cubic oscillator, *Journal of Physics A: Mathematical and General* **33**, 48, pp. 8771–8796, doi:10.1088/0305-4470/33/48/314, https://doi.org/10.1088/0305-4470/33/48/314.

Dirac, P. A. M. (1942). Bakerian Lecture. The physical interpretation of quantum mechanics, *Proceedings of the Royal Society A* **180**, pp. 1–40, doi:10.1098/rspa.1942.0023, https://doi.org/10.1098/rspa.1942.0023.

Doppler, J., Mailybaev, A. A., Böhm, J., Kuhl, U., Girschik, A., Libisch, F., Milburn, T. J., Rabl, P., Moiseyev, N., and Rotter, S. (2016). Dynamically encircling an exceptional point for asymmetric mode switching, *Nature* **537**, pp. 76–79, doi:10.1038/nature18605, https://doi.org/10.1038/nature18605.

Dorey, P., Dunning, C., Gliozzi, F., and Tateo, R. (2008). On the ODE/IM correspondence for minimal models, *Journal of Physics A: Mathematical and Theoretical* **41**, p. 132001, doi:10.1088/1751-8113/41/13/132001, https://doi.org/10.1088/1751-8113/41/13/132001.

Dorey, P., Dunning, C., Lishman, A., and Tateo, R. (2009). $\mathcal{PT}$ symmetry breaking and exceptional points for a class of inhomogeneous complex potentials, *Journal of Physics A: Mathematical and Theoretical* **42**, p. 465302, doi:10.1088/1751-8113/42/46/465302, https://doi.org/10.1088/1751-8113/42/46/465302.

Dorey, P., Dunning, C., Masoero, D., Suzuki, J., and Tateo, R. (2007a). Pseudo-differential equations, and the Bethe ansatz for the classical Lie algebras, *Nuclear Physics B* **772**, pp. 249–289, doi:10.1016/j.nuclphysb.2007.02.029, https://doi.org/10.1016/j.nuclphysb.2007.02.029.

Dorey, P., Dunning, C., and Tateo, R. (2000). Differential equations for general $su(n)$ Bethe ansatz systems, *Journal of Physics A: Mathematical and General* **33**, pp. 8427–8441, doi:10.1088/0305-4470/33/47/308, https://doi.org/10.1088/0305-4470/33/47/308.

Dorey, P., Dunning, C., and Tateo, R. (2001a). Spectral equivalence, Bethe ansatz, and reality properties in $\mathcal{PT}$-symmetric quantum mechanics, *Journal of Physics A: Mathematical and General* **34**, pp. 5679–5704, doi:10.1088/0305-4470/34/28/305, https://doi.org/10.1088/0305-4470/34/28/305.

Dorey, P., Dunning, C., and Tateo, R. (2001b). Supersymmetry and the spontaneous breakdown of $\mathcal{PT}$ symmetry, *Journal of Physics A: Mathematical and General* **34**, pp. L391–L400, doi:10.1088/0305-4470/34/28/102, https://doi.org/10.1088/0305-4470/34/28/102.

Dorey, P., Dunning, C., and Tateo, R. (2007b). The ODE/IM correspondence, *Journal of Physics A: Mathematical and Theoretical* **40**, pp. R205–R283, doi:10.1088/1751-8113/40/32/R01, https://doi.org/10.1088/1751-8113/40/32/R01.

Dorey, P., Dunning, C., and Tateo, R. (2012). Quasi-exact solvability, resonances and trivial monodromy in ordinary differential equations, *Journal of Physics A: Mathematical and Theoretical* **45**, p. 444013, doi:10.1088/1751-8113/45/44/444013, http://dfx.doi.org/10.1088/1751-8113/45/44/444013.

Dorey, P., Faldella, S., Negro, S., and Tateo, R. (2013). The Bethe ansatz and the Tzitzéica-Bullough-Dodd equation, *Philosophical Transactions of the Royal Society of London A* **371**, p. 20120052, doi:10.1098/rsta.2012.0052, https://doi.org/10.1098/rsta.2012.0052.

Dorey, P., Millican-Slater, A., and Tateo, R. (2005). Beyond the WKB approximation in $\mathcal{PT}$-symmetric quantum mechanics, *Journal of Physics A: Mathematical and General* **38**, pp. 1305–1332, doi:10.1088/0305-4470/38/6/010, https://doi.org/10.1088/0305-4470/38/6/010.

Dorey, P. and Tateo, R. (1999a). Anharmonic oscillators, the thermodynamic Bethe ansatz, and nonlinear integral equations, *Journal of Physics A: Mathematical and General* **32**, pp. L419–L425, doi:10.1088/0305-4470/32/38/102, https://doi.org/10.1088/0305-4470/32/38/102.

Dorey, P. and Tateo, R. (1999b). On the relation between Stokes multipliers and the T-Q systems of conformal field theory, *Nuclear Physics B* **563**, pp. 573–602, doi:10.1016/S0550-3213(99)00609-4, https://doi.org/10.1016/S0550-3213(99)00609-4, [Erratum: Nucl. Phys.B603,581(2001)].

Dorey, P. and Tateo, R. (2000). Differential equations and integrable models: the $su(3)$ case, *Nuclear Physics B* **571**, pp. 583–606, doi:10.1016/S0550-3213(99)00791-9, https://doi.org/10.1016/S0550-3213(99)00791-9.

Dorizzi, B., Grammaticos, B., and Ramani, A. (1983). A new class of integrable systems, *Journal of Mathematical Physics* **24**, pp. 2282–2288, doi:10.1063/1.525975, https://doi.org/10.1063/1.525975.

Dunne, G. V. and Feinberg, J. (1998). Self-isospectral periodic potentials and supersymmetric quantum mechanics, *Physical Review D* **57**, pp. 1271–1276, doi:10.1103/PhysRevD.57.1271, https://doi.org/10.1103/PhysRevD.57.1271.

Dutt, R., Khare, A., and Sukhatme, U. P. (1986). Exactness of supersymmetric WKB spectra for shape-invariant potentials, *Physics Letters B* **181**, pp. 295–298, doi:10.1016/0370-2693(86)90049-3, https://doi.org/10.1016/0370-2693(86)90049-3.

Dutt, R., Khare, A., and Varshni, Y. (1995). New class of conditionally exactly solvable potentials in quantum mechanics, *Journal of Physics A: Mathematical and General* **28**, pp. L107–L113, doi:10.1088/0305-4470/28/3/008, https://doi.org/10.1088/0305-4470/28/3/008.

Dyson, F. J. (1956). Thermodynamic behavior of an ideal ferromagnet, *Physical Review* **102**, pp. 1230–1244, doi:10.1103/PhysRev.102.1230, https://doi.org/10.1103/PhysRev.102.1230.

Egrifes, H. and Sever, R. (2005). Bound states of the Dirac equation for the $\mathcal{PT}$-symmetric generalized Hulthén potential by the Nikiforov-Uvarov method, *Physics Letters A* **344**, pp. 117–126, doi:10.1016/j.physleta.2005.06.061, https://doi.org/10.1016/j.physleta.2005.06.061.

El-Ganainy, R., Makris, K., Christodoulides, D., and Musslimani, Z. (2007). Theory of coupled optical $\mathcal{PT}$-symmetric structures, *Optics Letters* **32**, pp. 2632–2634, doi:10.1364/OL.32.002632, https://doi.org/10.1364/OL.32.002632.

El-Ganainy, R., Makris, K. G., Khajavikhan, M., Musslimani, Z. H., Rotter, S., and Christodoulides, D. N. (2018). Non-Hermitian physics and $\mathcal{PT}$ symmetry, *Nature Physics* **14**, pp. 11–19, doi:10.1038/nphys4323, https://doi.org/10.1038/nphys4323.

Eremenko, A. (2015). Entire functions, $\mathcal{PT}$-symmetry and Voros's quantization scheme, arXiv:1510.02504 [math-ph], https://arxiv.org/abs/1510.02504.

Eremenko, A. and Gabrielov, A. (2009a). Analytic continuation of eigenvalues of a quartic oscillator, *Communications in Mathematical Physics* **287**, pp. 431–457, doi:10.1007/s00220-008-0663-6, https://doi.org/10.1007/s00220-008-0663-6.

Eremenko, A. and Gabrielov, A. (2009b). Irreducibility of some spectral determinants, *ArXiv* 0904.1714, http://arxiv.org/abs/0904.1714.

Eremenko, A. and Gabrielov, A. (2011a). Quasi-exactly solvable quartic: elementary integrals and asymptotics, *Journal of Physics A: Mathematical and Theoretical* **44**, p. 312001, doi:10.1088/1751-8113/44/31/312001, https://doi.org/10.1088/1751-8113/44/31/312001.

Eremenko, A. and Gabrielov, A. (2011b). Singular perturbation of polynomial potentials with applications to $\mathcal{PT}$-symmetric families, *Moscow Mathematical Journal* **11**, pp. 473–503.

Eremenko, A. and Gabrielov, A. (2012a). Quasi-exactly solvable quartic: real algebraic spectral locus, *Journal of Physics A: Mathematical and Theoretical* **45**, 17, p. 175205, doi:10.1088/1751-8113/45/17/175205, https://doi.org/10.1088/1751-8113/45/17/175205.

Eremenko, A. and Gabrielov, A. (2012b). Two-parametric $\mathcal{PT}$-symmetric quartic family, *Journal of Physics A: Mathematical and Theoretical* **45**, 17, p. 175206, doi:10.1088/1751-8113/45/17/175206, https://doi.org/10.1088/1751-8113/45/17/175206.

Eremenko, A., Gabrielov, A., and Shapiro, B. (2008). High energy eigenfunctions of one-dimensional Schrödinger operators with polynomial potentials, *Computational Methods and Function Theory* **8**, pp. 513–529, doi: 10.1007/BF03321702, https://doi.org/10.1007/BF03321702.

Faddeev, L. and Yakubovskii, O. (2009). *Lectures on Quantum Mechanics for Mathematical Students* (American Mathematical Society, Providence, Rhode Island).

Fateev, V. and Lukyanov, S. (2006). Boundary RG flow associated with the AKNS soliton hierarchy, *Journal of Physics A: Mathematical and General* **39**, pp. 12889–12925, doi:10.1088/0305-4470/39/41/S10, hep-th/0510271, https://doi.org/10.1088/0305-4470/39/41/S10.

Fedorov, Y. N. and Gomez-Ullate, D. (2007). Dynamical systems on infinitely sheeted Riemann surfaces, *Physica D - Nonlinear Phenomena* **227**, pp. 120–134, doi:10.1016/j.physd.2007.02.001, http://doi.org/10.1016/j.physd.2007.02.001.

Feng, L., El-Ganainy, R., and Ge, L. (2017a). Non-Hermitian photonics based on parity-time symmetry, *Nature Photonics* **11**, p. 752, doi:10.1038/s41566-017-0031-1, https://doi.org/10.1038/s41566-017-0031-1.

Feng, L., El-Ganainy, R., and Ge, L. (2017b). Non-Hermitian photonics based on parity-time symmetry, *Nature Photonics* **11**, pp. 752–762, doi:10.1038/s41566-017-0031-1, https://doi.org/10.1038/s41566-017-0031-1.

Feng, L., Wong, Z. J., Ma, R.-M., Wang, Y., and Zhang, X. (2014). Single-mode laser by parity-time symmetry breaking, *Science* **346**, pp. 972–975, doi: 10.1126/science.1258479, https://doi.org/10.1126/science.1258479.

Fernández, F. M., Guardiola, R., Ros, J., and Znojil, M. (2005). A family of complex potentials with real spectrum, *Journal of Physics A: Mathematical and General* **32**, pp. 3105–3116, doi:10.1088/0305-4470/32/17/303, https://doi.org/10.1088/0305-4470/32/17/303.

Feynman, R. and Hibbs, A. (1965). *Quantum Mechanics and Path Integrals* (McGraw-Hill, New York).

Fisher, M. E. (1978). Yang-Lee edge singularity and ϕ^3 field theory, *Physical Review Letters* **40**, pp. 1610–1613, doi:10.1103/PhysRevLett.40.1610, https://doi.org/10.1103/PhysRevLett.40.1610.

Fleury, R., Khanikaev, A. B., and Alù, A. (2016). Floquet topological insulators for sound, *Nature Communications* **7**, p. 11744, doi:10.1038/ncomms11744, https://doi.org/10.1038/ncomms11744.

Fleury, R., Monticone, F., and Alù, A. (2015). Invisibility and cloaking: Origins, present, and future perspectives, *Physical Review Applied* **4**, p. 037001, doi:10.1103/PhysRevApplied.4.037001, https://doi.org/10.1103/PhysRevApplied.4.037001.

Fleury, R., Sounas, D., and Alù, A. (2014a). An invisible acoustic sensor based on parity-time symmetry, *Nature Communications* **6**, p. 5905, doi:10.1038/ncomms6905, https://doi.org/10.1038/ncomms6905.

Fleury, R., Sounas, D. L., and Alú, A. (2014b). Negative refraction and planar focusing based on parity-time symmetric metasurfaces, *Physical Review Letters* **113**, p. 023903, doi:10.1103/PhysRevLett.113.023903, https://doi.org/10.1103/PhysRevLett.113.023903.

Flügge, S. (1971). *Practical Quantum Mechanics* (Springer, Berlin).

Fring, A. (2006). A note on the integrability of non-Hermitian extensions of Calogero-Moser-Sutherland models, *Modern Physics Letters A* **21**, pp. 691–699, doi:10.1142/S0217732306019682, https://doi.org/10.1142/S0217732306019682.

Fring, A. (2007). $\mathcal{PT}$-symmetric deformations of the Korteweg-de Vries equation, *Journal of Physics A: Mathematical and Theoretical* **40**, pp. 4215–4224, doi:10.1088/1751-8113/40/15/012, https://doi.org/10.1088/1751-8113/40/15/012.

Fring, A. (2009). Particles versus fields in $\mathcal{PT}$-symmetrically deformed integrable systems, *Pramana-Journal of Physics* **73**, 2, pp. 363–373, doi:10.1007/s12043-009-0128-2, https://doi.org/10.1007/s12043-009-0128-2.

Fring, A. (2013). $\mathcal{PT}$-symmetric deformations of integrable models, *Philosophical Transactions of the Royal Society of London A: Mathematical, Physical and Engineering Sciences* **371**, 1989, p. 20120046, doi:10.1098/rsta.2012.0046, https://doi.org/10.1098/rsta.2012.0046.

Fring, A. and Smith, M. (2010). Antilinear deformations of Coxeter groups, an application to Calogero models, *Journal of Physics A: Mathematical and Theoretical* **43**, p. 325201, doi:10.1088/1751-8113/43/32/325201, https://doi.org/10.1088/1751-8113/43/32/325201.

Fring, A. and Smith, M. (2011). $\mathcal{PT}$ invariant complex E(8) root spaces, *International Journal of Theoretical Physics* **50**, pp. 974–981, doi:10.1007/s10773-010-0542-8, https://doi.org/10.1007/s10773-010-0542-8.

Fring, A. and Smith, M. (2012). Non-Hermitian multi-particle systems from complex root spaces, *Journal of Physics A: Mathematical and Theoretical* **45**, p. 085203, doi:10.1088/1751-8113/45/8/085203, https://doi.org/10.1088/1751-8113/45/8/085203.

Fring, A. and Znojil, M. (2008). $\mathcal{PT}$-symmetric deformations of Calogero models, *Journal of Physics A: Mathematical and Theoretical* **41**, p. 194010, doi:10.1088/1751-8113/41/19/194010, https://doi.org/10.1088/1751-8113/41/19/194010.

Gaiotto, D., Moore, G. W., and Neitzke, A. (2013). Wall-crossing, Hitchin systems, and the WKB approximation, *Advances in Mathematics* **234**, pp. 239–403, doi:10.1016/j.aim.2012.09.027, https://doi.org/10.1016/j.aim.2012.09.027.

Gao, T., Estrecho, E., Bliokh, K. Y., Liew, T. C. H., Fraser, M. D., Brodbeck, S., Kamp, M., Schneider, C., Höfling, S., Yamamoto, Y., Nori, F., Kivshar, Y. S., Truscott, A. G., Dall, R. G., and Ostrovskaya, E. A. (2015). Observation of non-Hermitian degeneracies in a chaotic exciton-polariton billiard, *Nature* **526**, pp. 554–558, doi:10.1038/nature15522, https://doi.org/10.1038/nature15522.

Ge, L., Chong, Y. D., and Stone, A. (2012a). Conservation relations and anisotropic transmission resonances in one-dimensional $\mathcal{PT}$-symmetric photonic heterostructures, Physical Review A **85**, p. 023802, doi:10.1103/PhysRevA.85.023802, https://doi.org/10.1103/PhysRevA.85.023802.

Ge, L., Chong, Y. D., and Stone, A. D. (2012b). Conservation relations and anisotropic transmission resonances in one-dimensional $\mathcal{PT}$-symmetric photonic heterostructures, Physical Review A **85**, p. 023802, doi:10.1103/PhysRevA.85.023802, https://doi.org/10.1103/PhysRevA.85.023802.

Ge, L., Makris, K. G., Christodoulides, D. N., and Feng, L. (2015). Scattering in $\mathcal{PT}$- and $\mathcal{RT}$-symmetric multimode waveguides: Generalized conservation laws and spontaneous symmetry breaking beyond one dimension, Physical Review A **92**, p. 062135, doi:10.1103/PhysRevA.92.062135, https://doi.org/10.1103/PhysRevA.92.062135.

Ge, L. and Stone, A. D. (2014). Parity-time symmetry breaking beyond one dimension: The role of degeneracy, Physical Review X **4**, p. 031011, doi:10.1103/PhysRevX.4.031011, https://doi.org/10.1103/PhysRevX.4.031011.

Gendenshtein, L. E. (1983). Derivation of exact spectra of the Schrödinger equation by means of supersymmetry, JETP Letters **38**, pp. 356–359.

Ginocchio, J. N. (1984). A class of exactly solvable potentials. I. One-dimensional Schrödinger equation, Annals of Physics **152**, pp. 203–219, doi:10.1016/0003-4916(84)90084-8, https://doi.org/10.1016/0003-4916(84)90084-8.

Ginocchio, J. N. (1985). A class of exactly solvable potentials II. The three-dimensional Schrödinger equation, Annals of Physics **159**, pp. 467–480, doi:10.1016/0003-4916(85)90120-4, https://doi.org/10.1016/0003-4916(85)90120-4.

Goldzak, T., Mailybaev, A. A., and Moiseyev, N. (2018). Light stops at exceptional points, Physical Review Letters **120**, p. 013901, doi:10.1103/PhysRevLett.120.013901, https://doi.org/10.1103/PhysRevLett.120.013901.

Gomez-Ullate, D., Kamran, N., and Milson, R. (2009). An extended class of orthogonal polynomials defined by a Sturm-Liouville problem, Journal of Mathematical Analysis and Applications **359**, pp. 352–367, doi:10.1016/j.jmaa.2009.05.052, https://doi.org/10.1016/j.jmaa.2009.05.052.

Gomez-Ullate, D., Kamran, N., and Milson, R. (2010a). Exceptional orthogonal polynomials and the Darboux transformation, Journal of Physics A: Mathematical and Theoretical **43**, p. 434016, doi:10.1088/1751-8113/43/43/434016, https://doi.org/10.1088/1751-8113/43/43/434016.

Gomez-Ullate, D., Kamran, N., and Milson, R. (2010b). An extension of Bochner's problem: Exceptional invariant subspaces, Journal of Approximation Theory **162**, pp. 987–1006, doi:10.1016/j.jat.2009.11.002, https://doi.org/10.1016/j.jat.2009.11.002.

Gomez-Ullate, D., Kamran, N., and Milson, R. (2012). On orthogonal polynomials spanning a non-standard flag, in Algebraic Aspects of Darboux Transformations, Quantum Integrable Systems and Supersymmetric Quantum Mechanics, Contemporary Mathematics, Vol. 563 (AMS), pp. 51–71, doi:10.1090/conm/563/11164, https://doi.org/10.1090/conm/563/11164.

Graefe, E.-M. (2012). Stationary states of a $\mathcal{PT}$ symmetric two-mode Bose–Einstein condensate, *Journal of Physics A: Mathematical and Theoretical* **45**, p. 444015, doi:10.1088/1751-8113/45/44/444015, https://doi.org/10.1088/1751-8113/45/44/444015.

Graefe, E.-M. and Jones, H. F. (2011). $\mathcal{PT}$-symmetric sinusoidal optical lattices at the symmetry-breaking threshold, *Physical Review A* **84**, p. 013818, doi:10.1103/PhysRevA.84.013818, https://doi.org/10.1103/PhysRevA.84.013818.

Graefe, E.-M., Korsch, H. J., and Rush, A. (2015a). Classical and quantum dynamics in the (non-Hermitian) Swanson oscillator, *Journal of Physics A: Mathematical and Theoretical* **48**, p. 055301, doi:10.1088/1751-8113/48/5/055301, https://doi.org/10.1088/1751-8113/48/5/055301.

Graefe, E.-M., Mudute-Ndumbe, S., and Taylor, M. (2015b). Random matrix ensembles for $\mathcal{PT}$-symmetric systems, *Journal of Physics A: Mathematical and Theoretical* **48**, p. 38FT02, doi:10.1088/1751-8113/48/38/38ft02, https://doi.org/10.1088/1751-8113/48/38/38ft02.

Graefe, E.-M. and Schubert, R. (2011). Wave-packet evolution in non-Hermitian quantum systems, *Physical Review A* **83**, p. 060101, doi:10.1103/physreva.83.060101, https://doi.org/10.1103/physreva.83.060101.

Graefe, E.-M. and Schubert, R. (2012). Complexified coherent states and quantum evolution with non-Hermitian Hamiltonians, *Journal of Physics A: Mathematical and Theoretical* **45**, p. 244033, doi:10.1088/1751-8113/45/24/244033, https://doi.org/10.1088/1751-8113/45/24/244033.

Grammaticos, B. and Ramani, A. (1997). Integrability and how to detect it, in *Integrability of Nonlinear Systems* (Springer), pp. 30–94, doi:10.1007/978-3-540-40962-5_3, https://doi.org/10.1007/978-3-540-40962-5_3.

Grod, A., Kuzhel, S., and Sudilovskaya, V. (2011). On operators of transition in Kreĭn spaces, *Opuscula Mathematica* **31**, pp. 49–59, doi:10.7494/OpMath.2011.31.1.49, https://doi.org/10.7494/OpMath.2011.31.1.49.

Günther, U. and Kuzhel, S. (2010). $\mathcal{PT}$-symmetry, Cartan decompositions, Lie triple systems and Kreĭn space-related Clifford algebras, *Journal of Physics A: Mathematical and Theoretical* **43**, p. 392002, doi:10.1088/1751-8113/43/39/392002, https://doi.org/10.1088/1751-8113/43/39/392002.

Guo, A., Salamo, G. J., Duchesne, D., Morandotti, R., Volatier-Ravat, M., Aimez, V., Siviloglou, G. A., and Christodoulides, D. N. (2009). Observation of $\mathcal{PT}$-symmetry breaking in complex optical potentials, *Physical Review Letters* **103**, p. 093902, doi:10.1103/PhysRevLett.103.093902, https://doi.org/10.1103/PhysRevLett.103.093902.

Gupta, S. N. (1950a). On the calculation of self-energy of particles, *Physical Review* **77**, pp. 294–295, doi:10.1103/PhysRev.77.294, https://doi.org/10.1103/PhysRev.77.294.

Gupta, S. N. (1950b). Theory of longitudinal photons in quantum electrodynamics, *Proceedings of the Physical Society. Section A* **63**, pp. 681–691, doi:10.1088/0370-1298/63/7/301, https://doi.org/10.1088/0370-1298/63/7/301.

Handy, C., Khan, D., Wang, X.-Q., and Tymczak, C. (2001). Multiscale reference function analysis of the $\mathcal{PT}$ symmetry breaking solutions for the $P^2 + iX^3 + i\alpha X$ Hamiltonian, *Journal of Physics A: Mathematical and General* **34**, pp. 5593–5602, doi:10.1088/0305-4470/34/27/309, https://doi.org/10.1088/0305-4470/34/27/309.

Handy, C. R. (2001). Generating converging bounds to the (complex) discrete states of the $p^2 + ix^3 + i\alpha x$ Hamiltonian, *Journal of Physics A: Mathematical and General* **34**, pp. 5065–5081, doi:10.1088/0305-4470/34/24/305, http://doi.org/10.1088/0305-4470/34/24/305.

Harms, B. C., Jones, S. T., and Tan, C.-I. (1980). New structure in the energy spectrum of Reggeon quantum mechanics with quartic couplings, *Physics Letters B* **91**, 2, pp. 291–295, doi:10.1016/0370-2693(80)90452-9, https://doi.org/10.1016/0370-2693(80)90452-9.

Hatano, N. and Nelson, D. R. (1996). Localization transitions in non-Hermitian quantum mechanics, *Physical Review Letters* **77**, pp. 570–573, doi:10.1103/PhysRevLett.77.570, https://doi.org/10.1103/PhysRevLett.77.570.

He, C., Zhang, X.-L., Feng, L., Lu, M.-H., and Chen, Y.-F. (2011). One-way cloak based on nonreciprocal photonic crystal, *Applied Physics Letters* **99**, p. 151112, doi:10.1063/1.3648112, https://doi.org/10.1063/1.3648112.

Herbst, I. (1979). Dilation analyticity in constant electric field I: The two body problem, *Communications in Mathematical Physics* **64**, pp. 279–298, doi:10.1007/BF01221735, https://doi.org/10.1007/BF01221735.

Hertog, T. and Horowitz, G. T. (2005). Holographic description of AdS cosmologies, *Journal of High Energy Physics*, p. 005doi:10.1088/1126-6708/2005/04/005, https://doi.org/10.1088/1126-6708/2005/04/005.

Hezaro, H. (2008). Complex zeros of eigenfunctions of 1D Schrödinger operators, *International Mathematics Research Notices* **2008**, p. rnm148, doi:10.1007/10.1093/imrn/rnm148, https://doi.org/10.1093/imrn/rnm148.

Hille, E. (1969). *Lectures on Ordinary Differential Equations* (Addison-Wesley, Reading, MA).

Hodaei, H., Hassan, A. U., Garcia-Gracia, H., Hayenga, W. E., Christodoulides, D. N., and Khajavikhan, M. (2017). Enhanced sensitivity in $\mathcal{PT}$-symmetric coupled resonators, in A. V. Kudryashov, A. H. Paxton, and V. S. Ilchenko (eds.), *Proceedings of SPIE: Laser Resonators, Microresonators, and Beam Control XIX*, Vol. 10090, p. 6 pages, doi:10.1117/12.2250047, https://doi.org/10.1117/12.2250047.

Hodaei, H., Miri, M.-A., Heinrich, M., Christodoulides, D. N., and Khajavikhan, M. (2014). Parity-time-symmetric microring lasers, *Science* **346**, pp. 975–978, doi:10.1126/science.1258480, https://doi.org/10.1126/science.1258480.

Hollowood, T. (1992). Solitons in affine Toda field theories, *Nuclear Physics B* **384**, 3, pp. 523–540, doi:10.1016/0550-3213(92)90579-Z, https://doi.org/10.1016/0550-3213(92)90579-Z.

Hook, J. and Hall, H. (1991). *Solid State Physics* (J Wiley & Sons, New York).

Horiuchi, N. (2017). View from... JSAP Spring Meeting: A marriage of materials and optics, *Nature Photonics* **11**, pp. 271–273, doi:10.1038/nphoton.2017.62, https://doi.org/10.1038/nphoton.2017.62.

Hsu, C. W., Zhen, B., Stone, A. D., Joannopoulos, J. D., and Soljačić, M. (2016). Bound states in the continuum, *Nature Reviews: Materials* **1**, p. 16048, doi:10.1038/natrevmats.2016.48, `https://doi.org/10.1038/natrevmats.2016.48`.

Hu, S., Ma, X., Lu, D., Yang, Z., Zheng, Y., and Hu, W. (2011). Solitons supported by complex $\mathcal{PT}$-symmetric gaussian potentials, *Physical Review A* **84**, p. 043818, doi:10.1103/PhysRevA.84.043818, `https://doi.org/10.1103/PhysRevA.84.043818`.

Humphreys, J. E. (1972). *Introduction to Lie Algebras and Representation Theory* (Springer, Berlin).

Infeld, L. and Hull, T. D. (1951). The factorization method, *Reviews of Modern Physics* **23**, pp. 21–68, doi:10.1103/RevModPhys.23.21, `https://doi.org/10.1103/RevModPhys.23.21`.

Ishkhanyan, A. (2015). Exact solution of the Schrödinger equation for the inverse square root potential $v_0/\sqrt{x}$, *Europhysics Letters* **112**, p. 10006, doi:10.1209/0295-5075/112/10006, `https://doi.org/10.1209/0295-5075/112/10006`.

Ishkhanyan, A. (2016). The Lambert-W step-potential – an exactly solvable confluent hypergeometric potential, *Physics Letters A* **380**, pp. 640–644, doi:10.1016/j.physleta.2015.12.004, `https://doi.org/10.1016/j.physleta.2015.12.004`.

Ishkhanyan, T. A. and Ishkhanyan, A. M. (2017). Solutions of the bi-confluent Heun equation in terms of the Hermite functions, *Annals of Physics* **383**, pp. 79–91, doi:10.1016/j.aop.2017.04.015, `https://doi.org/10.1016/j.aop.2017.04.015`.

Ito, K. and Locke, C. (2014). ODE/IM correspondence and modified affine Toda field equations, *Nuclear Physics B* **885**, pp. 600–619, doi:10.1016/j.nuclphysb.2014.06.007, `https://doi.org/10.1016/j.nuclphysb.2014.06.007`.

Ito, K. and Locke, C. (2015). ODE/IM correspondence and Bethe ansatz for affine Toda field equations, *Nuclear Physics B* **896**, pp. 763–778, doi:10.1016/j.nuclphysb.2015.05.016, `https://doi.org/10.1016/j.nuclphysb.2015.05.016`.

Ito, K. and Shu, H. (2017). ODE/IM correspondence for modified $B_2^{(1)}$ affine Toda field equation, *Nuclear Physics B* **916**, pp. 414–429, doi:10.1016/j.nuclphysb.2017.01.009, `https://doi.org/10.1016/j.nuclphysb.2017.01.009`.

Ivanov, E. A. and Smilga, A. V. (2007). Cryptoreality of nonanticommutative Hamiltonians, *Journal of High Energy Physics*, p. 036 doi:10.1088/1126-6708/2007/07/036, `https://doi.org/10.1088/1126-6708/2007/07/036`.

Jackson, J. (1999). *Classical Electrodynamics* (Wiley, Oxford).

Jahromi, A. K., Hassan, A. U., Christodoulides, D. N., and Abouraddy, A. F. (2017). Statistical parity-time-symmetric lasing in an optical fibre network, *Nature Communications* **8**, p. 1359, doi:10.1038/s41467-017-00958-x, `https://doi.org/10.1038/s41467-017-00958-x`.

Japaridze, G. S. (2002). Space of state vectors in $\mathcal{PT}$-symmetric quantum mechanics, *Journal of Physics A: Mathematical and General* **35**, pp. 1709–1718, doi:10.1088/0305-4470/35/7/315, `http://doi.org/10.1088/0305-4470/35/7/315`.

Jones, H. F. (1999). The energy spectrum of complex periodic potentials of the Kronig-Penney type, *Physics Letters A* **262**, pp. 242–244, doi:10.1016/S0375-9601(99)00672-6, `https://doi.org/10.1016/S0375-9601(99)00672-6`.

Jones, H. F. (2005). On pseudo-Hermitian Hamiltonians and their Hermitian counterparts, *Journal of Physics A: Mathematical and General* **38**, pp. 1741–1746, doi:10.1088/0305-4470/38/8/010, `https://doi.org/10.1088/0305-4470/38/8/010`.

Jones, H. F. (2008). Equivalent Hamiltonian for the Lee model, *Physical Review D* **77**, p. 065023, doi:10.1103/PhysRevD.77.065023, `https://doi.org/10.1103/PhysRevD.77.065023`.

Jones, H. F. (2012). Analytic results for a $\mathcal{PT}$-symmetric optical structure, *Journal of Physics A: Mathematical and Theoretical* **45**, p. 135306, doi:10.1088/1751-8113/45/13/135306, `https://doi.org/10.1088/1751-8113/45/13/135306`.

Jones, H. F. and Kulishov, M. (2016). Extension of analytic results for a $\mathcal{PT}$-symmetric structure, *Journal of Optics* **18**, p. 055101, doi:10.1088/2040-8978/18/5/055101, `https://doi.org/10.1088/2040-8978/18/5/055101`.

Jones, H. F., Kulishov, M., and Kress, B. (2016). Parity time-symmetric vertical cavities: Intrinsically single-mode regime in longitudinal direction, *Optics Express* **24**, pp. 17125–17137, doi:10.1364/OE.24.017125, `https://doi.org/10.1364/OE.24.017125`.

Jones, H. F. and Mateo, J. (2006). Equivalent Hermitian Hamiltonian for the non-Hermitian $-x^4$ potential, *Physical Review D* **73**, p. 085002, doi:10.1103/PhysRevD.73.085002, `https://doi.org/10.1103/PhysRevD.73.085002`.

Jones, H. F., Mateo, J., and Rivers, R. J. (2006). On the path-integral derivation of the anomaly for the Hermitian equivalent of the complex $\mathcal{PT}$-symmetric quartic Hamiltonian, *Physical Review D* **74**, p. 125022, doi:10.1103/PhysRevD.74.125022, `https://doi.org/10.1103/PhysRevD.74.125022`.

Jones-Smith, K. (2010). *Non-Hermitian Quantum Mechanics*, Ph.D. thesis, Case Western Reserve University, `http://rave.ohiolink.edu/etdc/view?acc_num=case1270231293`.

Jones-Smith, K. and Mathur, H. (2014). Relativistic non-Hermitian quantum mechanics, *Physical Review D* **89**, p. 125014, doi:10.1103/physrevd.89.125014, `https://doi.org/10.1103/physrevd.89.125014`.

Jones-Smith, K. and Mathur, H. (2016). Non-Hermitian neutrino oscillations in matter with $\mathcal{PT}$ symmetric Hamiltonians, *Europhysics Letters* **113**, p. 61001, doi:10.1209/0295-5075/113/61001, `https://doi.org/10.1209/0295-5075/113/61001`.

Junker, G. (1996). *Supersymmetric Methods in Quantum and Statistical Physics* (Springer, Berlin).

Junker, G. and Roy, P. (1997). Conditionally exactly solvable problems and non-linear algebras, *Physics Letters A* **232**, pp. 155–161, doi:10.1016/S0375-9601(97)00422-2, https://doi.org/10.1016/S0375-9601(97)00422-2.

Källén, G. and Pauli, W. (1955). On the mathematical structure of T. D. Lee's model of a renormalizable field theory, *Matematisk-Fysiske Meddelelser Konglige Danske Videnskabernes Selskab* **30**, p. 7, http://gymarkiv.sdu.dk/MFM/kdvs/mfm%2030-39/mfm-30-7.pdf.

Kamuda, A., Kuzhel, S., and Sudilovskaya, V. (2018). On dual definite subspaces in Krein space, *Complex Analysis and Operator Theory*, pp. 1–22, doi:10.1007/s11785-018-0838-x, https://doi.org/10.1007/s11785-018-0838-x.

Kevrekidis, P. G., Siettos, C. I., and Kevrekidis, Y. G. (2017). To infinity and some glimpses of beyond, *Nature Communications* **8**, p. 1562, doi:10.1038/s41467-017-01502-7, https://doi.org/10.1038/s41467-017-01502-7.

Khare, A. and Cooper, F. (1993). One-parameter family of soliton solutions with compact support in a class of generalized Korteweg-de Vries equations, *Physical Review E* **48**, 6, pp. 4843–4844, doi:10.1103/PhysRevE.48.4843, https://doi.org/10.1103/PhysRevE.48.4843.

Khare, A. and Mandal, B. P. (2000). A $\mathcal{PT}$-invariant potential with complex QES eigenvalues, *Physics Letters A* **272**, pp. 53–56, doi:10.1016/S0375-9601(00)00409-6, https://doi.org/10.1016/S0375-9601(00)00409-6.

Kim, K.-H., Hwang, M.-S., Kim, H.-R., Choi, J.-H., No, Y.-S., and Park, H.-G. (2016). Direct observation of exceptional points in coupled photonic-crystal lasers with asymmetric optical gains, *Nature Communications* **7**, p. 13893, doi:10.1038/ncomms13893, https://doi.org/10.1038/ncomms13893.

Klaiman, S., Günther, U., and Moiseyev, N. (2008). Visualization of branch points in $\mathcal{PT}$-symmetric waveguides, *Physical Review Letters* **101**, p. 080402, doi:10.1103/PhysRevLett.101.080402, https://doi.org/10.1103/PhysRevLett.101.080402.

Kleefeld, F. (2004). Non-Hermitian quantum theory and its holomorphic representation: Introduction and some applications, *ArXiv*, hep-th/0408028 http://arxiv.org/abs/hep-th/0408028.

Kleefeld, F. (2005). On (non-Hermitian) Lagrangeans in (particle) physics and their dynamical generation, *Czechoslovak Journal of Physics* **55**, pp. 1123–1134, doi:10.1007/s10582-005-0117-8, https://doi.org/10.1007/s10582-005-0117-8.

Kleefeld, F. (2006). Kurt Symanzik - a stable fixed point beyond triviality, *Journal of Physics A: Mathematical and General* **39**, pp. L9–L15, doi:10.1088/0305-4470/39/1/L02, https://doi.org/10.1088/0305-4470/39/1/L02.

Korteweg, D. J. and de Vries, G. (1895). On the change of form of long waves advancing in a rectangular canal, and on a new type of long stationary waves, *Philosophical Magazine* **39**, pp. 422–443, doi:10.1080/14786449508620739, https://doi.org/10.1080/14786449508620739.

Kretschmer, R. and Szymanowski, L. (2004). Quasi-Hermiticity in infinite-dimensional Hilbert spaces, *Physics Letters A* **325**, pp. 112–117, doi:10.1016/j.physleta.2004.03.044, https://doi.org/10.1016/j.physleta.2004.03.044.

Krichever, I., Babelon, O., Billey, E., and Talon, M. (1995). Spin generalization of the Calogero-Moser system and the matrix KP equation, in *Translations of the American Mathematical Society-Series 2*, Vol. 170 (American Mathematical Society), pp. 83–120.

Kruskal, M., Joshi, N., and Halburd, R. (2004). Analytic and asymptotic methods for nonlinear singularity analysis: A review and extensions of tests for the Painlevé property, in *Integrability of Nonlinear Systems*, Vol. 638 (Springer, Berlin), pp. 175–208, doi:10.1007/978-3-540-40962-5_6, https://doi.org/10.1007/978-3-540-40962-5_6.

Kulishov, M., Kress, B., and Jones, H. F. (2014). Novel optical characteristics of a Fabry-Perot resonator with embedded $\mathcal{PT}$-symmetrical grating, *Optics Express* **22**, pp. 23164–23181, doi:10.1364/OE.22.023164, https://doi.org/10.1364/OE.22.023164.

Kulishov, M., Laniel, J. M., Bélanger, N., Azaña, J., and Plant, D. V. (2005). Nonreciprocal waveguide Bragg gratings, *Optics Express* **13**, pp. 3068–3078, doi:10.1364/OPEX.13.003068, https://doi.org/10.1364/OPEX.13.003068.

Kuzhel, A. and Kuzhel, S. (1998). *Regular Extensions of Hermitian Operators* (VSP, Utrecht, The Netherlands).

Kuzhel, S. (2009). On pseudo-Hermitian operators with generalized C-symmetries, in *The Mark Kreĭn Centenary Conference Volume 1: Operator Theory and Related Topics, Operator Theory: Advances and Applications*, Vol. 190 (Springer), pp. 375–385, doi:10.1007/978-3-7643-9919-1_23, https://doi.org/10.1007/978-3-7643-9919-1_23.

Kuzhel, S. and Sudilovskaja, V. (2017). Towards theory of C-symmetries, *Opuscula Mathematica* **37**, pp. 65–80, doi:10.7494/OpMath.2017.37.1.65, https://doi.org/10.7494/OpMath.2017.37.1.65.

Kuzhel, S. O. and Patsyuk, O. M. (2012). On the theory of $\mathcal{PT}$-symmetric operators, *Ukrainian Mathematical Journal* **64**, pp. 35–55, doi:10.1007/s11253-012-0628-y, https://doi.org/10.1007/s11253-012-0628-y.

Lamb Jr, G. L. (1971). Analytical descriptions of ultrashort optical pulse propagation in a resonant medium, *Reviews of Modern Physics* **43**, pp. 99–124, doi:10.1103/RevModPhys.43.99, https://doi.org/10.1103/RevModPhys.43.99.

Langer, H. (1970a). On the maximal dual pairs of invariant subspaces of j-self-adjoint operators, *Mathematical notes of the Academy of Sciences of the USSR* **7**, pp. 269–271, doi:10.1007/BF01151700, https://doi.org/10.1007/BF01151700.

Langer, H. (1970b). On the maximal dual pairs of invariant subspaces of j-self-adjoint operators (Russian), *Mate matieheskie Zametki* **7**, pp. 443–447.

Langer, H. (1982). Spectral functions of definitizable operators in Kreĭn spaces, in D. Butković, H. Kraljević, and S. Kurepa (eds.), *Lecture Notes in Mathematics*, Vol. 948 (Springer-Verlag), pp. 1–46, doi:10.1007/BFb0069840.

Lax, P. D. (1968). Integrals of nonlinear equations of evolution and solitary waves, *Communications on Pure and Applied Mathematics* **21**, pp. 467–490 doi:10.1002/cpa.3160210503, https://doi.org/10.1002/cpa.3160210503.

Lee, T. D. (1954). Some special examples in renormalizable field theory, *Physical Review* **95**, p. 1329, doi:10.1103/PhysRev.95.1329, https://doi.org/10.1103/PhysRev.95.1329.

Lee, T. D. and Wick, G. C. (1969). Negative metric and the unitarity of the S-matrix, *Nuclear Physics B* **9**, pp. 209–243, doi:10.1016/0550-3213(69)90098-4, http://doi.org/10.1016/0550-3213(69)90098-4.

Lévai, G. (1989). A search for shape-invariant solvable potentials, *Journal of Physics A: Mathematical and General* **22**, pp. 689–702, doi:10.1088/0305-4470/22/6/020, https://doi.org/10.1088/0305-4470/22/6/020.

Lévai, G. (1991). A class of exactly solvable potentials related to the Jacobi polynomials, *Journal of Physics A: Mathematical and General* **24**, pp. 131–146, doi:10.1088/0305-4470/24/1/022, https://doi.org/10.1088/0305-4470/24/1/022.

Lévai, G. (1994a). Solvable potentials associated with SU(1,1) algebras: A systematic study, *Journal of Physics A: Mathematical and General* **27**, pp. 3809–3828, doi:10.1088/0305-4470/27/11/031, https://doi.org/10.1088/0305-4470/27/11/031.

Lévai, G. (1994b). Solvable potentials derived from supersymmetric quantum mechanics, in H. V. von Geramb (ed.), *Quantum inversion theory and applications, Lecture Notes in Physics*, Vol. 47 (Springer, Berlin), pp. 107–126, doi:10.1007/3-540-57576-6_7, https://doi.org/10.1007/3-540-57576-6_7.

Lévai, G. (2004). Supersymmetry without Hermiticity, *Czechoslovak Journal of Physics* **54**, pp. 1121–1124, doi:10.1023/B:CJOP.0000044013.61838.0d, https://doi.org/10.1023/B:CJOP.0000044013.61838.0d.

Lévai, G. (2006). On the pseudo-norm and admissible solutions of the $\mathcal{PT}$-symmetric Scarf I potential, *Journal of Physics A: Mathematical and General* **39**, pp. 10161–10169, doi:10.1088/0305-4470/39/32/S17, https://doi.org/10.1088/0305-4470/39/32/S17.

Lévai, G. (2007). Solvable $\mathcal{PT}$-symmetric potentials in higher dimensions, *Journal of Physics A: Mathematical and Theoretical* **40**, pp. F273–F280, doi:10.1088/1751-8113/40/15/F02, https://doi.org/10.1088/1751-8113/40/15/F02.

Lévai, G. (2008a). On the normalization constant of $\mathcal{PT}$-symmetric and real Rosen–Morse I potentials, *Physics Letters A* **372**, pp. 6484–6489, doi:10.1016/j.physleta.2008.08.073, https://doi.org/10.1016/j.physleta.2008.08.073.

Lévai, G. (2008b). Solvable $\mathcal{PT}$-symmetric potentials in 2 and 3 dimensions, *Journal of Physics: Conference Series* **128**, p. 012045, doi:10.1088/1742-6596/128/1/012045, https://doi.org/10.1088/1742-6596/128/1/012045.

Lévai, G. (2009). Spontaneous breakdown of $\mathcal{PT}$ symmetry in the complex Coulomb potential, *Pramana-Journal of Physics* **73**, pp. 329–335, doi:10.1007/s12043-009-0125-5, https://doi.org/10.1007/s12043-009-0125-5.

Lévai, G. (2011). Asymptotic properties of solvable $\mathcal{PT}$-symmetric potentials, *International Journal of Theoretical Physics* **50**, pp. 997–1004, doi:10.1007/s10773-010-0595-8, https://doi.org/10.1007/s10773-010-0595-8.

Lévai, G. (2012). Gradual spontaneous breakdown of symmetry in a solvable potential, *Journal of Physics A: Mathematical and Theoretical* **45**,

p. 444020, doi:10.1088/1751-8113/45/44/444020, `https://doi.org/10.1088/1751-8113/45/44/444020`.

Lévai, G. (2015). $\mathcal{PT}$ symmetry in Natanzon-class potentials, *International Journal of Theoretical Physics* **54**, pp. 2724–2736, doi:10.1007/s10773-014-2507-9, `https://doi.org/10.1007/s10773-014-2507-9`.

Lévai, G. (2017). Accidental crossing of energy eigenvalues in $\mathcal{PT}$-symmetric Natanzon-class potentials, *Annals of Physics* **380**, pp. 1–11, doi:10.1016/j.aop.2017.03.001, `https://doi.org/10.1016/j.aop.2017.03.001`.

Lévai, G., Baran, Á., Salamon, P., and Vertse, T. (2017). Analytical solutions for the radial Scarf II potential, *Physics Letters A* **381**, pp. 1936–1942, doi:10.1016/j.physleta.2017.04.010, `https://doi.org/10.1016/j.physleta.2017.04.0101`.

Lévai, G., Cannata, F., and Ventura, A. (2001). Algebraic and scattering aspects of a $\mathcal{PT}$-symmetric solvable potential, *Journal of Physics A: Mathematical and General* **34**, pp. 839–844, doi:10.1088/0305-4470/34/4/310, `https://doi.org/10.1088/0305-4470/34/4/310`.

Lévai, G., Cannata, F., and Ventura, A. (2002a). $\mathcal{PT}$-symmetric potentials and the SO(2, 2) algebra, *Journal of Physics A: Mathematical and General* **35**, pp. 5041–5057, doi:10.1088/0305-4470/35/24/305, `https://doi.org/10.1088/0305-4470/35/24/305`.

Lévai, G., Cannata, F., and Ventura, A. (2002b). $\mathcal{PT}$ symmetry breaking and explicit expressions for the pseudo-norm in the Scarf II potential, *Physics Letters A* **300**, pp. 271–281, doi:10.1016/S0375-9601(02)00779-X, `https://doi.org/10.1016/S0375-9601(02)00779-X`.

Lévai, G. and Magyari, E. (2009). The $\mathcal{PT}$-symmetric Rosen–Morse II potential: Effects of the asymptotically non-vanishing imaginary potential component, *Journal of Physics A: Mathematical and Theoretical* **42**, p. 195302, doi:10.1088/1751-8113/42/19/195302, `https://doi.org/10.1088/1751-8113/42/19/195302`.

Lévai, G., Siegl, P., and Znojil, M. (2009). Scattering in the $\mathcal{PT}$-symmetric Coulomb potential, *Journal of Physics A: Mathematical and Theoretical* **42**, p. 295201, doi:10.1088/1751-8113/42/29/295201, `https://doi.org/10.1088/1751-8113/42/29/295201`.

Lévai, G., Sinha, A., and Roy, P. (2003). An exactly solvable $\mathcal{PT}$ symmetric potential from the Natanzon class, *Journal of Physics A: Mathematical and General* **36**, pp. 7611–7623, doi:10.1088/0305-4470/36/27/313, `https://doi.org/10.1088/0305-4470/36/27/313`.

Lévai, G. and Znojil, M. (2000). Systematic search for $\mathcal{PT}$-symmetric potentials with real energy spectra, *Journal of Physics A: Mathematical and General* **33**, pp. 7165–7180, doi:10.1088/0305-4470/33/40/313, `https://doi.org/10.1088/0305-4470/33/40/313`.

Lévai, G. and Znojil, M. (2001). Conditions for complex spectra in a class of PT-symmetric potentials, *Modern Physics Letters A* **16**, pp. 1973–1981, doi:10.1142/S0217732301005321, `https://doi.org/10.1142/S0217732301005321`.

Lévai, G. and Znojil, M. (2002). The interplay of supersymmetry and $\mathcal{PT}$ symmetry in quantum mechanics: A case study for the Scarf II potential, *Journal of Physics A: Mathematical and General* **35**, pp. 8793–8804, doi:10.1088/0305-4470/35/41/311, https://doi.org/10.1088/0305-4470/35/41/311.

Li, K., Kevrekidis, P., Frantzeskakis, D., Rüter, C., and Kip, D. (2013). Revisiting the $\mathcal{PT}$-symmetric trimer: bifurcations, ghost states and associated dynamics, *Journal of Physics A: Mathematical and Theoretical* **46**, p. 375304, doi:10.1088/1751-8113/46/37/375304, https://doi.org/10.1088/1751-8113/46/37/375304.

Li, K., Kevrekidis, P., Malomed, B. A., and Günther, U. (2012). Nonlinear $\mathcal{PT}$-symmetric plaquettes, *Journal of Physics A: Mathematical and Theoretical* **45**, p. 444021, doi:10.1088/1751-8113/45/44/444021, https://doi.org/10.1088/1751-8113/45/44/444021.

Li, K. and Kevrekidis, P. G. (2011). $\mathcal{PT}$-symmetric oligomers: Analytical solutions, linear stability, and nonlinear dynamics, *Physical Review E* **83**, p. 066608, doi:10.1103/PhysRevE.83.066608, https://doi.org/10.1103/PhysRevE.83.066608.

Li, W., Jiang, Y., Li, C., and Song, H. (2016). Parity-time-symmetry enhanced optomechanically-induced-transparency, *Scientific Reports* **6**, p. 31095, doi:10.1038/srep31095, https://doi.org/10.1038/srep31095.

Lieb, E. H. (1967a). Exact solution of the f model of an antiferroelectric, *Physical Review Letters* **18**, pp. 1046–1048, doi:10.1103/PhysRevLett.18.1046, https://doi.org/10.1103/PhysRevLett.18.1046.

Lieb, E. H. (1967b). Exact solution of the problem of the entropy of two-dimensional ice, *Physical Review Letters* **18**, pp. 692–694, doi:10.1103/PhysRevLett.18.692, http://dx.doi.org/10.1103/PhysRevLett.18.692.

Lieb, E. H. (1967c). Exact solution of the two-dimensional Slater KDP model of a ferroelectric, *Physical Review Letters* **19**, pp. 108–110, doi:10.1103/PhysRevLett.19.108, https://doi.org/10.1103/PhysRevLett.19.108.

Lieb, E. H. (1967d). Residual entropy of square ice, *Physical Review* **162**, pp. 162–172, doi:10.1103/PhysRev.162.162, https://doi.org/10.1103/PhysRev.162.162.

Limonov, M. F., Rybin, M. V., Poddubny, A. N., and Kivshar, Y. S. (2017). Fano resonances in photonics, *Nature Photonics* **11**, pp. 543–554, doi:10.1038/nphoton.2017.142, https://doi.org/10.1038/nphoton.2017.142.

Lin, Z., Ramezani, H., Eichelkraut, T., Kottos, T., Cao, H., and Christodoulides, D. N. (2011). Unidirectional invisibility induced by $\mathcal{PT}$-symmetric periodic structures, *Physical Review Letters* **106**, p. 213901, doi:10.1103/PhysRevLett.106.213901, https://doi.org/10.1103/PhysRevLett.106.213901.

Liouville, J. (1855). Note sur l'intégration des équations différentielles de la dynamique, *Journal de Mathématiques pures et appliquées*, Vol. 20, pp. 137–138.

Liu, Z.-P., Zhang, J., Şahin Kaya Özdemir, Peng, B., Jing, H., Lü, X.-Y., Chun-Wen Li, L. Y., Nori, F., and xi Liu, Y. (2016). Metrology with $\mathcal{PT}$-symmetric cavities: Enhanced sensitivity near the $\mathcal{PT}$-phase transition,

Physical Review Letters **117**, p. 110802, doi:PhysRevLett.117.110802, https://doi.org/10.1103/PhysRevLett.117.110802.

Longhi, S. (2010a). $\mathcal{PT}$-symmetric laser absorber, *Physical Review A* **82**, p. 031801(R), doi:10.1103/PhysRevA.82.031801, https://doi.org/10.1103/ PhysRevA.82.031801.

Longhi, S. (2010b). Spectral singularities and Bragg scattering in complex crystals, *Physical Review A* **81**, p. 022102, doi:10.1103/PhysRevA.81.022102, https://doi.org/10.1103/PhysRevA.81.022102.

Longhi, S. (2011). Invisibility in $\mathcal{PT}$-symmetric complex crystals, *Journal of Physics A: Mathematical and Theoretical* **44**, p. 485302, doi:10.1088/ 1751-8113/44/48/485302, https://doi.org/10.1088/1751-8113/44/48/ 485302.

Longhi, S. (2015). Bound states in the continuum in $\mathcal{PT}$-symmetric optical lattices, *Optics Letters* **39**, pp. 1697–1700, doi:10.1364/OL.39.001697, https://doi.org/10.1364/OL.39.001697.

Longhi, S. and Della Valle, G. (2013). Transparency at the interface between two isospectral crystals, *European Physics Letters* **102**, p. 40008, doi:10.1209/ 0295-5075/102/40008, https://doi.org/10.1209/0295-5075/102/40008.

Longhi, S. and Della Valle, G. (2014). Optical lattices with exceptional points in the continuum, *Physical Review A* **89**, p. 052132, doi:10.1103/PhysRevA. 89.052132, https://doi.org/10.1103/PhysRevA.89.052132.

López-Ortega, A. (2015). New conditionally exactly solvable inverse power law potentials, *Physics Letters A* **90**, p. 085202, doi:10.1088/0031-8949/90/8/ 085202, https://doi.org/10.1088/0031-8949/90/8/085202.

Lü, X.-Y., Jing, H., Ma, J.-Y., and Wu, Y. (2015). $\mathcal{PT}$-symmetry-breaking chaos in optomechanics, *Physical Review Letters* **114**, p. 253601, doi:10. 1103/physrevlett.114.253601, https://doi.org/10.1103/physrevlett.114.253601.

Lukyanov, S. and Zamolodchikov, A. (2010). Quantum sine(h)-Gordon model and classical integrable equations, *Journal of High Energy Physics* **7**, p. 008, doi: 10.1007/JHEP07(2010)008, https://doi.org/10.1007/JHEP07(2010)008.

Lukyanov, S. L. (2013). ODE/IM correspondence for the Fateev model, *Journal of High Energy Physics* **12**, p. 012, doi:10.1007/JHEP12(2013)012, https://doi.org/10.1007/JHEP12(2013)012.

MacDonald, A. D. (1948). *Properties of the Confluent Hypergeometric Function* (Research Laboratory of Electronics, MIT).

Makris, K., El-Ganainy, R., Christodoulides, D., and Musslimani, Z. (2008). Beam dynamics in $\mathcal{PT}$ symmetric optical lattices, *Physical Review Letters* **100**, p. 103904, doi:10.1103/PhysRevLett.100.103904, https://doi.org/10.1103/PhysRevLett.100.103904.

Makris, K. G., El-Ganainy, R., Christodoulides, D. N., and Musslimani, Z. H. (2010). $\mathcal{PT}$-symmetric optical lattices, *Physical Review A* **81**, p. 063807, doi:10.1103/PhysRevA.81.063807, https://doi.org/10.1103/ PhysRevA.81.063807.

Makris, K. G., Musslimani, Z. H., Christodoulides, D. N., and Rotter, S. (2015). Constant-intensity waves and their modulation instability in non-Hermitian potentials, *Nature Communications* **6**, p. 7257, doi:10.1038/ncomms8257, https://doi.org/10.1038/ncomms8257.

Mathieu, P. (1988). Supersymmetric extension of the Korteweg–de Vries equation, *Journal of Mathematical Physics* **29**, 11, pp. 2499–2506, doi:10.1063/1.528090, https://doi.org/10.1063/1.528090.

Mezincescu, G. A. (2000). Some properties of eigenvalues and eigenfunctions of the cubic oscillator with imaginary coupling constant, *Journal of Physics A: Mathematical and General* **33**, pp. 4911–4916, doi:10.1088/0305-4470/33/27/308, https://doi.org/10.1088/0305-4470/33/27/308.

Midya, B. (2014). Supersymmetry-generated one-way-invisible $\mathcal{PT}$-symmetric optical crystals, *Physical Review A* **89**, p. 032116, doi:10.1103/PhysRevA.89.032116, https://doi.org/10.1103/PhysRevA.89.032116.

Midya, B. and Roy, B. (2013). Infinite families of (non)-Hermitian Hamiltonians associated with exceptional x_m Jacobi polynomials, *Journal of Physics A: Mathematical and Theoretical* **46**, p. 175201, doi:10.1088/1751-8113/46/17/175201, https://doi.org/10.1088/1751-8113/46/17/175201.

Miller Jr., W. (1968). *Lie Theory of Special Functions* (Academic Press, New York).

Millican-Slater, A. (2004). *Aspects of $\mathcal{PT}$-symmetric Quantum Mechanics*, Ph.D. thesis, Durham.

Milton, K. A. (2001). *The Casimir Effect: Physical Manifestations of Zero-Point Energy* (World Scientific, Singapore), ISBN 981-02-4397-9.

Milton, K. A. (2004). Anomalies in $\mathcal{PT}$-symmetric quantum field theory, *Czechoslovak Journal of Physics* **54**, pp. 85–91, doi:10.1023/B:CJOP.0000014372.21537.c0, https://doi.org/10.1023/B:CJOP.0000014372.21537.c0.

Miri, M.-A., Heinrich, M., and Christodoulides, D. N. (2013). Supersymmetry-generated complex optical potentials with real spectra, *Physical Review A* **87**, p. 043819, doi:10.1103/PhysRevA.87.043819, https://doi.org/10.1103/PhysRevA.87.043819.

Moffat, J. W. (2005). Charge conjugation invariance of the vacuum and the cosmological constant problem, *Physics Letters B* **627**, pp. 9–17, doi:10.1016/j.physletb.2005.09.012, https://doi.org/10.1016/j.physletb.2005.09.012.

Moffat, J. W. (2006). Positive and negative energy symmetry and the cosmological constant problem, *ArXiv*, hep–th/0610162, http://arxiv.org/abs/hep-th/0610162.

Moser, J. (1975). Three integrable Hamiltonian systems connected with isospectral deformations, *Advances in Mathematics* **16**, pp. 197–220.

Mostafazadeh, A. (2002a). Pseudo-Hermiticity for a class of nondiagonalizable Hamiltonians, *Journal of Mathematical Physics* **43**, pp. 6343–6352, doi:10.1063/1.1514834, https://doi.org/10.1063/1.1514834.

Mostafazadeh, A. (2002b). Pseudo-Hermiticity versus $\mathcal{PT}$ symmetry: The necessary condition for the reality of the spectrum of a non-Hermitian Hamiltonian, *Journal of Mathematical Physics* **43**, pp. 205–214, doi:10.1063/1.1418246, https://doi.org/10.1063/1.1418246.

Mostafazadeh, A. (2002c). Pseudo-Hermiticity versus PT-symmetry. II. A complete characterization of non-Hermitian Hamiltonians with a real spectrum, *Journal of Mathematical Physics* **43**, pp. 2814–2816, doi:10.1063/1.1461427, https://doi.org/10.1063/1.1461427.

Mostafazadeh, A. (2002d). Pseudo-Hermiticity versus PT-symmetry III: Equivalence of pseudo-Hermiticity and the presence of antilinear symmetries, *Journal of Mathematical Physics* **43**, pp. 3944–3951, doi:10.1063/1.1489072, https://doi.org/10.1063/1.1489072.

Mostafazadeh, A. (2002e). Pseudo-supersymmetric quantum mechanics and isospectral pseudo-Hermitian Hamiltonians, *Nuclear Physics B* **640**, pp. 419–434, doi:10.1016/S0550-3213(02)00347-4, https://doi.org/10.1016/S0550-3213(02)00347-4.

Mostafazadeh, A. (2003a). Erratum: Pseudo-Hermiticity for a class of nondiagonalizable Hamiltonians [J. Math. Phys. 43, 6343 (2002)], *Journal of Mathematical Physics* **44**, p. 943, doi:10.1063/1.1540714, https://doi.org/10.1063/1.1540714.

Mostafazadeh, A. (2003b). Exact PT-symmetry is equivalent to Hermiticity, *Journal of Physics A: Mathematical and General* **36**, pp. 7081–7092, doi:10.1088/0305-4470/36/25/312, https://doi.org/10.1088/0305-4470/36/25/312.

Mostafazadeh, A. (2003c). Pseudo-Hermiticity and generalized PT- and CPT-symmetries, *Journal of Mathematical Physics* **44**, pp. 974–989, doi:10.1063/1.1539304, https://doi.org/10.1063/1.1539304.

Mostafazadeh, A. (2005). Pseudo-Hermitian description of PT-symmetric systems defined on a complex contour, *Journal of Physics A: Mathematical and General* **38**, pp. 3213–3234, doi:10.1088/0305-4470/38/14/011, https://doi.org/10.1088/0305-4470/38/14/011.

Mostafazadeh, A. (2009). Spectral singularities of complex scattering potentials and infinite reflection and transmission coefficients at real energies, *Physical Review Letters* **102**, p. 220402, doi:10.1103/PhysRevLett.102.220402, https://doi.org/10.1103/PhysRevLett.102.220402.

Mostafazadeh, A. (2010). Pseudo-Hermitian representation of quantum mechanics, *International Journal of Geometric Methods in Modern Physics* **7**, pp. 1191–1306, doi:10.1142/S0219887810004816, https://doi.org/10.1142/S0219887810004816.

Mostafazadeh, A. (2014a). A dynamical formulation of one-dimensional scattering theory and its applications in optics, *Annals of Physics* **341**, pp. 77–85, doi:10.1016/j.aop.2013.11.008, https://doi.org/10.1016/j.aop.2013.11.008.

Mostafazadeh, A. (2014b). Unidirectionally invisible potentials as local building blocks of all scattering potentials, *Physical Review Letters* **90**, p. 023833, doi:10.1103/PhysRevA.90.023833, https://doi.org/10.1103/PhysRevA.90.023833.

Mostafazadeh, A. and Batal, A. (2004). Physical aspects of pseudo-Hermitian and PT-symmetric quantum mechanics, *Journal of Physics A: Mathematical and General* **37**, p. 11645, doi:10.1088/0305-4470/37/48/009, https://doi.org/10.1088/0305-4470/37/48/009.

Muga, J., Palao, J., Navarro, B., and Egusquiza, L. (2004). Complex absorbing potentials, *Physics Reports* **395**, 6, pp. 357–426, doi:10.1016/j.physrep. 2004.03.002, https://doi.org/10.1016/j.physrep.2004.03.002.

Musslimani, Z., Makris, K., El-Ganainy, R., and Christodoulides, D. (2008a). Optical solitons in $\mathcal{PT}$ periodic potentials, *Physical Review Letters* **100**, p. 030402, doi:10.1103/PhysRevLett.100.030402, https://doi.org/10.1103/PhysRevLett.100.030402.

Musslimani, Z. H., Makris, K. G., El-Ganainy, R., and Christodoulides, D. N. (2008b). Analytical solutions to a class of nonlinear Schrödinger equations with $\mathcal{PT}$-like potentials, *Journal of Physics A: Mathematical and Theoretical* **41**, p. 244019, doi:10.1088/1751-8113/41/24/244019, https://doi.org/10.1088/1751-8113/41/24/244019.

Naboko, S. (1984). Conditions for similarity to unitary and self-adjoint operators, *Functional Analysis and Its Applications* **18**, pp. 13–22, doi: 10.1007/BF01076357, https://doi.org/10.1007/BF01076357.

Nanayakkara, A. (2004a). Classical motion of complex 2-D non-Hermitian Hamiltonian systems, *Czechoslovak Journal of Physics* **54**, pp. 101–107, doi: 10.1023/B:CJOP.0000014374.61647.55, http://doi.org/10.1023/B:CJOP.0000014374.6164755.

Nanayakkara, A. (2004b). Classical trajectories of 1D complex non-Hermitian Hamiltonian systems, *Journal of Physics A: Mathematical and General* **37**, pp. 4321–4334, doi:10.1088/0305-4470/37/15/002, http://doi.org/10.1088/0305-4470/37/15/002.

Natanzon, G. A. (1971). The study of one-variable Schrödinger equation generated by hypergeometric equation, *Vestnik Leningradskogo Universiteta, Fizika i Khimiya* **2**, pp. 22–8.

Natanzon, G. A. (1979). General properties of potentials for which the Schrödinger equation can be solved by means of hypergeometric functions, *Theoretical and Mathematical Physics* **38**, pp. 146–153, doi:10.1007/BF01016836, https://doi.org/10.1007/BF01016836.

Negro, S. (2017). ODE/IM correspondence in Toda field theories and fermionic basis in sin(h)-Gordon model, *ArXiv e-prints* arXiv:1702.06657, http://arxiv.org/abs/1702.06657.

Nesemann, J. (2011). *$\mathcal{PT}$-Symmetric Schrödinger Operators with Unbounded Potentials* (Vieweg-Teubner Verlag, Heidelberg).

Nixon, S., Ge, L., and Yang, J. (2012a). Stability analysis for solitons in $\mathcal{PT}$-symmetric optical lattices, *Physical Review A* **85**, p. 023822, doi:10.1103/PhysRevA.85.023822, https://doi.org/10.1103/PhysRevA.85.023822.

Nixon, S., Zhu, Y., and Yang, J. (2012b). Nonlinear dynamics of wave packets in parity-time-symmetric optical lattices near the phase transition point, *Optics Letters* **37**, pp. 4874–4876, doi:10.1364/OL.37.004874, https://doi.org/10.1364/OL.37.004874.

Ogilvie, M. and Meisinger, P. (2007). $\mathcal{PT}$-symmetric matrix quantum mechanics, *ArXiv*, hep-th/0701207, http://arxiv.org/abs/hep-th/0701207.

Ohlsson, T. (2016). Non-Hermitian neutrino oscillations in matter with $\mathcal{PT}$ symmetric Hamiltonians, *Europhysics Letters* **113**, p. 61001, doi:10.1209/0295-5075/113/61001, https://doi.org/10.1209/0295-5075/113/61001.

Okamoto, K. (1986). Studies on the Painlevé equations. III. Second and fourth Painlevé equations PII and PIV. *Mathematische Annalen* **275**, pp. 221–256, http://eudml.org/doc/164145.

Olshanetsky, M. A. and Perelomov, A. M. (1977). Quantum completely integrable systems connected with semi-simple Lie algebras, *Letters in Mathematical Physics* **2**, pp. 7–13, doi:10.1007/BF00420664, https://doi.org/10.1007/BF00420664.

Olshanetsky, M. A. and Perelomov, A. M. (1981). Classical integrable finite dimensional systems related to Lie algebras, *Physics Reports* **71**, pp. 313–400, doi:10.1016/0370-1573(81)90023-5, https://doi.org/10.1016/0370-1573(81)90023-5.

Olver, F. (1974). *Asymptotics and Special Functions* (Taylor and Francis).

Painlevé, P. (1900). Mémoire sur les équations différentielles dont l'intégrale générale est uniforme, *Bulletin de la Société Mathématique de France* **28**, pp. 201–261, http://www.numdam.org/item?id=BSMF_1900__28__201_0.

Pauli, W. (1943). On Dirac's new method of field quantization, *Reviews of Modern Physics* **15**, pp. 175–207, doi:10.1103/RevModPhys.15.175, https://doi.org/10.1103/RevModPhys.15.175.

Pendry, J., Schurig, D., and Smith, D. (2006). Controlling electromagnetic fields, *Science* **312**, pp. 1780–1782, doi:10.1126/science.1125907, https://doi.org/10.1126/science.1125907.

Pendry, J. B. (2000). Negative refraction makes a perfect lens, *Physical Review Letters* **85**, pp. 3966–3969, doi:10.1103/PhysRevLett.85.3966, https://doi.org/10.1103/PhysRevLett.85.3966.

Peng, B., Şahin Kaya Özdemir, Lei, F., Monifi, F., Gianfreda, M., Long, G. L., Fan, S., Nori, F., Bender, C. M., and Yang, L. (2014). Parity-time-symmetric whispering-gallery microcavities, *Nature Physics* **10**, pp. 394–398, doi:10.1038/nphys2927, https://doi.org/10.1038/nphys2927.

Pile, D. F. P. (2017). Gaining with loss, *Nature Photonics* **11**, p. 742, doi:10.1038/s41566-017-0060-9, https://doi.org/10.1038/s41566-017-0060-9.

Poladian, L. (1996). Resonance mode expansions and exact solutions for nonuniform gratings, *Physical Review E* **54**, pp. 2963–2975, doi:10.1103/PhysRevE.54.2963, https://doi.org/10.1103/PhysRevE.54.2963.

Quesne, C. (2008). Exceptional orthogonal polynomials, exactly solvable potentials and supersymmetry, *Journal of Physics A: Mathematical and Theoretical* **41**, p. 392001, doi:10.1088/1751-8113/41/39/392001, https://doi.org/10.1088/1751-8113/41/39/392001.

Quesne, C. (2009). Solvable rational potentials and exceptional orthogonal polynomials in supersymmetric quantum mechanics, *SIGMA* **5**, p. 084, doi:10.3842/SIGMA.2009.084, https://doi.org/10.3842/SIGMA.2009.084.

Ralston, J. P. (2007). $\mathcal{PT}$ and $\mathcal{CPT}$ quantum mechanics embedded in symplectic quantum mechanics, *Journal of Physics A: Mathematical and Theoretical* **40**, pp. 9883–9904, doi:10.1088/1751-8113/40/32/013, https://doi.org/10.1088/1751-8113/40/32/013.

Ramezani, H., Kottos, T., El-Ganainy, R., and Christodoulides, D. N. (2010). Unidirectional nonlinear $\mathcal{PT}$-symmetric optical structures, *Physical Review*

A **82**, p. 043803, doi:10.1103/PhysRevA.82.043803, `https://doi.org/10.1103/PhysRevA.82.043803`.

Rechtsman, M. C., Zeuner, J. M., Tünnermann, A., Nolte, S., Segev, M., and Szameit, A. (2012). Strain-induced pseudomagnetic field and photonic Landau levels in dielectric structures, *Nature Photonics* **84**, pp. 153–158, doi:10.1038/nphoton.2012.302, `https://doi.org/10.1038/nphoton.2012.302`.

Regensburger, A., Bersch, C., Miri, M.-A., Onishchukov, G., Christodoulides, D. N., and Peschel, U. (2012). Parity-time synthetic photonic lattices, *Nature* **488**, pp. 167–171, doi:10.1038/nature11298, `https://doi.org/10.1038/nature11298`.

Ren, J., Hodaei, H., Harari, G., Hassan, A. U., Chow, W., Soltani, M., Christodoulides, D., and Khajavikhan, M. (2017a). Ultrasensitive microscale parity-time-symmetric ring laser gyroscope, *Optics Letters* **42**, pp. 1556–1559, doi:10.1364/OL.42.001556, `https://doi.org/10.1364/OL.42.001556`.

Ren, J., Hodaei, H., Harari, G., Hassan, A. U., Chow, W., Soltani, M., Christodoulides, D., and Khajavikhan, M. (2017b). Ultrasensitive microscale parity-time-symmetric ring laser gyroscope, *Optics Letters* **42**, pp. 1556–1559, doi:10.1364/OL.42.001556, `https://doi.org/10.1364/OL.42.001556`.

Rosenau, P. and Hyman, J. M. (1993). Compactons: Solitons with finite wavelength, *Physical Review Letters* **70**, 5, pp. 564–567, doi:10.1103/PhysRevLett.70.564, `https://doi.org/10.1103/PhysRevLett.70.564`.

Roychoudhury, R. and Roy, P. (2007). Construction of the $\mathcal{C}$ operator for a $\mathcal{PT}$ symmetric model, *Journal of Physics: Mathematical and Theoretical* **40**, pp. F617–F620, doi:10.1088/1751-8113/40/27/F06, `https://doi.org/10.1088/1751-8113/40/27/F06`.

Roychoudhury, R., Roy, P., Znojil, M., and Lévai, G. (2001). Comprehensive analysis of conditionally exactly solvable models, *Journal of Mathematical Physics* **42**, pp. 1996–2007, doi:10.1063/1.1362286, `https://doi.org/10.1063/1.1362286`.

Rüter, C. E., Makris, K. G., El-Ganainy, R., Christodoulides, D. N., Segev, M., and Kip, D. (2010). Observation of parity-time symmetry in optics, *Nature Physics* **6**, pp. 192–195, doi:10.1038/nphys1515, `https://doi.org/10.1038/nphys1515`.

Saff, E. B. and Snider, A. D. (1993). *Fundamentals of Complex Analysis for Mathematics, Science, and Engineering* (Prentice Hall, New Jersey), ISBN 978-0133274615, see Chapter 8, Section 8.5, problem 8.

Sakhdari, M., Farhat, M., and Chen, P.-Y. (2017). $\mathcal{PT}$-symmetric metasurfaces: wave manipulation and sensing using singular points, *New Journal of Physics* **19**, p. 065002, doi:10.1088/1367-2630/aa6bb9, `https://doi.org/10.1088/1367-2630/aa6bb9`.

Schindler, J., Li, A., Zheng, M. C., Ellis, F. M., and Kottos, T. (2011). Experimental study of active lrc circuits with $\mathcal{PT}$ symmetries, *Physical Review A* **84**, p. 040101(R), doi:10.1103/PhysRevA.84.040101, `https://doi.org/10.1103/PhysRevA.84.040101`.

Schindler, S. T. and Bender, C. M. (2018). Winding in non-Hermitian systems, *Journal of Physics A: Mathematical and Theoretical* **51**, p. 055201, doi:10.1088/1751-8121/aa9faf, https://dx.doi.org/10.1088/1751-8121/aa9faf.

Scholtz, F., Geyer, H., and Hahne, F. (1992). Quasi-Hermitian operators in quantum mechanics and the variational principle, *Annals of Physics* **213**, 1, pp. 74–101, doi:10.1016/0003-4916(92)90284-S, https://doi.org/10.1016/0003-4916(92)90284-S.

Scholtz, F. G. and Geyer, H. B. (2006a). Moyal products-a new perspective on quasi-Hermitian quantum mechanics, *Journal of Physics A: Mathematical and General* **39**, pp. 10189–10205, doi:10.1088/0305-4470/39/32/S19, https://doi.org/10.1088/0305-4470/39/32/S19.

Scholtz, F. G. and Geyer, H. B. (2006b). Operator equations and Moyal products–metrics in quasi-Hermitian quantum mechanics, *Physics Letters B* **634**, pp. 84–92, doi:10.1016/j.physletb.2006.01.022, https://doi.org/10.1016/j.physletb.2006.01.022.

Schomerus, H. (2010). Quantum noise and self-sustained radiation of $\mathcal{PT}$-symmetric systems, *Physical Review Letters* **104**, p. 233601, doi:10.1103/physrevlett.104.233601, https://doi.org/10.1103/physrevlett.104.233601.

Schrödinger, E. (1940a). Further studies on solving eigenvalue problems by factorization, *Proceedings of the Royal Irish Academy. Section A* **46**, pp. 183–206, http://www.jstor.org/stable/20490756.

Schrödinger, E. (1940b). A method of determining quantum-mechanical eigenvalues and eigenfunctions, *Proceedings of the Royal Irish Academy. Section A* **46**, pp. 9–16, http://www.jstor.org/stable/20490744.

Schrödinger, E. (1941). The factorization of the hypergeometric equation, *Proceedings of the Royal Irish Academy. Section A* **47**, pp. 53–54, http://www.jstor.org/stable/20488434.

Schurig, D., Mock, J. J., Justice, B. J., Cummer, S. A., Pendry, J. B., Starr, A. F., and Smith, D. R. (2006). Metamaterial electromagnetic cloak at microwave frequencies, *Science* **314**, pp. 977–980, doi:10.1126/science.1133628, https://doi.org/10.1126/science.1133628.

Schweber, S. S. (1961). *An Introduction to Relativistic Quantum Field Theory* (Harper & Row, New York).

Shapiro, B. and Tater, M. (2014). On spectral asymptotic of quasi-exactly solvable quartic and Yablonskii-Vorob'ev polynomials, *ArXiv* http://arxiv.org/abs/1412.3026.

Shapiro, B. and Tater, M. (2017). Asymptotics and monodromy of the algebraic spectrum of quasi-exactly solvable sextic oscillator, *Experimental Mathematics* **0**, 0, pp. 1–8, doi:10.1080/10586458.2017.1325792, https://doi.org/10.1080/10586458.2017.1325792.

Shin, K. C. (1998). Eigenvalues of complex Hamiltonians with $\mathcal{PT}$-symmetry. I, *Physics Letter A* **250**, pp. 25–28, doi:10.1016/S0375-9601(98)00791-9, https://doi.org/10.1016/S0375-9601(98)00791-9.

Shin, K. C. (2001). On the eigenproblems of $\mathcal{PT}$-symmetric oscillators, *Journal of Mathematical Physics* **42**, pp. 2513–2530, doi:10.1063/1.1366328, https://doi.org/10.1063/1.1366328.

Shin, K. C. (2002). On the reality of the eigenvalues for a class of $\mathcal{PT}$-symmetric oscillators, *Communications in Mathematical Physics* **229**, pp. 543–564, doi:10.1007/s00220-002-0706-3, https://doi.org/10.1007/s00220-002-0706-3.

Shin, K. C. (2004). On the shape of spectra for non-self-adjoint periodic Schrödinger operators, *Journal of Physics A: Mathematical and General* **37**, pp. 8287–8291, doi:10.1088/0305-4470/37/34/007, https://doi.org/10.1088/0305-4470/37/34/007.

Shin, K. C. (2005a). Eigenvalues of $\mathcal{PT}$-symmetric oscillators with polynomial potentials, *Journal of Physics A: Mathematical and General* **38**, pp. 6147–6166, doi:10.1088/0305-4470/38/27/005, https://doi.org/10.1088/0305-4470/38/27/005.

Shin, K. C. (2005b). The potential $(iz)^m$ generates real eigenvalues only, under symmetric rapid decay boundary conditions, *Journal of Mathematical Physics* **46**, p. 082110, doi:10.1063/1.2009667, https://doi.org/10.1063/1.2009667.

Sibuya, Y. (1975). *Global Theory of a Second Order Linear Ordinary Differential Equation with a Polynomial Coefficient* (North-Holland, Amsterdam).

Siegl, P. and Krejčiřík, D. (2012). On the metric operator for the imaginary cubic oscillator, *Physical Review D* **86**, p. 121702, doi:10.1103/PhysRevD.86.121702, https://doi.org/10.1103/PhysRevD.86.121702.

Simon, B. and Dicke, A. (1970). Coupling constant analyticity for the anharmonic oscillator, *Annals of Physics* **58**, 1, pp. 76–136, doi:10.1016/0003-4916(70)90240-X, https://doi.org/10.1016/0003-4916(70)90240-X.

Şimşek, M. and Eğrifes, H. (2004). The Klein–Gordon equation of generalized Hulthén potential in complex quantum mechanics, *Journal of Physics A: Mathematical and General* **37**, pp. 4379–4393, doi:10.1088/0305-4470/37/15/007, https://doi.org/10.1088/0305-4470/37/15/007.

Sinha, A. (2012). Scattering states of a particle, with position-dependent mass, in a $\mathcal{PT}$-symmetric heterojunction, *Journal of Physics A: Mathematical and Theoretical* **45**, p. 185305, doi:10.1088/1751-8113/45/18/185305, https://doi.org/10.1088/1751-8113/45/18/185305.

Sinha, A., Lévai, G., and Roy, P. (2004). $\mathcal{PT}$ symmetry of a conditionally exactly solvable potential, *Physics Letters A* **322**, pp. 78–83, doi:10.1016/j.physleta.2004.01.009, https://doi.org/10.1016/j.physleta.2004.01.009.

Smilga, A. V. (2008a). Cryptogauge symmetry and cryptoghosts for crypto-Hermitian Hamiltonians, *Journal of Physics A: Mathematical and Theoretical* **41**, p. 244026, doi:10.1088/1751-8113/41/24/244026, http://doi.org/10.1088/1751-8113/41/24/244026.

Smilga, A. V. (2008b). Physics of crypto-Hermitian and crypto-supersymmetric field theories, *Physical Review D* **77**, p. 061701(R), doi:10.1103/PhysRevD.77.061701, http://doi.org/10.1103/PhysRevD.77.061701.

Sorrell, M. (2007). Complex WKB analysis of a $\mathcal{PT}$-symmetric eigenvalue problem, *Journal of Physics A: Mathematical and Theoretical* **40**, pp. 10319–10336, doi:10.1088/1751-8113/40/33/023, https://doi.org/ 10.1088/1751-8113/40/33/023.

Sounas, D. L., Fleury, R., and Alú, A. (2015). Unidirectional cloaking based on metasurfaces with balanced loss and gain, *Physical Review Applied* **4**, p. 014005, doi:10.1103/PhysRevApplied.4.014005, https://doi.org/10. 1103/PhysRevApplied.4.014005.

Srivastava, S. and Jain, S. (2013). Pseudo-Hermitian random matrix theory, *Fortschritte der Physik* **61**, pp. 276–290, doi:10.1002/prop.201200107, https://doi.org/10.1002/prop.201200107.

Streater, R. F. and Wightman, A. S. (2000). *PCT, Spin and Statistics, and All That* (Princeton, New Jersey).

Strutt, J. W. (1878). *The Theory of Sound II* (Macmillan and Co, London).

Suchkov, S. V., Sukhorukov, A. A., Huang, J., Dmitriev, S. V., Lee, C., and Kivshar, Y. S. (2016). Nonlinear switching and solitons in $\mathcal{PT}$-symmetric photonic systems, *Laser and Photonics Reviews* **10**, pp. 177–213, doi:10. 1002/lpor.201500227, https://doi.org/10.1002/lpor.201500227.

Sudarshan, E. C. G. (1961). Quantum mechanical systems with indefinite metric. I, *Physical Review* **123**, pp. 2183–2193, doi:10.1103/PhysRev.123.2183, http://doi.org/10.1103/PhysRev.123.2183.

Sukhorukov, A. A., Xu, Z., and Kivshar, Y. S. (2010). Nonlinear suppression of time reversals in $\mathcal{PT}$-symmetric optical couplers, *Physical Review A* **82**, p. 043818, doi:10.1103/PhysRevA.82.043818, https://doi.org/10.1103/ PhysRevA.82.043818.

Sutherland, B. (1967). Exact solution of a two-dimensional model for hydrogen-bonded crystals, *Physical Review Letters* **19**, pp. 103–104, doi:10.1103/ PhysRevLett.19.103, https://doi.org/10.1103/PhysRevLett.19.103.

Sutherland, B. (1971a). Quantum many-body problem in one dimension: Ground state, *Journal of Mathematical Physics* **12**, pp. 246–250, doi:10.1016/0001-8708(75)90151-6, https://doi.org/10.1016/ 0001-8708(75)90151-6.

Sutherland, B. (1971b). Quantum many-body problem in one dimension: Thermodynamics, *Journal of Mathematical Physics* **12**, pp. 251–256, doi: 10.1063/1.1665585, https://doi.org/10.1063/1.1665585.

Suzuki, J. (2000). Functional relations in Stokes multipliers and solvable models related to $u_q(a^{(1)n})$, *Journal of Physics A: Mathematical and General* **33**, pp. 3507–3521, doi:10.1088/0305-4470/33/17/308, https://doi.org/ 10.1088/0305-4470/33/17/308.

Suzuki, J. (2001). Functional relations in Stokes multipliers: Fun with $x^6 + \alpha x^2$ potential, *Journal of Statistical Physics* **102**, pp. 1029–1047, doi:10.1023/A: 1004823608260, https://doi.org/10.1023/A:1004823608260.

Swanson, M. S. (2004). Transition elements for a non-Hermitian quadratic Hamiltonian, *Journal of Mathematical Physics* **45**, pp. 585–601, doi:10.1063/1. 1640796, http://doi.org/10.1063/1.1640796.

Symanzik, K. (1971a). Small-distance-behaviour analysis and Wilson expansions, *Communications in Mathematical Physics* **23**, pp. 49–86, doi:10. 1007/BF01877596, https://doi.org/10.1007/BF01877596.

Symanzik, K. (1971b). Small-distance behaviour in field theory, in *Strong Interaction Physics*, Vol. 57 (Springer, Berlin, Heidelberg), pp. 222–236.

Symanzik, K. (1973). Field-theory with cumputable large-moment behavior, *Lettere Al Nuovov Cimento* **6**, pp. 77–80, doi:10.1007/BF02788323, https://doi.org/10.1007/BF02788323.

Szameit, A., Rechtsman, M. C., Bahat-Treidel, O., and Segev, M. (2011). $\mathcal{PT}$-symmetry in honeycomb photonic lattices, *Physical Review A* **84**, p. 021806, doi:10.1103/physreva.84.021806, https://doi.org/10.1103/physreva.84.021806.

Takhtajan, L. (2008). *Quantum Mechanics for Mathematicians* (American Mathematical Society, Providence, Rhode Island).

Tanaka, T. (2006). General aspects of $\mathcal{PT}$-symmetric and $\mathcal{P}$-self-adjoint quantum theory in a Kreĭn space, *Journal of Physics: General and Mathematical* **39**, pp. 14175–14203, doi:10.1088/0305-4470/39/45/025, https://doi.org/10.1088/0305-4470/39/45/025.

ter Haar, D. (1975). *Problems in Quantum Mechanics* (Pion, London).

Tichy, V. and Skála, L. (2010). Analytic wave functions and energies for two-dimensional $\mathcal{PT}$-symmetric quartic potentials, *Central European Journal of Physics* **8**, pp. 519–522, doi:10.2478/s11534-009-0127-4, https://doi.org/10.2478/s11534-009-0127-4.

Titchmarsh, E. (1932). *The theory of functions* (OUP).

Trefethen, L. N. and Bau, D. (1997). *Numerical Linear Algebra* (Society for Industrial and Applied Mathematics, Philadelphia), ISBN 0-89871-361-7.

Trinh, D. T. (2005). Remarks on $\mathcal{PT}$-pseudo-norm in $\mathcal{PT}$-symmetric quantum mechanics, *Journal of Physics A: Mathematical and General* **38**, pp. 3665–3678, doi:10.1088/0305-4470/38/16/014, https://doi.org/10.1088/0305-4470/38/16/014.

Turbiner, A. V. (2016). One-dimensional quasi-exactly solvable Schrödinger equations, *Physics Reports* **642**, pp. 1–71, doi:10.1016/j.physrep.2016.06.002, https://doi.org/10.1016/j.physrep.2016.06.002.

Ushveridze, A. (1994). *Quasi-exactly solvable models in quantum mechanics* (IOP, Bristol).

Veselago, V. G. (1968). The electrodynamics of substances with simultaneously negative values of ϵ and μ, *Soviet Physics Uspekhi* **10**, pp. 509–514, doi:10.1070/PU1968v010n04ABEH003699, https://doi.org/10.1070/PU1968v010n04ABEH003699.

von Neumann, J. (1996). *Mathematical Foundations of Quantum Mechanics* (Princeton University Press, Princeton, New Jersey).

von Roos, O. (1983). Position-dependent effective masses in semiconductor theory, *Physical Review B* **27**, pp. 7547–7552, doi:10.1103/PhysRevB.27.7547, https://doi.org/10.1103/PhysRevB.27.7547.

Voros, A. (1982). Semi-classical correspondence and exact results : the case of the spectra of homogeneous Schrödinger operators, *Journal de Physique Lettres*

43, 1, pp. 1–4, doi:10.1051/jphyslet:019820043010100, https://doi.org/10.1051/jphyslet:019820043010100.

Voros, A. (1983). The return of the quartic oscillator. the complex WKB method, *Annales de l'I.H.P. Physique théorique* **39**, 3, pp. 211–338, http://eudml.org/doc/76217.

Voros, A. (1999). Exact resolution method for general 1D polynomial Schrödinger equation, *Journal of Physics A: Mathematical and General* **32**, pp. 5993–6007, doi:10.1088/0305-4470/32/32/311, https://doi.org/10.1088/0305-4470/32/32/311, [Corrigendum: J. Phys. A33 (2000) 5783-5784].

Weigert, S. (2003a). Completeness and orthonormality in PT-symmetric quantum systems, *Physical Review A* **68**, p. 062111, doi:10.1103/PhysRevA.68.062111, https://doi.org/10.1103/PhysRevA.68.062111.

Weigert, S. (2003b). PT-symmetry and its spontaneous breakdown explained by anti-linearity, *Journal of Optics B: Quantum and Semiclassical Optics* **5**, pp. S416–S419, doi:10.1088/1464-4266/5/3/380, https://doi.org/10.1088/1464-4266/5/3/380.

Weigert, S. (2005). An algorithmic test for diagonalizability of finite-dimensional PT-invariant systems, *Journal of Physics A: Mathematical and General* **39**, pp. 235–245, doi:10.1088/0305-4470/39/1/017, https://doi.org/10.1088/0305-4470/39/1/017.

Weigert, S. (2006). Detecting broken PT-symmetry, *Journal of Physics A: Mathematical and General* **39**, pp. 10239–10246, doi:10.1088/0305-4470/39/32/S22, https://doi.org/10.1088/0305-4470/39/32/S22.

Weimann, S., Kremer, M., Plotnik, Y., Lumer, Y., Nolte, S., Makris, K. G., Segev, M., Rechtsman, M. C., and Szameit, A. (2016). Topologically protected bound states in photonic parity-time-symmetric crystals, *Nature Materials* **16**, pp. 433–438, doi:10.1038/nmat4811, https://doi.org/10.1038/nmat4811.

Weiss, J., Tabor, M., and Carnevale, G. (1983). The Painlevé property for partial differential equations, *Journal of Mathematical Physics* **24**, pp. 522–526, doi:10.1063/1.525721, https://doi.org/10.1063/1.525721.

Witten, E. (1981). Dynamical breaking of supersymmetry, *Nuclear Physics B* **188**, pp. 513–554, doi:10.1016/0550-3213(81)90006-7, https://doi.org/10.1016/0550-3213(81)90006-7.

Witten, E. (2001). Quantum gravity in de Sitter space, *ArXiv*, hep-th/0106109, http://arxiv.org/abs/hep-th/0106109.

Wolfram Research (2015). NDEigensystem, http://reference.wolfram.com/language/ref/NDEigensystem.html.

Wong, Z. J., Xu, Y.-L., Kim, J., O'Brien, K., Wang, Y., Feng, L., and Zhang, X. (2016). Lasing and anti-lasing in a single cavity, *Nature Photonics* **10**, pp. 796–801, doi:10.1038/nphoton.2016.216, https://doi.org/10.1038/nphoton.2016.216.

Xu, H., Mason, D., Jiang, L., and Harris, J. G. E. (2016). Topological energy transfer in an optomechanical system with exceptional points, *Nature* **537**, pp. 80–83, doi:10.1038/nature18604, https://doi.org/10.1038/nature18604.

Zezyulin, D. A. and Konotop, V. V. (2012). Nonlinear modes in finite-dimensional $\mathcal{PT}$-symmetric systems, *Physical Review Letters* **108**, p. 213906, doi:10. 1103/PhysRevLett.108.213906, https://doi.org/10.1103/PhysRevLett. 108.213906.

Zhang, W., Wu, T., and Zhang, X. (2017). Tailoring eigenmodes at spectral singularities in graphene-based PT systems, *Scientific Reports* **7**, p. 11407, doi:10.1038/s41598-017-11231-y, https://doi.org/10.1038/ s41598-017-11231-y.

Zhen, B., Hsu, C. W., Igarashi, Y., Lu, L., Kaminer, I., Pick, A., Chua, S.-L., Joannopoulos, J. D., and Soljačić, M. (2015). Spawning rings of exceptional points out of Dirac cones, *Nature* **525**, pp. 354–358, doi: 10.1038/nature14889, https://doi.org/10.1038/nature14889.

Zhu, X., Feng, L., Zhang, P., Yin, X., and Zhang, X. (2013). One-way invisible cloak using parity-time symmetric transformation optics, *Optics Letters* **38**, pp. 2821–2824, doi:10.1364/OL.38.002821, https://doi.org/10.1364/OL. 38.002821.

Zhu, X., Ramezani, H., Shi, C., Zhu, J., and Zhang, X. (2014). $\mathcal{PT}$-symmetric acoustics, *Physical Review X* **4**, p. 031042, doi:10.1103/PhysRevX.4. 031042, https://doi.org/10.1103/PhysRevX.4.031042.

Znojil, M. (1999a). $\mathcal{PT}$-symmetric harmonic oscillators, *Physics Letters A* **259**, pp. 220–223, doi:10.1016/S0375-9601(99)00429-6, https://doi.org/10. 1016/S0375-9601(99)00429-6.

Znojil, M. (1999b). Exact solution for Morse oscillator in $\mathcal{PT}$-symmetric quantum mechanics, *Physics Letters A* **264**, pp. 108–111, doi:10.1016/S0375-9601(99)00805-1, https://doi.org/10.1016/S0375-9601(99)00805-1.

Znojil, M. (2001). $\mathcal{PT}$-symmetric square well, *Physics Letters A* **285**, pp. 7–10, doi:10.1016/S0375-9601(01)00301-2, https://doi.org/10.1016/ S0375-9601(01)00301-2.

Znojil, M. (2005a). $\mathcal{PT}$-symmetric quantum toboggans, *Physics Letters A* **342**, pp. 36–47, doi:10.1016/j.physleta.2005.05.029, https://doi.org/10.1016/ j.physleta.2005.05.029.

Znojil, M. (2005b). Solvable $\mathcal{PT}$-symmetric model with a tunable interspersion of nonmerging levels, *Journal of Mathematical Physics* **46**, p. 062109, doi: 10.1063/1.1925249, https://doi.org/10.1063/1.1925249.

Znojil, M. (2006). Coupled-channel version of the $\mathcal{PT}$-symmetric square well, *Journal of Physics A: Mathematical and General* **39**, pp. 441–455, doi: 10.1088/0305-4470/39/2/014, https://doi.org/10.1088/0305-4470/39/ 2/014.

Znojil, M. and Lévai, G. (2000). The Coulomb harmonic oscillator correspondence in $\mathcal{PT}$ symmetric quantum mechanics, *Physics Letters A* **271**, pp. 327–333, doi:10.1016/S0375-9601(00)00400-X, https://doi.org/10.1016/ S0375-9601(00)00400-X.

Znojil, M. and Lévai, G. (2001). Spontaneous breakdown of $\mathcal{PT}$ symmetry in the solvable square-well model, *Modern Physics Letters A* **16**, pp. 2273–2280, doi:10.1142/S0217732301005722, https://doi.org/10.1142/ S0217732301005722.

Znojil, M., Lévai, G., Roychoudhury, R., and Roy, P. (2001). Anomalous doublets of states in a $\mathcal{PT}$ symmetric quantum model, *Physics Letters A* **290**, pp. 249–254, doi:10.1016/S0375-9601(01)00676-4, https://doi.org/10.1016/S0375-9601(01)00676-4.

Znojil, M., Siegl, P., and Lévai, G. (2009). Asymptotically vanishing $\mathcal{PT}$-symmetric potentials and negative-mass Schrödinger equations, *Physics Letters A* **373**, pp. 1921–1924, doi:10.1016/j.physleta.2009.03.070, https://doi.org/10.1016/j.physleta.2009.03.070.

Znojil, M. and Tater, M. (2001). Complex Calogero model with real energies, *Journal of Physics A: Mathematical and General* **34**, pp. 1793–1803, doi:10.1088/0305-4470/34/8/321, https://doi.org/10.1088/0305-4470/34/8/321.

Zyablovsky, A. A., Vinogradov, A. P., Dorofeenko, A. V., Pukhov, A. A., and Lisyansky, A. A. (2014). Causality and phase transitions in $\mathcal{PT}$-symmetric optical systems, *Physical Review A* **89**, p. 033808, doi:10.1103/PhysRevA.89.033808, https://doi.org/10.1103/PhysRevA.89.033808.

Index

A

acoustical systems, ix, 17, 392
AKNS method, 308
analytic continuation, viii, 22–25,
  41–54, 75–82, 187, 196, 298–299
anharmonic oscillator, 20–25, 180.
  See also deformation
    classical deformed, 30–38,
      107–117
    cubic, vii–viii, 30–32, 175–207
    general classes of $\mathcal{PT}$-symmetric,
      30–31, 59–71, 186–188
    logarithmic, 63–65
    $\mathcal{PT}$ quartic, 45, 59–61
    $\mathcal{PT}$ sextic, 46, 49–50, 54, 59–60,
      62
    quantum, 45–50, 59–71
anomaly, quantum parity, 55, 57–59,
  122–123, 126, 146, 154, 155–157,
  166
antidamping, 12
anti-de Sitter cosmologies, 171
antilinear operator, 8, 40
Arnoldi algorithm, 71–72, 77–78
asymptotic freedom, 165

B

Bäcklund transformations, 314–317
basis functions and basis states
    biorthogonal basis, 84, 291–292
    orthogonal and orthonormal
      basis, 73, 84, 232, 265–266,
      290–292, 295, 297
    Riesz basis, 263, 266, 290–291
    Schauder basis, 266, 291–292
Baxter TQ relation, 41, 182, 185,
  219
Benjamin–Ono equation, 344
Berezin integral, 323
Bessel equation and Bessel functions,
  90, 153, 233, 257, 363, 376
beta function, 33
Bethe ansatz, 41, 184–185, 190–191,
  219
Bianchi diagrams, 315
Bianchi's theorem of permutability,
  315–316
bifurcation of optical beams, 355
binding energy, *see* bound states
birefringence, 354
Bloch functions, 356
Bloch sphere, 383
Bloch wave, 379, 381
Bohr–Sommerfeld quantization
    formula, 37, 43
Borel summation and summability,
  137–140, 144
bound states, 36, 44, 155–158
    in continuum, 381
    parity anomalies and, 55–57